D0138683

Ecological Methodology

Dedicated to the many biometricians working in ecology, fisheries, and wildlife who over the last 60 years have established a rigorous foundation for our ecological methods.

Ecological Methodology

Second Edition

Charles J. Krebs
University of British Columbia

An imprint of Addison Wesley Longman, Inc.

Menlo Park, California • Reading, Massachusetts • New York • Harlow, England
Don Mills, Ontario • Amsterdam • Madrid • Sydney • Mexico City

Publisher: Jim Green
Sponsoring Editor: Elizabeth Fogarty
Production Editor: Vivian McDougal
Copy Editor: Nick Murray
Proofreader: Jan McDearmon
Indexer: Nancy Ball, Nota Bene Indexing
Illustrators: G & S Typesetters, Inc.
Compositor: G & S Typesetters, Inc.

Library of Congress Cataloging-in-Publication Data

Krebs, Charles J.
Ecological methodology / Charles J. Krebs — 2nd ed.
p. cm.
Includes bibliographical references (p. 581) and index.
ISBN 0-321-02173-8
1. Ecology—Statistical methods. I. Title.
QH541.15.S72K74 1998
577'.07'27—dc21 98-8681
CIP

2 3 4 5 6 7 8 9 10—MA—02 01 00 99

Benjamin/Cummings
An imprint of Addison Wesley Longman
2725 Sand Hill Road
Menlo Park, CA 94025

Contents

Preface

This book attempts to present to ecologists in a coherent form the statistical methodology that is general to ecological field measurements. Scientific progress depends on good methods, and there are two components to progress in ecological methodology. The first component is biological and technical—you must have a good sampling device that catches the animals you wish to study or marks them with tags or bands that will not fall off. There has been great progress in designing better and better sampling devices in all areas of ecology, from mouse trapping to sampling the deep benthos of the oceans. The techniques and equipment used for sampling vary greatly for different groups of animals and plants and are the subject of many handbooks specific to a given area of ecology or specific group of plants or animals.

The second component of scientific progress is good statistical design. This component is general to all disciplines, and ecology is no exception. Over the past 60 years there has been superb development of statistical methods for increasing precision and avoiding bias in the estimation of ecological parameters like population size. Much of this has now penetrated ecological journals, but there have been few attempts to put these methods together in a textbook. Southwood (1978) has been the standard, followed by Ludwig and Reynolds (1988) and Sutherland (1996). Smaller books like Elliott (1977) cover some methods extremely well but are not comprehensive, and ecological methods books are still relatively scarce.

This book will not tell you what plankton sampler is best for oligotrophic lakes, but it will tell you how to design your plankton sampling in the most efficient way so that you need to do minimal work for maximal precision. All the methods presented here are well known to statisticians; that they are not always known to ecologists is one of the weaknesses of modern ecology. I hope that this book will assist field ecologists in the statistical design and analysis of ecological measurements.

Students using this book should have the basic knowledge of statistics presented in a one- or two-semester course in statistics. If your statistics are rusty, keep an introductory statistics text at hand to remind you of the basic ideas of standard errors, variances, and confidence limits. Since I am not a statistician and this is not a book for statisticians, I do not prove any theorems or present rigorous arguments about statistical expectations. I do attempt to translate the recommendations of statisticians into ecological English so that they can be used in the real world. I use approximations when they are needed, because in an ecological study an approximation is better than nothing.

To assist students, I work out numerous examples in boxes in each chapter, as well as in the text. For most of the statistical calculations I discuss, I have written FORTRAN programs to do the tedious arithmetic. Computer-literate students can do many of these calculations in EXCEL or other spreadsheets or in Visual Basic.

If you encounter mistakes in the book, I will be grateful to hear about them. I will maintain a website on the World Wide Web at *www.zoology.ubc.ca* to keep a list of the mistakes that come to my attention. Feedback can be delivered via e-mail to *krebs@zoology.ubc.ca*. One of the things I learned as I wrote this book was the frequency of errors of computation in the ecological literature. The computer should help us avoid mistakes of arithmetic, but of course we must be careful not to institutionalize mistakes in computer programs. Computers do not yet design experiments. If they did, you would not need this book.

I am grateful to my students at the University of British Columbia for helping me develop these thoughts on methodology, and to all those who have suggested improvements in this edition. In particular I thank James Bogart, Leslie Bowker, Robert Frederick, James Gilbert, C. Edward Miller, Jon Mendelson, John Yunger, Brad Anholt, and Lorne Rothman for their suggestions. I thank Nils Stenseth and Grant Singleton for arranging quiet time for me to work on this book, and Alice Kenney for assistance in getting all the material together efficiently.

A tremendous explosion of interest in the statistical problems of ecological methods has occurred during the last decade, and I hope that this book brings these to the attention of field ecologists trying to make the measurements that will help us understand the natural world.

Charles J. Krebs
8 October 1997

CHAPTER 1

Ecological Data

Ecologists, like other biologists, collect data to be used for testing hypotheses or describing nature. Modern science proceeds by conjecture and refutation, by hypothesis and test, by ideas and data, and it also proceeds by obtaining good descriptions of ecological events. Ecology is an empirical science that cannot be done solely on the blackboard or on the computer; it requires data from the real world. This book is about ecological data and how to wrestle it from the real world.

But data or ecological measurements are not all there is to ecology. At best, data may be said to be half of the science. Ecological hypotheses or ideas are the other half, and some ecologists feel that hypotheses are more important than data, while others argue the contrary. The central tenet of modern empirical science is that both are necessary. Hypotheses without data are not very useful, and data without hypotheses are wasted.

One problem that all the sciences face is what to measure, and the history of science is littered with examples of measurements that turned out not to be useful. Philosophers of science argue that we should measure only those things that theory dictates to be important. In abstract principle this is fine, but every field ecologist sees things he or she could measure about which current theory says nothing. Theory develops in a complex feedback loop with data, and an ecologist measuring the acidity of precipitation in 1950 would have been declared unfit for serious science. More typically, the mad ecologist is seen as a person who tries to measure everything. Do not try this, or you will waste much time and money collecting useless data. Data may be useless for several reasons. It may be unreliable or unrepeatable. It may be perfectly reliable and accurate but irrelevant to the problem at hand. It may be reliable, accurate, and very relevant but not collected at the right season of the year. Or the experimental design may be so hopeless that a statistical analysis is not possible. So start by recognizing the following rules.

Rule # 1 **Not everything that can be measured should be.**

Collect useful data and you have jumped the first hurdle of ecological research. But how do you know what data are useful? It is a mistake to think that statistical analysis by itself will give you any crisp insight into what data you should be collecting. Do not get the proverbial statistical cart in front of your ecological horse. Ecological theory and your ecological insight will give you the distinction between useful things to measure and useless ones, and you will not find this absolutely fundamental information in this book. So, before you do anything else, follow Rule # 2.

Rule # 2 **Find a problem and state your objectives clearly.**

Often your objective will be to answer a question, to test an ecological hypothesis. Do not labor under the false impression that a statistician can help you to find a problem that is ecologically important or to define appropriate objectives for your research. Many excellent ecology books can help you at this step. Start with Begon et al. (1996) or Krebs (1994) and move to Caughley and Sinclair (1994) or Diamond and Case (1986) for more advanced discussions of ecological theory. The key here is to find an important problem that, once solved, will have many ramifications in ecological theory or in the management and conservation of our resources.

When all the intellectually hard work is over, the statistician can be of great assistance. This book will try to lay out the ways in which some statistical knowledge can help answer ecological questions. We now proceed to describe in detail the statistical cart and forget about the ecological horse, but remember that the two must operate together.

Rule # 3 **Collect data that will achieve your objectives and make a statistician happy.**

Usually these two goals are the same, but if you ever find a dichotomy of purpose, achieve your objectives, answer your question, and ignore the statistician. In nearly all the cases ecologists have to deal with, a statistician's information can be vital to answering a question in a definitive way. This is a serious practical problem because all too often insufficient data are collected for reaching a firm conclusion. In some cases it is impossible to collect a sufficient amount of data, given normal budgetary constraints.

Some ecologists pick exceedingly interesting but completely impossible problems to study. So please beware of the following possible pitfall of the enthusiastic ecologist.

Rule # 4 **Some ecological questions are impossible to answer at the present time.**

You do not need to get depressed if you agree with this statement; realizing that adequate data cannot be obtained on some questions would save tax money, ulcers, and some marriages. Constraints may be technical, or it may be simply impossible to collect a large enough sample size. It might be interesting, for example, to map the movements of all the killer whales on the Pacific Coast in real time in order to analyze their social groupings, but it is not possible financially or technically to achieve this goal at this time.

We must always keep in mind that there are three populations out in the real world that keep getting confused.

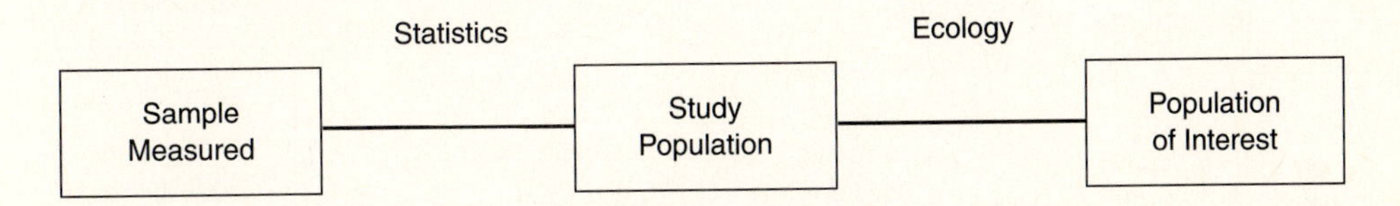

We actually measure a sample of animals or plants, and these are the primary data that an ecologist gathers. If we do our sampling correctly, this will be a representative sample of a *study population.* Often this is a random sample of the study population, and statisticians tell us that with a random sample, we can make valid inferences about the study population. In some cases the population studied is the population of interest, but the ecologist is often interested in yet another, broader population and would like to draw some conclusions about this population. Keep in mind the population of interest when designing your sampling methods.

Given these general warnings about the interface between statistics and ecology, we will review a few basic ideas about designing field studies and taking measurements. We will then consider a few problems in applying statistical inference to ecological data.

1.1 DESIGNING FIELD STUDIES

Is the event you wish to study controlled by you, or must you study uncontrolled events? This is the first and most important distinction you must make in designing your field studies. If you are studying the effects of natural forest fires on herb production, you are at the mercy of the recent fire season. If you are studying the effects of logging on herb production, you may be able to control where logging occurs and when. In this second case you can apply all the principles you learned in introductory statistics about replicated experimental plots and replicated control plots, and the analyses you must make are similar to those used in agricultural field research, the mother lode of modern statistics. If, on the other hand, you are studying uncontrolled events, you must use a different strategy based on sampling theory (Eberhardt and Thomas 1991). Figure 1.1 illustrates these two approaches to ecological field studies. Sampling studies are part of descriptive statistics, and they are appropriate for all ecological studies that attempt to answer the question *What happened?* Hypothesis testing may or may not be an important part of sampling studies, and the key question you should ask is what is your objective in doing these studies.

Figure 1.1 lists several types of statistical analysis that are often not familiar to ecologists; they are discussed in subsequent chapters of this book. *Intervention analysis* is a method for analyzing a time series in which, at some point in the series, an event like

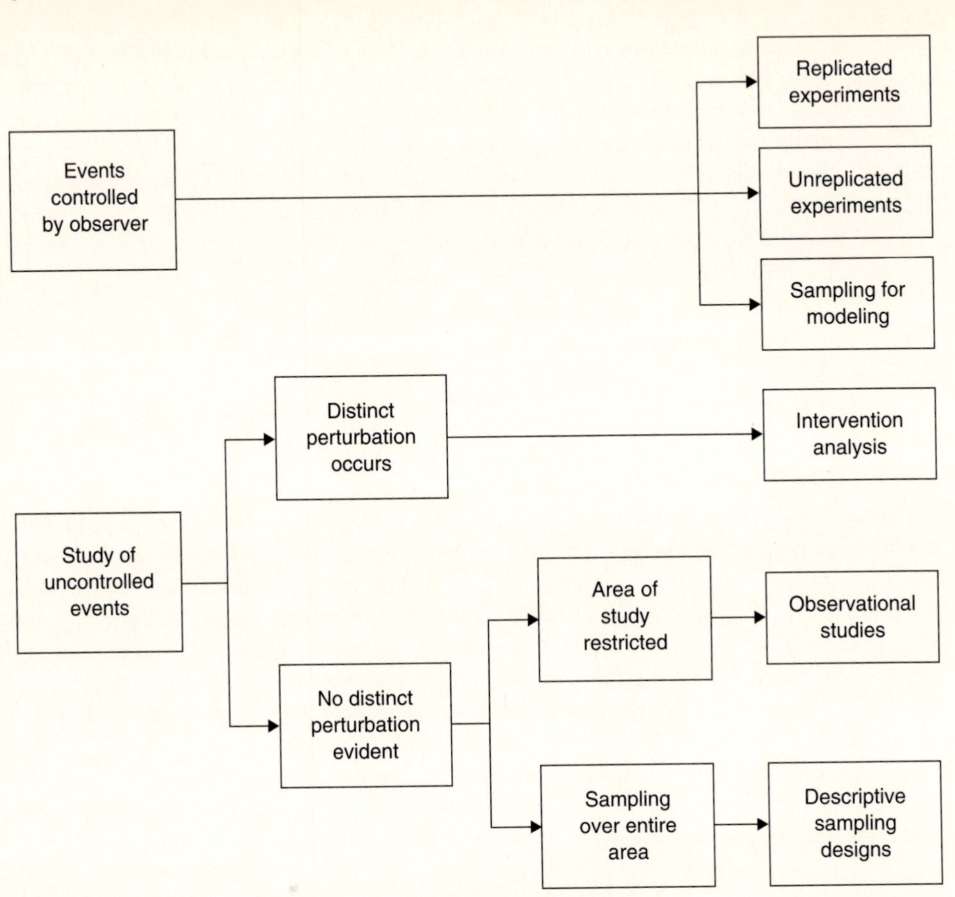

Figure 1.1 A classification of the methods used in ecological field studies. The key dichotomy is whether or not you are studying events you can control. (Modified after Eberhardt and Thomas 1991.)

a forest fire occurs, or a power plant is constructed. By comparing some environmental variable before and after the event, you may be able to detect an impact (Chapter 8). The remaining methods in Figure 1.1 are based on sampling, and they differ in one's objective in doing the work. Consider a study of the productivity of a complex forest ecosystem that contains many different communities. For an observational study, we may pick two forest communities and compare their production. Descriptive sampling could be applied to all the forest communities in the region to estimate productivity of the entire region, and analytical sampling could use these sampling data to test hypotheses about why one forest community differed in productivity from another. Chapters 8 and 9 discuss these sampling problems. Sampling may also be done to look for a pattern, which implies that one is interested, for example, in the pattern of geographical distribution of a species or a pollutant. We will discuss this type of sampling in Chapters 4 and 6.

Ecologists often attempt to determine the impact of a treatment applied to a population or community. One illustration of why it is important to think about experimental design before you begin is shown in Figure 1.2. Suppose that you are the manager of a nature reserve, and you wish to determine if fox and coyote control on the reserve will increase the nest success of ducks. If you do a single measurement before and after the fox

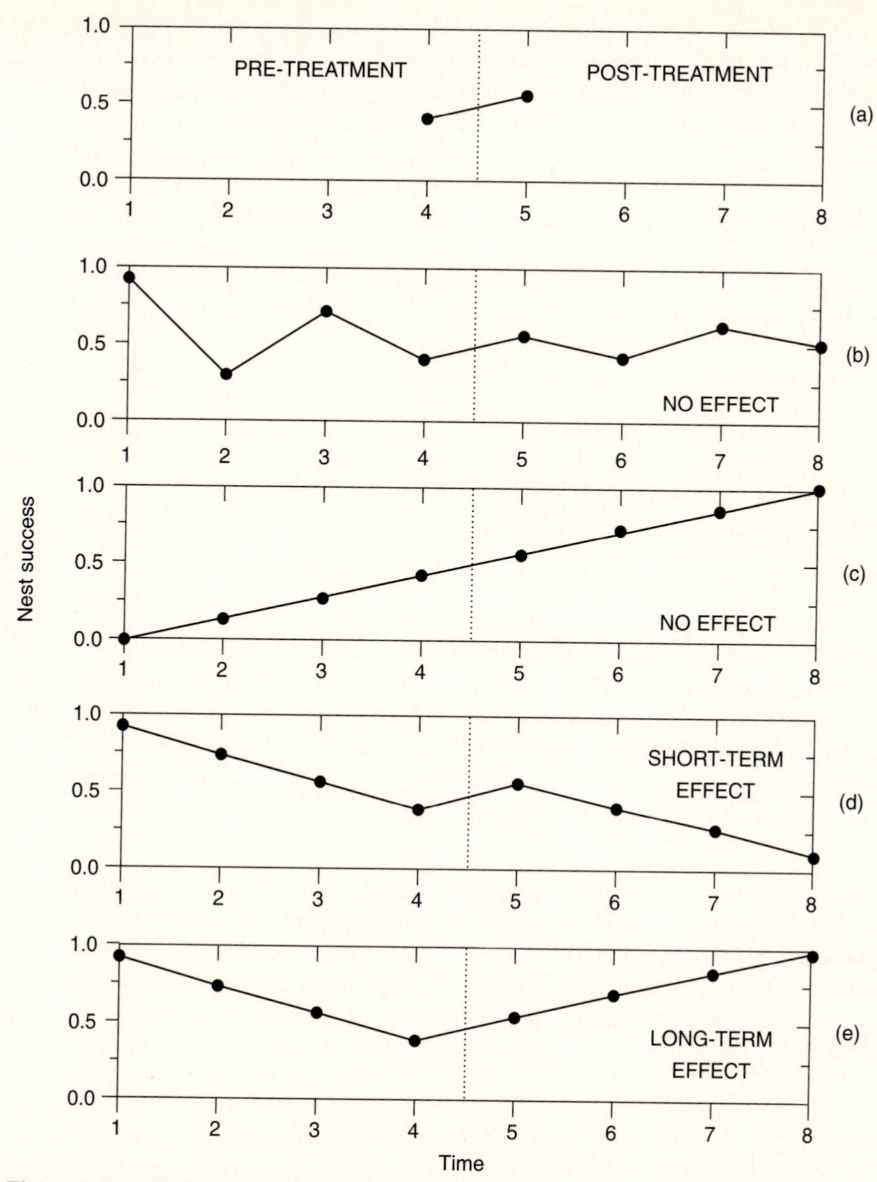

Figure 1.2 Advantages of a time-series experimental design. A manipulation is carried out between time 4 and 5 (dotted line). (a) A single pretest-posttest design with results that are impossible to interpret. (b)–(d) Four possible outcomes if additional pre- and post-manipulation data are available. By adding an unmanipulated control (not shown here), a much better experiment could be done. (Modified after Kamil 1988.)

and coyote removal, you might observe the data shown in Figure 1.2a. These results by themselves would be difficult to interpret. By collecting data for a longer time period, both before and after the experiment (the *time-series design*, Kamil 1988), you would be in a stronger position to draw the correct inference. As illustrated in Figure 1.2b–e, you might observe no effect, a temporary effect, or a long-term effect of the manipulation. You could further strengthen this study by using a control area, a refuge that is similar to the manipulated one but is not subject to fox and coyote control (the *multiple time-series design*, Kamil 1988). The key enemies that confound interpretations are random events, so we need to set up our studies to remove the impact of randomness from our conclusions.

The importance of pilot studies in both field and laboratory experiments cannot be overemphasized. Pilot studies lead you into more thorough studies and allow you to answer simple questions of technique like *Can we make this measurement on live plants?* and *Can we measure this many animals in one hour?* They are particularly important in designing experiments, since you can measure the variability in the animals or plants you are studying, and thus determine the sample sizes needed for a particular level of precision in your answers.

We will discuss sampling methods in detail in Chapters 7, 8, and 9, and experimental design in Chapter 10.

1.2 SCALES OF MEASUREMENT

Data may be collected on three basic scales of measurement.

(1) Nominal Scale Nominal data are attributes like sex or species, and represent measurement at its weakest level. We can determine if one object is different from another, and the only formal property of nominal scale data is *equivalence*. Nominal data are very common in ecology, and we often count individuals occurring in different classes. The colors of gastropods on a beach, or the names of different insect species collected in a light trap might be determined to provide nominal scale data.

(2) Ranking Scale Some biological variables cannot be measured on a numerical scale, but individuals can be ranked in relation to one another. Items in the diet, for example, might be ranked from more preferred to less preferred on the basis of cafeteria tests. As with the nominal scale, we have a series of classes, but now the classes bear some rank with respect to one another. Two formal properties occur in ranking data: *equivalence and greater than*. We can symbolize our ordered classes by the conventional number or letter order:

1, 2, 3, 4, . . . or A, B, C, D, E, . . .

I recommend using letters rather than numbers because the most common mistake in ranking data is to assume that they are measured on an absolute scale. For example, we might rank 5 grouse in a dominance hierarchy from 1 (low) to 5 (high) by means of their relative aggression. We might then be tempted to assume erroneously that a bird ranked 4 is really twice as dominant as a bird ranked 2 in the hierarchy. Do not confuse ranking numbers with absolute numbers.

Note that any quantitative measurement can also be expressed on a ranking scale. Clearly, if we weigh 14 fish, we can also rank these from lightest to heaviest. In some cases, we may deliberately adopt a ranking scale for ecological measurements because it is faster and cheaper than doing a precise quantitative measurement.

(3) Interval and Ratio Scales Interval and ratio scales have all the characteristics of the ranking scale, but we also know the distances between the classes. We must have a unit of measurement for interval and ratio data: cm, degrees, kg. If we have a unit of measurement and a true zero point, we have a ratio scale of measurement. A true zero point means that the variable being measured vanishes at zero. Thus fish length is a measurement on the ratio scale, and a 4 kg fish is twice as heavy as a 2 kg fish. But water temperature is a measurement on the interval scale because 0°C is not the lowest possible temperature, and 8°C is not twice as hot as 4°C. For statistical purposes, these two scales represent the highest form of measurement, and much of statistics deals with the analysis of data of this type. Most of the measurements we take in ecology are interval or ratio scale data. Height, weight, age, clutch size, population size—the list is endless. Data of this type can be subjected to the normal arithmetic operations of addition, subtraction, multiplication, and division because the unit of measurement is a constant—a centimeter is a centimeter is a centimeter.

Ratio scale data may be continuous or discrete. Discrete data are usually simple because they take on integer values only: 0, 1, 2, 3 . . . . Examples abound in ecology: number of plants in a quadrat, number of eggs in a nest, number of fish in a net. No intermediate values are possible with discrete data, so counts of this type are usually exact (or should be) with no error, at least when the numbers involved are small.

Continuous data may be measured to any degree of precision, so they do not represent such a simple situation. We must first distinguish *accuracy* and *precision*. *Accuracy* is the closeness of a measured value to its true value and is dependent on having a good measuring device or system. Some ecological data are very inaccurate. For example, estimates of plankton production in tropical ocean waters by the carbon-14 method may be an order of magnitude too low (Sheldon 1984). Vole population densities estimated from standard live traps may be only one-half the true density (Boonstra and Krebs 1978). The history of ecological measurement is largely the history of attempts to design better measuring techniques or better sampling devices that permit more accurate measurements (e.g., Cullen et al. 1986).

Precision is the closeness of repeated measurements to the same item. A ruler that has been marked off in the wrong places and is too short may give a very precise measurement of a fish's length because, if we are careful, every repeated measurement with this ruler will give nearly the same numerical value. But this precise measurement would be very inaccurate, and the ruler would give a biased measurement. Bias (Figure 1.3) is very important in many ecological measurements and is a recurrent problem with many estimation techniques. We will find that some ecological methods produce biased estimates, and it is important to try to find out the size of the bias. A biased estimate is better than no estimate at all, but we must be careful to remember when bias is present in our ecological estimates.

Continuous data are recorded to some fixed level of precision, and every measurement has its implied limits:

Measurement	Implied limits
67	66.5 to 67.5
67.2	67.15 to 67.25
67.23	67.225 to 67.235

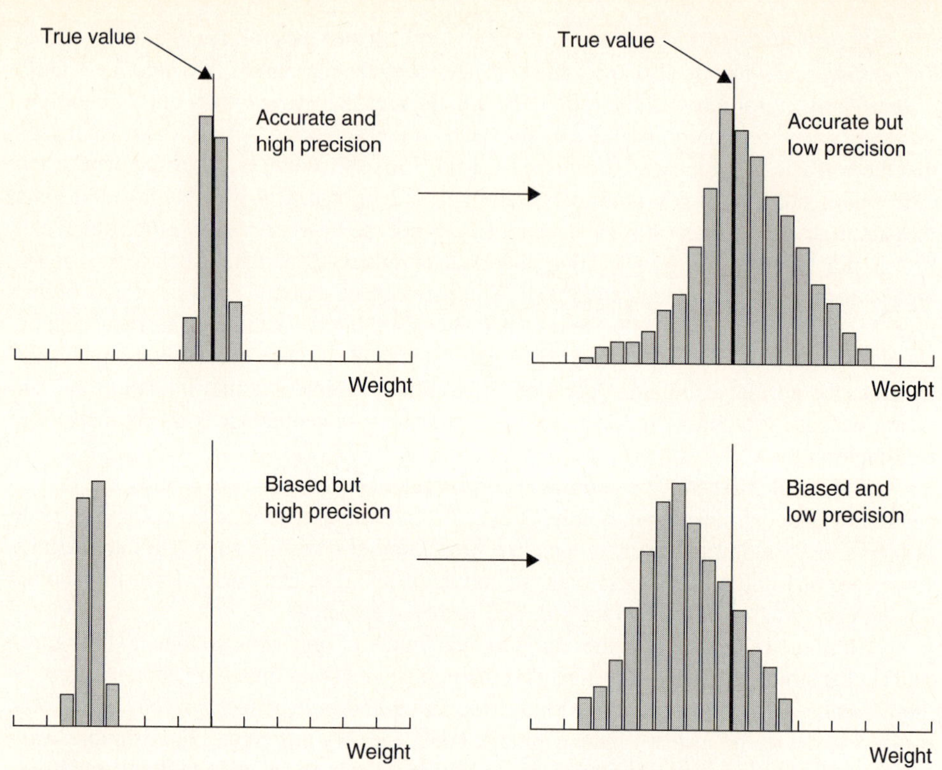

Figure 1.3 Illustration of accuracy and precision in ecological measurements. In each case a series of repeated measurements is taken on a single item (e.g. the weight of a single fish specimen).

Large numbers are often presented with less care than they should be, so that the level of precision is not always known. This deficiency can be overcome by either conventional rounding or the use of exponential notation.

Measurement	Implied limits
31,000	Not clear—could be any of the following.
3.1×10^4	3.05×10^4 to 3.15×10^4
3.10×10^4	3.095×10^4 to 3.105×10^4
3.100×10^4	3.0995×10^4 to 3.1005×10^4

Significant figures are defined as the digits in a number that denote the accuracy. In gathering ecological data we must often make some decision about how many significant figures to use. Sokal and Rohlf (1995, 14) make a practical rule of thumb, as follows: *The number of unit steps from the smallest to the largest measurement should be between 30 and 300.* For example, if we are measuring fish lengths and the smallest fish is 3.917 cm long and the largest fish is 38.142 cm long, we should record to the nearest centimeter (4 to 38 cm) to give 34 one-centimeter steps. There is no point in recording these fish lengths to 5 significant figures (see Box 1.1). We thus reach the next rule of ecological measurement.

Box 1.1 Methods for Determining the Number of Significant Figures to Record in Your data

Sokal and Rohlf (1995) Method

1. Determine the *range* for your data:

$$\text{Range} = \text{Maximum value} - \text{minimum value}$$

2. Divide the range into unit steps numbering between 30 and 300:

$$\frac{\text{Range}}{30} = \text{Minimal desired level of measurement}$$

$$\frac{\text{Range}}{300} = \text{Maximal desired level of measurement}$$

Example: $n = 100$, $\bar{x} = 173.86$ mm; $s = 12.26$; $s_{\bar{x}} = 1.226$.

$$\text{Maximum value observed} = 210.64 \text{ mm}$$

$$\text{Minimum value observed} = 143.21 \text{ mm}$$

$$\text{Range} = 210.64 - 143.21 = 67.43 \text{ mm}$$

$$\text{Minimal level of measurement} = \frac{67.43}{30} = 2.25 \text{ mm}$$

$$\text{Maximal level of measurement} = \frac{67.43}{300} = 0.225 \text{ mm}$$

If you record your data to the nearest 1 mm, you will have about 67 possible values, and if you record your data to the nearest 2 mm, you will have about 34 possible values. Both of these would fall within the Sokal and Rohlf rule of 30 to 300 steps, and thus you should record your data to the nearest 1 or 2 mm. If you record to the nearest 0.1 mm, you are being overly precise (because you will have over 600 possible values), and if you record to the nearest 5 mm, you are being too imprecise.

Barford (1985) Method

1. Determine the relative accuracy of the standard error of your data ($s_{\bar{x}}$):

$$\text{Relative accuracy of } s_{\bar{x}} \cong \frac{1}{\sqrt{n-2}}$$

2. Determine the range of probable error of the standard error:

$$\text{Probable error of standard error} = \pm(s_{\bar{x}})(\text{relative accuracy of } s_{\bar{x}})$$

continues

3. Round the standard error to the precision set by the probable error limits, and measure to the same number of decimal points.

Example: $n = 100, \bar{x} = 173.86$ mm; $s = 12.26$; $s_{\bar{x}} = 1.226$.

$$\text{Relative accuracy of } (s_{\bar{x}}) \cong \frac{1}{\sqrt{100 - 2}} = 0.1010$$

$$\text{Probable error of } s_{\bar{x}} = \pm(1.226)(0.1010) = \pm 0.1238 \text{ mm}$$

Hence the standard error could probably range from $1.226 + 0.1238 = 1.3498$ to $1.226 - 0.1238 = 1.1022$ and so is precise at most to one decimal point (1.2). Thus the original lengths should be measured at most to one decimal point, or 0.1 mm.

In any data set involving 10 or fewer measurements, there is no point in giving the standard error to more than one significant figure. Thus in this example, if $n = 10$, the probable error of the standard error would be ± 0.433, and clearly the standard error could easily range from 0.8 to 1.7, so $s_{\bar{x}} = 1$, and your lengths should be measured to the nearest 1 mm only.

In determining the number of significant figures to record in your data, Barford's method is slightly more conservative than Sokal and Rohlf's minimal level in recommending more measuring precision.

These calculations can be carried out in program-group EXTRAS, listed in Appendix 2.

Rule # 5 **With continuous data, save time and money by deciding on the number of significant figures needed in the data *before* you start an experiment.**

Some ecological measurements are used as part of an equation to estimate a derived variable. For example, the product of the number of locusts per square meter and their rate of food intake per individual per day will give us an estimate of the total food consumption of a locust population. Errors multiply in any calculation of this type, and for this reason we may want to record data to one more significant digit than we recommended above. Note, however, that any chain of measurement is only as strong as its weakest link and if we have an estimate with only two significant figures for locust density, there is no point to having an estimate of food intake accurate to five significant figures. Achieving a balanced set of measurements is one of the most difficult of the ecological arts.

Ratio or interval data can also be recorded as ranking data, and it is important to decide before you start exactly what detail of measurement your study requires.

1.3 STATISTICAL INFERENCE

Ecological statistics differ in emphasis from most types of statistics because the problems of estimation and the problems of sampling are much more difficult in ecology than they are in other biological disciplines. Descriptive statistics occupy a short part of most statistics books, and random sampling is often discussed only briefly. This book is largely concerned with the descriptive statistics of ecology and the types of sampling schemes that ecologists can adopt. For example, the conceptually simple variable *population size* is difficult to measure, and yet is vital for most types of ecological work. Because ecological variables may be difficult to measure, we often tend to forget the following basic rule of descriptive statistics.

Rule # 6 **Never report an ecological estimate without some measure of its possible error.**

This elementary statistical rule is violated daily, and we must be humble enough to realize that, even though we may spend two months' hard work obtaining one estimate, that estimate is still subject to error.

In contrast, hypothesis testing is much the same in ecology as it is in other biological disciplines. Consequently, we will not repeat the analysis of variance in this book but will assume that you can obtain details of statistical tests from other excellent books such as Sokal and Rohlf (1995), Underwood (1997), or Zar (1996). Statistical tests that are unique to ecological types of data will be given here.

Statistical inference is particularly difficult in ecology. Statistical populations are not biological populations, and the unfortunate dualism of the word *population* is a constant source of confusion. For any valid statistical inference, we must specify the statistical population, the "target" population we are studying, and for most ecological studies we cannot do this very rigorously. Biological populations and communities change in space and time in such a complex manner that if we specify a statistical population very broadly, we cannot sample it in a random manner (Hagood 1970; Scheiner 1993). This fundamental problem undercuts the very foundation of normal statistical inference in ecology (Morrison and Henkel 1970b).

Ecologists face a second problem in hidden variables that confound our interpretations of data. If you count rabbits in one year, then remove foxes and count the rabbits in the second year, you may find that rabbit numbers have increased. This increase could be due to reduced fox predation (your favorite interpretation), but it could also be caused by better food in the second year, or a reduction in disease or parasites, or other variables you did not measure. Before-after experiments are often difficult to interpret because of confounding.

A third major difficulty in ecological statistics is that statistical methods can cope only with random errors. In real situations systematic errors, or bias, may be more important, and no statistical test can detect biased data. We try to minimize bias in our measurements, and in this manner to reduce this source of possible error. But the net result of these weak points is to provide a cautionary note.

Rule # 7 **Be skeptical about the results of statistical tests of significance.**

The conventional approach to a statistical test is too often presented as a black-or-white decision whether to accept or reject the null hypothesis. We would be better off if we viewed this problem of statistical decisions as an area of shades of gray with no pure blacks or pure whites.

We all wish to have our ecological studies recognized as important contributions to ecological understanding. This desire can lead us into one of the most common traps in modern science: using statistical tests of significance as a measure of importance. To avoid this trap, remember the following rule.

Rule # 8 ***Never*** **confuse statistical significance with biological significance.**

The greatest mistake any ecologist can make in the use of routine statistics is to confuse the concept of statistical significance with that of biological significance. Statistical significance is commonly thought to be achieved in any statistical test in which the probability of obtaining the given results under the null hypothesis is less than 5%. Biological significance is not a mechanical concept like statistical significance. It refers to the importance of a particular set of measurements in the context of a theoretical hypothesis. Small effects can be important in some ecological processes. For example, a difference in survival rate of 3% per year between males and females of a seabird population may be very significant biologically but not statistically significant unless one has large sample sizes. Conversely, by measuring 10,000 whitefish in two lakes, one might establish beyond any statistical doubt that whitefish in lake A are 0.03 grams heavier than whitefish in lake B. A difference may be biologically trivial but highly significant statistically. In a very real sense the null hypothesis of no difference is irrelevant to ecological statistics because we know as ecologists that each biological population and community will differ from all others. To demonstrate a difference statistically is trivial and often gets in the way of the real ecological question: *How different are the two populations or communities?* And, secondly, are the differences large enough to be ecologically relevant? We come back to the central theme of this chapter, that ecological hypotheses, ecological insights, and ecological theory must be the arbiters of what we measure and how we interpret our results. And we must strive in ecology to build strong feedback loops between theory and data as our science matures.

1.4 DATA RECORDS

The most mundane aspect of data recording is to decide how to write the data records. Little is said of this technical problem, and one learns by bitter experience how not to record data. Every ecological problem is unique, so no one can give you a universal recipe for a good data form. Research budgets differ, and access to computers is not uniform. But the trend is very clear, and the gradually improving access to computing equipment dictates the following prescription.

Rule # 9 **Code all your ecological data and enter it on a computer in some machine-readable format.**

To achieve this goal, you must set up proper data sheets, decide on significant digits for each type of measurement, and in general organize your data-collection procedure. If you do only this and neglect to put data into computer files subsequently, you will have most of the benefit of this exercise. If you do not at this stage know what a computer file is and how it works, you may wish to find out. There are many database management programs and spreadsheet programs available for personal computers—EXCEL, DBASE, FOXPRO, ACCESS—and the list will grow every year. You may save yourself an enormous amount of time later by systematic coding now in one of these programs. The advantages of putting data into computer files may not be large for ecological work that involves only a few observations on a few animals or plants, but I have never seen any ecological data that suffered by coding. With small samples, data analysis by computer may require the same time and effort as pencil and paper or your pocket calculator. With large samples, there is no other way to handle ecological data than by computer. Another important advantage of computer analysis is that you know the answers are mathematically correct. Hand calculations sometimes err.

You should store your computer data as raw data so that you can recalculate everything once you find a new method of analysis. If you store only the mean values, you will not be able to recover the raw data when you discover, for example, that you should have done a logarithmic transform before your analysis. Similarly, keep study areas separate in your data coding, since you can combine them later if needed.

Personal computers now have available a large set of statistical programs that can be used directly on data coded in computer files. Ecological data are commonly evaluated using standard statistical packages like SAS, SYSTAT, JMP, NCSS, and a variety of packages that improve rapidly from year to year. If you are a novice, it will pay you to search for evaluations of statistical packages in the literature before you decide which to use.[1] Thus, you do not need to be a computer operator or a computer programmer to make use of many standard statistical tests. An additional advantage is that you can easily recalculate new parameters several years after you collected the data if you suddenly get a bright new idea. If you use computer programs to analyze your data, you should always be aware that some programs may contain "bugs" that produce errors. It pays to check computer programs with calculations for which you already know the answer.

The computer is just another labor-saving device that can be used or abused by ecologists. If we do not take careful measurements, or we shortcut proper techniques, we will be no better off with a computer than without it. You can ignore all the assumptions of any statistical test, and the computer will still grind out an answer. The following ancient rule applies with particular force to "computerized" ecology and to all ecological statistics.

Rule # 10 Garbage in, garbage out.

Garbage, or poor data, may result from purely technical problems like a balance that is not weighing accurately or from human problems of inattentiveness or a lack of proper instruction in how to take field measurements. A good experimental design can build in data-

[1] The *Bulletin of the Ecological Society of America* contains a section on technological tools that can be useful, as does the *Wildlife Society Bulletin*.

checking procedures by repeated measurements or the injection of known samples. But a good ecologist will always be alert to possible errors in data collection and strive to eliminate them. Simple procedures can be highly effective. Persons recording data in the field can reduce errors by repeating data back to those doing the measurements. Ecological data collected carefully may be difficult enough to interpret without adding unnecessary noise.

SELECTED READING

Barford, N. C. 1985. *Experimental Measurements: Precision, Error, and Truth.* John Wiley and Sons, New York.

Berger, J. O., and Berry, D. A. 1988. Statistical analysis and the illusion of objectivity. *American Scientist* 76: 159–165.

Cleveland, W. S. 1994. *The Elements of Graphing Data.* Hobart Press, Summit, New Jersey.

Eberhardt, L. L., and Thomas, J. M. 1991. Designing environmental field studies. *Ecological Monographs* 61: 53–73.

James, F. C., and McCulloch, C. E. 1985. Data analysis and the design of experiments in ornithology. *Current Ornithology* 2: 1–63.

Morrison, D. E., and Henkel, R. E., (eds). 1970. *The Significance Test Controversy.* Butterworth, London.

Sokal, R. R., and Rohlf, F. J. 1995. *Biometry.* Chapter 2, Data in biology, and Chapter 3, The handling of data. W. H. Freeman and Co., New York.

Stevens, S. S. 1946. On the theory of scales of measurement. *Science* 103: 677–680.

Tukey, J. W. 1960. Conclusions vs. decisions. *Technometrics* 2: 423–433.

Yoccoz, N. G. 1991. Use, overuse, and misuse of significance tests in evolutionary biology and ecology. *Bulletin of the Ecological Society of America* 72: 106–111.

QUESTIONS AND PROBLEMS

1.1. Turk (1978) reports on an experiment involving six intertidal areas in which he removed starfish from two areas, added starfish to two areas, and used two areas as unmanipulated controls. He measured the abundance of the snail *Tegula eiseni* (a prey of the starfish). He measured snail abundance at approximately 2-month intervals for one year, then did the experimental removals and additions, and measured snail abundance on all areas for an additional year. Do you consider the ecological experiment relatively simple? Is the analysis of the data statistically simple? Read the analyses of this experiment by Finney (1978) and Van Belle and Zeisig (1978). How might you improve on this experiment?

1.2. The diameters of 16 pollen grains were obtained as follows:

12.478	12.475	12.504	12.457
12.482	12.473	12.492	12.501
12.470	12.499	12.509	12.477
12.490	12.502	12.482	12.512

Use the two methods given in Box 1.1 to determine how many significant figures you should record for future data sets. Would these recommendations change if you had measured 100 grains instead of 16? 1000 grains?

1.3. Go to an introductory statistics text and find out what a statistician means by Type I and Type II errors. Suppose you were testing the hypothesis that a particular type of logging operation was not affecting the breeding performance of a grouse species. Describe in ecological English the meaning of Type I and II errors for this example.

1.4. You have been given the task of weighing plant samples, and a preliminary analysis like that in Box 1.1 has shown that you should be weighing the samples to the nearest 1.0 g. Unfortunately you have available only a spring scale that weighs to the nearest 5 grams. What do you lose statistically by continuing to weigh your samples to the nearest 5 g?

1.5. How would you know if you were violating Rule # 1?

1.6. Compare the ten rules given in this chapter with the ten principles given in Chapter 2 of R. H. Green's (1979) book *Sampling Design and Statistical Methods for Environmental Biologists.*

1.7. Deming (1975) states, "We do not perform an experiment to find out if two varieties of wheat or two drugs are equal. We know in advance, without spending a dollar on an experiment, that they are not equal." Do you agree with this statement? What cautionary note does it inject into ecological statistics?

1.8 After clear-cut logging in the Pacific Northwest, a variety of deciduous trees and shrubs typically begin forest succession. Many experiments have been completed to try to increase the growth of the desired coniferous trees (like Douglas fir) by removing the deciduous trees from early succession. In about half of these studies the conifers grow better on the treated plots; in the other half, growth is equal on the treated and the untreated plots. Given these facts, what would you conclude if you were responsible for forest management, and what further actions would you suggest?

PART ONE

Estimating Abundance in Animal and Plant Populations

How many are there? This question is the central question of many ecological studies. If you wish to harvest a lake trout population, one helpful bit of information is the size of the population you wish to harvest. If you are trying to decide whether to spray a pesticide for aphids on a crop, you may wish to know how many aphids are living on your crop plants. If you wish to measure the impact of lion predation on a zebra population, you will need to know the size of both the lion and the zebra populations.

The cornerstone of many ecological studies is an estimate of the abundance of a particular population. This is true for both population ecology, in which interest centers on individual species, and community ecology, in which interest centers on groups of species, which makes the estimation problem more complex. Note that some ecological studies do not require an estimate of abundance, so before you begin the task of estimating the size of a population, you should have a clear idea why you need these data. Estimates of abundance themselves are not valuable, and a large book filled with estimates of the abundance of every species on earth as of January 1, 1998 would be a good conversation piece but not science.

Abundance can be measured in two ways. *Absolute density* is the number of organisms per unit area or volume. A red deer density of four deer per square kilometer is an absolute density. *Relative density* is the density of one population relative to that of another population. Blue grouse may be more common in a block of recently burned woodland than in a block of mature woodland. Relative density estimates are usually obtained with some biological index that is correlated with absolute density. Relative density may be adequate for many ecological problems, and should always be used when adequate because it is much easier and cheaper to determine than absolute density.

Absolute density must be obtained for any detailed population study in which you attempt to relate population density to reproductive rate, or any other vital statistic. The analysis of harvesting strategies may demand information on absolute numbers. All community studies that estimate energy flow or nutrient cycles require reliable estimates of absolute density.

The sequence of decisions by which we decide how to estimate absolute density is outlined in Figure A. Many factors—ecological, economic, and statistical—enter into a decision about how to proceed in estimating population size. Figure A thus gives relative guidance rather than absolute rules. Quadrat counts and spatial distribution methods are usually chosen in plant studies. Many vertebrate studies use mark-recapture techniques. Fish and wildlife populations that are exploited can be estimated by a special set of techniques applicable to harvested populations. There is a rapidly accumulating store of ecological wisdom for estimating populations of different animal and plant species, and if you are assigned to study elephant populations, you should begin by finding out how other elephant ecologists have estimated population size in this species. But do not stop there. Read the next four chapters and you may find that the conventional wisdom needs updating. Fresh approaches and new improvements in population estimation occur often enough that ecologists cannot yet become complacent about knowing the best way to estimate population size in each species.

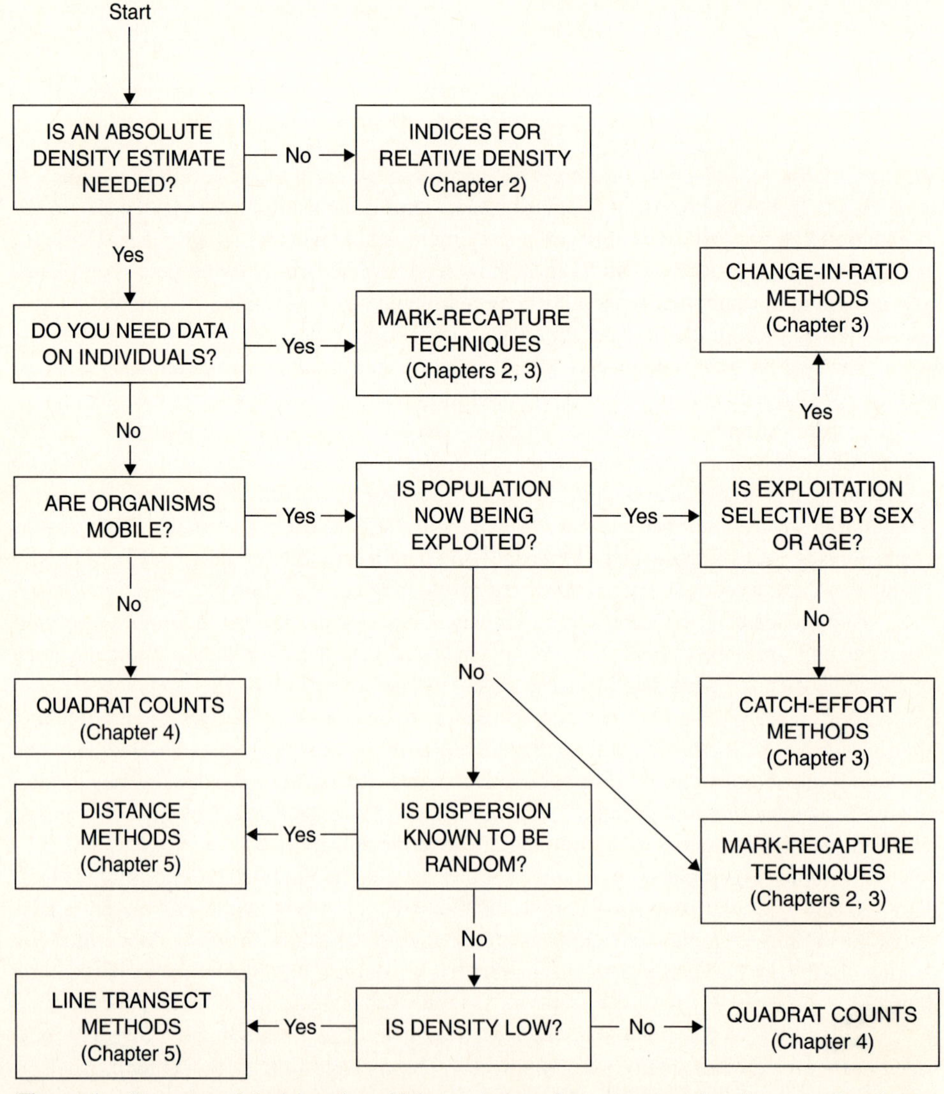

Figure A Sequence of decisions by which a technique for estimating abundance can be chosen. (Modified from Caughley 1977.)

CHAPTER 2

Estimating Abundance: Mark-Recapture Techniques

One way to estimate the size of a population is to capture and mark individuals from the population and then resample to see what fraction of individuals carry marks. John Graunt first used this simple principle to estimate the human population of London in 1662. The first ecological use of mark-recapture was carried out by the Danish fisheries biologist C. G. J. Petersen in 1896 (Ricker 1975). Tagging of fish was first used to study movements and migration of individuals, but Petersen realized that tagging could also be used to estimate population size and to measure mortality rates. Fisheries biologists were well advanced over others in applying these methods. Lincoln (1930) used mark-recapture to estimate the abundance of ducks from band returns, and Jackson (1933) was the first entomologist to apply mark-recapture methods to insect populations. This chapter concentrates on the mark-recapture techniques that are used most often when data are required on individual organisms that are highly mobile. The strength of mark-recapture techniques is that they can provide information on birth, death, and movement rates in addition to information on absolute abundance. The weakness of these techniques is that they require considerable time and effort to get the required data and, to be accurate, they require a set of very restrictive assumptions about the properties of the population being studied. Seber (1982), Otis et al. (1978), and Pollock et al. (1990) have described mark-recapture methods in great detail, and this chapter is an abstract of the more common methods they describe.

Mark-recapture techniques may be used for *open* or *closed* populations. A *closed* population is one that does not change in size during the study period; that is, the effects of births, deaths, and movements are negligible. Thus populations are typically closed over only a short period of time. An *open* population is the more usual case, a population that changes in size and composition from births, deaths, and movements. Different methods must be applied to open and closed populations, and I will discuss three cases in this chapter:

Closed populations

1. Single marking, single recapture: *Petersen method*
2. Multiple markings and recaptures: *Schnabel method*

Open populations

1. Multiple census: *Jolly-Seber method*

2.1 PETERSEN METHOD

The Petersen method is the simplest mark-recapture method because it is based on a single episode of marking animals, and a second single episode of recapturing individuals. The basic procedure is to mark a number of individuals over a short time, release them, and then to recapture individuals to check for marks. All individuals can be marked in the same way. The second sample must be a *random* sample for this method to be valid; that is, all individuals must have an equal chance of being captured in the second sample, regardless of whether they are marked or not. The data obtained are

M = Number of individuals marked in the first sample

C = Total number of individuals captured in the second sample

R = Number of individuals in second sample that are marked

From these three variables, we need to obtain an estimate of

N = Size of population at time of marking

By a proportionality argument, we obtain

$$\frac{N}{M} = \frac{C}{R}$$

or, transposing

$$\hat{N} = \frac{CM}{R} \tag{2.1}$$

where $\hat{N}$ = Estimate of population size at time of marking,* and the other terms are as defined above

This formula is the "Petersen estimate" of population size and has been widely used because it is intuitively clear. Unfortunately, formula (2.1) produces a *biased* estimator of population size, tending to overestimate the actual population. This bias can be large for small samples, and several formulas have been suggested to reduce this bias. Seber (1982) recommends the estimator

$$\hat{N} = \frac{(M+1)(C+1)}{(R+1)} - 1 \tag{2.2}$$

which is unbiased if $(M + C) > N$ and nearly unbiased if there are at least seven recaptures of marked animals ($R > 7$). This formula assumes sampling *without* replacement (see page 263) in the second sample, so any individual can only be counted once.

In some ecological situations, the second sample of a Petersen series is taken *with* replacement so that a given individual can be counted more than once. For example, animals may be merely observed at the second sampling and not captured. For these cases, the size of the second sample (C) can be even larger than total population size (N) because individuals might be sighted several times. In this situation we must assume that the chances of sighting a marked animal are on the average equal to the chances of sighting an unmarked animal. The appropriate estimator from Bailey (1952) is

$$\hat{N} = \frac{M(C+1)}{(R+1)} \tag{2.3}$$

which differs only very slightly from equation (2.2) and is nearly unbiased when the number of recaptures (R) is 7 or more.

2.1.1 Confidence Intervals

How reliable are these estimates of population size? To answer this critical question, a statistician constructs *confidence intervals* around the estimates. A *confidence interval* is a range of values that is expected to include the true population size a given percentage of the time. Typically the given percentage is 95%, but you can construct 90% or 99% confidence intervals, or any range you wish. The high and low values of a confidence interval are called the *confidence limits*. Clearly, we want confidence intervals to be as small as possible, and the statistician's job is to recommend confidence intervals of minimal size consistent with the assumptions of the data at hand.

Confidence intervals are akin to gambling. You can state that the chances of flipping a coin and getting "heads" is 50%, but after the coin is flipped, it is either "heads" or

*A ^ over a variable means "an estimate of."

"tails." Similarly, after you have estimated population size by the Petersen method and calculated the confidence interval, the true population size (unfortunately not known to you) will either be inside your confidence interval or outside it. You cannot know which, and all the statistician can do is tell you that *on the average* 95% of confidence intervals will cover the true population size. Alas, you only have one estimate, and *on the average* does not tell you whether your one confidence interval is lucky or unlucky.

Confidence intervals are an important guide to the precision of your estimates. If a Petersen population estimate has a very wide confidence interval, you should not place too much faith in it. If you wish, you can take a larger sample next time and narrow the confidence limits. But remember that even when the confidence interval is narrow, the true population size may *sometimes* be outside the interval. Figure 2.1 illustrates the variability of Petersen population estimates from artificial populations of known size, and shows that some random samples by chance produce confidence intervals that do not include the true value.

Several techniques of obtaining confidence intervals for Petersen estimates of population size are available, and the particular one to use for any specific set of data depends upon the size of the population in relation to the samples we have taken. Seber (1982) gives the following general guide:

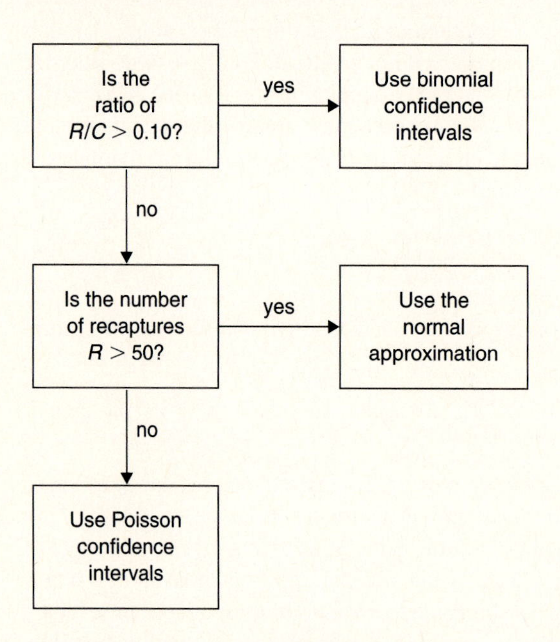

Poisson Confidence Intervals We discuss the Poisson distribution in detail in Chapter 4 (Section 4.2.1). We are concerned here only with the mechanics of determining confidence intervals for a Poisson variable.

Table 2.1 provides a convenient listing of values for obtaining 95% confidence intervals based on the Poisson distribution. An example will illustrate this technique. If I mark

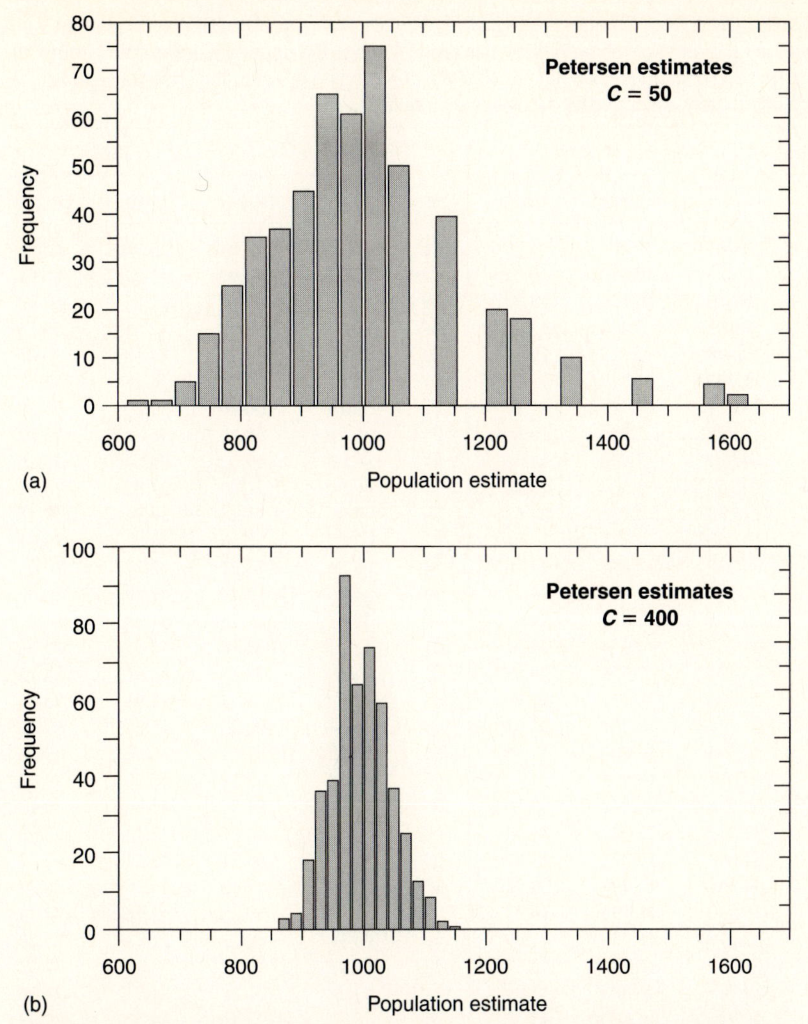

Figure 2.1 Petersen population estimates for an artificial population of $N = 1000$. Five hundred replicate samples were drawn. In both cases $M = 400$ individuals were marked in the first sample. (a) Samples of $C = 50$ were taken repeatedly for the second sample. A total of 13 estimates out of 500 did not include the known population size of 1000 (estimates below 705 or above 1570). (b) Samples of $C = 400$ were taken for the second sample. A total of 22 estimates out of 500 did not include the known population size (estimates below 910 or above 1105). Note the wide range of estimates of population size when the number of animals recaptured is small.

600 (M) and then recatch a total of 200 (C) animals, 13 (R) of which are marked, we have from Table 2.1:

$$\text{Lower 95\% confidence limit of } R \text{ when } R \text{ is } 13 = 6.686$$

$$\text{Upper 95\% limit of } R \text{ when } R \text{ is } 13 = 21.364$$

TABLE 2.1 CONFIDENCE LIMITS FOR A POISSON FREQUENCY DISTRIBUTION. Given the number of organisms observed (x), this table provides the upper and lower limits from the Poisson distribution. Do not use this table unless you are sure the observed counts are adequately described by a Poisson distribution.

	95%		99%			95%		99%	
x	Lower	Upper	Lower	Upper	x	Lower	Upper	Lower	Upper
0	0	3.285	0	4.771	46	34.05	60.24	29.90	65.96
1	0.051	5.323	0.010	6.914	47	34.66	61.90	31.84	66.81
2	0.355	6.686	0.149	8.727	48	34.66	62.81	31.84	67.92
3	0.818	8.102	0.436	10.473	49	36.03	63.49	32.55	69.83
4	1.366	9.598	0.823	12.347	50	37.67	64.95	34.18	70.05
5	1.970	11.177	1.279	13.793	51	37.67	66.76	34.18	71.56
6	2.613	12.817	1.785	15.277	52	38.16	66.76	35.20	73.20
7	3.285	13.765	2.330	16.801	53	39.76	68.10	36.54	73.62
8	3.285	14.921	2.906	18.362	54	40.94	69.62	36.54	75.16
9	4.460	16.768	3.507	19.462	55	40.94	71.09	37.82	76.61
10	5.323	17.633	4.130	20.676	56	41.75	71.28	38.94	77.15
11	5.323	19.050	4.771	22.042	57	43.45	72.66	38.94	78.71
12	6.686	20.335	4.771	23.765	58	44.26	74.22	40.37	80.06
13	6.686	21.364	5.829	24.925	59	44.26	75.49	41.39	80.65
14	8.102	22.945	6.668	25.992	60	45.28	75.78	41.39	82.21
15	8.102	23.762	6.914	27.718	61	47.02	77.16	42.85	83.56
16	9.598	25.400	7.756	28.852	62	47.69	78.73	43.91	84.12
17	9.598	26.306	8.727	29.900	63	47.69	79.98	43.91	85.65
18	11.177	27.735	8.727	31.839	64	48.74	80.25	45.26	87.12
19	11.177	28.966	10.009	32.547	65	50.42	81.61	46.50	87.55
20	12.817	30.017	10.473	34.183	66	51.29	83.14	46.50	89.05
21	12.817	31.675	11.242	35.204	67	51.29	84.57	47.62	90.72
22	13.765	32.277	12.347	36.544	68	52.15	84.67	49.13	90.96
23	14.921	34.048	12.347	37.819	69	53.72	86.01	49.13	92.42
24	14.921	34.665	13.793	38.939	70	54.99	87.48	49.96	94.34
25	16.768	36.030	13.793	40.373	71	54.99	89.23	51.78	94.35
26	16.77	37.67	15.28	41.39	72	55.51	89.23	51.78	95.76
27	17.63	38.16	15.28	42.85	73	56.99	90.37	52.28	97.42
28	19.05	39.76	16.80	43.91	74	58.72	91.78	54.03	98.36
29	19.05	40.94	16.80	45.26	75	58.72	93.48	54.74	99.09
30	20.33	41.75	18.36	46.50	76	58.84	94.23	54.74	100.61
31	21.36	43.45	18.36	47.62	77	60.24	94.70	56.14	102.16
32	21.36	44.26	19.46	49.13	78	61.90	96.06	57.61	102.42
33	22.94	45.28	20.28	49.96	79	62.81	97.54	57.61	103.84
34	23.76	47.02	20.68	51.78	80	62.81	99.17	58.35	105.66
35	23.76	47.69	22.04	52.28	81	63.49	99.17	60.39	106.12
36	25.40	48.74	22.04	54.03	82	64.95	100.32	60.39	107.10
37	26.31	50.42	23.76	54.74	83	66.76	101.71	60.59	108.61
38	26.31	51.29	23.76	56.14	84	66.76	103.31	62.13	110.16
39	27.73	52.15	24.92	57.61	85	66.76	104.40	63.63	110.37
40	28.97	53.72	25.83	58.35	86	68.10	104.58	63.63	111.78
41	28.97	54.99	25.99	60.39	87	69.62	105.90	64.26	113.45
42	30.02	55.51	27.72	60.59	88	71.09	107.32	65.96	114.33
43	31.67	56.99	27.72	62.13	89	71.09	109.11	66.81	114.99
44	31.67	58.72	28.85	63.63	90	71.28	109.61	66.81	116.44
45	32.28	58.84	29.90	64.26	91	72.66	110.11	67.92	118.33

TABLE 2.1 *Continued*

	95%		99%			95%		99%	
x	Lower	Upper	Lower	Upper	x	Lower	Upper	Lower	Upper
92	74.22	111.44	69.83	118.33	97	78.73	116.93	73.20	124.16
93	75.49	112.87	69.83	119.59	98	79.98	118.35	73.62	125.70
94	75.49	114.84	70.05	121.09	99	79.98	120.36	75.16	127.07
95	75.78	114.84	71.56	122.69	100	80.25	120.36	76.61	127.31
96	77.16	115.60	73.20	122.78					

Source: Crow and Gardner 1959.

When $x > 100$ use the normal approximation:
95% confidence limits of x:

Lower limit $= x - 0.94 - 1.96\sqrt{x - 0.02}$

Upper Limit $= x + 1.94 + 1.96\sqrt{x + 0.98}$

99% confidence limits of x:

Lower limit $= x - 1.99 - 2.576\sqrt{x + 0.33}$

Upper Limit $= x + 2.99 + 2.576\sqrt{x + 1.33}$

and we obtain the 95% confidence interval for estimated population size (sampling without replacement) by using these values of R in equation (2.2):

$$\text{Lower 95\% confidence limit on } \hat{N} = \frac{(601)(201)}{21.364 + 1} - 1 = 5402$$

$$\text{Upper 95\% confidence limit on } \hat{N} = \frac{(601)(201)}{6.686 + 1} - 1 = 15{,}716$$

Normal Approximation Confidence Intervals This method is essentially a "large sample" method that obtains a confidence interval on the fraction of marked animals in the second catch (R/C). It should be used only when R is above 50. The confidence interval for (R/C) is defined by the following formula:

$$\frac{R}{C} \pm \left\{ z_\alpha \left[\sqrt{\frac{(1 - f)(R/C)(1 - R/C)}{(C - 1)}} \right] + \frac{1}{2C} \right\} \tag{2.4}$$

where f = fraction of total population sampled in the second sample $= \dfrac{R}{M}$

$\dfrac{1}{2C}$ = correction for continuity

z_α = standard normal deviate for $(1 - \alpha)$ level of confidence
= 1.96 (for 95% confidence limits)
= 2.576 (for 99% confidence limits)

For large samples and a large population size, both the *finite population correction* $(1 - f)$ and the correction for continuity are negligible, and this formula for the normal approximation to the binomial simplifies to

$$\frac{R}{C} \pm z_\alpha \sqrt{\frac{(R/C)(1 - R/C)}{(C - 1)}} \tag{2.5}$$

The constant z_α defines 100 $(1 - \alpha)$ percent confidence limits, and values can be substituted from tables of the standard normal distribution (z); see Zar (1996, app19). For example, for 80% confidence limits, replace z_α with the constant 1.2816.

One example will illustrate this method. If I mark 1800 animals (M) and catch at the second sampling a total of 800 (C), of which 73 (R) are already tagged, then from formula (2.4) for 95% confidence limits,

$$\frac{73}{800} \pm \left\{ 19.6 \left[\sqrt{\frac{(1 - 73/1800)(73/800)(1 - 73/800)}{(800 - 1)}} \right] + \frac{1}{2(800)} \right\}$$

$$= 0.09125 \pm 0.020176$$

and the 95% confidence interval for R/C is 0.07107 to 0.111426. To obtain a 95% confidence interval for the estimated population size, we use these limits for R/C in equation (2.1):

$$\hat{N} = \frac{CM}{R}$$

$$\text{Lower 95\% confidence limit on } \hat{N} = \frac{1}{0.111426}(1800) = 16{,}154$$

$$\text{Upper 95\% confidence limit on } \hat{N} = \frac{1}{0.07107}(1800) = 25{,}326$$

Binomial Confidence Intervals Binomial confidence intervals for the fraction of marked animals (R/C) can be obtained most easily graphically from Figure 2.2. The resulting confidence interval will be approximate but should be adequate for most ecological data. For example, suppose I mark 50 birds (M), and then capture 22 (C) birds of which 14 (R) are marked. The fraction of marked animals (R/C) is 14/22, or 0.64. Move along the x-axis (*Sample Proportion*) to 0.64, and then move up until you intercept the first sample size line of C, or 22. Then read across to the y-axis (*Population Proportion*) to obtain 0.40, the lower 95% confidence limit for R/C. Now repeat the procedure to intercept the second sample size line of C, or 22. Reading across again, you find on the y-axis 0.83, the upper 95% confidence limit for R/C.

These confidence limits can be converted to confidence limits for population size (N) by the use of these limits for R/C in formula (2.1), exactly as described above (page 26). We use these limits for R/C in equation (2.1):

$$\hat{N} = \frac{CM}{R}$$

$$\text{Lower 95\% confidence limit on } \hat{N} = \frac{1}{0.83}(50) = 60$$

$$\text{Upper 95\% confidence limit on } \hat{N} = \frac{1}{0.40}(50) = 125$$

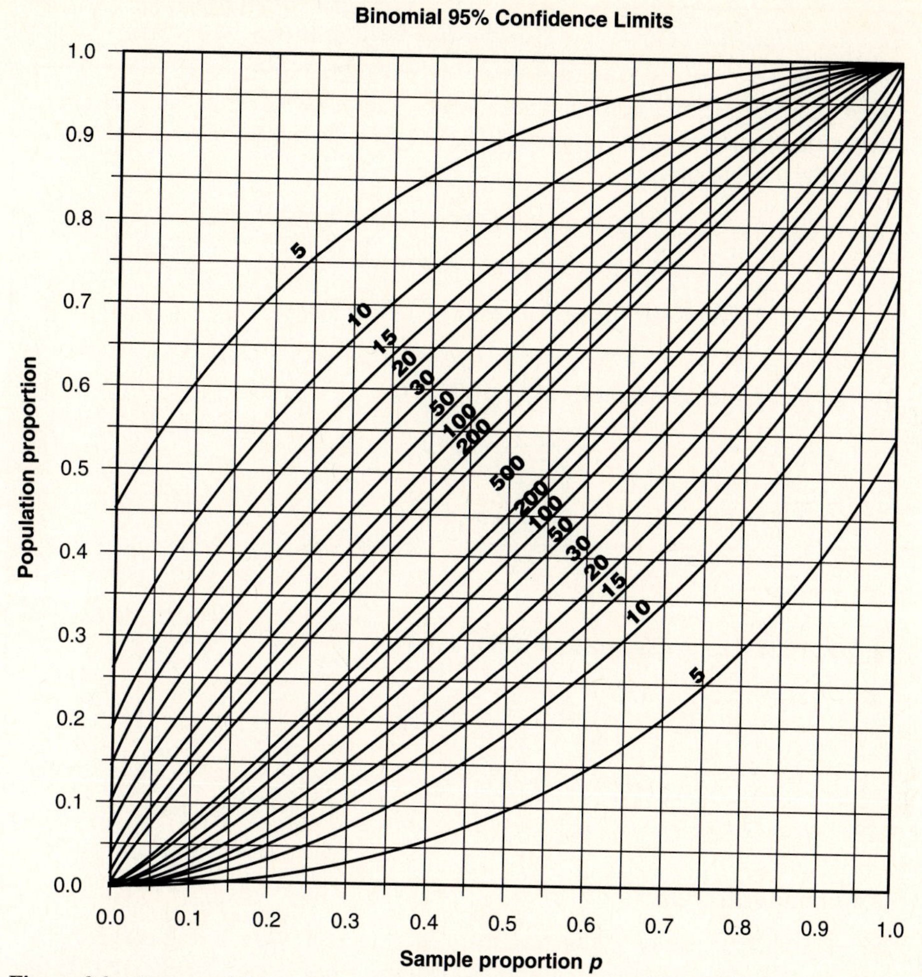

Figure 2.2 Upper and lower 95% confidence limits for a population proportion. Confidence limits are read off the *y*-axis for an observed value of *p* on the *x*-axis. Sample sizes are marked on the contour lines.

Alternatively, binomial confidence limits can be calculated using program-group EXTRAS (see Appendix 2), which uses the formulas given in Zar (1996, 524). Binomial confidence intervals can also be read from extensive tables, like those of Burnstein (1971), but for most ecological applications the precision of extensive tables of the binomial distribution is not required. For the bird example above, the tables provide the exact 95% confidence limits of 0.407 to 0.828, compared with the slightly less accurate 0.40 to 0.83 from Figure 2.2.

Program-group MARK-RECAPTURE (see Appendix 2) computes the Petersen estimate of population size and the appropriate confidence interval according to the recommendations of Seber (1982). Box 2.1 illustrates how the Petersen calculations are done.

Box 2.1 Petersen Method of Population Estimation

Green and Evans (1940) estimated the size of a snowshoe hare population at Lake Alexander, Minnesota in the winter of 1932–33 from these live-trapping data (sampling without replacement):

M = 948 hares caught and tagged in first sample

C = 421 total caught in second sample

R = 167 marked hares in second sample

Biased Estimator (equation [2.1])

$$\hat{N} = \frac{CM}{R}$$

$$= \frac{(421)(948)}{167} = 2390 \text{ hares}$$

Unbiased Estimator (equation [2.2])

$$\hat{N} = \frac{(M + 1)(C + 1)}{(R + 1)} - 1$$

$$= \frac{(421 + 1)(948 + 1)}{167 + 1} - 1 = 2383 \text{ hares}$$

Confidence Interval

R/C = 0.3967, so we use a binomial confidence interval. We can read the approximate confidence limits from Figure 2.2, and this provides visual estimates of 0.35 to 0.45 as 95% confidence limits.

We thus obtain a confidence interval for population size as:

$$\text{Lower 95\% confidence limit on } \hat{N} = \frac{C}{R}M = \frac{1}{0.45}(948) = 2107$$

$$\text{Upper 95\% confidence limit on } \hat{N} = \frac{C}{R}M = \frac{1}{0.35}(948) = 2709$$

If we use the more exact equations for binomial confidence limits given by Zar (1996, 524), we obtain 95% confidence limits of 0.369 to 0.425 for the ratio of R/C, and narrower confidence limits of 2231 to 2569 individuals.

These calculations can be done in program-group MARK-RECAPTURE (Appendix 2).

2.1.2 Sample Size Estimation

Let us now turn the Petersen method upside down and ask, given a rough estimate of population size (N), *How large a sample do I need to take to get a good estimate of abundance?* This is a key question for a great deal of ecological work that uses mark-recapture techniques, and should be asked *before* a study is done, not afterward. We need two preliminary bits of information before we can answer the sample size question precisely:

1. Initial estimate of population size (N)
2. Accuracy desired in Petersen estimate

The first requirement is the usual statistical one of "knowing the answer before you do any of the work," but this can be a very rough, order-of-magnitude guess. The only rule is to guess on the high side if you wish to be conservative.

For the second requirement, we need a quantitative definition of *accuracy*. We want our estimate of population size to be within a certain range of the true value, and we call this the accuracy (A):

$$A = \pm 100 \left(\frac{\text{Estimated population size} - \text{true population size}}{\text{True population size}} \right)$$

where A = Accuracy of an estimate (as a percentage)

Thus we may wish our estimate to be within $\pm 10\%$ accuracy. We cannot guarantee this accuracy all the time, and we must allow a probability (α) of not achieving our desired accuracy. Robson and Regier (1964) suggest using an α of 0.05 and three standard levels for A:

1. *Preliminary surveys*: $A = \pm 50\%$; these cases require only a rough idea of population size.
2. *Management work*: $A = \pm 25\%$; a moderate level of accuracy is desired.
3. *Research work*: $A = \pm 10\%$; these cases require quite accurate data on population size.

Figure 2.3 gives sample sizes needed for small populations for the three levels of accuracy, and Figure 2.4 gives sample sizes needed for large populations. Both these graphs are to be read in the same way. The contour lines are estimated population sizes (N). The vertical and horizontal axes of the graphs are values of M, number marked in the first Petersen sample, and C, number caught in the second Petersen sample. From the contour line corresponding to your estimated population size, read off pairs of M, C values. All of these pairs of values are suitable for obtaining the accuracy specified, and you can pick any set you like.

EXAMPLE 1

$A = \pm 25\%$, rough population estimate, $\hat{N} = 20$; reading from the contour line of $N = 20$ in Figure 2.3(b):

M	C	
1	20	Any one of these
10	15	four pairs will
15	10	produce the
17	8	desired accuracy.

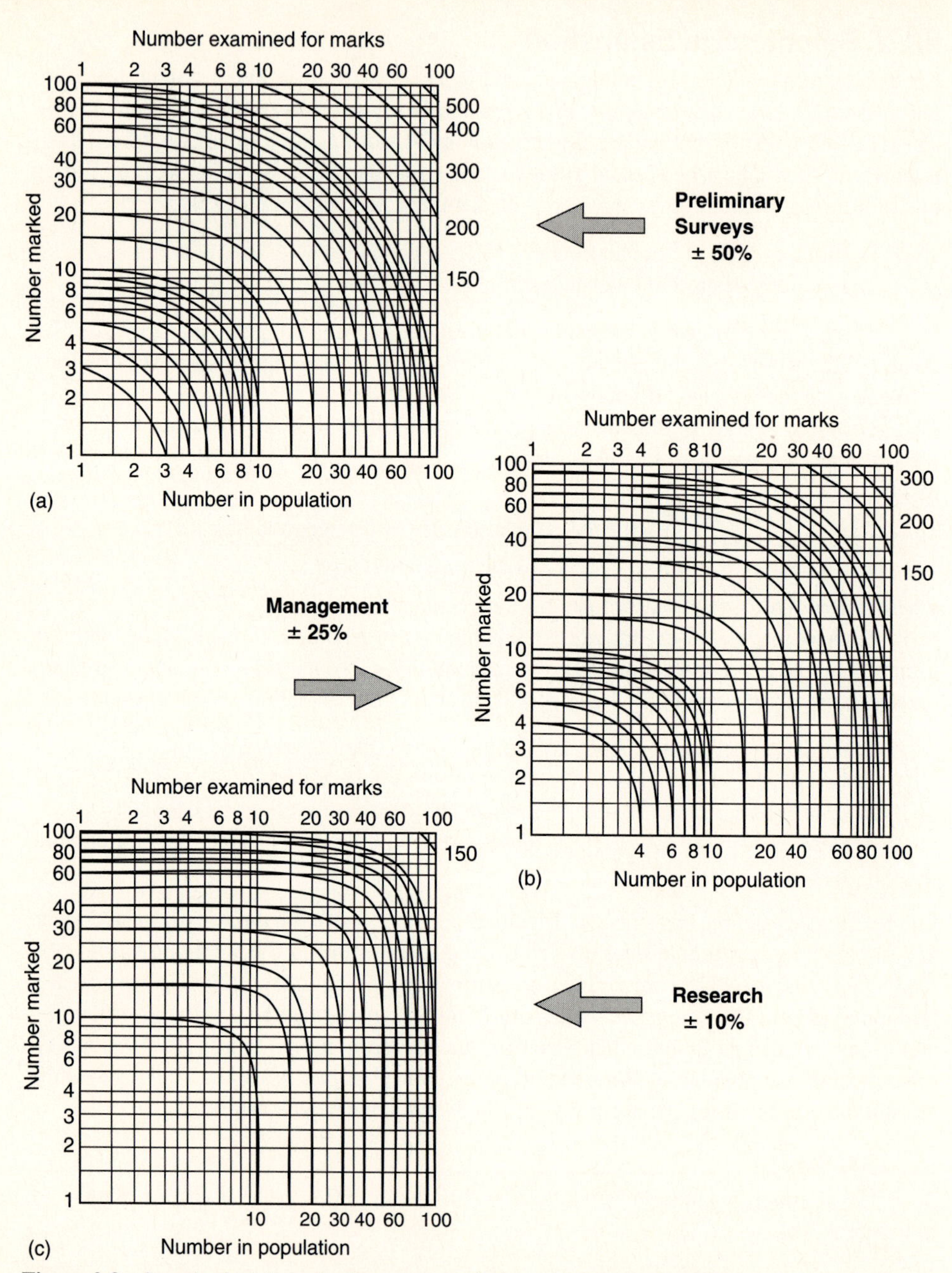

Figure 2.3 Sample size charts for Petersen population estimates for small populations; α is 5% in all cases. (a) For preliminary surveys, an accuracy of ±50% is recommended; (b) for management studies, ±25%; and (c) for research, ±10%. (After Robson and Regier 1964.)

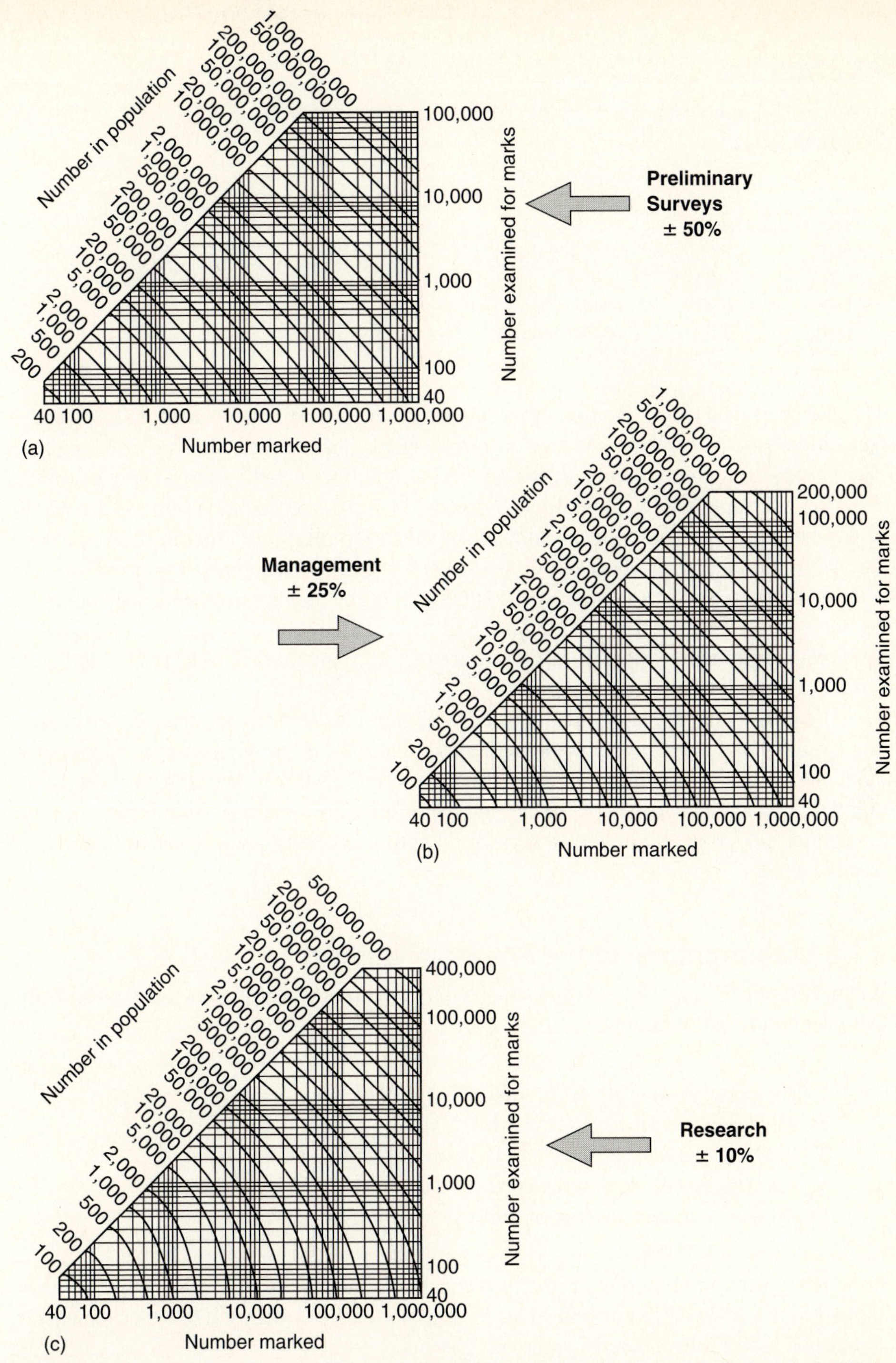

Figure 2.4 Sample size charts for Petersen population estimates for *large* populations; α is 5% in all cases. (a) For preliminary surveys, an accuracy of ±50% is recommended; (b) for management studies, ±25%; and (c) for research, ±10%. These charts are a continuation of those in Figure 2.3. (After Robson and Regier 1964.)

EXAMPLE 2

$A = \pm 10\%$, rough estimate $\hat{N} = 200{,}000$; reading from the contour line of $N = 200{,}000$ in Figure 2.4(c):

M	*C*	
125,000	200	Any one of these
92,000	400	four pairs will
30,000	2000	produce the
7,000	10,000	desired accuracy.

The choice of which combination of marking (M) and capturing (C) to use will depend in part on the ease of marking and ease of sampling for marked individuals. For example, it may be very easy to mark individuals, but very difficult to recapture large samples. In this situation in Example 2, we might decide to take $M = 92{,}000$ and $C = 400$. Alternatively, it may be very expensive to mark animals and cheap to recapture them in large numbers, and we might take $M = 7000$ and $C = 10{,}000$ for the same example. In normal circumstances we try to equalize sample sizes so that $M \approx C$, the number marked is approximately the same as the number caught in the second Petersen sample. This balance of $M \approx C$ is the point of minimum effort in terms of numbers of animals that need to be handled to achieve a given level of precision.

I cannot overemphasize the importance of going through this sample size estimation procedure *before you start* an ecological study that uses mark-recapture. To obtain even moderate accuracy in population estimation, we often need to mark 50% or more of the individuals in a population (Roff 1973), and the information available from Figures 2.3 and 2.4 can tell you how much work you must do and whether the project you are starting is feasible at all.

2.1.3 Assumptions of the Petersen Method

If $\hat{N}$ in formula (2.2) or (2.3) is to be an accurate estimate of population size the following five assumptions must hold:

1. The population is *closed*, so that N is constant.
2. All animals have the same chance of getting caught in the first sample.
3. Marking individuals does not affect their catchability.
4. Animals do not lose marks between the two sampling periods.
5. All marks are reported upon discovery in the second sample.

If the first assumption is to hold in a biological population, the Petersen method must be applied over a short period of time. This is an important consideration in experimental design, because if a long period of sampling is needed to get the size of sample required, the method may lose its precision.

There are several simple ways in which the first assumption can be violated without affecting the validity of the Petersen estimate. If accidental deaths occur during the first sampling and marking, the estimate of N is valid but refers to the number of individuals *alive* in the population after the first sample is released. Natural mortality may occur between the first and second Petersen samples without affecting the estimation if marked and unmarked animals have an equal chance of dying between the first and second samples. The question of whether mortality falls equally on marked and unmarked animals is difficult to answer for any natural population, and this assumption is best avoided if possible.

Population size (N) in a mark-recapture estimation always refers to the *catchable population*, which may or may not be the entire population. For example, nets will catch only fish above a certain size, and smaller individuals are thus ignored in the Petersen calculations. The recruitment of young animals into the catchable population between the first and second samples tends to decrease the proportion of marked animals in the second sample and thus inflate the Petersen estimate of N above the true population size at time 1. When there is recruitment but no mortality, $\hat{N}$ will be a valid estimate of population size at time 2. Fishery scientists have developed tests for recruitment and subsequent corrections that can be applied to Petersen estimates when there is recruitment (Seber 1982), but the best advice is still the simplest: Avoid recruitment of new individuals into the population by sampling over a short time interval.

One of the crucial assumptions of mark-recapture work is that marked and unmarked animals are equally catchable. This assumption is necessary for all the methods discussed in this chapter, and I defer discussion of methods for testing this assumption until Section 2.4.

Random sampling is critical if we are to obtain accurate Petersen estimates, but it is difficult to achieve in field programs. We cannot number all the animals in the population and select randomly M individuals for marking and later C individuals for capture. If all animals are equally catchable, we can approximate a random sample by sampling *areas* at random with a constant effort. We can divide the total area into equal subareas, and allot sampling effort to the subareas selected from a random number table. All points selected must be sampled with the same effort.

Systematic sampling (Chapter 8, Section 8.4) is often used in place of random sampling. Systematic sampling will provide adequate estimates of population size only if there is uniform mixing of marked and unmarked animals and all individuals are equally catchable. Uniform mixing is unlikely among most animals that show territorial behavior or well-defined home ranges. Where possible you should aim for a random sample and avoid the dangerous assumption of uniform mixing that must be made after systematic samples are taken.

I will not review here the various marking methods applied to animals and plants. Seber (1982), Southwood (1978), and Sutherland (1996) give general references regarding marking methods for fish, birds, mammals, and insects. Experienced field workers can offer much practical advice on marking specific kinds of organisms, not all of which is contained in books. New metals, plastics, and dyes are constantly being developed as well. The importance of these technological developments in marking cannot be overemphasized because of the problems of lost marks and unreported marks. Poor marking techniques will destroy the most carefully designed and statistically perfect mark-recapture scheme, and it is important to use durable tags that will not interfere with the animal's life cycle.

Tag losses can be estimated easily by giving all the M individuals in the first Petersen sample two types of tags, and then recording the following in the second sample:

R_A = Number of tagged animals in second sample with only an A-tag (i.e., they have lost their B-tag)

R_B = Number of tagged animals in second sample with only a B-tag (i.e., they have lost their A-tag)

R_{AB} = Number of tagged animals in second sample with both tags present.

Clearly, if R_A and R_B are both zero, you are probably not having a tag-loss problem. The total number of recaptures is defined as

$$R = R_A + R_B + R_{AB} + \begin{Bmatrix} \text{Number of individuals losing both tags} \\ \text{and thus being classed as unmarked} \end{Bmatrix}$$

Seber (1982, 95) shows that if we define

$$k = \frac{R_A R_B}{(R_A + R_{AB})(R_B + R_{AB})}$$

and

$$c = \frac{1}{1 - k}$$

then we can estimate the total number of recaptures as

$$\hat{R} = c(R_A + R_B + R_{AB}) \tag{2.6}$$

For example, if we mark 500 beetles with dots of cellulose paint (A) and reflecting paint (B) and obtain in the second sample

$$R_A = 23$$

$$R_B = 6$$

$$R_{AB} = 127$$

then we obtain

$$k = \frac{(23)(6)}{(23 + 127)(6 + 127)} = 0.006917$$

$$c = \frac{1}{1 - 0.006917} = 1.00697$$

$$\hat{R} = 1.00697(23 + 6 + 127) = 157.09$$

Thus we observe $R = 156$ and estimate $\hat{R} = 157$, so that only one insect is estimated to have lost both marks during the experiment. From this type of experiment, we can calculate the probabilities of losing marks of a given type:

$$\left\{\begin{matrix}\text{Probability of losing a tag of type A} \\ \text{between samples 1 and 2}\end{matrix}\right\} = \frac{R_B}{(R_B + R_{AB})} \tag{2.7}$$

$$\left\{\begin{matrix}\text{Probability of losing a tag of type B} \\ \text{between samples 1 and 2}\end{matrix}\right\} = \frac{R_A}{(R_A + R_{AB})} \tag{2.8}$$

In this example the probability of losing a cellulose dot is only 0.045, but the chances of losing a reflecting dot are 0.153, more than a threefold difference between tag types.

The failure to report all tags recovered may be important for species sampled by hunting or commercial fishing in which the individuals recovering the tags are not particularly interested in the data obtained. Tag returns are affected by the size of the tag reward and the ease of visibility of the tags. Paulik (1961) has discussed methods of testing for incomplete tag returns in large-scale mark-recapture experiments.

2.2 SCHNABEL METHOD

Schnabel (1938) extended the Petersen method to a series of samples in which there is a 2nd, 3rd, 4th . . . *n*th sample. Individuals caught at each sample are first examined for marks, then marked and released. Marking occurs at each of the sampling times. Only a single type of mark need be used, since throughout a Schnabel experiment we need to distinguish only two types of individuals: *marked* = caught in one or more prior samples; and *unmarked* = never caught before. We determine for each sample time (t):

C_t = Total number of individuals caught in sample t

R_t = Number of individuals already marked when caught in sample t

U_t = Number of individuals marked for first time and released in sample t

Normally,

$$C_t = R_t + U_t$$

but if there are accidental deaths involving either marked or unmarked animals, these are subtracted from the U_t value.* The number of marked individuals in the population continues to accumulate as we add further samples, and we define

$$M_t = \left\{\begin{matrix}\text{Number of marked individuals in the} \\ \text{population just before sample } t \text{ is taken}\end{matrix}\right\}$$

$$M_t = \sum_{i=1}^{t-1} U_i$$

so that, for example, $M_6 = U_1 + U_2 + U_3 + U_4 + U_5$

*The number of accidental deaths is assumed to be small.

TABLE 2.2 MARK-RECAPTURE DATA OBTAINED FOR A SCHNABEL-TYPE ESTIMATE OF POPULATION SIZE

Date, t	Number of fish caught C_t	Number of recaptures[a] R_t	Number newly marked (less deaths)[b]	Marked fish at large[c] M_t
June 2	10	0	10	0
June 3	27	0	27	10
June 4	17	0	17	37
June 5	7	0	7	54
June 6	1	0	1	61
June 7	5	0	5	62
June 8	6	2	4	67
June 9	15	1	14	71
June 10	9	5	4	85
June 11	18	5	13	89
June 12	16	4	10	102
June 13	5	2	3	112
June 14	7	2	4	115
June 15	19	3	—	119
Totals	162	24	119	984

Source: S.D. Gerking (1953) marked and released sunfish in an Indiana lake for 14 days and obtained these data.

[a] The number of fish already marked when taken from the nets.

[b] Note that there were two accidental deaths on June 12 and one death on June 14.

[c] Number of marked fish assumed to be alive in the lake in the instant just before sample t is taken.

Table 2.2 gives an example of some Schnabel-type data obtained on a sunfish population.

Given these counts of marked and unmarked individuals, we need to derive an estimate of population size N for a closed population. Ricker (1975) gives several methods, and Seber (1982, Chap. 4) discusses this estimation problem in detail. We will describe here two methods of estimation: the original Schnabel method and the Schumacher and Eschmeyer method, which is the most robust and useful ecological model according to Seber (1982).

Schnabel Method Schnabel (1938) treats the multiple samples as a series of Petersen samples. She obtained a population estimate as a weighted average of Petersen estimates:

$$\hat{N} = \frac{\Sigma_t (C_t M_t)}{\Sigma_t R_t} \tag{2.9}$$

If the fraction of the total population that is caught in each sample $(C_t/\hat{N})$ *and* the fraction of the total population that is marked $(M_t/\hat{N})$ is always less than .1, a better estimate is

$$\hat{N} = \frac{\Sigma_t (C_t M_t)}{\Sigma_t R_t + 1} \tag{2.10}$$

The variance of the Schnabel estimator is calculated on the reciprocal of N:

$$\text{Variance}\left(\frac{1}{\hat{N}}\right) = \frac{\Sigma\, R_t}{(\Sigma\, C_t M_t)^2} \tag{2.11}$$

$$\text{Standard error of } \frac{1}{\hat{N}} = \sqrt{\text{Variance}\left(\frac{1}{\hat{N}}\right)} \tag{2.12}$$

Schumacher and Eschmeyer Method Schumacher and Eschmeyer (1943) pointed out that if we plot on graph paper

x-axis: M_t, number of individuals previously marked (before time t)

y-axis: R_t/C_t, proportion of marked individuals in the t-th sample

the plotted points should lie on a straight line of slope ($1/N$) passing through the origin. Thus one could use linear regression techniques to obtain an estimate of the slope ($1/N$) and thus an estimate of population size. The appropriate formula for this estimation is

$$\hat{N} = \frac{\sum_{t=1}^{s} (C_t M_t^2)}{\sum_{t=1}^{s} (R_t M_t)} \tag{2.13}$$

where s = Total number of samples

The variance of the Schumacher estimator is obtained from linear regression theory as the variance of the slope of the regression (Zar 1996, 330; Sokal and Rohlf 1995, 471). In terms of mark-recapture data,

$$\text{Variance of}\left(\frac{1}{\hat{N}}\right) \frac{\Sigma\,(R_t^2/C_t) - [(\Sigma\, R_t M_t)^2/\Sigma\, C_t M_t^2]}{s - 2} \tag{2.14}$$

where s = Number of samples included in the summations

The standard error of the slope of the regression is obtained as follows:

$$\text{Standard error of}\left(\frac{1}{\hat{N}}\right) = \sqrt{\frac{\text{Variance of } (1/\hat{N})}{\Sigma\,(C_t M_t^2)}} \tag{2.15}$$

2.2.1 Confidence Intervals

If the total number of recaptures ($\Sigma\, R_t$) is less than 50, confidence limits for the Schnabel population estimate should be obtained from the Poisson distribution (Table 2.1). These confidence limits for $\Sigma\, R_t$ from Table 2.1 can be substituted into equations (2.9) or (2.10) as the denominator, and the upper and lower confidence limits for population size estimated.

For the Schnabel method, if the total number of recaptures ($\Sigma\, R_t$) is above 50, use the normal approximation derived by Seber (1982, 142). This large-sample procedure uses the standard error and a t-table to get confidence limits for ($1/\hat{N}$) as follows:

$$\frac{1}{\hat{N}} \pm t_\alpha \text{S.E.} \tag{2.16}$$

where S.E. = standard error of $1/N$ (equation [2.12] or [2.15])

t_α = value from Student's t-table for $(100 - \alpha)\%$ confidence limits

Enter the t-table with $(s - 1)$ degrees of freedom for the Schnabel method and $(s - 2)$ degrees of freedom for the Schumacher and Eschmeyer methods, where s is the number of samples. Invert these limits to obtain confidence limits for $\hat{N}$. Note that this method (equation [2.16]) is used for all Schumacher-Eschmeyer estimates, regardless of the number of recaptures. This procedure is an approximation, but the confidence limits obtained are sufficiently accurate for all practical purposes.

We can use the data in Table 2.1 to illustrate both these calculations. From that data we obtain

$$\sum C_t M_t = 10{,}740$$

$$\sum (C_t M_t^2) = 970{,}296$$

$$\sum R_t M_t = 2294$$

$$\sum (R_t^2 / C_t) = 7.7452$$

For the Schnabel estimator, from equation (2.9), we have

$$\hat{N} = \frac{10{,}740}{24} = 447.5 \text{ sunfish}$$

A 95% confidence interval for this estimate is obtained from the Poisson distribution because there are only 24 recaptures. From Table 2.1, with $\Sigma R_t = 24$ recaptures, the 95% confidence limits on ΣR_t are 14.921 and 34.665. Using equation (2.9) with these limits, we obtain

$$\text{Lower 95\% confidence limit} = \frac{\Sigma (C_t M_t)}{\Sigma R_t} = \frac{10{,}740}{34.665} = 309.8 \text{ sunfish}$$

$$\text{Upper 95\% confidence limit} = \frac{\Sigma (C_t M_t)}{\Sigma R_t} = \frac{10{,}740}{14.921} = 719.8 \text{ sunfish}$$

The 95% confidence limits for the Schnabel population estimate are 310 to 720 for the data in Table 2.1.

For the Schumacher-Eschmeyer estimator, from equation (2.13), we have

$$\hat{N} = \frac{970{,}296}{2294} = 423 \text{ sunfish}$$

The variance of this estimate, from equation (2.14), is

$$\text{Variance of } \left(\frac{1}{\hat{N}}\right) = \frac{7.7452 - [(2294)^2/970{,}296]}{14 - 2} = 0.1934719$$

$$\text{Standard error } \left(\frac{1}{\hat{N}}\right) = \sqrt{\frac{0.1934719}{970{,}296}} = 0.0004465364$$

The confidence interval from equation (2.16) is

$$\frac{1}{423} \pm (2.179)(0.0004465364)$$

or 0.0013912 to 0.0033372. Taking reciprocals, the 95% confidence limits for the Schumacher-Eschmeyer estimator are 300 and 719 sunfish, very similar to those obtained from the Schnabel method.

Seber (1982) recommends the Schumacher-Eschmeyer estimator as the most robust and useful one for multiple censuses on closed populations. Program-group MARK-RECAPTURE (Appendix 2) computes both these estimates and their appropriate confidence intervals. Box 2.2 illustrates an example of the use of the Schnabel estimator.

2.2.2 Assumptions of the Schnabel Method

The Schnabel method makes all the same assumptions that the Petersen method makes: that the population size is constant without recruitment or losses, that sampling is random, and that all individuals have an equal chance of capture in any given sample. Our previous discussion of the limitations these assumptions impose are thus all relevant here.

The major advantage of the multiple sampling in a Schnabel experiment is that it is easier to pick up violations of these assumptions. A regression plot of the proportion of marked animals (Y) on the number previously marked (X) will be linear if these assumptions are true, but will become curved when the assumptions are violated. Figure 2.5 illustrates one data set from cricket frogs that fits the expected straight line and a second data set from red-backed voles that is curvilinear. Unfortunately, there is no unique interpretation possible for curvilinear plots, and they signal that one or more assumptions are violated without telling us which assumption or how it is violated. For example, the downward curve in Figure 2.5b could be due to the immigration of unmarked animals into the study area, or to the fact that marked voles are less easily caught in live traps than unmarked voles.

When a curvilinear relationship is present, you may still be able to obtain a population estimate by the use of *Tanaka's model* (see Seber 1982, 145). The procedure is to plot the following on log-log paper:

x-axis: M_t (number marked at large)

y-axis: C_t/R_t (number caught/number of recaptures)

Figure 2.6 shows this graph for the same vole data plotted in Figure 2.5b. If Tanaka's model is applicable, this log-log plot should be a straight line, and the x-intercept is an estimate of population size. Thus, as a first approximation, a visual estimate of N can be made by drawing the regression line by eye and obtaining the x-axis intercept, as shown in Figure 2.6. The actual formulas for calculating $\hat{N}$ by the Tanaka method are given in Seber (1982, 147) and are not repeated here.

The Schnabel and Tanaka methods both assume that the number of accidental deaths or removals is negligible, so that population size is constant. If a substantial fraction of the

Box 2.2 Schnabel Method of Population Estimation

Turner (1960) obtained data on a frog population sampled over 5 days:

Sample #	No. caught, C_t	No. recaptures, R_t	No. newly marked (less deaths)	No. marked frogs at large, M_t
1	32	0	32	0
2	54	18	36	32
3	37	31	6	68
4	60	47	13	74
5	41	36	5	87

$$\sum C_t M_t^2 = (32)(0^2) + (54)(32^2) + \cdots + (41)(87^2) = 865{,}273$$

$$\sum R_t M_t = (0)(0) + (18)(32) + \cdots + (36)(87) = 9294$$

$$\sum \frac{R_t^2}{C_t} = \frac{(18)^2}{54} + \frac{(31)^2}{37} + \cdots = 100.3994$$

To estimate population size using the Schumacher-Eschmeyer method, we obtain from equation (2.13)

$$\hat{N} = \frac{\sum_{t=1}^{s} (C_t M_t^2)}{\sum_{t=1}^{s} (R_t M_t)} = \frac{865{,}273}{9294} = 93.1 \text{ frogs}$$

To obtain confidence intervals, we can use the normal approximation method defined in equations (2.14) and (2.15) as follows:

$$\text{Variance of } \left(\frac{1}{\hat{N}}\right) = \frac{\Sigma\, (R_t^2/C_t) - [(\Sigma\, R_t M_t)^2/\Sigma\, (C_t M_t^2)]}{s - 2}$$

$$\text{Variance of } \left(\frac{1}{\hat{N}}\right) = \frac{100.3994 - [(9294)^2/865{,}273]}{5 - 2} = 0.1904809$$

$$\text{Standard error of } \left(\frac{1}{\hat{N}}\right) = \sqrt{\frac{\text{Variance of } (1/\hat{N})}{\Sigma\, (C_t M_t^2)}}$$

$$\text{Standard error of } \left(\frac{1}{\hat{N}}\right) = \sqrt{\frac{0.1904809}{865{,}273}} = 0.0004692$$

Note that it is important to maintain an excessive number of decimal places in these intermediate calculations to preserve accuracy in the final estimates.

The 95% confidence interval is, from equation (2.16),

$$\frac{1}{\hat{N}} \pm t_\alpha \text{S.E.}$$

since $t_{.025}$ for 3 d.f. = 3.182, we obtain:

$$\frac{1}{93.1} \pm (3.182)(0.0004692)$$

or 0.09248 to 0.012234. Taking reciprocals, we obtain the 95% confidence limits on population size of 82 to 108. Note that these confidence limits are *not* symmetrical about $\hat{N}$.

These calculations can be done by program-group MARK-RECAPTURE (Appendix 2).

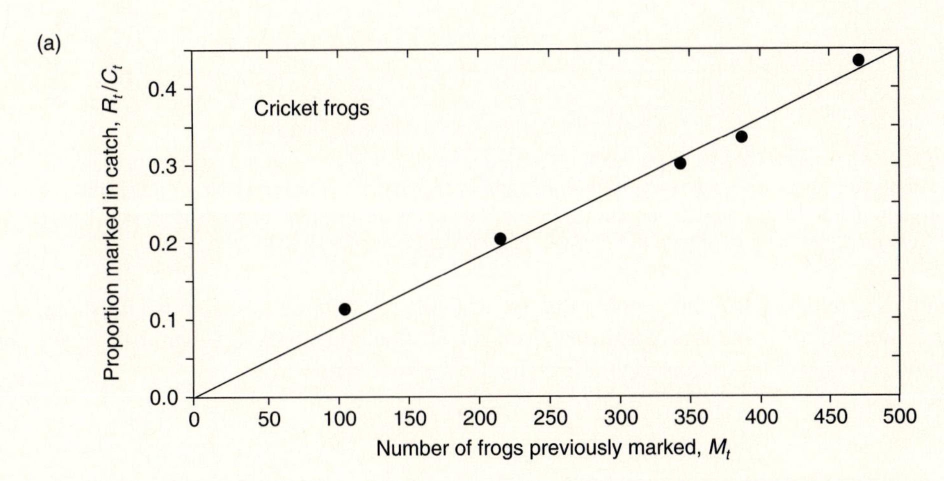

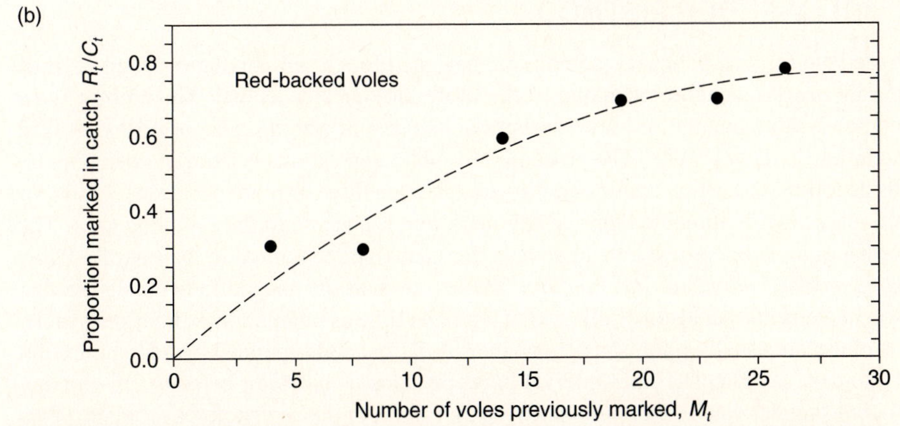

Figure 2.5 Schnabel method of population estimation: (a) cricket frogs; (b) red-backed voles. A plot of the accumulated number of marked animals (M_t) against the proportion of marked animals in each sample (R_t/C_t) will be linear (a) if the assumptions underlying the method are fulfilled. A curvilinear plot (b) shows that the assumptions are violated, and either the population is not closed or catchability is not constant. [Data from Pyburn 1958 for (a) and from Tanaka 1951 for (b).]

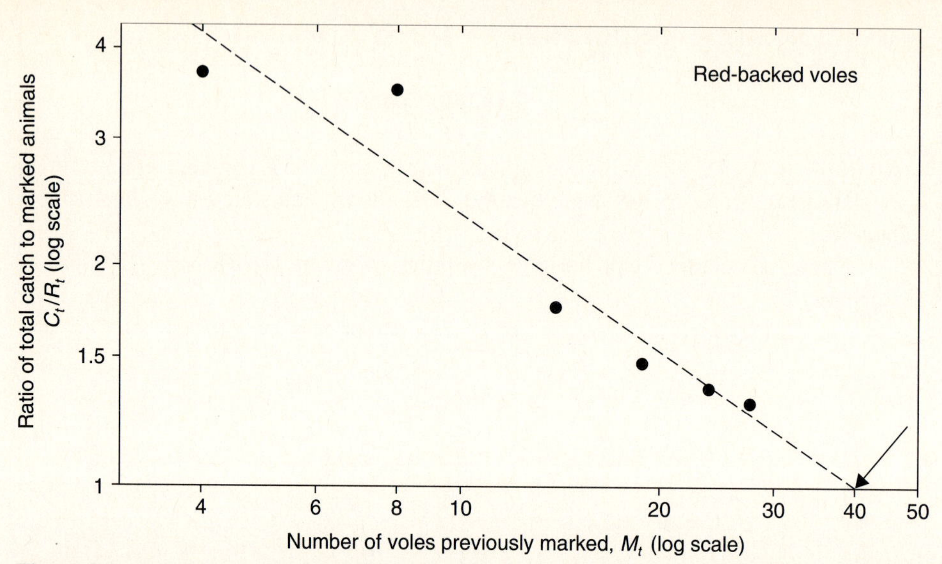

Figure 2.6 Tanaka's model of population estimation: red-backed voles. A log-log plot of the number of marked animals at large (M_t) against the ratio of total catch to marked catch (C_t/R_t) should be a straight line to fit this model, and the *x*-intercept (arrow) is the estimate of population size. These data are the same as those plotted in Figure 2.5b. (Data from Tanaka 1951.)

samples is removed from the population, for example, by hunting or commercial fishing, corrections to the population estimates given above should be applied. Seber (1982, 152) provides details of how to correct these estimates for known removals.

2.3 JOLLY-SEBER METHOD

Both the Petersen and Schnabel methods are designed for closed populations, and we now extend the mark-recapture technique to the more biologically realistic situation of *open* populations. Most populations are constantly changing in size because of births, deaths, immigration, and emigration. The procedure for obtaining estimates from an open population is as follows. Mark-recapture samples are taken on three or more occasions. Individuals are marked with numbered tags or any mark that is specific to the sampling time. The important point here is to be able to answer, for each marked animal in the sample: *When was this marked individual last captured?* Often animals are tagged individually so that data on movements can also be collected at the same time as population estimation is done. The samples are usually point samples of short duration, and separated by a long duration from the next sample. The time interval between samples need not be constant, and any number of samples can be accommodated, so that series of data extending over many years can be used in this method.

Before we discuss the estimation procedures, let us get a good grasp of the data used in the calculations. All the animals in the first sample must be unmarked by definition. For the second and all subsequent samples, the total catch can be subdivided into two fractions: *marked animals* and *unmarked animals*. For marked individuals, we ask one important

TABLE 2.3 MARK-RECAPTURE DATA FOR A SERIES OF 11 SAMPLES OF A FIELD VOLE (*Microtus pennsylvanicus*) POPULATION IN THE SOUTHWESTERN YUKON

Time of last capture	Time of capture											
	1	2	3	4	5	6	7	8	9	10	11	
1		15	1	0	0	0	0	0	0	0	0	
2			15	0	1	0	0	0	0	0	0	←Z_6
3				37	2	0	0	0	0	0	0	
4					61	4	1	1	0	0	0	
5						75	3	2	0	0	0	
6					m_6↗		77	4	0	0	0	
7						R_6↗		69	0	0	0	
8									8	1	0	
9										14	0	
10											19	
Total marked (m_t)	0	15	16	37	64	79	81	76	8	15	19	
Total unmarked (u_t)	22	26	32	45	25	22	26	15	11	12	3	
Total caught (n_t)	22	41	48	82	89	101	107	91	19	27	22	
Total released (s_t)	21	41	46	82	88	99	106	90	19	26	22	

Note: Each sample was captured over 2 days, and samples were usually 2 weeks apart. The data are cast in a Method B table as a preliminary to the estimation of population size by the Jolly-Seber method.

question: *When was this individual* last *captured?* Leslie (1952) showed that this was the most informative question to ask about marked individuals—better, for example, than asking when it was first captured. The answers to this question are tallied in a Method B table (Leslie and Chitty 1951). Table 2.3 gives a Method B table for a series of 11 samples from a field vole population. In the second sample all of the 15 marked voles must have been last caught in the first sample. But in the third sample marked animals may have been last caught at time 2 (15) or time 1 (1). Marked individuals may thus evade capture for one or more sampling periods. In an intensive mark-recapture program, most of the marked animals caught will have been last caught at the previous sampling, and consequently will appear along the subdiagonal of the Method B table. Conversely, when populations are very large or sampling is less intensive, more and more recaptures will appear above the subdiagonal of the Method B table, since marked individuals will typically evade capture for several sampling periods in this situation.

Given a Method B table, we can now define the following variables using the terminology of Jolly (1965):

m_t = Number of marked animals caught in sample t

u_t = Number of unmarked animals caught in sample t

n_t = Total number of animals caught in sample t

$= m_t + u_t$

s_t = Total number of animals released after sample t

$= (n_t -$ accidental deaths or removals)

m_{rt} = Number of marked animals caught in sample t last caught in sample r

All these variables are symbols for the data written in the Method B table. For example, Table 2.3 shows that in column 6

$$m_6 = 75 + 4 + 0 + 0 + 0 = 79 \text{ voles}$$

We require two more variables for our calculations,

R_t = Number of the s_t individuals released at sample t and caught again in some later sample

Z_t = Number of individuals marked *before* sample t, not caught in sample t, but caught in some sample after sample t

These last two variables are more easily visualized than described (see Table 2.3). For example, as shown in the table, along row 6

$$R_6 = 77 + 4 + 0 + 0 + 0 = 81 \text{ voles}$$

The Z_t are those animals that missed getting caught in sample t, and survived to turn up later on. In an intensive mark-recapture program, the Z_t values will approach zero. In this table for example in columns 7–11, rows 1–5

$$Z_6 = 3 + 2 + 1 + 1 = 7 \text{ individuals}$$

We can now proceed to estimate population size following Jolly (1965) from the simple relationship

$$\text{Population size} = \frac{\text{Size of marked population}}{\text{Proportion of animals marked}}$$

The proportion of animals marked is estimated as

$$\hat{\alpha}_t = \frac{m_t + 1}{n_t + 1} \tag{2.17}$$

where the "+ 1" is a correction for bias in small samples (Seber 1982, 204). The size of the marked population is more difficult to estimate because there are two components of the marked population at any sampling time: (1) marked animals actually caught; and (2) marked animals present but not captured in sample t. Seber (1982) showed that the sizes of the marked population could be estimated by:

$$\hat{M}_t = \frac{(s_t + 1)Z_t}{R_t + 1} + m_t \tag{2.18}$$

where M_t = Estimated size of the marked population just before sample time t

We can now estimate population size:

$$\hat{N}_t = \frac{\hat{M}_t}{\hat{\alpha}_t} \tag{2.19}$$

where $\hat{N}_t$ = Estimated population size just before sample time t

TABLE 2.4 POPULATION ESTIMATES DERIVED FROM DATA IN TABLE 2.3 BY USE OF THE JOLLY-SEBER MODEL OF POPULATION ESTIMATION. Eleven samples from a field vole population in the southwestern Yukon

Sample	Proportion marked	Size of marked population	Population estimate	Probability of survival	Number joining	Standard errors of N_t	Standard errors of φ_t	Standard errors of B_t
1	.000	0.0	—[a]	.832	—	—	.126	—
2	.381	17.5	45.9	.395	31.4	6.1	.077	2.7
3	.347	17.2	49.5	.862	47.9	4.0	.055	3.6
4	.458	40.7	88.8	.824	24.6	5.3	.043	3.3
5	.722	70.5	97.7	.925	22.1	5.3	.032	2.6
6	.784	87.5	111.6	.853	27.3	5.8	.043	2.8
7	.759	91.7	120.8	.651	12.8	7.1	.046	1.7
8	.837	76.0	90.8	.104	11.4	6.5	.033	1.8
9	.450	9.3	20.7	.738	10.9	3.6	.101	1.1
10	.571	15.0	26.2	—	—	4.4	—	—
11	.870	—	—	—	—	—	—	—

[a]— means that no estimate can be made of this parameter from the data available.

Table 2.4 gives these estimates for the data in Table 2.3, and a few sample calculations follow:

$$\hat{\alpha}_5 = \frac{64 + 1}{89 + 1} = 0.7222$$

$$\hat{\alpha}_6 = \frac{79 + 1}{101 + 1} = 0.7843$$

$$\hat{M}_2 = \frac{(41 + 1)(1)}{16 + 1} + 15 = 17.47$$

$$\hat{M}_6 = \frac{(99 + 1)(7)}{81 + 1} + 79 = 87.54$$

$$\hat{N}_6 = \frac{87.54}{0.7843} = 111.6$$

We can derive an additional bonus from the use of the Jolly-Seber Model of population estimation—an estimate of the loss rate and the addition rate of the population. Consider first the loss rate. We define

$$\Phi_t = \text{Probability of survival from sample time } t \text{ to sample time } t + 1$$
$$= \frac{\text{Size of marked population at start of sample time } t + 1}{\text{Size of marked population at end of sample time } t}$$

The marked population is added to during each sampling time as new individuals are marked and released. Thus the size of the marked population at the end of sample t consists of the marked individuals alive at the start of t plus the new individuals marked during sampling time t, or in the symbols defined above,

$$\hat{\Phi}_t = \frac{\hat{M}_{t+1}}{\hat{M}_t + (s_t - m_t)} \tag{2.20}$$

This formula corrects for all accidental deaths or removals at time t. Note that the probability of survival is determined by sampling the *marked population* only. Survival in this context means staying alive on the study area. Individuals that emigrate are counted as losses in the same way as individuals that die.

An addition rate to the population is called the *dilution rate*,* since it includes both additions by births and by immigration:

$$\lambda_t = \text{Dilution rate from sample time } t \text{ to sample time } t + 1$$

$$= \frac{\text{Actual population size at sample time } t + 1}{\text{Expected population size at sample time } t + 1 \text{ if no additions occurred}}$$

The expected population size in the absence of any gains is clearly

$$\left\{\begin{matrix}\text{Expected population} \\ \text{size at time } t + 1\end{matrix}\right\} = \left(\begin{matrix}\text{Probability of survival} \\ \text{from } t \text{ to } t + 1\end{matrix}\right)\left(\begin{matrix}\text{Population size} \\ \text{at time } t\end{matrix}\right)$$

Once we correct for accidental deaths, we obtain

$$\hat{\lambda}_t = \frac{\hat{N}_{t+1}}{\hat{\Phi}_t[\hat{N}_t - (n_t - s_t)]} \tag{2.21}$$

Thus if there are no additions at all, we expect the dilution rate to be 1.0, its theoretical minimum value.

From a series of population estimates we can define

$$\text{Finite rate of population change} = \frac{N_{t+1}}{N_t}$$

if there are no accidental deaths at time t. By rearranging formula (2.21) we see for the ideal case of no accidental losses ($n_t = s_t$)

$$\Phi_t \lambda_t = \frac{N_{t+1}}{N_t} = \text{Finite rate of population change} \tag{2.22}$$

Thus when there are no losses ($\Phi_t = 1.0$) and no additions ($\lambda_t = 1.0$), the population remains constant.

The addition rate to the population can also be expressed as a *number* of individuals, if we define

$$B_t = \left\{\begin{matrix}\text{Number of new animals joining the population between} \\ \text{time } t \text{ and } t + 1 \textit{ and } \text{still alive at time } t + 1\end{matrix}\right\}$$

$$\hat{B}_t = \hat{N}_{t+1} - \hat{\Phi}_t[\hat{N}_t - (n_t - s_t)] \tag{2.23}$$

This formula is clearly just a rearrangement of formula (2.21).

*Note that the *dilution rate* defined here follows Jolly (1965) and differs from that of Leslie et al. (1953), who defined a dilution factor that is zero when there is no dilution.

It is important to keep in mind that these estimates of N_t, Φ_t, λ_t, and B_t are *not* independent estimates of what is happening in a population, but they are all interconnected, so that when one estimate is poor, they are all poor.

Estimates of population size (N_t) and dilution rate (λ_t) cannot be obtained for the first sample, nor can any of these parameters be estimated for the last sample (see Table 2.4). In addition, the probability of survival cannot be estimated for the next to last sample (Φ_{s-1}). For this reason, population studies using the Jolly-Seber model should be planned to start before the time periods of interest and extend at least two samples beyond the last time periods of interest.

2.3.1 Confidence Intervals

Variances for the estimates of population size, probability of survival, and dilution rate are given in Jolly (1965) but these are valid only for large samples. Manly (1971) and Roff (1973) have shown with computer simulation that the large-sample variances are not reliable for getting confidence intervals because the estimated confidence belt is too narrow. Manly (1984) suggested an alternative method of obtaining confidence limits for Jolly-Seber estimates. Manly's methods have been criticized as being slightly arbitrary by Pollock et al. (1990), but since they work reasonably well, we will present them here.

Confidence Limits for Population Size Transform the estimates of N_t as follows:

$$T_1(\hat{N}_t) = \log_e(\hat{N}_t) + \log_e\left[\frac{1 - (p_t/2) + \sqrt{1 - p_t}}{2}\right] \tag{2.24}$$

where $\hat{p}_t = \dfrac{n_t}{\hat{N}_t} = \dfrac{\text{Total caught at time } t}{\text{Estimated population size at time } t}$

The variance of this transformation is given by

$$Var[T_1(\hat{N}_t)] = \left(\frac{\hat{M}_t - m_t + s_t + 1}{\hat{M}_t + 1}\right)\left(\frac{1}{R_t + 1} - \frac{1}{s_t + 1}\right) + \frac{1}{m_t + 1} + \frac{1}{n_t + 1} \tag{2.25}$$

The upper and lower 95% confidence limits for T_1 are given by

$$T_{1L} = T_1(\hat{N}_t) - 1.6\sqrt{\hat{V}ar[T_1(\hat{N}_t)]} \tag{2.26}$$

$$T_{1U} = T_1(\hat{N}_t) + 2.4\sqrt{\hat{V}ar[T_1(\hat{N}_t)]} \tag{2.27}$$

where T_{1L} = lower confidence limit for T_1
T_{1U} = upper confidence limit for T_1

The confidence limits for population size are then

$$\frac{(4L + n_t)^2}{16L} < \hat{N}_t < \frac{(4U + n_t)^2}{16U} \tag{2.28}$$

where $L = e^{T_{1L}}$
$U = e^{T_{1U}}$
$e = 2.71828 \ldots$ (the base of natural logarithms)

These confidence limits will not be symmetrical about $\hat{N}_t$.

Confidence Limits for Probability of Survival Transform the estimates of Φ_t as follows:

$$T_2(\hat{\Phi}_t) = \log_e\left(\frac{1 - \sqrt{1 - A_t\Phi_t}}{1 + \sqrt{1 + A_t\Phi_t}}\right) \tag{2.29}$$

where $A_t = \dfrac{C_t}{B_t + C_t}$

B_t and C_t are defined below. The variance of this transformed value of Φ_t is

$$\hat{V}ar[T_2(\Phi_t)] = B_t + C_t \tag{2.30}$$

$$\text{where } \hat{B}_t = \left[\frac{(\hat{M}_{t+1} - m_{t+1} + 1)(\hat{M}_{t+1} - m_{t+1} + s_{t+1} + 1)}{(\hat{M}_{t+1} + 1)^2}\right]$$

$$\times\left(\frac{1}{R_{t+1} + 1} - \frac{1}{s_{t+1} + 1}\right) + \left(\frac{\hat{M}_t - m_t + 1}{\hat{M}_t - m_t + s_t + 1}\right)$$

$$\times\left[\left(\frac{1}{R_t + 1} - \frac{1}{s_t + 1}\right)\right]$$

$$\hat{C}_t = \frac{1}{\hat{M}_{t+1} + 1}$$

The upper and lower 95% confidence limits for T_2 are given by

$$\hat{T}_{2L} = T_2(\hat{\Phi}_t) - 1.9\sqrt{\hat{V}ar[T(\hat{\Phi}_t)]} \tag{2.31}$$

$$\hat{T}_{2U} = T_2(\hat{\Phi}_t) + 2.1\sqrt{\hat{V}ar[T(\hat{\Phi}_t)]} \tag{2.32}$$

The confidence limits for the probability of survival are then

$$\frac{1}{A_t}\left[1 - \frac{(1 - L)^2}{(1 + L)^2}\right] < \hat{\Phi}_t < \frac{1}{A_t}\left[1 - \frac{(1 - U)^2}{(1 + U)^2}\right] \tag{2.33}$$

where $L = e^{T_{2L}}$

$U = e^{T_{2U}}$

$A_t = \dfrac{C_t}{B_t + C_t}$ as defined above

There are at present no comparable estimates for the confidence limits that should be used for the dilution rate.

These formulas can be calculated by a computer program. Program-group MARK-RECAPTURE in Appendix 2 computes the Jolly-Seber estimators of population size, probability of survival, and dilution rate, along with the confidence limits from Manly (1984).

The Jolly-Seber model can be considerably simplified if some of the parameters of the model are constant. Jolly (1982) discusses four versions of the Jolly-Seber model:

1. **Model A**—the full Jolly-Seber model as discussed above, in which all parameters may change from sampling period to sampling period.

2. **Model B**—the constant survival model, which assumes that survival rates are constant over the entire period of sampling, so that only one survival estimate must be calculated.
3. **Model C**—the constant capture model, which assumes that the probability of capture remains constant for all animals over all the entire sampling periods.
4. **Model D**—the constant survival and constant capture model, which assumes both constant survival and constant capture probabilities during the study period. This is the simplest model.

If any of these reduced models, B, C, or D is appropriate for your data, you can gain considerably in precision by adopting the reduced model. All these models involve maximum likelihood estimation, so that they can be solved only with the help of a computer. Pollock et al. (1990) describe Program JOLLY, which will carry out these calculations on an IBM PC.

2.3.2 Assumptions of the Jolly-Seber Method

The Jolly-Seber method is designed for open populations, so unlike the earlier methods discussed in this chapter we do not need to assume the absence of recruitment and mortality. Random sampling becomes the crucial assumption, and we assume the following:

1. Every individual has the same probability (α_t) of being caught in the t-th sample, regardless whether it is marked or unmarked.
2. Every marked individual has the same probability (Φ_t) of surviving from the t-th to the $(t + 1)$th sample.
3. Individuals do not lose their marks, and marks are not overlooked at capture.
4. Sampling time is negligible in relation to intervals between samples.

The critical assumption of equal catchability for marked and unmarked individuals must be tested before you can rely on estimates provided by the Jolly-Seber model or indeed by any mark-recapture model.

2.4 TESTS OF EQUAL CATCHABILITY

A number of tests of equal catchability have been proposed. Unequal catchability may be due to three general causes (Eberhardt 1969):

1. The behavior of individuals in the vicinity of the trap.
2. Learning by animals already caught to come to or to avoid traps (trap-addicted or trap-shy animals).
3. Unequal opportunity to be caught because of trap positions.

Some forms of unequal catchability cannot be tested for statistically, and we must rely on biological intuition to recognize them. For example, access to traps may be controlled by social position in many vertebrates, so that subordinate or juvenile animals are rarely captured. In extreme cases an individual's catchability is zero, and the presence of such individuals can be detected only by a change in the capturing technique (e.g., Boonstra and Krebs 1978). Catchability of individual snowshoe hares is affected by the number of traps located within the individual's home range (Boulanger and Krebs 1996). The second form of unequal catchability can be detected by the statistical techniques I am about to describe, and can be alleviated by a change in capturing technique or possibly by a reduction in trapping frequency. The third form of unequal catchability can be eliminated by random sampling of trapping stations, and is thus a statistical defect rather than a biological problem. Whatever the causes, unequal catchability seems to be more the rule than the exception for biological populations (Caughley 1977a), and hence it is essential that every mark-recapture census include as a first priority a test of this assumption.

Figure 2.7 gives a flow chart to guide you in deciding which test of equal catchability you should employ. There is a gradient of simple to more complex tests from top to bottom in Figure 2.7. Seber (1982) and Otis et al. (1978) provide detailed discussion of these and other tests of equal catchability.

Several techniques are available in mark-recapture studies for testing the assumption that all individuals in the population have an equal probability of capture. These tests assume that each individual in the population has a property called "catchability" and that this property is not affected by the individual's previous capture history. When there are only two or three periods of capture, it is impossible to test the equal catchability assump-

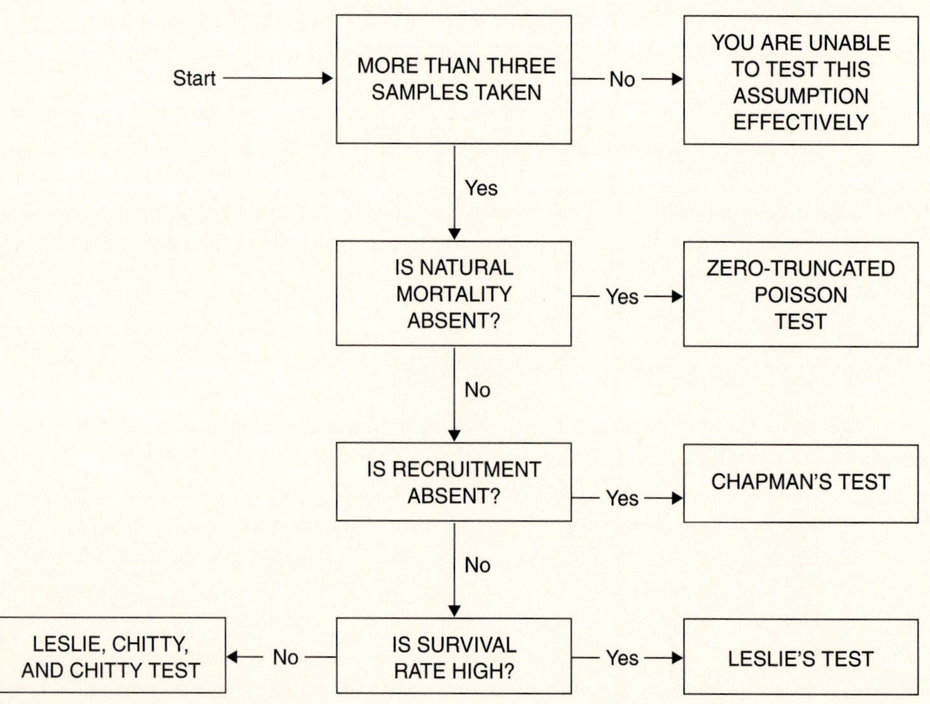

Figure 2.7 Sequence of decisions for choosing a statistical test for the equal catchability assumption of mark-recapture models.

tion with any statistical power. Cormack (1966) provides a test that can be used in Petersen-type marking experiments in which there are three sampling times, but the power of Cormack's test is low, and it seems to be rarely used. All other tests of equal catchability require four or more periods of capture, and the power of the test typically increases as the number of sampling periods gets larger and the average probability of capture increases.

2.4.1 Zero-Truncated Poisson Test

If the time period between the first and last samples in a multiple mark-recapture series is short enough to ensure little or no mortality, an efficient test of equal catchability is available (Caughley 1977a). The basic data for this test are provided as a frequency distribution of the number of animals caught once in a series of samples, the number caught twice, and so on. Box 2.3 gives some sample data.

To calculate the expected frequency distribution—expected on the basis of the null hypothesis of equal catchability—we need know only the mean of the observed data:

$$\bar{x} = \frac{\Sigma f_x x}{\Sigma f_x} \tag{2.34}$$

where x = number of times captured (1, 2, 3 . . .)
f_x = number of animals caught exactly x times

Box 2.3 Zero-Truncated Poisson Test for Equal Catchability

Keith and Meslow (1968) captured snowshoe hares during 7 days with the following results:

Number of captures, x	No. of hares caught, f_x	Expected frequencies (calculated below)
1	184	174.6
2	55	66.0
3	14	16.7
4	4	3.7
5	4	
6	0	
7	0	

The mean number of captures per individual hare is obtained from equation (2.34):

$$\bar{x} = \frac{\Sigma f_x x}{\Sigma f_x} = \frac{(184)(1) + (55)(2) + (14)(3) + (4)(4) + (4)(5)}{184 + 55 + 14 + 4 + 4} = 1.4253 \text{ captures}$$

Use this observed mean of the truncated distribution to estimate m as follows, using equation (2.35):

$$\bar{x} = \frac{m}{1 - e^{-m}}$$

continues

Provisionally estimating $\hat{m} = 1.3$, we get for this equation

$$1.4253 \overset{?}{=} \frac{1.3}{1 - e^{-1.3}} = \frac{1.3}{0.727} = 1.787$$

so our estimate of 1.3 is too high. After trial and error we find $\hat{m} = 0.7563$ because

$$1.4253 \overset{?}{=} \frac{0.7563}{1 - e^{-0.7563}} = \frac{0.7563}{0.5306} = 1.42537$$

Now we calculate the expected values from equation (2.36):

$$\begin{Bmatrix}\text{Expected number of} \\ \text{hares caught once}\end{Bmatrix} = \sum f_x \left(\frac{e^{-m}}{1 - e^{-m}} \frac{m^1}{1!} \right)$$

$$= 261\left(\frac{e^{-0.7563}}{1 - e^{-0.7563}} \frac{0.7563}{1} \right) = 174.6 \text{ hares}$$

$$\begin{Bmatrix}\text{Expected number of} \\ \text{hares caught twice}\end{Bmatrix} = \sum f_x \left(\frac{e^{-m}}{1 - e^{-m}} \frac{m^2}{2!} \right)$$

$$= 261\left(\frac{e^{-0.7563}}{1 - e^{-0.7563}} \frac{0.7563^2}{2} \right) = 66.0 \text{ hares}$$

$$\begin{Bmatrix}\text{Expected number of} \\ \text{hares caught three times}\end{Bmatrix} = \sum f_x \left(\frac{e^{-m}}{1 - e^{-m}} \frac{m^3}{3!} \right)$$

$$= 261\left(\frac{e^{-0.7563}}{1 - e^{-0.7563}} \frac{0.7563^3}{3!} \right) = 16.7 \text{ hares}$$

Of the 261 hares total, we now have accounted for

$$174.6 + 66.0 + 16.7 = 257.3 \text{ hares}$$

so there are only 3.7 hares left in the remainder of the distribution.

The χ^2 goodness-of-fit test compares the observed (O) and expected (E) values above in the usual way:

$$\chi^2 = \sum \left[\frac{(O - E)^2}{E} \right]$$

$$= \frac{(184 - 174.6)^2}{174.6} + \frac{(55 - 66.0)^2}{66.0} + \frac{(14 - 16.7)^2}{16.7} + \frac{(8 - 3.7)^2}{3.7}$$

$$= 7.77 \qquad \text{d.f.} = (n - 2) = (4 - 2) = 2$$

The observed χ^2 value of 7.77 exceeds the critical value of χ^2 for 2 d.f., $\alpha = 0.05$ ($\chi^2_{0.05} = 5.99$), and thus the null hypothesis of equal catchability of hares within this 7-day period is rejected.

Note that this method assumes that no deaths occur within the sampling period. Program-group MARK-RECAPTURE (Appendix 2) will do these calculations.

If the ratio $\bar{x}/s$, where s is the total number of sampling times, is less than about 0.25, a Poisson distribution can be fitted as an approximation to the data. If this ratio is too large, fit a binomial distribution as described by Seber (1982, 169). To fit a Poisson (with zero class omitted, since we do not know how many animals missed capture altogether—if we knew *that* we would not be doing all this work!), calculate m from the following equation:

$$\bar{x} = \frac{m}{1 - e^{-m}} \tag{2.35}$$

This equation can be solved by trial-and-error by inserting values of m less than $\bar{x}$ and testing the equality. Given m, we can obtain the expected frequency distribution:

$$\left\{ \begin{matrix} \text{Expected number of animals} \\ \text{captured } x \text{ times} \end{matrix} \right\} = \sum f_x \left(\frac{e^{-m} m^x}{(1 - e^{-m})x!} \right) \tag{2.36}$$

Box 2.3 illustrates these calculations. The observed and expected frequency distributions can be compared with a χ^2 goodness-of-fit test:

$$\chi^2 = \sum \left[\frac{(\text{Observed frequency} - \text{expected frequency})^2}{\text{Expected frequency}} \right]$$

combining the smaller frequencies so that the expected frequencies are all above 1.

Program-group MARK-RECAPTURE (Appendix 2) does these calculations for a zero-truncated Poisson test of equal catchability.

2.4.2 Chapman's Test

When there is no recruitment into a population (no births or immigration), we can test the validity of the assumption that all individuals have equal chances of capture by the following nonparametric test devised by Chapman (1952). The critical variable to be tallied for this test is the *time of first capture* of each marked animal in a series of samples. We define:

b_{ij} = number of marked animals caught in sample j *first* tagged in sample i

n_i = total number of individuals caught in sample i

u_i = number of unmarked animals caught in sample i

We can now set up the following array:

$$\begin{matrix} \dfrac{b_{12}}{u_1 n_2} & \dfrac{b_{13}}{u_1 n_3} & \dfrac{b_{14}}{u_1 n_4} & \cdots & \dfrac{b_{1s}}{u_1 n_s} \\ & \dfrac{b_{23}}{u_2 n_3} & \dfrac{b_{24}}{u_2 n_4} & \cdots & \dfrac{b_{2s}}{u_2 n_s} \\ & & \cdots & & \cdots \\ & & & & \dfrac{b_{s-1,s}}{u_{s-1} n_s} \end{matrix}$$

where s = Total number of samples taken.

TABLE 2.5 SIGNIFICANCE TABLE FOR CHAPMAN'S TEST OF EQUAL CATCHABILITY

Number of sampling periods, s	Observed number of negative differences, X							
	0	1	2	3	4	5	6	7
5	0.0035	0.0590	0.3056	0.6944	0.9410	0.9965	1.0000	1.0000
6	—	0.0012	0.0172	0.1052	0.3392	0.6608	0.8948	0.9828
7	—	—	0.0001	0.0020	0.0166	0.0627	0.2010	—
8	—	—	—	—	0.0001	0.0009	0.0323	0.1103

Source: Chapman (1952).

Note: Values in the table are probabilities, and the table is entered with the number of samples (s) down the side and the observed number of negative differences (X) along the top. The probabilities are the cumulative probabilities of getting this many or fewer negative signs under the null hypothesis of equal catchability.

Under the null hypothesis each element in the different rows is an independent estimate of the expected value ($1/\hat{N}$). If the null hypothesis of equal catchability is valid, we expect these estimates to bounce around at random. But if individuals are not equally catchable, the numbers in this array will tend to fall from left to right. To test for this trend, we take successive differences of all the observations across the rows (first observation minus second, second minus third, etc.) and keep track of the number of *negative* differences. We define the test statistic X as follows:

$$X = D_1 + D_2 + \cdots + D_{s-2} \tag{2.37}$$

where D_i = Number of negative differences in row i

Significant values of X could be either suspiciously low or suspiciously high values of X, and Table 2.5 gives critical values of the X statistic. Box 2.4 gives an example to illustrate Chapman's test.

2.4.3 Leslie's Test

Chapman's test can distinguish variation in catchability between the marked and the unmarked segments of the population. Leslie's test and the next one can distinguish unequal catchability only *within the marked segment* of the population. Leslie et al. (1953) showed that it was possible to have unequal catchability between marked and unmarked animals but still to have equal catchability within the marked segment of the population. This distinction is important because the estimation of total population size depends on equal catchability within the entire population, but the estimation of probability of survival depends only on equal catchability within the marked segment of the population.

Leslie's test of equal catchability can be applied to any cohort of animals known to be alive over a group of sampling periods. Schematically, we form a group of individuals all of which have a capture history like the following:

Samples of interest

Animal caught and marked sometime before i	i $i+1$ $i+2$ $i+3$ $i+4$ $i+5$ . . .	Animal caught sometime after last sample of interest

Box 2.4 Chapman's Test of Equal Catchability

Leslie, Chitty, and Chitty (1953) obtained these data for a vole (*Microtus agrestis*) population in Wales during a winter period when no births were occurring:

	Time of capture					
	1	2	3	4	5	6
Time of first capture						
1		14	8	7	4	2
2			5	4	2	1
3				3	2	4
4					14	7
5						12
No. unmarked (u_i)	193	25	29	38	34	21
Total catch (n_i)	193	39	42	52	56	47

Leslie (1952) called this grouping according to the time of first capture the *Method C* table. From this table, we calculate Chapman's table directly as the matrix of elements

$$\frac{b_{ij}}{u_i n_j}$$

where b_{ij} = Number of marked animals caught in sample j and first caught in sample i

Thus the first element is

$$\frac{b_{12}}{u_1 n_2} = \frac{14}{(193)(39)} = 0.00186$$

The second element in row 1 is

$$\frac{b_{13}}{u_1 n_3} = \frac{8}{(193)(42)} = 0.00099$$

We end up with this table:

	Time of capture				
	2	3	4	5	6
Time of first capture					
1	0.00186	0.00099	0.00070	0.00037	0.00022
2		0.00476	0.00308	0.00143	0.00085
3			0.00199	0.00123	0.00392
4				0.00658	0.00392
5					0.00751

continues

Across the rows of the table we subtract successive elements and record only the sign (+ or −) of the difference. Thus for the first row:

$$\text{(First element} - \text{second)} = 0.00186 - 0.00099 = +0.00087 \quad \oplus$$

$$\text{(Second element} - \text{third)} = 0.00099 - 0.00070 = +0.00029 \quad \oplus$$

$$\text{(Third element} - \text{fourth)} = 0.00070 - 0.00037 = +0.00033 \quad \oplus$$

$$\text{(Fourth element} - \text{fifth)} = 0.00037 - 0.00022 = +0.00015 \quad \oplus$$

We record 4 + and 0 − for the first row, so $D_1 = 0$ negative differences. Similarly for the other rows:

$$D_2 = 0 \qquad (3\ +,\ 0\ -)$$

$$D_3 = 1 \qquad (1\ +,\ 1\ -)$$

$$D_4 = 0 \qquad (1\ +,\ 0\ -)$$

Since the fifth row has only one element, we cannot obtain a difference for it. We calculate

$$X = \text{Sum of negative differences} = D_1 + D_2 + D_3 + D_4 = 1$$

for our six samples. Referring to Table 2.5 for $s = 6$ samples, we read the probability values directly:

$$\begin{Bmatrix} \text{Probability of getting 1 negative sign} \\ \text{under the null hypothesis when } s = 6 \end{Bmatrix} = 0.0012$$

Thus under the null hypothesis of equal catchability, this is a very unlikely outcome, and we can reject the null hypothesis at $p < 0.01$. The explanation in this case seems to be that marked and unmarked voles differ in their chances of entering a trap, and this leads to a steady influx of untagged voles through the winter, just as if breeding or immigration were going on.

At least 3 and preferably 5 or more samples should be used. We define the total cohort of individuals that satisfy the above criterion as G. We first tally the number of these G individuals caught in each of the samples (g_i) and the frequency distribution of captures (see Table 2.6). The test statistic is

$$\chi^2 = \frac{\Sigma f_x (x - \bar{x})^2}{\bar{x} - (\Sigma g_i^2 / G^2)} \qquad \text{d.f.} = (G - 1) \tag{2.38}$$

where $\bar{x} = \dfrac{\Sigma g_i}{G}$ = Mean number of captures per individual

Leslie suggests that at least $G = 20$ animals are needed to make this a satisfactory test. If the number of degrees of freedom is too large for the tabled values of χ^2 (d.f. > 100), you can convert the χ^2 value to the standard normal deviation by the approximate formula

$$z = \sqrt{2\chi^2} - \sqrt{(2G - 3)} \tag{2.39}$$

where G = Total number of animals

TABLE 2.6 DATA FOR LESLIE'S TEST OF EQUAL CATCHABILITY[a]

Year	Number of birds recaptured, g_i	Number of captures, x	Number of birds, f_x
1947	7	0	15
1948	7	1	7
1949	6	2	7
1950	4	3	2
1951	7	4	1
Total recaptures =	31	5	0
		Total birds =	32

[a]These data were taken from a cohort of shearwaters ($G = 32$) marked in 1946 and known to be alive in 1952 or later years. Recapture samples were taken annually, and thus the 5 years 1947 to 1951 are available for the test. Two frequency distributions are tallied. From these data we need to calculate:

$$\bar{x} = \textit{Mean number of recaptures per bird}$$
$$= \textit{Total recaptures/Total birds} = 31/32 = 0.9687$$
$$\Sigma g_i^2 = 7^2 + 7^2 + 6^2 + 4^2 + 7^2 = 199$$

For the data in Table 2.6,

$$\chi^2 = \frac{15(0 - 0.9687)^2 + 7(1 - 0.9687)^2 + 7(2 - 0.9687)^2 + \cdots}{0.9687 - (199/32^2)}$$

$$= \frac{38.97}{0.77436} = 50.32 \qquad \text{d.f.} = (32 - 1) = 31$$

At $\alpha = .05$, the critical value of χ^2 is 44.98 for 31 degrees of freedom, and thus $p < .05$. We reject the null hypothesis of equal catchability at the 5% level.

Box 2.5 gives another example of Leslie's test. The biggest drawback of Leslie's test is that it uses only a small fraction of the data collected. Carothers (1971) has extended

TABLE 2.7 METHOD B TABLE OF RE-RECAPTURES FOR A FIELD VOLE (*Microtus pennsylvanicus*) POPULATION IN THE SOUTHWESTERN YUKON

Time of last capture	Time of capture 1	2	3	4	5	6	7	8	9
1			5	0	1	0	0	0	0
2				12	1	0	0	0	0
3					30	2	0	0	0
4						55	3	1	0
5							62	4	0
6								57	0
7									8
8									8
Total marked	0	0	5	12	32	57	65	62	8
Total unmarked	—	15	11	25	32	22	16	14	0
Total caught	—	15	16	37	64	79	81	76	8
Total released	—	15	16	37	64	78	80	75	8

Note: These data form the basis of the calculations for the Leslie, Chitty and Chitty test of equal catchability. The compilation is exactly the same as the usual Method B table except that we define unmarked animals as those caught only once and marked animals as those caught twice or more. Note that because this is a compilation of re-recaptures, effectively no data are obtained from the first sample. From these data we can calculate the estimates as explained earlier with the results shown in Table 2.7.

Box 2.5 Leslie's Test of Equal Catchability

Manly and Parr (1968) marked and released moths (*Zygaema filipendula*) over 5 days. They marked the wings with dots of cellulose paint and used this color code:

Day 1: Green (g)
Day 2: White (w)
Day 3: Blue (b)
Day 4: Orange (o)

They obtained the following set of captures:

Day 1 (19 July): 57 marked green and released
Day 2 (20 July): 52 captures; 25 g, 27 unmarked
Day 3 (21 July): 52 captured; 8 g, 9 w, 11 gw, 24 unmarked
Day 4 (22 July): 31 captured: 2 g, 3 w, 4 b, 5 gb, 1 wb, 2 gwb, 14 unmarked
Day 5 (24 July): 54 captured: **1 g**, 2 w, 7 b, 5 o, **4 gw**, **2 gb**, **2 go**, 4 wb, 1 wo, 1 bo, **5 gbo**, **1 gwbo**, 19 unmarked

Of this total data set, we can use only the 15 moths (boldface) caught in the fifth sample and also caught in the first sample. These are identified as the 15 green marks in the day 5 sample. For these 15 moths, we record the following capture history with respect to the samples taken on days 2, 3, and 4:

Day	No. of moths recaptured, g_i
2	5 (4 gw, 1 gwbo)
3	8 (2 gb, 5 gbo, 1 gwbo)
4	8 (2 go, 5 gbo, 1 gwbo)
	Total recaptures = 21

No. of captures, x	No. of moths, f_x
0	1 (1 g)
1	8 (4 gw, 2 gb, 2 go)
2	5 (5 gbo)
3	1 (1 gwbo)
	Total moths = 15

$$\bar{x} = \text{Mean number of recaptures per moth} = \frac{\text{Total recaptures}}{\text{Total moths}} = \frac{21}{15} = 1.4$$

$$\sum g_i^2 = 5^2 + 8^2 + 8^2 = 153$$

From equation (2.38),

$$\chi^2 = \frac{\Sigma f_x(x - \bar{x})^2}{\bar{x} - (\Sigma g_i^2/G^2)} \qquad \text{d.f.} = (G - 1)$$

$$\chi^2 = \frac{1(0 - 1.4)^2 + 8(1 - 1.4)^2 + 5(2 - 1.4)^2 + \cdots}{1.4 - (153/15^2)}$$

$$= \frac{7.6}{0.72} = 10.56 \qquad \text{d.f.} = (15 - 1) = 14$$

The critical value of χ^2 at 14 d.f., $\alpha = 0.05$, is 23.68, and hence we accept the null hypothesis of equal catchability for these moths. Because of the small sample size, we should be cautious and try to repeat this test on a larger data set if possible.

Leslie's test to make it more efficient in the use of data from long chains of samples. Arnason and Baniuk (1980) have assembled a computer program to do Carothers (1971) test. The next test can also be applied to data of this sort.

2.4.4 Leslie, Chitty, and Chitty Test

Leslie et al. (1953) developed a clever test of the assumption of equal catchability within the marked segment of the population. This method can be used in open populations subject to births and deaths and is thus one of the most general of the tests for equal catchability. This test is usefully applied to data designed for the Jolly-Seber model of estimation.

The trick Leslie used to analyze this situation was to take one step back and, considering only the marked animals, define individuals caught only once as "unmarked" animals and individuals caught twice or more as "marked" animals. We apply exactly the same arguments we made above in developing the Jolly-Seber model and obtain an estimate of the number of new individuals joining the population over each sampling interval (called B_t in equation [2.23]). But because of the peculiar definitions of "unmarked" and "marked," we *know* what this number of new individuals is, and we can compare our *known* values with the *estimated* values. In Leslie's notation, we have the known values:

$$Z'_t = \left\{ \begin{array}{c} \text{Number of new additions to the marked population} \\ \text{that were made at sampling time } t \end{array} \right\}$$

$$= s_t - m_t$$

(using the terminology set out on page 43).

We get the estimated values as follows. First, we tally a Method B table for our marked animals, using the above definitions of "marked" and "unmarked" animals. This is thus a table of *re-recaptures*, and an example is given in Table 2.7. We proceed by the methods outlined on page 44 to estimate the size of the "marked" population by

$$\hat{M}_t = \frac{(s_t + 1)Z_t}{R_t + 1} + m_t \tag{2.40}$$

For example, from Table 2.7 the size of the "marked" population at time 6 is

$$\hat{M}_6 = \frac{(78 + 1)(4)}{66 + 1} + 57 = 61.72$$

The proportion of "marked" animals is obtained from

$$\hat{\alpha}_t = \frac{m_t + 1}{n_t + 1} \tag{2.41}$$

For example, from Table 2.7

$$\hat{\alpha}_6 = \frac{57 + 1}{79 + 1} = 0.725$$

Having calculated these values, as shown in Table 2.8, we can obtain the desired estimate:

$$\hat{Z}''_t = \left(\frac{\hat{M}_t - m_t + s_t}{\alpha_{t+1}}\right) - \frac{\hat{M}_t}{\hat{\alpha}_t} + d_t \tag{2.42}$$

where d_t = Number caught in sample t and not returned to the population (accidental deaths or removals)

For example, from the data in Table 2.8,

$$\hat{Z}''_6 = \left(\frac{61.72 - 57 + 78}{0.8049}\right) - \frac{61.72}{0.725} + 1 = 18.6 \text{ voles}$$

The completed calculations are given in Table 2.8. We can now compare the actual Z'_t values with the estimated values. The actual Z'_t values can be obtained from Table 2.3 as

$$Z'_t = s_t - m_t$$

and are respectively 30, 45, 24, 20, 25, and 14 for samples 3 to 8. Thus over this period of weeks we observe a total of 158 new individuals entering the marked population, and we predict a total of 129.0 new individuals, an 18% underestimate. This is a relatively large bias, and it suggests for these animals unequal catchability within the marked population. If socially dominant individuals are easier to trap, this sort of bias could arise.

TABLE 2.8 ESTIMATES OBTAINED FOR THE LESLIE, CHITTY, AND CHITTY TEST OF EQUAL CATCHABILITY FOR THE FIELD VOLE DATA IN TABLE 2.7. These estimates of the number of "new" animals marked can be compared to the observed values of Z'_t as discussed in the text.

Sample	Proportion "marked", α_t	Size of "marked" population, $\hat{M}_t$	Number of "new" animals marked for first time, $\hat{Z}''_t$	Standard error of $\hat{Z}''_t$
2	0.0000	—	—	—
3	0.3529	6.21	32.7	13.4
4	0.3421	14.30	35.6	13.8
5	0.5077	34.20	24.0	10.5
6	0.7250	61.72	18.6	8.2
7	0.8049	72.11	17.9	7.6
8	0.8182	62.00	0.2	4.1
9	1.0000	—	—	—

Seber (1982, 225) gives the large sample variance for the estimated Z''_t values. The square root of the variance of Z''_t can be used to judge the significance of the differences between observed and estimated Z'_t values (Leslie et al. 1953, 149). For most cases, however, a comparison of the sum of the observed and the sum of the estimated number of newly marked animals (as described above) will give the best information about the validity of the assumption of equal catchability.

Pollock et al. (1985a) have proposed a goodness-of-fit test for the Jolly-Seber model that compares the observed and expected numbers of individuals captured with each possible capture history. They provide a computer program to do these tests, which are similar to the Leslie, Chitty, and Chitty test.

2.5 PLANNING A MARK-RECAPTURE STUDY

The most important message this chapter should give you is that you must plan a mark-recapture study carefully in advance to avoid being disappointed by the precision of your results. Pollock et al. (1990) have provided an excellent discussion of how to plan marking experiments, and I summarize here some of their important suggestions.

The equal catchability assumption is the Achilles heel of all estimation that utilizes marked animals. In some cases you can adjust your study to minimize violations in this assumption. For example, if juvenile animals have a much lower chance of capture, you can separate adults and juveniles in your data. If a shortage of traps limits catchability, you might increase the number of traps available. One design that is useful to consider is the robust capture-recapture design first suggested by Pollock (1982). The robust design involves the use of two methods of estimation: an open method like the Jolly-Seber model and a closed method like the Schnabel method or the CAPTURE model described in the next chapter. Figure 2.8 illustrates the robust design. Primary sample periods might be weeks or months, and within these periods are secondary periods of capture of short duration. For example, primary periods for rodents might be months and secondary periods a series of 6 daily samples. For large birds, primary periods might be years and secondary periods a series of 5 weekly samples. The data for the Jolly-Seber calculations are obtained by pooling the data from the secondary periods. Recaptures within the secondary periods make no difference for the Jolly-Seber calculations but are essential information for closed population estimators like the Schnabel model.

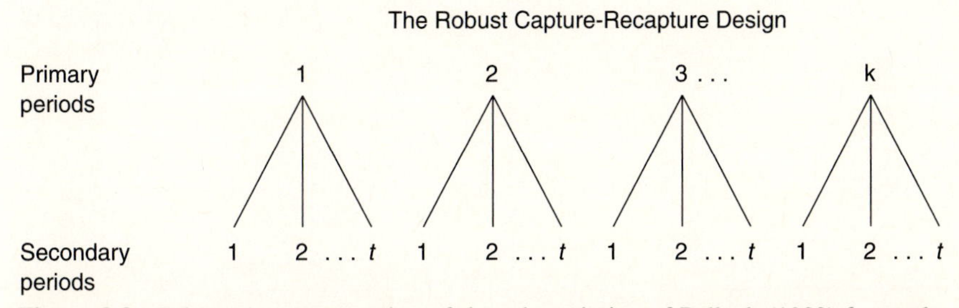

Figure 2.8 Schematic representation of the robust design of Pollock (1982) for mark-recapture studies. A primary period might be one week, and within the week, secondary periods might be the animals captured each day. Closed population estimators can be used for the secondary periods, and open estimators like the Jolly-Seber model can be used for the capture data grouped into the primary periods.

The strength of the robust design is in the fact that survival rates estimated by the Jolly-Seber model are insensitive to variations in catchability, and population estimates from some closed models like CAPTURE can take unequal catchability into account to get a relatively unbiased estimate of population size. Given these two parameters, we can use equation (2.23) to estimate the number of births in the population. Population estimates derived from the Jolly-Seber model are very sensitive to deviations from equal catchability, and the robust design avoids relying on this model for an unbiased estimate of population size.

To use the robust design, it is useful to determine what level of sampling intensity is needed to provide reliable estimates. As a measure of precision, we will use the coefficient of variation. For population estimates, this is

$$\mathrm{CV}(\hat{N}) = \frac{\sqrt{\mathrm{Var}(\hat{N})}}{\hat{N}} \tag{2.43}$$

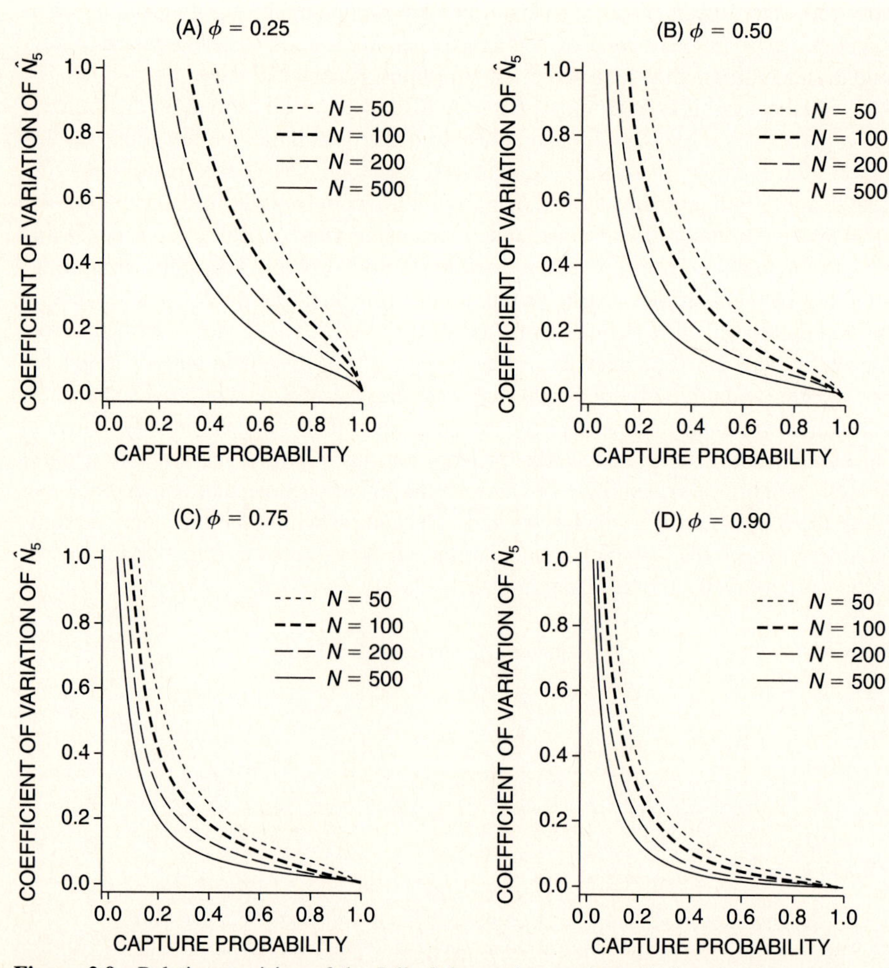

Figure 2.9 Relative precision of the Jolly-Seber estimate of population size for a range of population sizes from 50 to 500 when there are 10 sampling periods and the survival rate varies from 0.25 to 0.90. (From Pollock et al. 1990.)

The smaller the coefficient of variation, the higher the precision. A rough rule of thumb is that a CV of 0.20 (or 20%) is reasonable for ecological work. Pollock et al. (1990) have provided a series of figures to assist ecologists in planning how to sample to achieve a CV of 20%. Figure 2.9 illustrates one of these figures for a range of population sizes from 50 to 500 and a range of survival rates from low to high. Four general principles are illustrated in these graphs:

1. Precision increases as the capture probability of individuals increases.
2. As the number of sampling times increases, precision increases.
3. If survival rates are higher, precision is also higher.
4. Precision increases as population size increases, for fixed survival rates, number of samples, and capture probability.

For Jolly-Seber estimates, the precision of estimating survival probabilities is nearly the same as that of estimating population size, but it is more difficult to estimate birth rates. Figure 2.10 illustrates this difference for one set of population parameters. Capture probabilities must typically be above 0.80 to achieve high precision in estimates of the birth rate (λ_t or B_t) for populations of 200 or less.

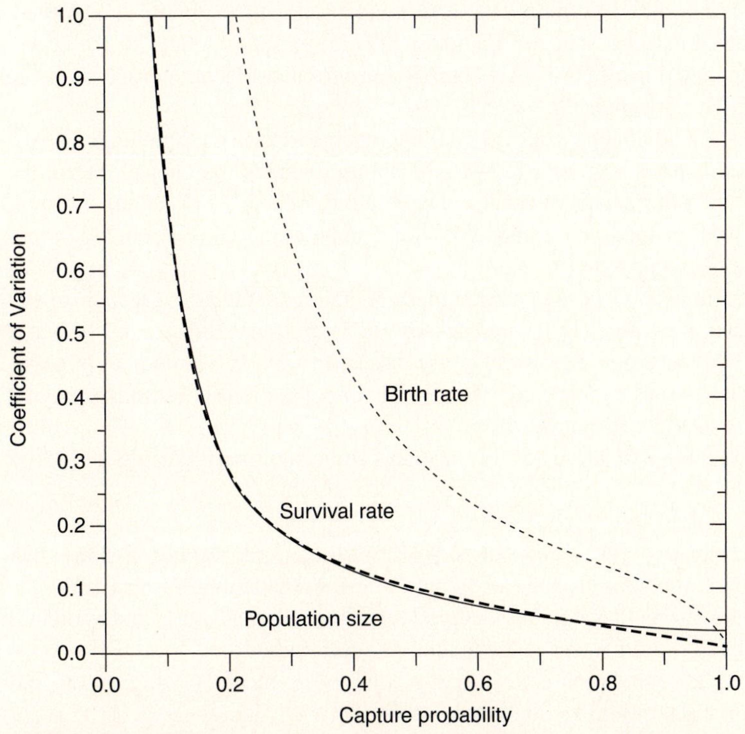

Figure 2.10 Relative precision of Jolly-Seber estimates of population size, survival rate, and birth rate for a hypothetical population of 200 individuals, 10 sample periods, and a survival rate of 0.75. The population size and survival rate curves are nearly identical. The birth rate is more difficult to estimate with high precision. (From Pollock et al. 1990.)

2.6 WHAT TO DO IF NOTHING WORKS

Estimating population size by mark-recapture may be your goal, but it may be unachievable. The assumptions you must make to use any method of estimation may be violated, and the question is what to do in these cases. Four general strategies can be used. First, you can examine more closely the assumptions as they apply to your particular population and try to design new ways of satisfying them. Perhaps random sampling is needed rather than systematic sampling (Chapter 8, Sections 8.1 and 8.4). Perhaps a second type of sampling gear can be used to catch some of the samples. Perhaps the interval between sampling should be longer or shorter. Perhaps the two sexes should be separated in the analysis, or even sampled with different kinds of traps. I call this the *tinkering strategy*, since the aim is to alter your techniques to satisfy the existing methods.

A second strategy is to use a very general, empirical model of population estimation. Caughley (1977a, 152) has reviewed this approach and emphasized its value for practical work. The major assumption of these general models is that the population is closed, so there is no mortality during the marking experiment, and consequently they are useful only for intensive short-term sampling. The basic data of these models are the number of individuals caught once, twice, three times, and so on (frequency-of-capture approach). These data form a zero-truncated frequency distribution of captures, and the missing zero-class represents the unknown number of individuals that were never caught. A variety of statistical distributions can be fitted to these data, and we stop once a "good fit" is achieved. Figure 2.11 illustrates this approach, and Caughley (1977a) provides details for the necessary calculations, included in a FORTRAN computer program available in program-group MARK-RECAPTURE (Appendix 2).

A third strategy is to change your original approach and use other methods of estimation included in Chapters 3, 4, and 5. Mark-recapture methods may not be useful for many populations, and other forms of census may be more reliable. Perhaps quadrat sampling can be employed or line transect methods. Some populations can be counted *in toto* if enough effort can be employed.

The fourth and final strategy is to use the mark-recapture methods and recognize that the resulting estimates are biased (Jolly and Dickson 1983). Biased estimates may be better than no estimates, but you should be careful to use these estimates only as *indices* of population size. If the bias is such as to be consistent over time, your biased estimates may be reliable indicators of changes in a population.

The principal recommendations of this chapter can be summarized as three rules of procedure:

- **Rule 1**: Evaluate your objectives *before* starting, and do *not* assume that mark-recapture methods are the easiest path to valid population estimates.
- **Rule 2**: Pick your mark-recapture method *before* starting field work, and build into your sampling program a test of the model's assumptions.
- **Rule 3**: Treat all population estimates and confidence intervals with caution, and recheck your assumptions as often as possible.

There is a considerable literature on more specialized techniques of mark-recapture estimation. Seber (1982), Ricker (1975), and Pollock et al. (1990) cover all these problems in more detail.

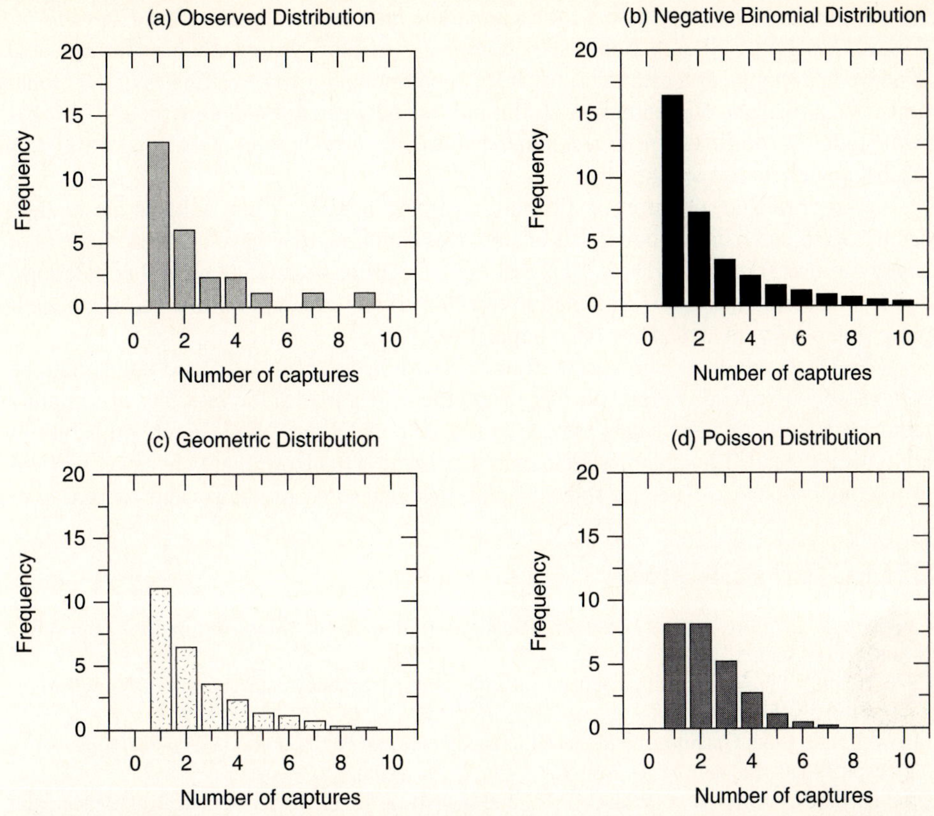

Figure 2.11 Frequency-of-capture models of population estimation. An observed distribution of frequency of captures (a) is used as a template, and several theoretical statistical distributions are fitted: (b), (c), and (d). The theoretical distribution with the closest fit is used to extrapolate to the zero class, those animals that missed getting caught. In this case the geometric distribution fits best because it has the smallest deviation of (observed − expected) values, and a population estimate of 45 is obtained with 19 animals in the zero class ("unseen"). Data on 26 male agamid lizards are from Caughley (1977a, 153).

2.7 SUMMARY

In both population and community ecology the single most important variable that must be estimated is the size of a particular population or series of populations. One way to do this is to use *mark-recapture* methods. Two general types are available, depending on whether the population is *closed* or *open*.

Closed populations do not change during the sampling period, and thus sampling must be done over a short time period. When only one period of marking and recapture is available, the *Petersen* method is appropriate. When several samples are taken with marking and recapturing, the *Schnabel* method or the *Schumacher-Eschmeyer* method can be used. The assumptions that are most critical are that sampling is random and that marking animals does not affect their catchability. A test of these assumptions is critical in any mark-recapture study.

Open populations change in size continuously, and the *Jolly-Seber* model is the best for analysis. The critical question for each marked individual caught is, *When was this individual* ***last*** *caught?* Populations can be sampled over many years with this approach, and in addition to population estimates, the Jolly-Seber model provides estimates of the probability of survival and the recruitment (dilution) rate between the sampling times. For long-term studies a *robust design* is recommended, in which both open and closed estimators can be applied to the recapture data.

The critical assumption of all mark-recapture models is that animals are equally catchable, so that marked individuals at any given sampling time have the same chances of capture as unmarked individuals. This assumption is often violated in natural populations, and four statistical tests are presented to test this critical assumption in a variety of sampling situations with closed and open populations.

Mark-recapture techniques are difficult to use in the real world, and you should be certain they are needed before starting to use them. A guide to the intensity of sampling required to attain reliable data is provided for open populations. High capture probabilities are typically needed to obtain reliable data. For many organisms, there are easier ways of estimating abundance. The next three chapters describe some alternative approaches.

SELECTED READING

Caughley, G. 1977. *Analysis of Vertebrate Populations*. Chapter 10, "Mark-Recapture." John Wiley and Sons, London.

Chao, A. 1988. Estimating animal abundance with capture frequency data. *Journal of Wildlife Management* 52: 295–300.

Manly, B. F. J. 1984. Obtaining confidence limits on parameters of the Jolly-Seber model for capture-recapture data. *Biometrics* 40: 749–758.

Otis, D. L., Burnham, K. P., White, G. C., and Anderson, D. R. 1978. Statistical inference from capture data on closed animal populations. *Wildlife Monographs* 62: 1–135.

Pollock, K. H. 1982. A capture-recapture sampling design robust to unequal catchability. *Journal of Wildlife Management* 46: 752–757.

Pollock, K. H., J. D. Nichols, C. Brownie, and J. E. Hines. 1990. Statistical inference for capture-recapture experiments. *Wildlife Monographs* 107: 1–97.

Ricker, W. E. 1975. *Computation and Interpretation of Biological Statistics of Fish Populations*. Fisheries Research Board of Canada, Bulletin 191.

Rosenberg, D. K., W. S. Overton, and R. G. Anthony. 1995. Estimation of animal abundance when capture probabilities are low and heterogeneous. *Journal of Wildlife Management* 59: 252–261.

Seber, G. A. F. 1982. *The Estimation of Animal Abundance and Related Parameters*. 2d ed. London: Griffin.

QUESTIONS AND PROBLEMS

2.1. Andersen (1962) marked roe deer with leather collars, and "recaptured" them over 3 months by binocular sightings of the colored collars. Individual deer could thus be observed (or "recaptured") several times. He obtained these data:

number marked and released = 74

number of sightings over 3 months = 462

number of these sightings that were marked deer = 340

Calculate a population estimate and its 95% confidence limits. What assumptions must you make to calculate this estimate? Read Strandgaard (1967) for a discussion of the reliability of this method.

2.2. Wood (1963) sampled climbing cutworms at night, marked them with a fluorescent powder, and resampled the population the next night. He obtained these data for 1962:

$M = 1000$ first night

$C = 1755$ second night

$R = 41$

Calculate the estimated population size and its 95% confidence interval for these data.

2.3. Hayne (1949b) marked and released meadow voles in Michigan over five days with the following results:

Date	Total catch	No. marked	No. accidental deaths
July 19 P.M.	8	0	0
July 20 A.M.	19	0	0
July 20 P.M.	10	2	1
July 21 A.M.	23	8	0
July 21 P.M.	9	0	0
July 22 A.M.	14	9	0
July 22 P.M.	9	7	1
July 23 A.M.	21	13	0

Calculate an appropriate estimate of population size and justify your choice of methods.

2.4. Dunnet (1963) sampled quokkas (small marsupials) on Rottnest Island off Western Australia over three sampling periods with these results for 1957:

	Mature animals	Immature animals
Number marked at time 1	32	35
Number marked at time 1 and never seen again	22	20
Number marked caught at times 1 and 2 only	5	1
Number marked caught at times 1 and 3 only	4	11
Number marked caught at times 1, 2, and 3	1	3

Recast these data in the form of a Schnabel sample, and estimate the size of the mature population and the immature population of this marsupial. What assumptions must you make to do these calculations? Do the estimates obtained by the Schumacher-Eschmeyer model differ significantly from those of the Schnabel model? Can you test for unequal catchability in these data?

2.5. Leslie et al. (1953) tallied the following Method B table for a population of field voles in Wales. No breeding occurred between October 1948 and April 1949. Estimate the population parameters for these voles, and discuss the biological implications of these estimates. Compare your analysis with that in Leslie et al. (1953, 144).

Time of last capture		Time of capture										
		June	July	Sept.	Oct.	Nov.	March	April	May	June	July	Sept.
1948	June		12	7	4	0	1	1	0	0	0	0
	July			10	0	1	0	0	0	0	0	0
	Sept.				19	8	4	3	0	0	0	0
	Oct.					11	9	1	2	0	0	0
	Nov.						14	6	2	1	1	0
1949	March							46	11	4	0	0
	April								34	18	0	0
	May									34	3	1
	June										40	5
	July											56
Total marked		0	12	17	23	20	28	57	49	57	44	62
Total unmarked		107	45	85	69	67	106	125	99	117	98	127
Total released		96	41	82	64	64	104	121	89	92	95	127

2.6. Bustard (1969) reported the following summary of recaptures of marked lizards in eastern Australia during 1963–1965:

No. of times recaptured	No. of lizards
1	238
2	91
3	46
4	33
5	15
6	9
7	9
8	10
9	2
10	4
11	1
12	0
13	1
14	3
15	2
19	2

Bustard tested the null hypothesis of equal catchability by fitting a Poisson distribution to these data. Do this test. What do you conclude? Is this test valid? Why or why not?

2.7. Rose and Armentrout (1974) marked and recaptured aquatic larvae of the tiger salamander in temporary ponds in Texas. For a rough estimate of 1000 salamanders, estimate from Figure 2.4 the sample size they would need for detailed research data on this population.

They obtained this sample for one pond:

No. tagged on days 1–3 = 552

No. captured on day 4 = 312

No. tagged Individuals on day 4 = 200

Calculate the population estimate for these data, and compare with the known value of 875 salamanders for this pond. Was the estimated population size close to the true value?

2.8. The outer-bound method of population estimation derives an estimate of population size from "total" counts when no animals could be marked and recaptured. For example, grizzly bears could be counted from an airplane in a block of habitat, and the count repeated on several days. Review the outer-bound method of population estimation in Giles (1978, 44–45), and the analysis of the large biases that make this superficially attractive method impossible to use in practice (Routledge 1982).

2.9. Since all field populations of animals are open populations, why would you ever use closed population estimators to determine population size?

CHAPTER 3

Estimating Abundance: Removal Methods and Resight Methods

Several methods have been developed for population estimation in which the organisms need to be captured only one time. The first set of these were developed for exploited populations from which individuals were removed as a harvest from the population, and they are loosely described as *removal methods*. These methods were first developed in the 1940s, so that wildlife and fisheries managers could get estimates of the population under harvest. A discussion of removal methods forms the first section of this chapter.

A second set of methods are of much more recent development and are based on the principle of resighting animals that have been marked. They require an individual animal to be captured only once, and then all subsequent "recaptures" are from sighting records only—the individual never has to be physically captured again. These newer methods were developed for animals with radio-collars but can be used with any kind of mark that is visible to a distant observer. We shall discuss these resighting methods in the second section of this chapter.

Mark-recapture methods have developed into very complex statistical methods during the last 10 years. While many of these methods are beyond the scope of this book, we can still use them for field populations once we understand their assumptions. The third section of this chapter provides an overview of computer-intensive methods that can be used for mark-recapture estimation.

3.1 EXPLOITED POPULATION TECHNIQUES

A special set of techniques has been developed for estimating population size in exploited populations. Many of these techniques are highly specific for exploited fish populations (Ricker 1975; Seber 1982), but some are of general interest because they can be applied to wildlife and fisheries problems as well as other field situations. I will briefly describe two types of approaches to population estimation that can be used with exploited populations.

3.1.1 Change-in-Ratio Methods

The idea that population size could be estimated from field data on the change in sex ratio during a hunting season was first noted by Kelker (1940). When only one sex is hunted, and the sex ratio before and after hunting is known, as well as the total kill, Kelker showed that one could calculate a population estimation from the simple ratio

$$\left\{\begin{matrix}\text{Fraction of males}\\\text{in population}\\\text{after hunting}\\\text{removals}\end{matrix}\right\} = \frac{\begin{pmatrix}\text{Number of males}\\\text{before hunting}\end{pmatrix} - \begin{pmatrix}\text{Number of males}\\\text{removed}\end{pmatrix}}{\begin{pmatrix}\text{Total population size}\\\text{before hunting season}\end{pmatrix} - \begin{pmatrix}\text{Total number of}\\\text{animals killed}\\\text{by hunters}\end{pmatrix}}$$

Several investigators discovered and rediscovered this approach during the last 50 years (Hanson 1963), and only recently has the general theory of change-in-ratio estimators been pulled together (Paulik and Robson 1969; Seber 1982). All change-in-ratio estimators are based on two critical assumptions:

1. The population is composed of *two types* of organisms, such as males and females, or adults and young.
2. A *differential change* in the numbers of the two types of organisms occurs during the observation period.

I will use here the general terminology of Paulik and Robson (1969), calling the two types of organisms *x-types* and *y-types*. To make the situation more concrete, you can think of x-types as females and y-types as males.

We define the following symbols:

N_1 = Total population size at time 1

N_2 = Total population size at time 2

X_1, X_2 = Number of x-type organisms in the population at times 1 and 2

Y_1, Y_2 = Number of y-type organisms in the population at times 1 and 2

$$p_1 = \frac{X_1}{N_1} = \text{Proportion of } x\text{-types in population at time 1}$$

$$p_2 = \frac{X_2}{N_2} = \text{Proportion of } x\text{-types in population at time 2}$$

$R_x = X_2 - X_1$ = Net change in numbers of x-type organisms between times 1 and 2 (may be + or −)

$R_y = Y_2 - Y_1$ = Net change in numbers of y-type organisms between times 1 and 2 (may be + or −)

$R = R_x + R_y$ = Net addition (+) to or net removal (−) from the total population between times 1 and 2

Given these symbols, we can restate the verbal model of Kelker (1940) given above by the formula:

$$\hat{N}_1 = \frac{R_x - \hat{p}_2 R}{\hat{p}_2 - \hat{p}_1} \tag{3.1}$$

This is the generalized change-in-ratio estimator of population size. We can illustrate this method with some hypothetical data on ringed-neck pheasants used by Paulik and Robson (1969). During a preseason survey, 800 of 1400 adult birds were females, and after hunting was over, 1800 of 2000 birds were females. The total kill was estimated to be 8000 male pheasants and 500 females. Thus:

$$\hat{p}_1 = \frac{800}{1400} = 0.571428$$

$$\hat{p}_2 = \frac{1800}{2000} = 0.900000$$

$$R_x = -500$$

$$R_y = -8000$$

$$R = R_x + R_y = -8500$$

Thus, from equation (3.1),

$$\hat{N}_1 = \frac{-500 - (0.900000)(-8500)}{(0.900000) - (0.571428)} = 21{,}761 \text{ pheasants}$$

Because of the structure of this formula, it is desirable to keep the estimated proportions to 6 or more decimal places to minimize rounding errors.

It follows that

$$\hat{X}_1 = \hat{p}_1 \hat{N}_1$$
$$= (0.571428)(21{,}761) = 12{,}435 \text{ female pheasants at time 1}$$

and

$$\hat{N}_2 = \hat{N}_1 + R$$
$$= 21{,}761 + (-8500) = 13{,}261 \text{ pheasants alive at time 2}$$

Confidence intervals for this population estimate can be calculated in two different ways, depending on sample size (Paulik and Robson 1969).

Large samples When approximately 500 or more individuals are sampled to estimate p_1 and an equally large number to estimate p_2, you should use the normal approximation:

$$\text{Variance}(\hat{N}_1) = \frac{\hat{N}_1^2\,[\text{variance}(\hat{p}_1)] + \hat{N}_2^2\,[\text{variance}(\hat{p}_2)]}{(\hat{p}_1 - \hat{p}_2)^2} \tag{3.2}$$

where

$$\text{Variance}(\hat{p}_1) = \frac{\hat{p}_1(1 - \hat{p}_1)}{n_1} \tag{3.3}$$

$$\text{Variance}(\hat{p}_2) = \frac{\hat{p}_2(1 - \hat{p}_2)}{n_2} \tag{3.4}$$

where n_1 = Total sample size used to estimate the ratio p_1 at time 1
n_2 = Total sample size used to estimate the ratio p_2 at time 2

This variance formula assumes binomial sampling with replacement and is a reasonable approximation to sampling without replacement when less than 10% of the population is sampled (Seber 1982, 356). It also assumes that R_x and R_y (the removals) are known exactly with no error.

For the pheasant example above,

$$\text{Variance of } \hat{p}_1 = \frac{(0.571428)(1 - 0.571428)}{1400} = 0.0001749$$

$$\text{Variance of } \hat{p}_2 = \frac{(0.9)(1 - 0.9)}{2000} = 0.000045$$

Thus,

$$\text{Variance}(\hat{N}_1) = \frac{(21{,}761)^2(0.0001749) + (13{,}261)^2(0.000045)}{(0.428571 - 0.100000)^2}$$

$$= 840{,}466$$

The standard error is given by:

$$\text{Standard error of } \hat{N}_1 = \sqrt{\text{Variance}(\hat{N}_1)}$$

$$= \sqrt{840{,}446} = 916.8$$

The 95% confidence limits follow from the normal distribution:

$$\hat{N}_1 \pm 1.96[\text{S.E.}(\hat{N}_1)]$$

For these data, we have

$$21{,}761 \pm 1.96(916.8)$$

or 19,964 to 23,558 pheasants.

Small samples When less than 100 or 200 individuals are sampled to estimate p_1 and p_2, you should use an alternative method suggested by Paulik and Robson (1969). This method obtains a confidence interval for the reciprocal of $\hat{N}_1$ as follows:

$$\text{Variance of}\left(\frac{1}{\hat{N}_1}\right) = \frac{(R_x - \hat{p}_1 R)^2}{(R_x - \hat{p}_2 R)^4}\begin{pmatrix}\text{Variance}\\ \text{of } \hat{p}_2\end{pmatrix} + \frac{1}{(R_x - p_2 R)^2}\begin{pmatrix}\text{Variance}\\ \text{of } \hat{p}_1\end{pmatrix} \quad (3.5)$$

where Variance of p_1 = as defined in equation (3.3)
Variance of $\hat{p}_2$ = as defined in equation (3.4)

and all other terms are as defined above.

This formula assumes that R_x and R_y are known without error. If we apply this formula to the pheasant data used above, we have

$$\text{Variance of}\left(\frac{1}{\hat{N}_1}\right) = \frac{[-500 - (0.571428)(-8500)]^2}{[-500 - (0.90)(-8500)]^4}(0.000045)$$

$$+ \frac{1}{[-500 - (0.90)(-8500)]^2}(0.0001749)$$

$$= 3.7481 \times 10^{-12}$$

$$\text{Standard error of}\left(\frac{1}{\hat{N}_1}\right) = \sqrt{\text{Variance}(1/\hat{N}_1)}$$

$$= \sqrt{3.7481 \times 10^{-12}} = 1.9360 \times 10^{-6}$$

The 95% confidence interval is thus:

$$\frac{1}{\hat{N}_1} \pm 1.96\left[\text{S.E.}\left(\frac{1}{\hat{N}_1}\right)\right]$$

$$\frac{1}{21{,}761} \pm 1.96(1.9360 \times 10^{-6})$$

or 4.2159×10^{-5} to 4.9748×10^{-5}

Inverting these limits to get confidence limits for $\hat{N}_1$,

$$\text{Lower 95\% confidence limit} = \frac{1}{4.9748 \times 10^{-5}} = 20{,}101 \text{ pheasants}$$

$$\text{Upper 95\% confidence limit} = \frac{1}{4.2159 \times 10^{-5}} = 23{,}720 \text{ pheasants}$$

Note that these confidence limits are asymmetrical about $\hat{N}_1$ and are slightly wider than those calculated above using the (more appropriate) large-sample formulas on these data.

The program-group MARK-RECAPTURE (Appendix 2) can do these calculations for the generalized change-in-ratio estimator of population size.

Planning Change-in-Ratio Studies If you propose to use the change-in-ratio estimator (equation [3.1]) to estimate population size, you should use the approach outlined by Paulik and Robson (1969) to help plan your experiment. Five variables must be guessed at to do this planning:

1. $\Delta p = p_1 - p_2$ = Expected change in the proportion of x-types during the experiment.
2. u = Rate of exploitation = R/N_1, which is the fraction of the whole population that is removed.
3. $f = R_x/R$ = fraction of x-types in the removals.
4. Acceptable limits of error for the $\hat{N}_1$ estimate; ±25% might be usual (see page 29).
5. Probability $(1 - \alpha)$ of achieving the acceptable limits of error defined in (4); 90% or 95% might be the usual values here.

Figure 3.1 shows the sample sizes required for *each* sample (n_1, n_2) for combinations of these parameters with the limits of error set at ±25% and $(1 - \alpha)$ as 90%. To use Figure 3.1, proceed as follows:

1. Estimate two of the three variables plotted: the change in proportion of x-types, the rate of exploitation, and the fraction of x-types in the removal.
2. From these two variables, locate your position on the graph in Figure 3.1, and read the nearest contour line to get the sample size required for each sample.

For example, if 60% of the whole population will be harvested and you expect 75% of the harvested animals to be x-types, Figure 3.1 shows that sample size should be about 100 for

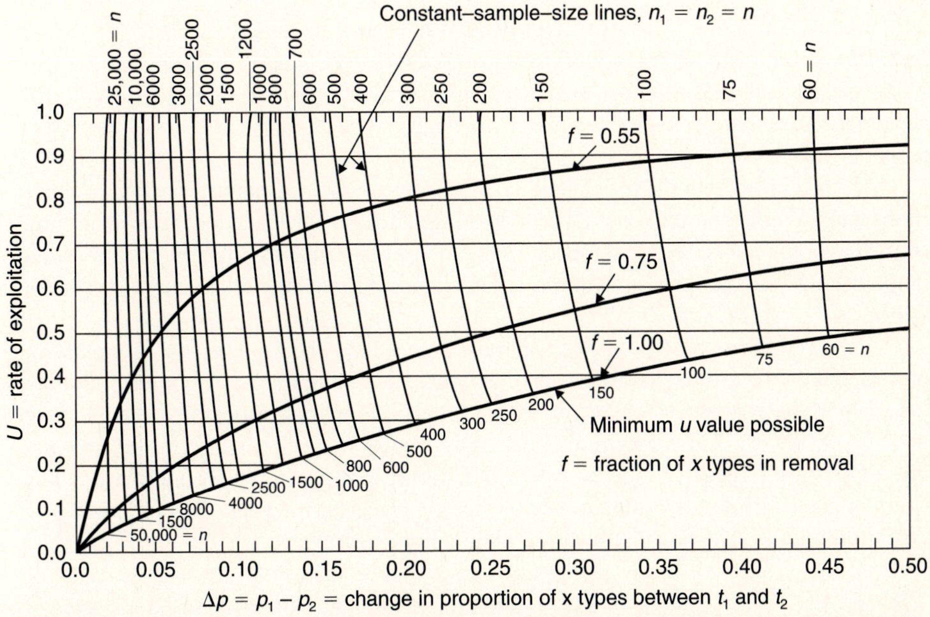

Figure 3.1 Sample sizes required for change-in-ratio estimation of population size with acceptable error limits of ±25% with $(1 - \alpha)$ of 0.90 when the initial proportion of x-types is 0.50. The contour lines represent the sample sizes needed in *each* of the time 1 and time 2 samples. The required sample sizes are only slightly affected by changes in the rate of exploitation (u) but are greatly affected by changes in Δp. (From Paulik and Robson 1969.)

the first sample and 100 for the second sample (to achieve error limits of ±25% with a $1 - \alpha$ of 0.90).

Two general points can be noted by inspecting Figure 3.1, and we can formulate them as rules of guidance:

1. Δp, the expected change in proportions, is the critical variable affecting the required sample sizes of change-in-ratio experiments. The rate of exploitation (u) is of minor importance, as is the fraction of x-types in the removals (f).
2. For Δp less than 0.05, this method requires enormous sample sizes and in practice is not useful for field studies. If p is less than 0.10, large sample sizes are needed, and it is especially critical to test all the assumptions of this approach.

Using a series of graphs like that of Figure 3.1, Paulik and Robson (1969) synthesized Figure 3.2, from which one can read directly the required sample sizes for change-in-ratio experiments. We can illustrate the utility of this graph with an example.

Suppose you are planning a change-in-ratio experiment for a deer population and you expect a change in the sex ratio by at least 0.30 over the hunting season. If you desire to estimate population size with an accuracy of ±25%, from Figure 3.2 you can estimate the necessary sample size to be approximately 200–250 deer to be sexed in *both* the before-hunting sample *and* the after-hunting sample. If you want ±10% accuracy, you must pay for it by increasing the sample size to 1050–1400 deer in each sample. Clearly, if your budget will allow you to count and sex only 40 deer in each sample, you will get an estimate of N_1 accurate only to ±100%. If this is inadequate, you need to reformulate your research goals.

Figures 3.1 and 3.2 are similar in general approach to Figures 2.3 and 2.4, which allow one to estimate samples required in a Petersen-type mark-recapture experiment. Paulik and Robson (1969) show that Petersen-type marking studies can be viewed as a special case of a change-in-ratio estimator in which x-type animals are *marked* and y-type are *unmarked*.

All the above methods for change-in-ratio estimation are designed for *closed* populations and are based on the assumption that all individuals have an equal chance of being sampled in both the first and in the second samples. Seber (1982, Chapter 9) discusses more complex situations for open populations and techniques for testing the assumptions of this method.

If the x-types and the y-types of animals do not have the same probability of being counted, one of the critical assumptions of the change-in-ratio method is violated. Pollock et al. (1985b) have proposed a new procedure for estimating population size when there is unequal catchability. This procedure operates by having two periods of removal in which only x-types are removed in the first period and only y-types in the second period. This occurs in some states, for example, where there is a season for male deer followed by a season for females. The procedure can be diagrammed as follows:

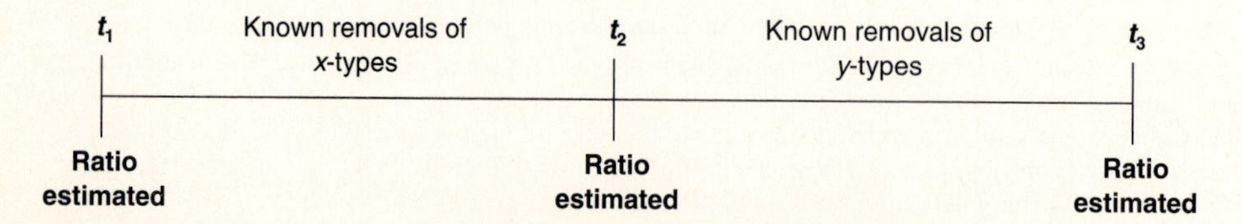

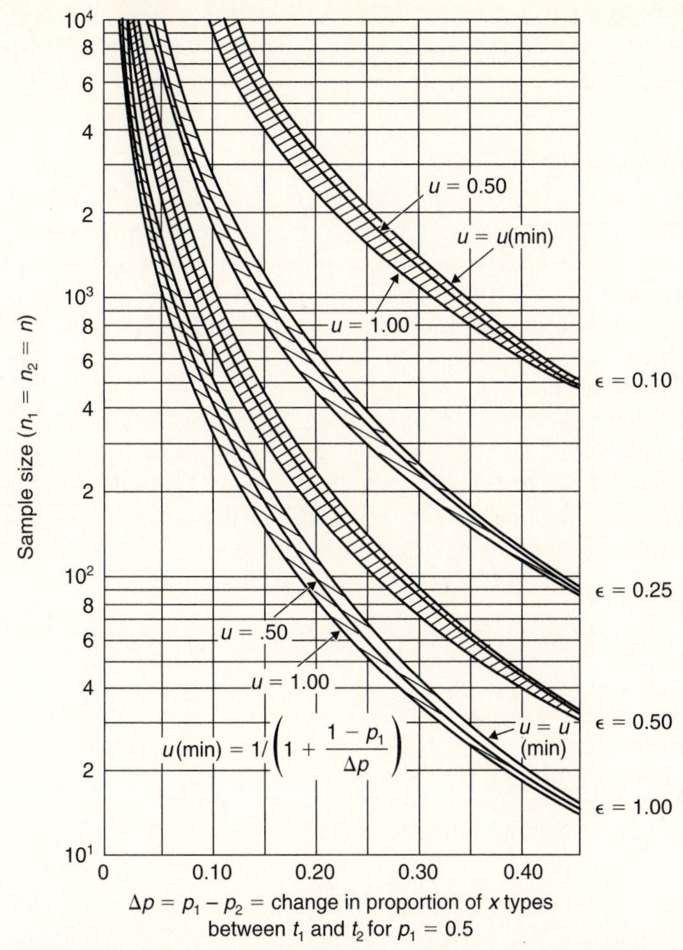

Figure 3.2 Sample sizes required for a change-in-ratio estimate of population size at several possible acceptable error limits (ϵ as a proportion of the population size) with $(1 - \alpha)$ of 0.95. The initial proportion of x-types is assumed to be 0.50. Error limits (ϵ) ranging from $\pm 10\%$ of the population size up to $\pm 100\%$ are indicated on the right side of the graph. These estimates are affected only very slightly by changes in the initial proportion of x-types between 0.05 and 0.50, and the critical variable is Δp (x-axis). The upper limits of the shaded zones are for situations with minimal exploitation, and the lower limits of the shaded zones are for situations with maximal exploitation. By reading sample sizes off the top of the curves, one can operate conservatively. (From: Paulik and Robson 1969.)

Given that you can estimate the proportion of x-types in the population at these three times, Pollock et al. (1985b) show how to estimate the initial population size. One advantage of this procedure is that the proportion of x-types in the population need not change from time 1 to time 3 (it must obviously change at time 2 if the method is to work!). But the main advantage of the Pollock design is that it is robust to unequal catchability of the two types of animals. It clearly involves more work however to obtain an estimate of population size.

3.1.2 Eberhardt's Removal Method

A simpler use of removal data to estimate population size was suggested by Eberhardt (1982) and has been called the "index-removal" method. If an index of population size (like roadside counts) can be made before and after the removal of a known number of individuals, it is possible to use the indices to estimate absolute density. This method does not require you to classify individuals into x-types and y-types as does the change-in-ratio method, and there is no need to identify individuals. Eberhardt (1982) discusses how this removal method compares with methods based on mark-recapture.

To use Eberhardt's method, an index of population size is obtained before and after the removals. This index is assumed to have some constant but unknown relationship to population size. For example, a roadside count may see 15% of the deer in an area. If you know the number of animals removed from the population after the first index is taken, you can calculate an estimate of population size by the following equation:

$$\hat{N} = \frac{x_1 R}{x_1 - x_2} \tag{3.6}$$

where $\hat{N}$ = Estimated population size at time 1
x_1 = Index count at time 1
x_2 = Index count at time 2
R = Number of animals removed

The proportion of animals removed can be estimated from the ratio

$$\hat{p} = \frac{(x_1 - x_2)}{x_1} \tag{3.7}$$

where $\hat{p}$ = Estimated proportion of animals removed

Eberhardt (1982) derives a variance estimate for population size as follows:

$$s_{\hat{N}}^2 = \left(\frac{1 - \hat{p}}{\hat{p}}\right)^2 \left(\frac{1}{x_1} + \frac{1}{x_2}\right)(\hat{N}^2) \tag{3.8}$$

where $s_{\hat{N}}^2$ = Variance of estimated population size

and the other terms are as defined above. From this variance, you can construct the 95% confidence limits in the usual manner:

$$\hat{N} \pm 1.96[\text{S.E.}(\hat{N})]$$

where $\text{S.E.}(\hat{N}) = \sqrt{s_{\hat{N}}^2}$ as defined in equation (3.8)

As with mark-recapture estimates, Eberhardt's removal method works best when a high fraction of the population is seen and a high fraction is removed. Table 3.1 gives the expected coefficient of variation of population estimates made by Eberhardt's method. Unless

TABLE 3.1 COEFFICIENTS OF VARIATION OF EBERHARDT'S INDEX REMOVAL POPULATION ESTIMATE FOR VARIOUS VALUES OF THE PROPORTION OF THE POPULATION COUNTED AND THE PROPORTION REMOVED

	Proportion seen			
Percentage removed	0.2	0.4	0.6	0.8
10%	0.92	0.65	0.53	0.46
20%	0.42	0.30	0.25	0.21
30%	0.26	0.18	0.15	0.13
40%	0.17	0.12	0.10	0.09
50%	0.12	0.09	0.07	0.06
60%	0.09	0.06	0.05	0.04
70%	0.06	0.05	0.04	0.03
80%	0.04	0.03	0.03	0.02

Source: Eberhardt (1982)

the percentage of the population seen is above about 40% and the percentage removed above 20%, the method is not very precise.

Box 3.1 illustrates the use of Eberhardt's index removal method.

3.1.3 Catch-Effort Methods

In exploited populations it may be possible to estimate population size by the decline in catch-per-unit-effort with time. This possibility was first recognized by Leslie and Davis (1939), who used it to estimate the size of a rat population that was being exterminated by trapping. DeLury (1947) and Ricker (1975) discuss the method in more detail. This method is highly restricted in its use because *it will work only if a large enough fraction of the population is removed so that there is a decline in the catch-per-unit-effort*. It will not work if the population is large relative to the removals. The following assumptions are also critical for this method:

1. The population is closed.
2. Probability of each individual being caught in a trap is constant throughout the experiment.
3. All individuals have the same probability of being caught in sample i.

The data required for catch-effort models are as follows:

c_i = Catch or number of individuals removed at sample time i

K_i = Accumulated catch from the start up to the beginning of sample time i

f_i = Amount of trapping effort expended in sample time i

F_i = accumulated amount of trapping effort from the start up to the beginning of time i

Table 3.2 gives an example of such data for a fishery operating on blue crabs.

Box 3.1 Eberhardt's Index-Removal Method of Population Estimation

Feral horses were counted before and after a removal program in Oregon with the following results:

$$x_1 = 301 \text{ horses counted before removals}$$

$$x_2 = 76 \text{ horses counted after removals}$$

$$R = 357 \text{ horses removed}$$

Assuming that the visual counts are in direct proportion to population size, we can use equation (3.6) to estimate population size at time 1:

$$\hat{N} = \frac{x_1 R}{x_1 - x_2} = \frac{(301)(357)}{301 - 76} = 477.6 \text{ horses}$$

In this example an estimated 63% of the total population were counted in the first index count (301/477.6). The percentage removed is also high (75%). From Table 3.1, the standard error of this population estimate should be about 3–4% of the estimate, so the confidence interval should be approximately ±6–8% of the population size.

The variance of this estimate is, from equation (3.8),

$$s_{\hat{N}}^2 = \left(\frac{1 - \hat{p}}{\hat{p}}\right)^2 \left(\frac{1}{x_1} + \frac{1}{x_2}\right)(\hat{N}^2)$$

$$= \left(\frac{1 - 0.74686}{0.74686}\right)^2 \left(\frac{1}{301} + \frac{1}{76}\right)(477.6)^2$$

$$= 432.566$$

From this variance we obtain the 95% confidence limits as

$$\hat{N} \pm 1.96[\text{S.E.}(\hat{N})]$$

$$477.6 \pm 1.96[\sqrt{432.566}]$$

$$\text{or } 437 \text{ to } 518$$

These confidence limits are ±8.5% of population size.

Eberhardt's removal method is most useful in management situations in which a controlled removal of animals is undertaken and it is not feasible or economic to mark individuals. It must be feasible to count a significant fraction of the whole population to obtain results that have adequate precision.

Under the assumptions listed above, the catch-per-unit-effort is directly proportional to the existing population size. Leslie and Davis (1939) showed that, because the population must be declining from time to time by an amount equal to the catch, a regression plot of:

x-axis accumulated catch (K_i)

y-axis catch-per-unit-effort (c_i/f_i)

TABLE 3.2 CATCH-EFFORT DATA FOR A POPULATION OF MALE BLUE CRABS (*Callinectes sapidus*) FOR A 12-WEEK PERIOD

Week no., i	Catch (pounds), c_i	Effort (lines per day), f_i	Catch per unit effort, $c_{i}/f_i = Y_i$	Accumulated catch, K_i	Accumulated effort, F_i
1	33,541	194	172.9	0	0
2	47,326	248	190.8	33,541	194
3	36,460	243	150.0	80,867	442
4	33,157	301	110.2	117,327	685
5	29,207	357	81.8	150,484	986
6	33,125	352	94.1	179,691	1343
7	14,191	269	52.8	212,816	1695
8	9,503	244	38.9	227,007	1964
9	13,115	256	51.2	236,510	2208
10	13,663	248	55.1	249,625	2464
11	10,865	234	46.4	263,288	2712
12	9,887	227	43.6	274,153	2946

Source: Data from Fischler 1965.

should be a straight line. Figure 3.3 illustrates this plot for the blue crab data in Table 3.2. This graph is easy to grasp because the x-intercept (the point where $y = 0$, or the catch-per-unit-effort falls to zero) is the initial population size (N), since it represents the exhaustion of the catch. Secondly, the slope of the line is an estimate of the *catchability* of the individuals, the probability that a given individual will be caught with one unit of effort. With

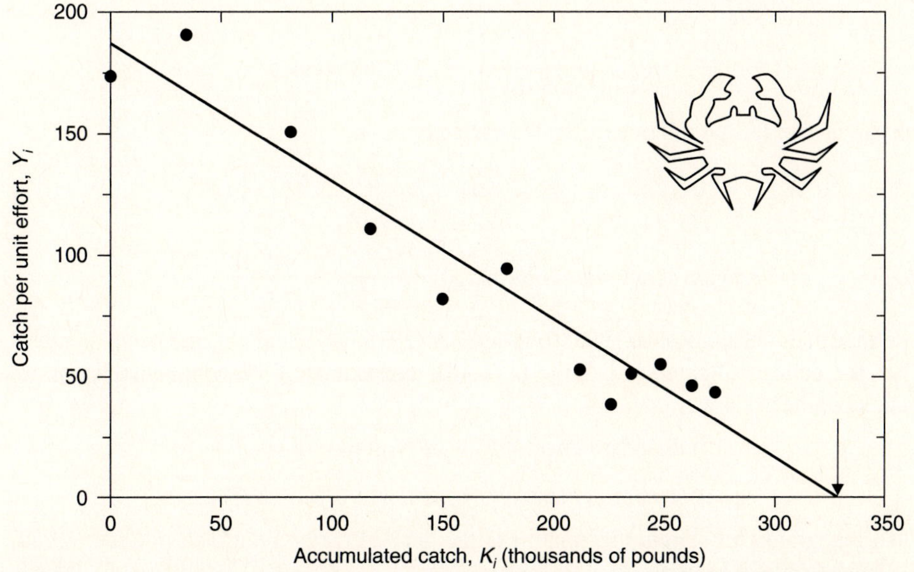

Figure 3.3 Leslie plot of catch-effort data of Fischler (1965) for a population of male blue crabs. Using the Leslie model, one can estimate initial population size (N) by extrapolating the linear regression to the x-axis. In this case (arrow) $\hat{N}$ is about 330×10^3 pounds. (Original data in Table 3.2.)

this regression (Figure 3.3) we can estimate these parameters by eye, or to be more precise, use linear regression techniques as follows:

$$\text{Catchability} = \hat{C} = \frac{-\sum_{i=1}^{s} Y_i(K_i - \bar{K})}{\sum_{i=1}^{s} (K_i - \bar{K})^2} \tag{3.9}$$

$$\text{Population size} = \hat{N} = \bar{K} + \left(\frac{\bar{Y}}{\hat{C}}\right) \tag{3.10}$$

where Y_i = Catch-per-unit-effort = c_i/f_i

$\bar{K}$ = Mean value of K_i (accumulated catch) = $\dfrac{(\Sigma K_i)}{s}$

s = Total number of samples ($i = 1, 2, 3 \ldots s$)

For example, using the data in Table 3.2,

$$\bar{Y} = \frac{\Sigma Y_i}{s} = \frac{172.9 + 190.8 + \cdots}{12} = 90.65$$

$$\bar{K} = \frac{\Sigma K_i}{s} = \frac{0 + 33{,}541 + 80{,}867 + \cdots}{12} = 168{,}775.75$$

Thus,

$$\hat{C} = \frac{-[172.9(0 - 168{,}775.75) + 190.8(33{,}541 - 168{,}775.75) + \cdots}{[(0 - 168{,}775.75)^2 + (33{,}541 - 168{,}775.75)^2 + \cdots}$$

$$= 0.0005614$$

$$\hat{N} = 168{,}775.75 + \left[\frac{90.65}{(0.0005614)}\right] = 330{,}268 \text{ pounds}$$

The variance of this population estimate is given by

$$\text{Variance of } (\hat{N}) = \frac{s_{yx}^2}{\hat{C}^2}\left[\frac{1}{s} + \frac{(\hat{N} - \bar{K})^2}{\Sigma (K_i - \bar{K})^2}\right] \tag{3.11}$$

where s_{yx}^2 = Variance about regression = $\sum \dfrac{[Y_i - \hat{C}(\hat{N} - K_i)]^2}{(s - 2)}$

as defined in Sokal and Rohlf (1995, 471) and Zar (1996, 327), and s = number of samples. When the number of samples is large ($s > 10$), approximate 95% confidence limits are obtained in the usual way:

$$\text{Standard error of } \hat{N} = \sqrt{\text{Variance of } \hat{N}}$$

$$95\% \text{ confidence limits} = \hat{N} \pm 1.96[\text{S.E.}(\hat{N})]$$

When the number of samples is small, use the general method outlined in Seber (1982, 299) to get confidence limits on $\hat{N}$.

The plot shown in Figure 3.3 provides a rough visual check on the assumptions of this model. If the data do not seem to fit a straight line, or if the variance about the regres-

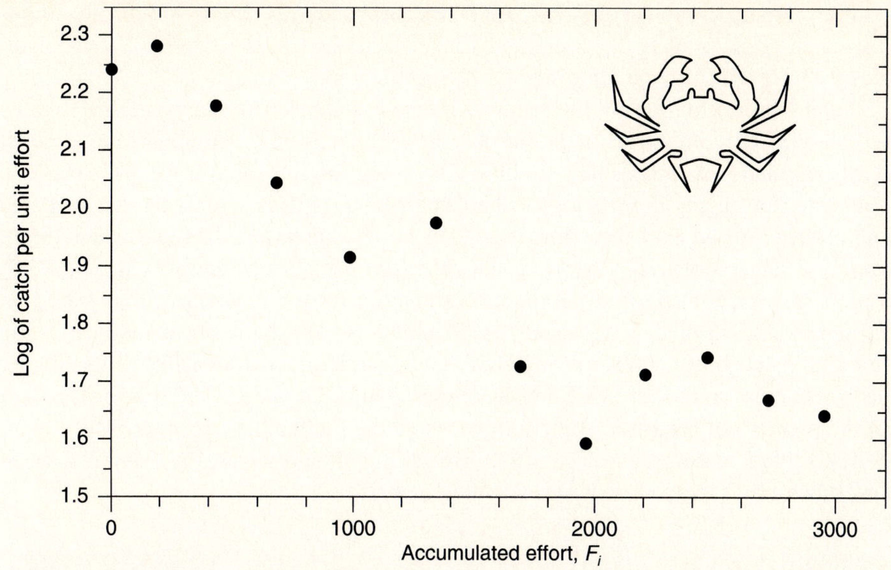

Figure 3.4 Ricker plot of catch-effort data of Fischler (1965) for male blue crabs. This is a semilogarithmic plot of the data in Table 3.2, in which the y-axis is the log of the catch-per-unit-effort. The graph shows a curved line and not a straight line, and consequently the Ricker catch-effort model is not supported for these data.

sion line is not constant, the data violate some or all of the assumptions, and this model is not appropriate. Ricker (1975) discusses how to deal with certain cases in which the assumptions of the model are not fulfilled.

DeLury (1947) and Ricker (1975) provide two alternative models for analyzing catch-effort data based on a semilogarithmic relationship between the log of the catch-per-unit-effort (y-axis) and the accumulated fishing effort (F_i) on the x-axis (e.g., Figure 3.4). The calculations are again based on linear regression techniques. For the more general Ricker model, defining $z_i = \log(Y_i)$ we have

$$\log(1 - \hat{C}) = \frac{\sum_{i=1}^{s} z_i(F_i - \bar{F})}{\sum_{i=1}^{s} (F_i - \bar{F})^2} \tag{3.12}$$

and

$$\log \hat{N} = \bar{F}\log(1 - \hat{C}) - \log \hat{C} \tag{3.13}$$

Parameter estimates of catchability ($\hat{C}$) and population size ($\hat{N}$) are obtained by taking antilogs. Seber (1982, 302) and Ricker (1975) show how confidence intervals may be calculated for these semilogarithmic models.

Since the Leslie model (equation [3.10]) and the Ricker model (equation [3.13]) are based on somewhat alternative approaches, it is desirable to plot both the Leslie regression (e.g., Figure 3.3) and the Ricker semilog regression (Figure 3.4) as checks on whether the underlying assumptions may be violated.

The program-group MARK-RECAPTURE (Appendix 2) can do these calculations for the catch-effort models of Leslie and Ricker described in this section; it calculates confidence intervals for population size for both estimators.

Population estimation by the removal method is subject to many pitfalls. Braaten (1969) showed by computer simulation that the DeLury (1947) semilogarithmic model produces population estimates that are biased by being too low on average. This bias can be eliminated by regressing the log of the catch-per-unit-effort against the accumulated fishing effort (F_i) *plus* half the effort expended in the *i*-interval ($1/2\ f_i$) (Braaten 1969). The more serious problem of variable catchability has been discussed by Schnute (1983), who proposes a new method of population estimation for the removal method based on maximum-likelihood models. In particular, Schnute (1983) provides a method of testing for constant catchability and for fitting an alternate model in which catchability is higher in the first sample, and lower in all subsequent samplings. Otis et al. (1978) also discuss in detail the problem of removal estimation, in which the probability of capture varies with time. The critical assumption of constant catchability should always be examined when you use catch-effort methods to estimate population size.

3.2 RESIGHT METHODS

In recent work on vertebrate population, individuals are often marked with radio transmitters, and for such studies a range of new population estimators are available (Arnason et al. 1991; Neal et al. 1993; White 1996). Radio-tagged individuals are often expensive to capture and tag, but after they are released, they can be easily resighted using radio-telemetry equipment (Kenward 1987; White and Garrott 1990). The methods that have developed are an important extension of Petersen-type population estimators for closed populations. Resight methods do not require radio-telemetry, of course, and any marking method that allows one to identify an individual at a distance could be used with this approach.

Four estimators of population size are available for mark-resight data, and White (1996) has implemented these in Program NOREMARK. I discuss here only two of these four estimators, the joint hypergeometric maximum-likelihood estimator (JHE) and Bowden's estimator. The statistical theory behind the JHE estimator is beyond the scope of this book, but I provide a general understanding of how maximum-likelihood estimators are derived, since they form the core of much of modern estimation theory.

Maximum likelihood is a method of statistical inference in which a particular model is evaluated with reference to a set of data. In this case we have a set of observations based on the sighting frequencies of marked and unmarked animals. If we assume that resightings occur at random, we can connect through probability theory the observed data and the likely value of population size N. For simple situations, such as the Petersen method, in which only two sampling times are involved, the resulting estimator for N will be a simple equation like equation (2.2). For more complex situations involving several sampling times, the resulting estimator cannot be written as a simple equation with a solution, and we must search to find the best value of N by trial and error. In mathematical jargon, we search for the maximum value of a function called the likelihood function (Edwards 1972). Figure 3.5 illustrates the maximum-likelihood approach for equation (3.14) and the data used in Box 3.2.

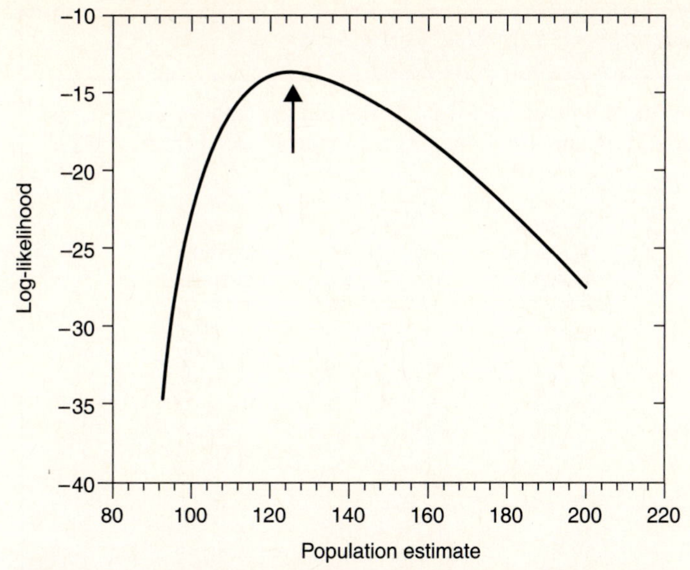

Figure 3.5 Likelihood ratio for the mountain sheep example given in Box 3.2. Given the observed data, estimated population sizes are substituted in equation (3.14) and the likelihood calculated. The most probable value of N is that which gives the highest likelihood (arrow), in this case $\hat{N} = 127$, at which point the log-likelihood is -13.6746 (likelihood $= 1.1513 \times 10^{-6}$). Because likelihoods cover a broad numerical range, it is easier to graph them as log-likelihoods ($\log_e$ or ln) as we have done here.

For the JHE estimator used by White (1996), we search for the value of N that maximizes the following function:

$$L(N|M, n_i, m_i) = \prod_{i=1}^{k} \frac{\binom{M}{m_i}\binom{N-M}{n_i-m_i}}{\binom{N}{n_i}} \tag{3.14}$$

where $L(N|M, n_i, m_i)$ = Likelihood of N conditional on the observed values of M, n_i and m_i, and

N = Population size

M = Number of radio-marked animals in the study zone

n_i = Total number of animals seen in the i-th sample survey (i = i, 2, 3 . . . k)

m_i = Number of marked animals sighted in the i-th survey

and the operator $\binom{M}{m_i}$ indicates the number of possible combinations of M items taken m_i at a time $= M!/[m_i!(M - m_i)!]$.

Box 3.2 Resight Method of Population Estimation

Neal et al. (1993) report the following data from radio-collared bighorn sheep (*Ovis canadensis*) from Trickle Mountain, Colorado, from 14 January to 1 February 1989:

Date	Number of marked sheep, M_i	Number animals seen during survey, n_i	Marked animals seen, m_i
January 14	25	40	9
January 17	25	63	11
January 18	25	66	14
January 19	25	57	11
January 27	25	52	10
January 28	25	61	9
February 1	25	87	19

There were a total of 426 recaptures in the sampling period, of which 83 sheep had radio-collars and 343 sheep were not marked. We can use each of these daily samples as Petersen estimates and obtain a series of Petersen estimates from equation (2.3). For example, from January 19 with $M = 25$, $C = 57$, and $R = 11$, using equation (2.3) for sampling with replacement, we obtain a Petersen estimate as follows:

$$\hat{N} = \frac{M(C + 1)}{(R + 1)} = \frac{25(57 + 1)}{(11 + 1)} = 120.8 \text{ sheep}$$

To use the radio method, we compute the likelihood of a range of possible values of N and find the one that has the maximum likelihood. This is tedious, and I illustrate only one calculation. Use as a preliminary estimate $\hat{N} = 95$. From equation (3.14),

$$L(N|M, n_i, m_i) = \prod_{i=1}^{k} \frac{\binom{M}{m_i}\binom{N - M}{n_i - m_i}}{\binom{N}{n_i}}$$

Since $M = 25$, we have seven terms to evaluate, as follows:

$$L(95|25, n_i, m_i) = \frac{\binom{25}{9}\binom{95 - 25}{40 - 9}}{\binom{95}{40}} \cdot \frac{\binom{25}{11}\binom{95 - 25}{63 - 11}}{\binom{95}{63}} \cdot \frac{\binom{25}{14}\binom{95 - 25}{66 - 14}}{\binom{95}{66}} \cdot \cdots$$

$$= (0.14632)\,(0.00522)\,(0.04817) \cdots = 1.2318 \times 10^{-13}$$

To avoid these very small numbers, we can express these likelihoods as log-likelihoods:

$$\log_e (1.2318 \times 10^{-13}) = -29.7251$$

Repeating these calculations for a range of values of N from 96 to 200 gives us the data plotted in Figure 3.5. The maximum likelihood occurs at $\hat{N} = 127$, which is the best estimate of population size given these resighting data. The 95% confidence interval for this estimate from Program NOREMARK is 112 to 146.

For the Bowden estimator, we require the sighting frequencies of each of the 25 mountain sheep that had radio-transmitters. For these 25 animals, these values (f_i) were: 6, 4, 3, 6, 5, 3, 0, 5, 4, 6, 1, 0, 0, 5, 6, 0, 6, 1, 5, 5, 2, 2, 0, 4, and 4. From the usual statistical calculations and the variance equation above,

$$\bar{f} = 3.320 \text{ resightings per animal}$$

$$\text{VAR}(f_j) = 4.8576$$

From equation (3.15) we obtain for these data

$$\hat{N} = \frac{\left(\frac{(u_T + m_T)}{\bar{f}}\right) + \left(\frac{s_f^2}{\bar{f}^2}\right)}{\left[1 + \left(\frac{s_f^2}{T\bar{f}^2}\right)\right]} = \frac{\left(\frac{343 + 83}{3.320}\right) + \left(\frac{4.8576}{(3.320)^2}\right)}{\left[1 + \left(\frac{4.8576}{25(3.320)^2}\right)\right]} = 126.52$$

with the following variance from equation (3.16):

$$V\hat{a}r(\hat{N}) = \frac{\hat{N}^2\left(\frac{1}{T} - \frac{1}{\hat{N}}\right)\left(\frac{s_f^2}{\bar{f}^2}\right)}{\left(1 + \frac{s_f^2}{T\bar{f}^2}\right)^2} = \frac{126.52^2\left(\frac{1}{25} - \frac{1}{126.52}\right)\left(\frac{4.8576}{3.320^2}\right)}{\left(1 + \frac{4.8576}{25(3.320^2)}\right)^2} = 218.658$$

The 95% confidence limits from equations (3.17) and (3.18) are, with $t_\alpha = 2.064$ for 24 degrees of freedom,

$$\text{Lower confidence limit} = \frac{\hat{N}}{\exp[t_\alpha \text{CV}(\hat{N})]} = \frac{126.52}{e^{(2.064[\sqrt{218.658}/126.52])}} = 99$$

$$\text{Upper confidence limit of } \hat{N} = (\hat{N})\{\exp[t_\alpha \text{CV}(\hat{N})]\} = (126.52)e^{(2.064[\sqrt{218.658}/126.52])}$$

$$= 161$$

Note that these confidence limits are slightly wider than those given for the JHE estimator above. This is the price we pay for the less restrictive assumption that individual sheep may have different probabilities of being sighted on the study area.

These calculations are done by program-group MARK-RECAPTURE (Appendix 2) and by Program NOREMARK from White (1994).

Equations of this type for many maximum-likelihood estimators can be solved only by trial and error, by substituting values of N into the equation and finding the value that maximizes the likelihood. For this reason all these methods are computer-intensive and were not feasible before the advent of modern computers.

Radio estimators can be viewed as a weighted overall average of a series of Petersen estimates for each sampling period. Box 3.2 illustrates the application of the radio method to sample data on radio-collared mountain sheep.

Confidence intervals for the JHE estimator can be determined by the profile-likelihood method described by Venzon and Moolgavkar (1988). These methods are computer-intensive and will not be described here.

The radio method assumes that all the marked animals remain on the study area during the surveys. Thus the number of marked animals is constant, although the probability of resighting need not be constant from sample to sample. Day-to-day variation in sightability can thus be accommodated in this model.

Bowden and Kufeld (1995) developed an estimate for population size with radio data that allows us to relax the assumption that all individuals in the population have the same probability of resighting. To compute this estimator, we need to have the resighting frequency for each radio-tagged individual in the population. Some animals may not be seen at all, and some may be seen at every one of the sampling times. The Bowden estimator of population size is given by

$$\hat{N} = \frac{\left(\frac{(u_T + m_T)}{\bar{f}}\right) + \left(\frac{s_f^2}{\bar{f}^2}\right)}{\left[1 + \left(\frac{s_f^2}{T\bar{f}^2}\right)\right]} \tag{3.15}$$

where u_T = Total number of sightings of unmarked animals over all time periods

m_T = Total number of sightings of marked animals over all time periods

$\bar{f}$ = Mean resighting frequency of marked animals $= \dfrac{m_T}{T}$

$T = M$ = Number of marked animals in the population at the time of the surveys

s_f^2 = Variance of sighting frequencies of marked animals $= \dfrac{\sum_{j=1}^{T} (f_j - \bar{f})^2}{T}$

Note that the variance estimate for sighting frequencies has T in the denominator (and not $T - 1$) because it is a complete sample of all the population of radio-tagged animals.

This estimate (3.15) has the following variance:

$$\text{Var}(\hat{N}) = \frac{\hat{N}^2\left(\frac{1}{T} - \frac{1}{\hat{N}}\right)\left(\frac{s_f^2}{\bar{f}^2}\right)}{\left(1 + \frac{s_f^2}{T\bar{f}^2}\right)^2} \tag{3.16}$$

Confidence intervals for the Bowden estimator are obtained from a log-transformation as

$$\text{Lower confidence limit} = \frac{\hat{N}}{\exp[t_\alpha \text{CV}(\hat{N})]} \tag{3.17}$$

$$\text{Upper confidence limit of } \hat{N} = (\hat{N})\{\exp[t_\alpha \text{CV}(\hat{N})]\} \tag{3.18}$$

where t_α = t-value with $(T - 1)$ degrees of freedom for the specified confidence level, and

$$\text{CV}(\hat{N}) = \text{coefficient of variation of } (\hat{N}) = \frac{\sqrt{\text{Var}(\hat{N})}}{\hat{N}} \tag{3.19}$$

The Bowden estimator usually gives somewhat wider confidence limits for the population estimate because it does not make the restrictive assumption that all individuals have equal sightability. It can be computed with program-group MARK-RECAPTURE (Appendix 2) or in Program NOREMARK from White (1996).

3.3 COMPUTER PROGRAMS FOR POPULATION ESTIMATORS

Estimation methods for closed populations have been unified by the approaches outlined in the monograph by Otis et al. (1978). These methods complement the estimators discussed in Chapter 2 for closed populations, but they are more restrictive because they assume that every marked animal can be individually recognized and that at least three sampling periods were used. These methods are all computer-intensive and have been codified in Program CAPTURE and Program MARK for the IBM PC.* I will discuss here only the basic outline of these methods and indicate the data needed for these calculations.

The simplest form of data input is in the form of an **X** matrix. The rows of this matrix represent the individual animals that were captured in the study, and the columns of the matrix represent the time periods of capture. In each column a 0 (zero) indicates that the individual in question was not caught during this sampling time, and a 1 (one) indicates that the individual was captured. A sample **X** matrix is as follows:

Tag number	Time 1	Time 2	Time 3	Time 4	Time 5	Time 6	Time 7
3455	1	0	0	1	1	1	0
3456	1	1	1	0	0	0	1
3458	1	0	0	0	0	0	0
3462	0	1	1	1	1	0	0
3463	0	0	0	1	0	1	0
3476	0	0	1	0	1	0	1
3488	0	0	0	0	0	0	1

This indicates that animal tag number 3455 was caught in the first trapping session but not caught in the second or third session, caught in the fourth, fifth, and sixth sessions but not

*These programs are currently available from Colorado State University at *www.cnr.colostate.edu/~gwhite/software.html.*

in the seventh. In a normal study these catching sessions might represent one day but they could be one hour or one week or whatever sampling time unit is appropriate to your animals.

Given the matrix of 0's and 1's, it is possible to use probability theory to ask how these should be arranged under several different models of capture behavior. The simplest model is the *null model*, in which the captures occur completely at random with respect to all individuals and all individuals have equal chances of being caught at any time. The null model in Program CAPTURE is essentially the equivalent of the Petersen and Schnabel models discussed in Chapter 2. If the null model does not fit the observed data well, there are three primary sources of variation that can cause changes in capture probabilities:

1. **Time**: The probability of capture varies with the time of capture. If it rains one day, the capture rate may be lower than usual (or higher depending on the species).
2. **Heterogeneity**: Individual animals may differ in the propensity to be caught, so that some individuals are trap-happy and some are trap-shy. Alternatively, some animals move around more and are exposed to more traps. This individual variation in chances of capture is called *heterogeneity*, and it is an important source of violation of the equal catchability assumption that many mark-recapture models make.
3. **Behavior**: Individual animals may change their behavior after they are caught once, so that the chances of capture may be quite different for the first capture and all subsequent captures. This source of variation is also common and is labeled *behavior* because it arises in general as a behavioral response of animals to the trapping devices.

These sources of variation can be combined to produce a model that includes both *time* and *heterogeneity*, and the shorthand used by Otis et al. (1978) is to label this Model M_{TH}. There are thus eight possible models that might be used in Program CAPTURE:

M_0: the null model

M_T: the time model (Darroch)

M_H: the heterogeneity model (Jackknife)

M_B: the behavior model (Zippin)

M_{TH}: the time and heterogeneity model

M_{TB}: the time and behavior model

M_{BH}: the behavior and heterogeneity model (Generalized Removal)

M_{TBH}: the full model with time, heterogeneity, and behavior varying

The more complicated models, as you might guess, are harder to fit to observed data, and at present no one has solved the full model M_{TBH}; therefore, it cannot be utilized.

The key problem remaining is which of these models to use on your particular data. Otis et al. (1978) have devised a series of chi-squared tests to assist in this choice, but these do not give a unique answer with most data sets. Work is continuing in this area to develop better methods of model selection.

The details of the models and the calculations are presented in Otis et al. (1978). For ecologists, a general description of the procedures used and some sample data runs are given in White et al. (1982). Maximum-likelihood methods are used in all the estimation procedures in CAPTURE, similar to those described above for the RESIGHT method, and I will illustrate only one method here, the null model M_0. The best estimate of $\hat{N}$ for model M_0 is obtained from the following maximum-likelihood equation:

$$L(\hat{N}_0, \hat{p}|X) = \ln\left(\frac{N!}{(N-M)!}\right) + (n)\ln(n) + (tN - n)\ln(tN - n) - (tN)\ln(tN) \quad (3.20)$$

where $\hat{N}_0$ = Estimated population size from the null model of CAPTURE
N = Provisional estimate of population size
$\hat{p}$ = Probability of capture
M = Total number of different individuals captured in the entire sampling period
n = Total number of captures during the entire sampling period
t = Number of samples (e.g., days)
ln = Natural log ($\log_e$)
L = Log-likelihood of the estimated value $\hat{N}_0$ and p, given the observed **X** matrix of captures

This equation is solved by trial and error to determine the value of N that maximizes the log-likelihood (as in Figure 3.5), and this value of N is the best estimate of population size.

Once the value of $\hat{N}_0$ has been determined, the probability of capture can be obtained from

$$\hat{p} = \frac{n}{tN_0} \quad (3.21)$$

and the variance of the estimated population size is obtained from

$$\text{Var}(\hat{N}_0) = \frac{\hat{N}_0}{(1-\hat{p})^{-t} - (t/1-\hat{p}) + t - 1} \quad (3.22)$$

The standard error of this population estimate is the square root of this variance, and hence the confidence limits for the estimated population size are given by the usual formula:

$$\hat{N}_0 \pm z_\alpha \sqrt{\hat{V}ar(\hat{N}_0)} \quad (3.23)$$

where z_α = Standard normal deviate (i.e., 1.960 for 95% confidence limits, 2.576 for 99% limits, or 1.645 for 90% limits)

Because these confidence limits are based on the normal distribution, there is a tendency for confidence intervals of population estimates to be more narrow than they ought to be (Otis et al. 1978, 105).

The null model M_0 has a simple form when there are only two sampling periods (as in a Petersen sample). For this situation, equation (3.20) simplifies to

$$\hat{N}_0 = \frac{(n_1 + n_2)^2}{4m} \quad (3.24)$$

where n_1 = Number of individuals captured in the first sample and marked
n_2 = Number of individuals captured in the second sample
m = Number of recaptured individuals in the second sample

Box 3.3 Null Model of Program Capture to Estimate Population Size for Humpback Whales

Palsbøll et al. (1997) reported on mark-recapture studies of humpback whales (*Megaptera novaeangliae*) in the North Atlantic Ocean. For the Gulf of St. Lawrence subpopulation, they captured 65 genetically distinct individuals, and over 5 sampling periods they tallied 86 sightings (sampling with replacement).

To illustrate how Program CAPTURE calculates a population estimate, we use these data with the null model, which assumes no heterogeneity in capture probabilities, no time variation in catchability, and no behavioral changes as a result of the original capture of a genetic sample for identification.

Given a provisional range of estimates of population size, the method of maximum likelihood is used to find the most likely estimate of N. From equation (3.20),

$$L(\hat{N}_0, \hat{p}|\mathbf{X}) = \ln\left(\frac{N!}{(N-M)!}\right) + (n)\ln(n) + (tN - n)\ln(tN - n) - (tN)\ln(tN)$$

where $\hat{N}_0$ = Estimated population size from the null model of CAPTURE
N = Provisional estimate of population size
$\hat{p}$ = Probability of capture
M = Total number of different individuals captured in the entire sampling period
n = Total number of captures during the entire sampling period
t = Number of samples (e.g., days)
ln = Natural log ($\log_e$)
L = Log-likelihood of the estimated value $\hat{N}_0$ and p, given the observed data

A great deal of computation goes into finding the maximum of this function, and I illustrate here only one set of calculations to show in principle how the method works.

Use 112 for a provisional estimate of population size (N). Calculate the log-likelihood as

$$L(\hat{N}_0, \hat{p}|\mathbf{X}) = \ln\left(\frac{112!}{(112-65)!}\right) + (86)\ln(86) + (5[112] - 86)\ln(5[112] - 86) - (5[112])\ln(5[112]) = 42.789$$

By repeating this calculation for other provisional estimates of N, you can determine

$$\text{for } N = 117,\ L(\hat{N}_0, \hat{p}|\mathbf{X}) = 42.885$$

$$\text{for } N = 121,\ L(\hat{N}_0, \hat{p}|\mathbf{X}) = 42.904$$

$$\text{for } N = 126,\ L(\hat{N}_0, \hat{p}|\mathbf{X}) = 42.868$$

and the maximum likelihood occurs at $\hat{N}$ of 121 whales.

Note that in practice you would use Program CAPTURE to do these calculations and also to test whether more complex models involving variation in probability of capture due to time or behavior might be present in these data.

The probability of an individual whale being sighted and sampled for DNA at any given sample period can be determined from equation (3.21):

$$\hat{p} = \frac{n}{t\hat{N}_0} = \frac{86}{5(121)} = 0.142$$

Given this probability, we can now estimate the variance of population size from equation (3.22)

$$\text{Var}(\hat{N}_0) = \frac{\hat{N}_0}{(1 - \hat{p})^{-t} - (t/1 - \hat{p}) + t - 1}$$

$$= \frac{121}{(1 - 0.142)^{-5} - [5/(1 - 0.142)] + 5 - 1} = 373.5$$

and the resulting 90% confidence interval from equation (3.23) is

$$\hat{N}_0 \pm z_\alpha \sqrt{\hat{V}ar(\hat{N}_0)}$$

$$121 \pm 1.645\sqrt{373.5} \text{ or } 121 \pm 32$$

These calculations, including confidence limits, can be done by Program CAPTURE, and this method is discussed in detail by Otis et al. (1978).

For example, from the data in Box 2.1, the null model estimate of population size is

$$\hat{N}_0 = \frac{(n_1 + n_2)^2}{4m} = \frac{(948 + 421)^2}{4(167)} = 2806 \text{ hares}$$

This estimate is 18% higher than the Petersen estimate calculated in Box 2.1. The null model tends to be biased towards overestimation when the number of sampling times (t) is less than 5, unless the proportion of marked animals is relatively high. For this reason the Petersen method is recommended for data gathered over two sampling periods, as in this example. Program CAPTURE and Program MARK become most useful when there are at least 4–5 sampling times in the mark-recapture data, and like all mark-recapture estimators, they provide better estimates when a high fraction of the population is marked.

Box 3.3 gives a sample set of calculations for the null model from Program CAPTURE.

3.4 ENUMERATION METHODS

Why estimate when you can count the entire population? This is one possible response to the problems of estimating density by mark-recapture, removal, or resight methods. All of these methods have their assumptions, and you could avoid all this statistical hassle by counting all the organisms in the population. In most cases this simple solution is not possible either physically or financially, and you must rely on a sampling method of some type, but in a few situations, enumeration is a possible strategy.

Enumeration methods have been used widely in small mammal studies (Krebs 1966) where they have been called the *minimum-number-alive* method (MNA). The principle of this method is simple. Consider a small example from a mark-recapture study (0 = not caught, 1 = caught):

Tag number	Time 1	Time 2	Time 3	Time 4	Time 5
A34	1	0	1	1	0
A38	1	1	1	0	1
A47	1	0	0	1	1
A78	1	1	0	1	1
A79	0	0	1	0	1
A83	1	1	1	0	1
No. caught	5	3	4	3	5
Minimum number alive	5	5	6	6	5

Individuals miss being caught at one or more sample periods, but if they are known to be present before and after the time of sampling, we assume they were present but not caught during the intervening sample times. These individuals are added to the number actually caught to give the minimum number alive—simple arithmetic, no complicated equations, and apparently an ecologist's dream!

Enumeration methods have been used for birds, small mammals, reptiles, and amphibians (Pollock et al. 1990), and their simplicity makes them popular. But there is one overwhelming complaint about enumeration methods—they all suffer from a negative bias. Numbers enumerated are always less than or equal to the true population size (Jolly and Dickson 1983). If catchability is high, so that most individuals are caught in most of the sampling periods, this negative bias is small. But if catchability is low, the negative bias is very large.

The net result is that enumeration methods should not be used when it is practical to use one of the unbiased mark-recapture methods. At the very least we should compute both, for example, the Jolly-Seber population estimates and the MNA estimates, and show that these are nearly the same. Jolly-Seber estimates can never be below MNA estimates, and these two will converge only when all animals are caught each sampling time.

In a few cases enumeration methods are needed and can be justified on the principle that a negatively biased estimate of population size is better than no estimate. In some cases the numbers of animals in the population are so low that recaptures are rare. In cases with endangered species, it may not be possible to mark and recapture animals continuously without possible damage, so sample sizes may be less than the 5–10 individuals needed for most estimators to be unbiased. In these cases it is necessary to ask whether a population estimate is needed, or whether an index of abundance is sufficient for management purposes. It may also be possible to use the methods presented in the next two chapters for rare and endangered species. The important point is to recognize that complete enumeration is rarely a satisfactory method for estimating population size.

The fact that most animals do not satisfy the randomness-of-capture assumption of mark-recapture models is often used to justify the use of enumeration as an alternative estimation procedure. But Pollock et al. (1990) show that even with heterogeneity of capture, the Jolly-Seber model is less biased than MNA-based enumeration procedures. A closed population estimator or the Jolly-Seber model should be used in almost all cases instead of enumeration.

3.5 ESTIMATING DENSITY

We have been discussing methods of estimating abundance, and yet most ecologists want to know the *density* of the population, the number of individuals per unit area or unit volume. If you know the size of the study area, it is easy to determine the density once you know the abundance of a species. But in many cases the transition from abundance to density is far from simple. Consider, for example, a grouse population in an extensive area of forest or moorland. Individuals captured and marked on the study area will move on and off the area, and the home ranges of vertebrates may be only partly on the study area. In practice this means that the size of the study area is larger than the physical area trapped, and the true population density is less than the biased density estimate obtained by dividing the population estimate by the physical area of study. Figure 3.6 illustrates the problem

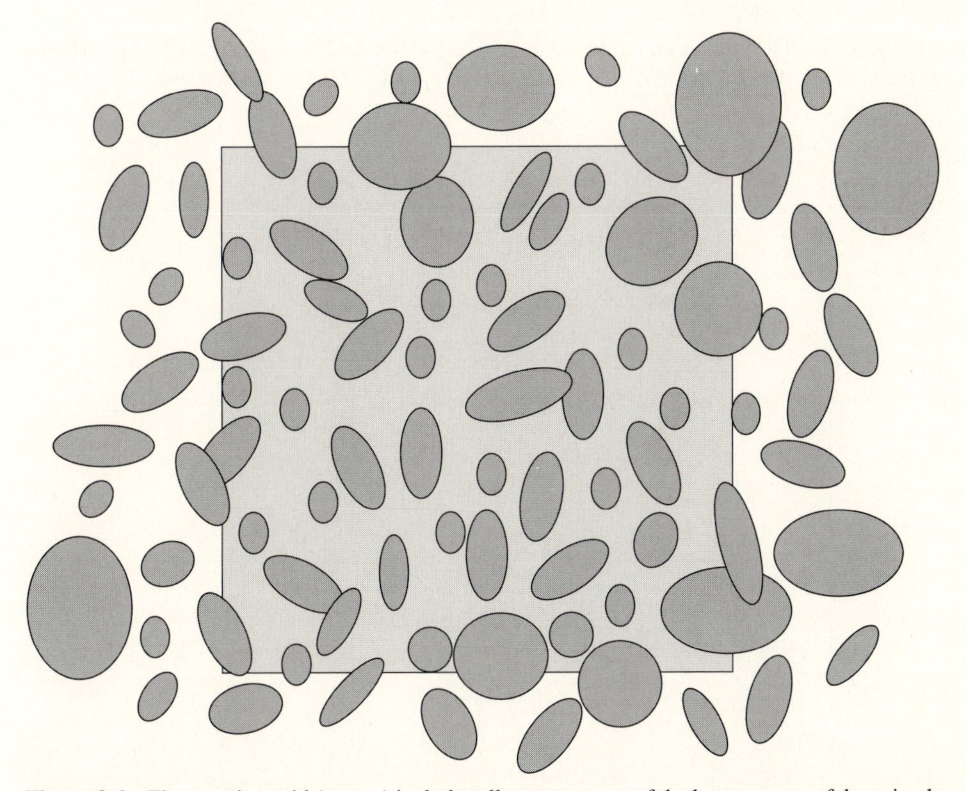

Figure 3.6 The trapping grid (square) includes all, part, or none of the home ranges of the animals in the study zone. Some of the animals whose home ranges overlap the trapping grid will be captured, marked, and released. The effective size of the trapping grid is thus larger than its physical area. If home ranges are very small, the effective grid area is only slightly larger than its physical area. (Modified from White et al. 1983.)

schematically. How can we determine how large an effective area we are studying so that we can obtain estimates of true density? Several methods have been employed to estimate the effective size of a trapping area.

3.5.1 Boundary Strip Methods

The simplest methods add a boundary strip around the trapping area to estimate the effective size of area trapped. Figure 3.7 illustrates the principle of a boundary strip. The width of the boundary strip can be estimated in several ways (Stenseth and Hansson 1979). The simplest procedure is to add a strip one-half the movement radius of the animals under study. For example, for mammals, we can determine the average distance moved between trap captures and use this as a likely estimate of movement radius. The problem with this approach is that it is highly dependent on the spacing between the capture points and the number of recaptures. While this approach is better than ignoring the problem of a boundary strip, it is not completely satisfactory (Otis et al. 1978).

A more sophisticated boundary strip method was devised by Bondrup-Nielsen (1983), who used home range size to estimate effective size of a sampling area. The general principle is to try to minimize the boundary strip effect by defining the required size of the sampling area. This method proceeds in three steps:

1. Calculate the average home range size for the animal under study. Methods for doing this are presented in Kenward (1987) and in White and Garrott (1990).

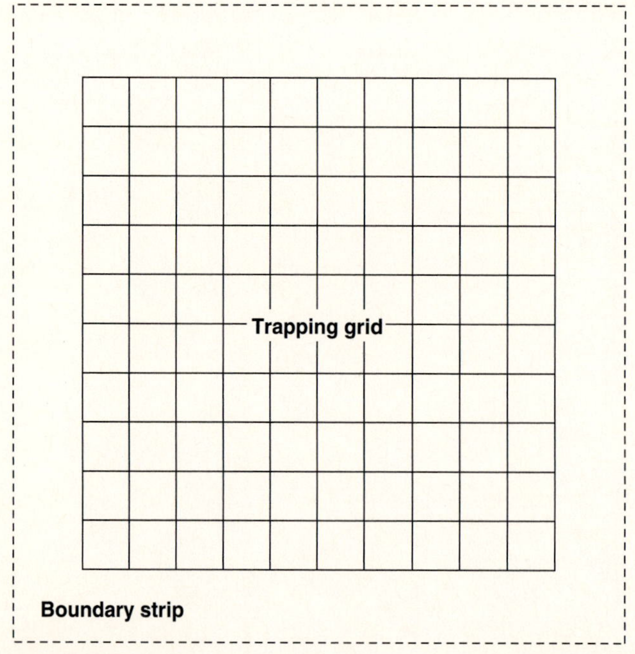

Figure 3.7 Illustration of a study grid of 10 × 10 checkerboard shape with a boundary strip added to delineate the effective study area. Population size would be estimated for the checkerboard, and the total area, including the boundary strip, would be used to estimate population density.

2. Compute the ratio of grid size to area of the average home range. In a hypothetical world with square home ranges that do not overlap, the overestimation of density is shown by Bondrup-Nielsen (1983) to be

$$\frac{\text{Estimated density}}{\text{True density}} = \frac{(\sqrt{A} + 1)^2}{A} \tag{3.25}$$

$$\text{where} \quad A = \frac{\text{Area of study grid}}{\text{Average home range size}} \tag{3.26}$$

3. Assume that the home range is elliptical, and use a computer to "throw" home ranges at random on a map that delineates the study zone as part of a large area of habitat.

Figure 3.8 shows the results of this simulation. On the basis of this graph, Bondrup-Nielsen (1983) suggested that the trapping grid size should be at least 16 times the size of the average home range of the species under study to minimize the edge effect.

Given that we have data on study area size and average home range size, we can use Figure 3.8 as a guide to estimating true density of a population. For example, in a study of house mouse populations in wheat fields, live traps were set out on a 0.8 ha grid, and the Petersen population estimate was 127 mice. The average home range size of these house

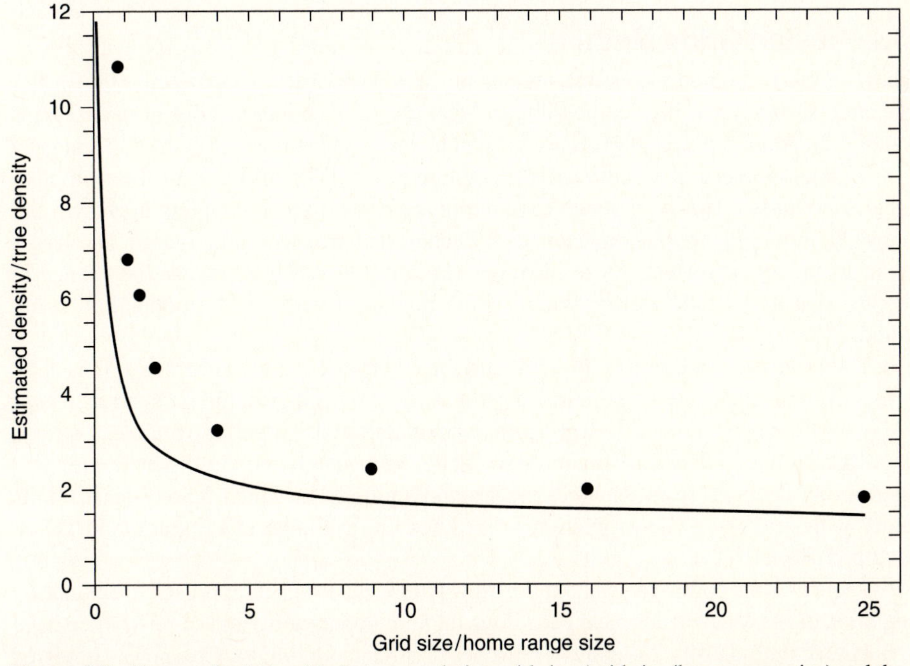

Figure 3.8 Expected relationship between relative grid size (grid size/home range size) and the overestimation of density if no boundary strip is used to calculate population density. The theoretical curve is shown for square home ranges that do not overlap, and the points (•) show the average results of computer simulations with elliptical home ranges that are allowed to overlap. (Modified from Bondrup-Nielsen 1983.)

mice was 0.34 ha. We can calculate as follows from equation (3.26):

$$A = \frac{\text{Grid size}}{\text{Home range size}} = \frac{0.8}{0.34} = 2.35$$

From Figure 3.8 or equation (3.25) the expected overestimate is

$$\frac{\text{Estimated density}}{\text{True density}} = \frac{(\sqrt{A} + 1)^2}{A} = \frac{(\sqrt{2.35} + 1)^2}{2.35} = 2.73$$

Thus the biased density estimated from the area of the grid is

$$\text{Biased density estimate} = \frac{\text{Population estimate}}{\text{Size of study area}} = \frac{127 \text{ mice}}{0.8 \text{ ha}}$$

$$= 159 \text{ mice per ha}$$

$$\text{Corrected density estimate} = \frac{159}{2.73} = 58 \text{ mice per ha}$$

This correction factor is only approximate because home ranges are not exactly elliptical, and there is some overlap among individuals, but the adjusted density estimate is closer to the true value than is the biased estimate that does not take into account the effective grid size.

3.5.2 Nested Grids Method

A more rigorous method for estimating density was developed by Otis et al. (1978). If a large area is sampled, it may be possible to break the data up into a series of nested grids (Figure 3.9). At least 4 nested grids are needed to get good estimates, and this will mean a 15×15 checkerboard of sample traps or capture points. The principle of the method is simple: The biased density estimate (population estimate/size of sampling area) will decline as the sampling area increases in size. Each nested grid will have a larger and larger area of boundary strip as grid size increases (Figure 3.9), and we can use the change in observed density to estimate the width of the boundary strip for the population being studied.

The estimation problem is difficult, and the details are not given here (see Otis et al. 1978). Given the area and the population estimate for each subgrid, the problem is to estimate two parameters: boundary strip width and population density. If the boundary strip is very wide, the biased density estimates will change very slowly with grid size. Figure 3.10 illustrates an example from Richardson's ground squirrels in which biased density falls rapidly with grid size. The estimation procedures are available in Program CAPTURE (described above).

There is some controversy about the utility of the nested grids approach to estimating population density. Few trials have been done on field populations, and some of these have provided accurate density estimates (Wilson and Anderson 1985a). Computer-simulated populations however typically show a large positive bias in the density estimate because the strip width is underestimated (Wilson and Anderson 1985b).

The nested grids approach to density estimation is very data-intensive, and it works best when a high fraction of the population can be marked and recaptured and when density

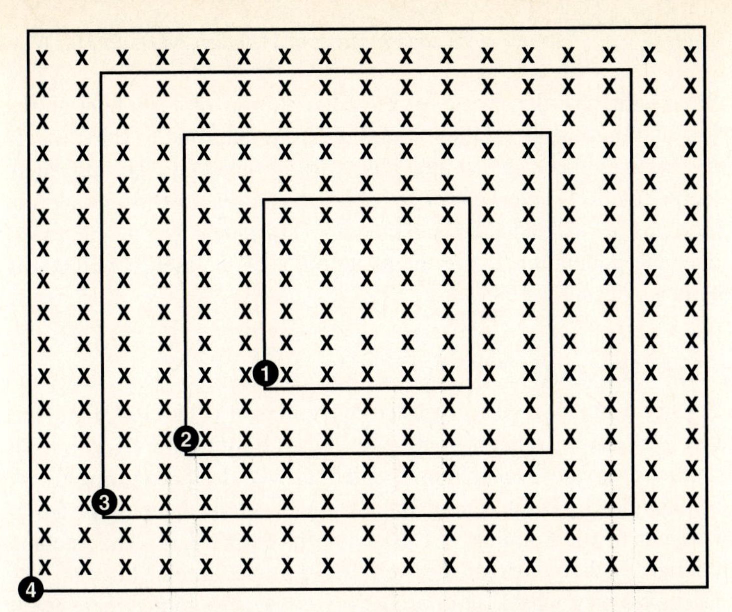

Figure 3.9 An example of a set of nested grids for population density estimation. The entire study area is a 17 × 17 checkerboard, and 4 nested subgrids are shown by the lines. The nested grids are 5 × 5, 9 × 9, 13 × 13, and 17 × 17, and if the sample points are 10 m apart (for example), the areas of these four subgrids would be 0.16 ha, 0.64 ha, 1.44 ha, and 2.56 ha.

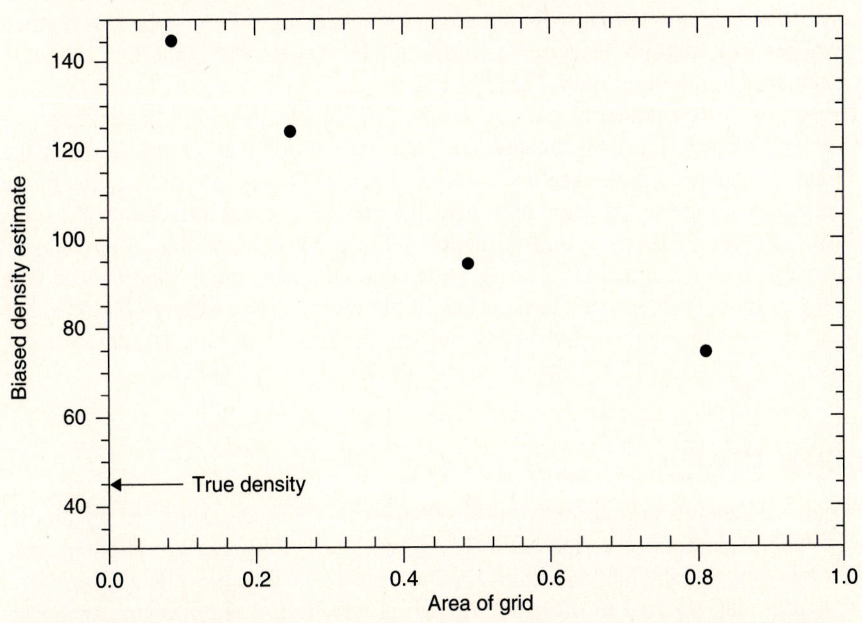

Figure 3.10 Illustration of the approach to density estimation from nested grids. Data on Richardson's ground squirrels live-trapped on a 10 × 10 checkerboard grid with 10 m spacing. Four nested subgrids of 4 × 4, 6 × 6, 8 × 8 and 10 × 10 were used to generate four separate estimates of population size. The biased density estimate (population estimate/area trapped) declines with grid size. The best estimate of true density for these data was 45 squirrels per hectare, with a boundary strip width of 12.4 m. These estimates were obtained using Program CAPTURE. (Data from White et al. 1983.)

is moderate to high. It assumes that the density of the population is uniform over all the area of the nested grids, and that the population is closed during the time of sampling. If the traps used are attractive to animals, so that individuals are attracted into the study area, the assumption that the population is closed will be violated, and this method should not be used. If a removal study is being conducted, and the removal extends over a long period so that immigrants enter the population, again the assumption of a closed population is violated, and this method is not useful.

3.5.3 Trapping Web Method

A third approach to density estimation was suggested by Anderson et al. (1983) and has been called the *trapping web design*. This method is described only briefly here, since it depends on distance methods that are discussed in Chapter 5. It was designed primarily for small mammals that are sampled by live-trapping, but it could be adapted to any species of limited mobility. The layout of the trapping web is shown in Figure 3.11. This method assumes that all individuals at the center of the web are captured. Radial lines of 15–20 traps are laid out as spokes on a wheel, and trap density is adjusted so that there are at least 8–12 traps in every home range at the center of the web. The critical assumption is that every individual in the center of the web is captured.

The data obtained from this design are the numbers of individuals captured in each of the trapping rings. As you move away from the center of the web, the area of each successive ring is larger, and consequently the trap density is lower. The expectation is that the number of individuals caught will increase in the outer rings of the web. This method is a type of removal method, because each individual is counted only the first time it is caught and marked, and all recaptures are ignored.

The details of the calculations for the density estimate are given in Chapter 5 and are covered in Anderson et al. (1983). In addition to the assumption that all individuals in the center of the web are caught, this method assumes that the distances from the center of the web to each trap are measured accurately, and that animals are not attracted to the traps from outside the web. If there is some attraction to the traps, it is possible to throw away the data from the outer 2 or 3 rings. The suggested configuration of the trapping web is to have 16 lines with at least 16 traps in each line of the web. Approximately 60 individuals or more are needed to obtain a reasonably precise estimate of density (Anderson et al. 1983).

3.6 SUMMARY

Three special techniques are presented in this chapter that may be useful for population estimation with exploited populations. If there are two types of organisms in a population, like males and females, and hunting removes more of one type than the other, the resulting *change-in-ratio* may be used to estimate population size. Large samples are usually required for change-in-ratio estimators, and you need to plan your sampling program thoroughly. Alternatively, if removals are not sex- or age-specific, you can use Eberhardt's index-removal method to estimate population size. This method depends on index counts that must enumerate about 20% or more of the total population to achieve precise estimates of population size.

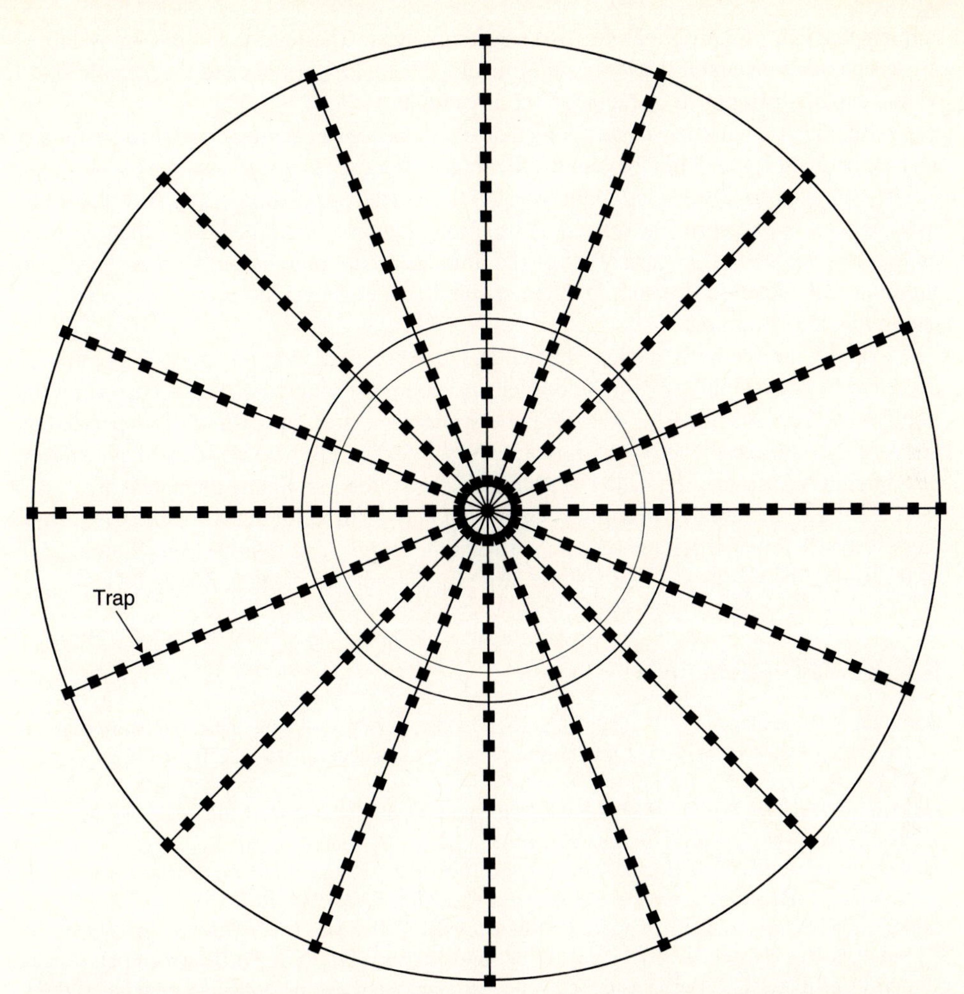

Figure 3.11 Trapping web design for estimating population density. Traps are laid out in a radial pattern. Sixteen lines are shown in this example, with 17 traps in each line, equally spaced from the center of the web. One ring is illustrated that includes the 7th ring of traps. The area of each ring grows larger as you move from the center of the web to the outermost traps. The basic data obtained are the numbers of individuals captured in each of the 17 rings of traps. (Modified after Anderson et al. 1983.)

If harvesting removes a reasonable fraction of the population, the *catch-per-unit-effort* may decline with time. If catchability is constant and this decline is linear, regression methods may be used to estimate population size at the time when exploitation begins. Fisheries managers have used this method extensively, but the assumption of constant catchability is vital if the population estimates are to be accurate.

Recent developments in mark-recapture methods have led to a series of models for closed population estimation. These models can accommodate variation in catchability over time as well as among individuals, and thus circumvent the randomness-of-capture requirement of simpler methods. These new methods are computer-intensive and have been implemented in two programs for personal computers. They are restrictive in assuming closed populations and at least 3 (and typically 5) capture periods, and they require individual capture histories to be recorded for all animals.

Enumeration methods have a long history in population ecology, and their simplicity makes them attractive. But all methods of enumeration suffer from a negative bias—population estimates are always less than or equal to the true population—and they should be used only as a last resort. The critical assumption of enumeration methods is that you have captured or sighted all or nearly all of the animals in the population, so that the bias is minimal. Enumeration methods may be required for endangered species that cannot be captured without trauma.

All of the methods in this chapter and the previous chapter estimate population size, and to convert this to population density, you must know the area occupied by the population. This is simple for discrete populations on islands but difficult for species that are spread continuously across a landscape. Nested grids may be used to estimate density, or a boundary strip may be added to a sampling area to approximate the actual area sampled. For small mammals, a trapping web design may be used for density estimation. Density estimates are more reliable when the sampling area is large relative to the home range of the animal being studied and when a large fraction of the individuals can be captured.

SELECTED READING

Burnham, K. P., Anderson, D. R., and White, G. C. 1995. Selection among open population capture-recapture models when capture probabilities are heterogeneous. *Journal of Applied Statistics* 22: 611–624.

Lebreton, J. D. 1995. The future of population dynamics studies using marked individuals: A statistician's perspective. *Journal of Applied Statistics* 22: 1009–1030.

Matlock, R. B. J., Welch, J. B., and Parker, F. D. 1996. Estimating population density per unit area from mark, release, recapture data. *Ecological Applications* 6: 1241–1253.

Miller, S. D., White, G. C., Sellers, R. A., Reynolds, H. V., Schoen, J. W., Titus, K., Barnes, V. G., Smith, R. B., Nelson, R. R., Ballard, W. B., and Schwartz, C. C. 1997. Brown and black bear density estimation in Alaska using radiotelemetry and replicated mark-resight techniques. *Wildlife Monographs* 61: 1–55.

Rosenberg, D. K., Overton, W. S., and Anthony, R. G. 1995. Estimation of animal abundance when capture probabilities are low and heterogeneous. *Journal of Wildlife Management* 59: 252–261.

Seber, G. A. F. 1992. A review of estimating animal abundance, part 2. *International Statistical Review* 60: 129–166.

White, G. C. 1996. NOREMARK: Population estimation from mark-resighting surveys. *Wildlife Society Bulletin* 24: 50–52.

Wilson, K. R., and Anderson, D. R. 1985. Evaluation of two density estimators of small mammal population size. *Journal of Mammalogy* 66: 13–21.

QUESTIONS AND PROBLEMS

3.1. For the catch-effort regression estimation technique of Leslie, discuss the type of bias in estimated population size when there is

(a) Change in catchability during the experiment.

(b) Natural mortality during the experiment.

(c) Mortality caused by the marking procedure or the tag itself.
(d) Emigration of animals from the population.

3.2. Leslie and Davis (1939) set 210 rat traps (break-back type) in 70 houses in Freetown, Sierra Leone, in 1937. They caught, over 18 days,

Day	No. of *Rattus rattus* caught
1	49
2	32
3	31
4	34
5	16
6	33
7	22
8	27
9	17
10	19
11	18
12	16
13	18
14	12
15	14
16	12
17	17
18	7

They set 210 traps each day. Estimate the total population size of *Rattus rattus* at the start of the study. What sources of error might affect this particular estimate?

3.3. During a severe winter in 1938–39 in Utah, mule deer populations suffered heavy losses from starvation. The ratio of fawns to adults was 83 fawns to 100 adults before the severe weather (n=69 and 83 observed), and 53 fawns and 100 adults after the winter (n=38 and 72). Over the study area, 248 dead fawns and 60 dead adults were found, and these were believed to represent the whole of the losses (Rasmussen and Doman 1943). Estimate population size for this deer population at the start and end of this winter. Calculate 95% confidence intervals for your estimates.

3.4. Dunnet (1963) sampled quokkas (small marsupials) on Rottnest Island off Western Australia over three sampling periods with these results for 1957:

	Mature animals	Immature animals
Number marked at time 1	32	35
Number marked at time 1 and never seen again	22	20
Number marked caught at times 1 and 2 only	5	1
Number marked caught at times 1 and 3 only	4	11
Number marked caught at times 1, 2, and 3	1	3

Recast these data in the form of an **X** matrix, and use Program CAPTURE to estimate the size of the mature population and the immature population of this marsupial. What model is selected by Program CAPTURE, and how different are the estimates from the different models available in this program? Does the test in Program CAPTURE suggest any evidence for unequal catchability in these data?

3.5. Plan a study that will use the change-in-ratio estimator of population size. You would like to get an estimate within $\pm 10\%$ of the true value at an α of 5%. You expect a change in the proportion of males in your population to be about 10–15%. Describe your proposed sampling program in detail. How is it affected by uncertainty about the initial sex ratio (which might be from 50% males to 70% males)? How is it affected by the rate of exploitation?

3.6. White-footed mice were live-trapped in northern Michigan by Blair (1942). He trapped a large area of 7.4 ha. The population estimate was 39 males and 31 females for this trapping area. Male white-footed mice have home ranges that average 0.93 ha, while females have smaller ranges (0.56 ha). What are the biased density estimates for the two sexes of mice for this particular area, and what are the best estimates of the true density for each sex? What size of live-trapping area would you recommend for this species?

3.7. An aerial index count of feral water buffalo in the Northern Territory of Australia was carried out before and after a culling program in 1984. In the first index count 2786 buffalo were counted, and in the second count 1368 were seen. The cull occurred over two months, and 12,890 buffalo were shot. Can you estimate population size for this population? What proportion of the population was culled? What assumptions must you make to use these estimates?

3.8. During 7 aerial flights to locate 19 radio-collared caribou in southern British Columbia over a 2-week period, a total of 144 caribou were seen, of which 54 had radio-collars and the remaining 90 were not marked. The frequency of resighting of the 19 collared caribou was as follows: 7, 5, 1, 2, 3, 0, 5, 3, 1, 3, 0, 2, 4, 6, 1, 5, 1, 3, and 2. Estimate the population size of this caribou herd and the 90% confidence limits for this estimate.

3.9. Discuss from a practical point of view the advantages and disadvantages of using a trapping web design versus a nested grid design to estimate population density for a small mammal. What information might be lost if you adopted one design over the other?

outside the area to be counted? This decision is often biased by keen ecologists who prefer to count an organism rather than ignore it. Edge effects thus often produce a positive bias. The general significance of possible errors of counting at the edge of a quadrat cannot be quantified because it is organism- and habitat-specific, and can be reduced by training. If edge effects are a significant source of error, you should prefer a quadrat shape with less edge/area. Figure 4.1 illustrates one way of recognizing an edge effect problem. Note that there is no reason to expect any bias in *mean abundance* estimated from a variety of quadrat sizes and shapes. If there is no edge effect bias, we expect in an ideal world to get the same mean value regardless of the size or shape of the quadrats used, if the mean is expressed in the same units of area. This is important to remember: Quadrat size and shape are *not* about biased abundance estimates but are about narrower confidence limits. If you find a relationship like that in Figure 4.1 in your data, you should immediately disqualify the smallest quadrat size from consideration to avoid bias from the edge effect.

The second problem regarding quadrat shape is that nearly everyone has found that long, thin quadrats are better than circular or square ones of the same area. The reason for this is habitat heterogeneity. Long quadrats cross more patches. Areas are never uniform,

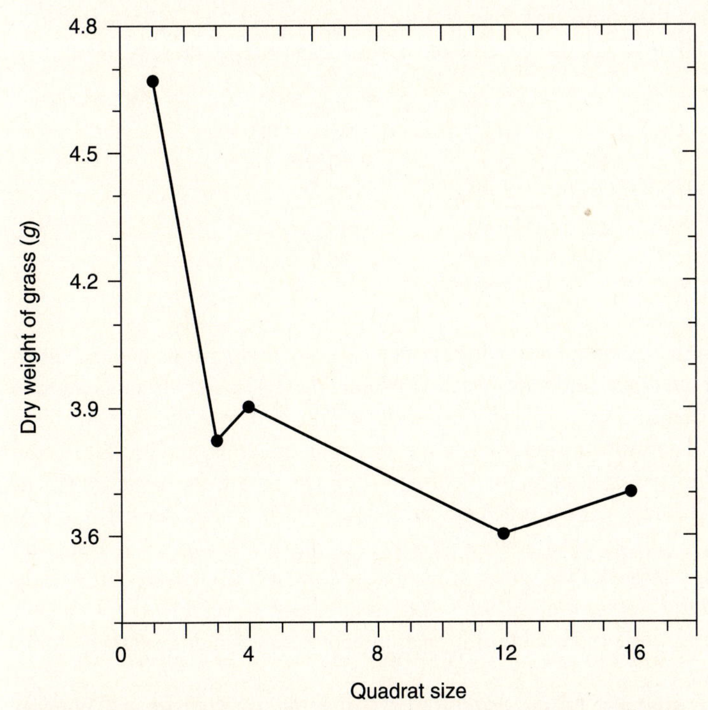

Figure 4.1 Edge effect bias in small quadrats. The estimated mean dry weight of grass (per 0.25 m^2) is much higher in quadrats of size 1 (0.016 m^2) than in all other quadrat sizes. This suggests an overestimation bias due to edge effects, and that quadrats of size 1 should not be used to estimate abundance of these grasses. Estimates of mean values (per unit area) should not be affected by quadrat size. (Data of Wiegert 1962, Table 1.)

TABLE 4.2 EFFECT OF PLOT SIZE ON STANDARD DEVIATION FOR MEASUREMENTS OF BASAL AREA OF TREES IN AN OAK-HICKORY FOREST IN NORTH CAROLINA

(1) Plot size (m)	(2) Observed standard deviation (per 4 m^2)	(3) Sample size[b]	(4) Standard error of mean for sample size in column (3)
4 × 4[a]	50.7	70	6.06
4 × 10	47.3	28	8.94
4 × 20	44.6	14	11.92
4 × 70	41.3	4	20.65
4 × 140	34.8	2	24.61

Source: Bormann 1953, Table IV.

[a] If equal total areas are sampled with each plot size, the best plot size is clearly 4 × 4 m.

[b] Number of quadrats of a given size needed to sample 1120 m^2.

and organisms are usually distributed somewhat patchily within the overall sampling zone. Clapham (1932) counted the number of *Prunella vulgaris* plants in 1 m^2 quadrats of two shapes: 1 m × 1 m and 4 m × 0.25 m. He counted 16 quadrats and got these results:

	Mean	Variance	S.E.	95% confidence interval
1 × 1 m	24	565.3	5.94	±12.65
4 × 0.25 m	24	333.3	4.56	±9.71

Clearly in this situation the rectangular quadrats are more efficient than square ones. Given that only two shapes of quadrats were tried, we do not know if even longer, thinner quadrats might be still more efficient.

Not all sampling data show this preference for long, thin quadrats, and for this reason each situation should be analyzed on its own. Table 4.2 shows data from basal area measurements on trees in a forest stand studied by Bormann (1953). The observed standard deviation almost always falls as quadrat area increases, as shown in Table 4.2. But if an equal total *area* is being sampled, the highest precision (= lowest S.E.) will be obtained by taking 70 4 × 4 m quadrats rather than two 4 × 140 m quadrats. If, on the other hand, an equal *number* of quadrats were to be taken for each plot size, one would prefer the long, thin quadrat shape.

Two methods are available for choosing the best quadrat size statistically. Wiegert (1962) proposed a general method that can be used to determine optimal size or shape.* Hendricks (1956) proposed a more restrictive method for estimating optimal size of quadrats. In both methods it is essential that data from all quadrats be standardized to a single

* Following the analysis by Cochran (1953, Chap. 9).

unit area—for example, per square meter. This conversion is simple for means, standard deviations, and standard errors: divide by the relative area. For example,

$$\text{Mean number per sq. meter} = \frac{\text{Mean number per } 0.25 \text{ m}^2}{0.25}$$

$$\text{Standard deviation per sq. meter} = \frac{\text{Standard deviation per } 4 \text{ m}^2}{4}$$

For variances, the square of the conversion factor is used:

$$\text{Variance per sq. meter} = \frac{\text{Variance per } 9 \text{ m}^2}{9^2}$$

For both Wiegert's and Hendricks' methods, you should standardize all data to a common base area before testing for optimal size or shape of quadrat. They both assume further that you have tested for and eliminated quadrat sizes that give an edge effect bias (see Figure 4.1).

4.1.1 Wiegert's Method

Wiegert (1962) proposed that two factors were of primary importance in deciding on optimal quadrat size or shape: *relative variability* and *relative cost*. In any field study, *time* or *money* would seem to be the limiting resource, and we must consider how to optimize with respect to sampling time. We will assume that time = money, and in the formulas that follow, either unit may be used. Costs of sampling have two components (in a simple world):

$$C = C_0 + C_x$$

where C = Total cost for one sample
C_0 = Fixed costs or overhead
C_x = Cost for taking one sample quadrat of size x

Fixed costs involve the time spent walking or flying between sampling points and the time spent locating a random point for the quadrat; these costs may be trivial in an open grassland or enormous when sampling the ocean in a large ship. The cost for taking a single quadrat may or may not vary with the size of the quadrat. Consider a simple example from Wiegert (1962) of grass biomass in quadrats of different sizes:

	Quadrat size (area)				
	1	3	4	12	16
Fixed cost ($)	10	10	10	10	10
Cost per sample ($)	2	6	8	24	32
Total cost for one quadrat ($)	12	16	18	34	42
Relative cost for one quadrat	1	1.33	1.50	2.83	3.50

We need to balance these costs against the relative variability of samples taken with quadrats of different sizes:

	Quadrat size (area)				
	1	3	4	12	16
Observed variance per 0.25 m^2	0.97	0.24	0.32	0.14	0.15

The operational rule is, *Pick the quadrat size that minimizes the product of* (Relative cost) (Relative variability).

In this example we begin by disqualifying quadrat size 1 because it showed a strong edge effect bias (Figure 4.1). Figure 4.2 shows that the optimal quadrat size for grass in this particular study is 3, although there is relatively little difference between the products for quadrats of size 3, 4, 12, and 16. In this case again the size 3 quadrat gives the maximum precision for the least cost.

Box 4.1 illustrates the application of Wiegert's procedures for deciding on the optimal size or the optimal shape of quadrat for counting animals or plants.

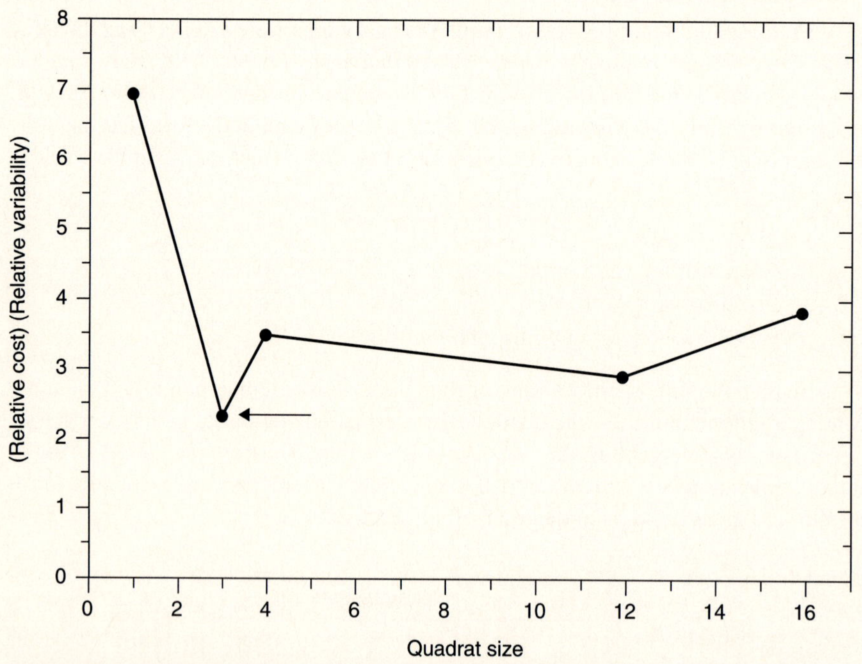

Figure 4.2 Determination of the optimal quadrat size for sampling. The best quadrat size is that which gives the minimal value of the product of relative cost and relative variability, which is quadrat size 3 (arrow) in this example. Quadrat size 1 data are plotted here for illustration, even though this quadrat size is disqualified because of edge effects. Data here are dry weights of grasses in quadrats of variable size. (After Wiegert 1962.)

Box 4.1 Wiegert's Method to Determine Optimal Quadrat Size for Biomass Estimates of the Seaweed *Chondrus crispus* ("Irish moss").

Quadrat size (m)	Sample size[a]	Mean biomass[b] (g)	Standard deviation[b]	S.E. of mean[b]	Time to take one sample (min)
0.5 × 0.5	79	1524	1022	115	6.7
1 × 1	20	1314	963	215	12.0
1.25 × 1.25	13	1037	605	168	13.2
1.5 × 1.5	9	1116	588	196	11.4
1.73 × 1.73	6	2021	800	327	33.0
2 × 2	5	1099	820	367	23.0

Source: Data from Pringle 1984.

[a] Sample sizes are approximately the number needed to make a total sampling area of 20 m^2.

[b] All expressed per square meter.

If we neglect any fixed costs and use the time to take one sample as the total cost, we can calculate the relative cost for each quadrat size as

$$\text{Relative cost} = \frac{\text{Time to take one sample of a given size}}{\text{Minimum time to take one sample}}$$

In this case the minimum time = 6.7 minutes for the 0.5 × 0.5 m quadrat. We can also express the variance [= (standard deviation)^2] on a relative scale:

$$\text{Relative variance} = \frac{(\text{Standard deviation})^2}{(\text{Minimum standard deviation})^2}$$

In this case the minimum variance occurs for quadrats of 1.5 × 1.5 m. We obtain for these data:

Quadrat size (m)	(1) Relative variance	(2) Relative cost	Product of (1) × (2)
0.5 × 0.5	3.02	1.00	3.02
1 × 1	2.68	1.79	4.80
1.25 × 1.25	1.06	1.97	2.09
1.5 × 1.5	1.00	1.70	1.70
1.73 × 1.73	1.85	4.93	9.12
2 × 2	1.94	3.43	6.65

The operational rule is to pick the quadrat size with minimal product of (cost) × (variance), and this is clearly 1.5 × 1.5 m, which is the optimal quadrat size for this particular sampling area.

There is a slight suggestion of a positive bias in the mean biomass estimates for the smallest quadrats, but this was not tested for by Pringle and does not appear to be statistically significant.

The optimal quadrat shape could be decided in exactly the same way.

Program-group QUADRAT SAMPLING (Appendix 2) can do these calculations.

4.1.2 Hendricks' Method

Hendricks (1956) noted that the variance (per standard area) usually decreased with larger quadrat sizes. In a simple situation the log of the variance will fall linearly with the log of quadrat size, and one critical parameter is how fast the variance declines with quadrat size. Hendricks assumes that the slope of this line will be between 0 and -1. This is a very restrictive assumption, and if the slope is not within this range, the method cannot be used.

In many cases the amount of time or money needed to count quadrats of different sizes is directly proportional to quadrat size. Thus a 2 m^2 quadrat will require twice as much time to count as a 1 m^2 quadrat. Hendricks makes a second assumption that this is true for all quadrat sizes. In this simple situation the critical cost is the fixed cost for each quadrat, the cost to select and move to and locate a single, new, random quadrat.

Given these simplifying assumptions, Hendricks shows that optimal quadrat size is determined as follows:

$$\hat{A} = \left(\frac{a}{1 - a}\right)\left(\frac{C_0}{C_x}\right) \tag{4.1}$$

where $\hat{A}$ = Estimate of optimal quadrat size
a = Absolute value of the slope of the regression of log (variance) on log (quadrat size), assumed to be between 0 and 1
C_0 = Cost of locating one additional quadrat
C_x = Cost of measuring one unit area of sample

Because Hendricks' method makes so many simplifying assumptions, it cannot be applied as generally as Wiegert's method. And because it is concerned only with quadrat size, it cannot answer questions about quadrat shape. But Hendricks' method can be a simple yet rigorous way of deciding on what size of quadrat is best for a particular sampling program.

Program-group QUADRAT SAMPLING (Appendix 2) does the calculations for Wiegert's method and for Hendricks' method for estimating the optimal quadrat size or shape.

Considerations of quadrat size and shape are complicated by seasonal changes within a population, and by the fact that in many cases several different species are being counted in the same quadrat. There is no reason why quadrat size cannot change seasonally or change from species to species. One simple way to do this is to use nested quadrats:

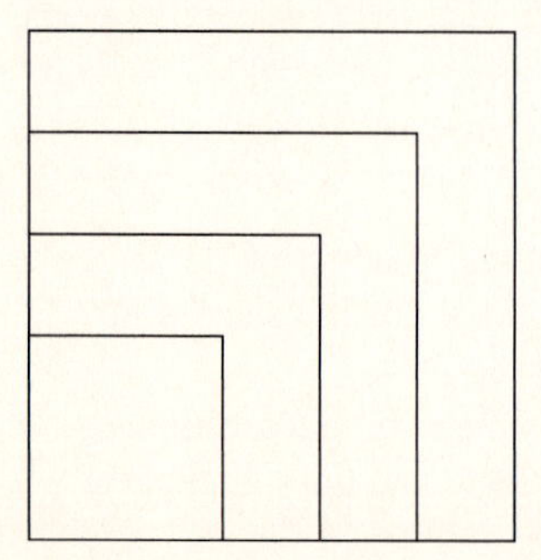

Plant ecologists use nested quadrats to define a species-area curve for a plant community. The number of species typically rises with quadrat size but then plateaus at a quadrat

TABLE 4.3 SAMPLE PLOTS OF DIFFERENT SIZES AND SHAPES

Area	Circular plots (radius)	Square plots (side)	Rectangular quadrats (ratio of sides)		
			1:2	1:5	1:10
1	0.56	1.00	0.71 × 1.41	0.44 × 2.20	0.32 × 3.16
2	0.80	1.41	1.00 × 2.00	0.63 × 3.16	0.45 × 4.47
3	0.98	1.73	1.22 × 2.44	0.78 × 3.86	0.55 × 5.48
4	1.13	2.00	1.41 × 2.82	0.89 × 4.45	0.63 × 6.32
5	1.26	2.24	1.58 × 3.16	1.00 × 5.00	0.71 × 7.07
10	1.78	3.16	2.24 × 4.47	1.41 × 7.07	1.00 × 10.00
20	2.52	4.47	3.16 × 6.32	2.00 × 10.00	1.41 × 14.14
30	3.09	5.48	3.94 × 7.88	2.45 × 12.25	1.73 × 17.32
40	3.57	6.32	4.47 × 8.94	2.83 × 14.15	2.00 × 20.00
50	3.99	7.07	5.00 × 10.00	3.16 × 15.81	2.24 × 22.36
100	5.64	10.00	7.07 × 14.14	4.47 × 22.36	3.16 × 31.62
200	7.98	14.14	10.00 × 20.00	6.32 × 31.62	4.47 × 44.72
300	9.77	17.32	12.25 × 24.50	7.74 × 38.70	5.48 × 54.77
400	11.28	20.00	14.14 × 28.28	8.94 × 44.70	6.32 × 63.24
500	12.62	22.36	15.81 × 31.62	10.00 × 50.00	7.07 × 70.71
1000	17.84	31.62	22.36 × 44.72	14.14 × 70.71	10.0 × 100.0

Source: Table courtesy of E. A. Johnson, University of Calgary.

size that determines the minimal area of a community. Species-area curves can be used to define quadrat size for plant community studies (Goldsmith and Harrison 1976). Nested quadrats may be square, circular, or any other shape. Table 4.3 can be useful in designing nested quadrats.

The message is that no single quadrat size or shape can be universally recommended. In any long-term study it is desirable to do a pilot study to gather the means, variances, and costs for quadrats of different sizes and shapes, and then to decide objectively on the size and shape of the sampling unit you will use. Such a pilot study can save you a great deal of time and money, and should be a required part of every experimental design.

4.1.3 When Should You Ignore These Recommendations?

Sampling plant and animal populations with quadrats is done for many different reasons, and it is important to ask when you might be advised to ignore the recommendations for quadrat size and shape that arise from Wiegert's method or Hendricks' method.

There are two common situations when you may wish to ignore these recommendations. In some cases you will wish to compare your data with older data gathered with a specific quadrat size and shape. Even if the older quadrat size and shape are inefficient, you may be advised for comparative purposes to continue using the old quadrat size and shape. In principle this is not required, as long as no bias is introduced by a change in quadrat size. But in practice many ecologists are more comfortable using the same quadrats as the earlier studies. This can become a more serious problem in a long-term monitoring program in which a poor choice of quadrat size in the early stages could condemn the whole project to wasting time and resources in inefficient sampling procedures. In such cases you will have to decide between the Concorde strategy (continuing with the older, inefficient procedure) and the Revolutionary strategy (sweeping the decks clean with proper methods). Ecological wisdom is making the right choice in these matters.

If you are sampling several habitats, sampling for many different species, or sampling over several seasons, you may find that these procedures result in a recommendation for a different size and shape of quadrat for each situation. It may be impossible to do this for logistical reasons, and thus you may have to compromise. For example, sampling plankton from a ship requires a standard size of net, and while some variation is possible, cost will prevent many implementations of the most efficient quadrat sizes. In multispecies studies it may be possible to use nested quadrats, but again only a restricted array of sizes and shapes may be feasible logistically. The human element is important to recognize here, and very complex sampling designs may be self-defeating if they are difficult to implement in the field and tiring for the observers. It is important to keep in mind that an average sampling strategy that is adequate but not optimal for all species or habitats may still save you time and energy.

4.2 STATISTICAL DISTRIBUTIONS

Plants or animals in any geographical area are scattered about in one of three spatial patterns: *uniform*, *random*, or *aggregated*,* as shown in Figure 4.3. We can describe degrees of uniformity and aggregation, so it is possible to say that redwood trees are *more* or *less* aggregated in certain forest stands. But randomness is randomness, and it is *not* correct to say that one pattern is more random than another.

The simplest view of spatial patterning can be obtained by adopting an individual orientation, and asking the question, *Given the location of one individual, what is the probability that another individual is nearby?* There are three possibilities:

1. This probability is increased—*aggregated* pattern
2. This probability is reduced—*uniform* pattern
3. This probability is unaffected—*random* pattern

Statisticians have employed many different statistical distributions to describe these three basic spatial patterns in populations, so that the jargon of ecology and statistics has become enmeshed and somewhat confused. Following Pielou (1977), I shall try to keep the terminology distinct for the following words:

- **Distribution**
  (a) In **statistics**, a mathematical frequency distribution that defines the frequency of counts in quadrats of a given size (see Figure 4.4). In this book *distribution* will always mean a statistical frequency distribution.
  (b) In **ecology**, the observed geographic dispersion or patterning of individuals (as in Figure 4.3). In this book I will use *spatial pattern* or *pattern* to describe the geographic distribution of individuals.

*The terminology for these patterns can be most confusing. Some synonyms follow: **uniform** = regular = even = negative contagion = under-dispersed; **aggregated** = contagious = clustered = clumped = patchy = positive contagion = over-dispersed.

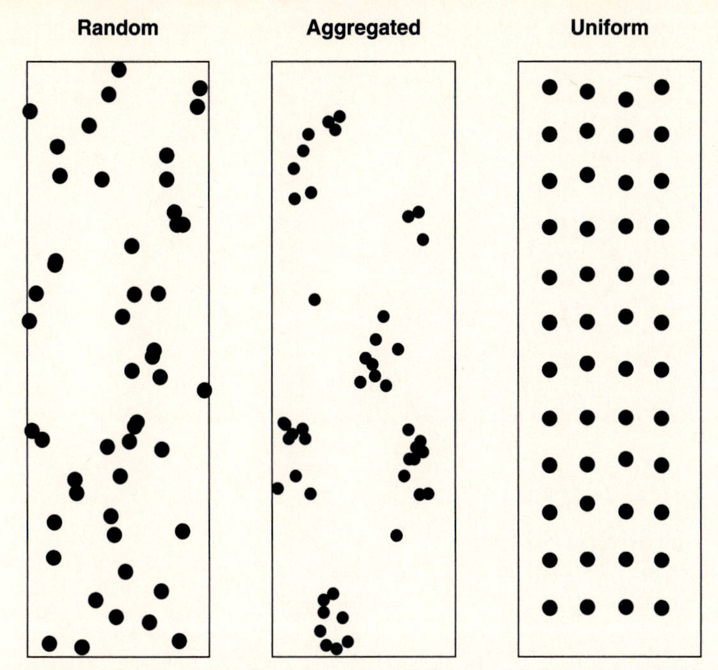

Figure 4.3 Three possible types of spatial patterning of individual animals or plants in a population.

- **Population**
  (a) In **statistics**, the universe of items under study is called the population. I will distinguish this concept as the *statistical population.*
  (b) In **ecology**, a population is a defined group of organisms of one species living in a particular area at a particular time. I will distinguish this concept as the *biological population.*

There is no end to the confusion that these two simple words have sown in ecological statistics, and you should try to use them only in a clear context.

Why should we be concerned about statistical distributions? For any particular kind of ecological data, the proper statistical distribution will determine how to calculate confidence limits correctly and how to set up a sampling program for a proper experimental design. Because quadrat sampling has an explicit spatial framework, we need to know about spatial patterns to estimate abundance properly. We begin by considering the simplest case: random spatial patterns and the corresponding statistical distribution, the Poisson.

4.2.1 Poisson Distribution

The approach to the statistical analysis of spatial patterns in biological populations is simple and straightforward. First, a frequency distribution is tallied of the counts in randomly placed quadrats of a given size, as in Figure 4.4. Then a statistician asks what this

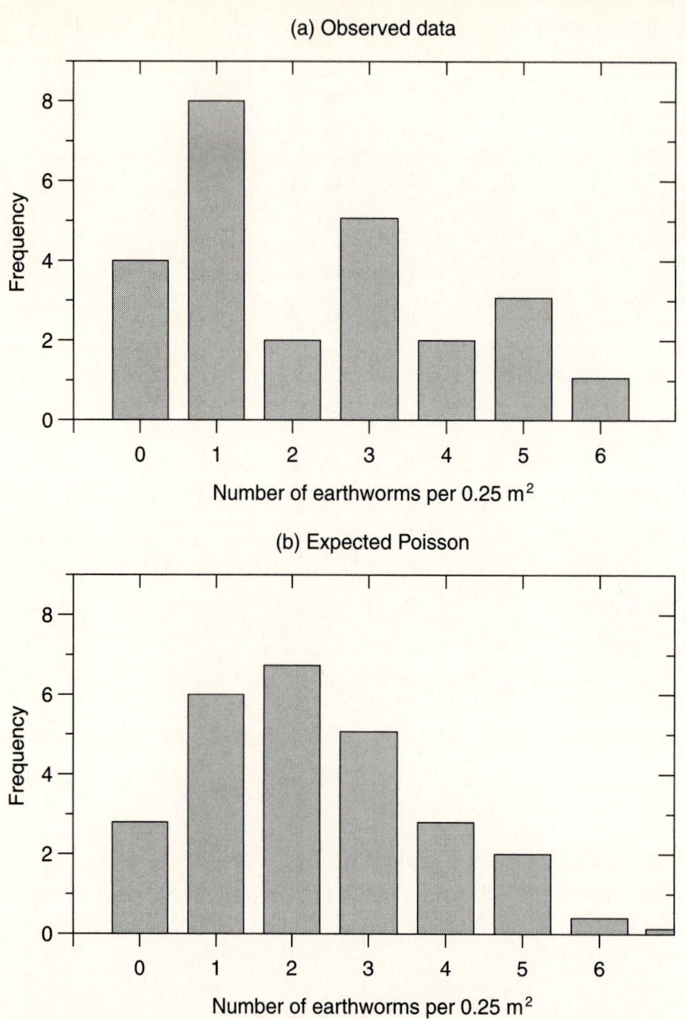

Figure 4.4 (a) Observed frequency distribution of the number of earthworms (*Lumbricus terrestris*) counted on 25 quadrats of 0.25 m^2 in Cairnshee pasture, Scotland, July 1977. $\bar{x} = 2.24$ worms per 0.25 m^2; $s^2 = 3.28$. (b) Expected frequency distribution based on the Poisson.

frequency distribution would look like if it were generated by completely random processes. In biological terms this is equivalent to asking if the organisms are distributed randomly in space. By spatial randomness we mean organisms whose x- and y-coordinates in geographical space are taken from a random number table.

If randomness of spatial pattern prevails, the Poisson distribution is the appropriate statistical descriptor of the data. The Poisson distribution is a discrete frequency distribution that is mathematically very simple because it depends on only one parameter: the *mean*. The terms of the Poisson distribution are defined as follows: (relative frequency = proportion = probability)

$$\text{Probability of observing zero individuals in a quadrat} = e^{-\mu}$$

$$\text{Probability of observing one individual in a quadrat} = e^{-\mu}\left(\frac{\mu}{1}\right)$$

$$\text{Probability of observing two individuals in a quadrat} = e^{-\mu}\left(\frac{\mu^2}{2}\right)$$

$$\text{Probability of observing three individuals in a quadrat} = e^{-\mu}\left(\frac{\mu^3}{3!}\right)$$

It is simpler to write out one general term to describe the Poisson distribution:

$$P_x = e^{-\mu}\left(\frac{\mu^x}{x!}\right) \tag{4.2}$$

where P_x = probability of observing x individuals in a quadrat
x = an integer counter; 0, 1, 2, 3 . . .
μ = true mean of the distribution
$x! = (x)(x-1)(x-2)\ldots(1)$ and $0! = 1$ by definition

In fitting the Poisson distribution to observed data, we need only to estimate the mean, which we do in the usual way by assuming

$$\bar{x} = \mu$$

The Poisson distribution assumes that the expected number of organisms is the same in all quadrats and is equal to μ. Thus it assumes a uniform world with no habitat patchiness, no "good" and "poor" quadrats.

Let us consider one example. Earthworms (*Lumbricus terrestris*) were counted on 25 quadrats in a Scottish pasture with these results: the numbers in each quadrat were

3	4	1	1	3	0	0	1	2	3	4	5	0
1	3	5	5	2	6	3	1	1	1	0	1	

For these data, $n = 25$, $\bar{x} = 2.24$, $s = 1.809$. The observed data are plotted in Figure 4.4(a). The terms of the theoretical Poisson distribution for these data can be calculated as follows:

e = Base of natural logarithms = 2.71828 . . .

P_0 = Proportion of quadrats expected to have no earthworms

$$= e^{-2.24} = 0.1065$$

P_1 = Proportion of quadrats expected to have 1 earthworm

$$= e^{-2.24}\left(\frac{2.24}{1}\right) = 0.2385$$

P_2 = Proportion of quadrats expected to have 2 earthworms

$$= e^{-2.24}\left(\frac{2.24^2}{2}\right) = 0.2671$$

$$P_3 = \text{Proportion of quadrats expected to have 3 earthworms}$$
$$= e^{-2.24}\left(\frac{2.24^3}{(3)(2)}\right) = 0.1994$$

Similarly,

$$P_4 = e^{-2.24}\left(\frac{2.24^4}{(4)(3)(2)}\right) = 0.1117$$

$$P_5 = e^{-2.24}\left(\frac{2.24^5}{(5)(4)(3)(2)}\right) = 0.0500$$

$$P_6 = e^{-2.24}\left(\frac{2.24^6}{(6)(5)(4)(3)(2)}\right) = 0.0187$$

Note that we *could* continue these calculations forever, but the proportions grow small very quickly. We know that

$$P_0 + P_1 + P_2 + P_3 + P_4 + P_5 + \cdots = 1.000$$

In this case

$$P_0 + P_1 + P_2 + P_3 + P_4 + P_5 + P_6 = 0.992$$

so that all the remaining terms $P_7 + P_8 + P_9 \ldots$ must sum to 0.008. We can calculate that $P_7 = 0.006$, so we can for convenience put all the remainder in P_8 and say

$$P_8 \approx 0.002$$

to round off the calculations. These calculations can be done by program-group QUADRAT SAMPLING (Appendix 2). We have calculated the *proportions* in the expected Poisson distribution, and the final step to tailor it to our earthworm data is to multiply each proportion by the number of quadrats sampled, 25 in this case.

Expected number of quadrats with no earthworms =

$$(P_0)(\text{total number of quadrats}) = (0.1065)(25) = 2.66$$

Expected number of quadrats with one earthworm =

$$(P_1)(\text{total number of quadrats}) = (0.2385)(25) = 5.96$$

And similarly,

No. worms, x	Expected number of quadrats
2	6.68
3	4.99
4	2.79
5	1.25
6	0.47
7	0.20

for a grand total of 25.00 expected. These expected numbers are plotted in Figure 4.4b. The observed and expected frequency distributions in Figure 4.4 are fairly similar, and we need a statistical technique to test the null hypothesis that the Poisson distribution provides an adequate fit to the observed data. This statistical hypothesis is inferred to be testing the ecological hypothesis that the spatial pattern observed is a random one.

Tests for Goodness-of-Fit There are two methods for testing the goodness-of-fit of the Poisson distribution: the index of dispersion test and the chi-squared goodness-of-fit test.

Index of Dispersion Test. We define an index of dispersion I to be

$$I = \frac{\text{Observed variance}}{\text{Observed mean}} = \frac{s^2}{\bar{x}} \tag{4.3}$$

For the theoretical Poisson distribution, the variance equals the mean, so the expected value of I is always 1.0 in a Poisson world. The simplest test statistic for the index of dispersion is a chi-squared one:

$$\chi^2 = I(n - 1) \tag{4.4}$$

where I = Index of dispersion (as defined in equation 4.3)
n = Number of quadrats counted
χ^2 = value of chi-squared with $(n - 1)$ degrees of freedom.

For the earthworm data with mean 2.24 and variance 3.27,

$$I = \frac{s^2}{\bar{x}} = \frac{3.27}{2.24} = 1.46$$

and

$$\chi^2 = I(n - 1) = 1.46(25 - 1) = 35.0$$

The index of dispersion test is a two-tailed chi-squared test because there are two possible directions of deviations: if organisms are uniformly spaced (Figure 4.3), the variance will be much less than the mean, and the index of dispersion will be close to zero. If organisms are aggregated, the observed variance will be greater than the mean, and the index of dispersion will be much larger than 1.0. Figure 4.5 gives the critical values of chi-squared for this two-tailed test. In more formal terms the decision rule is as follows: Provisionally accept the null hypothesis of an adequate fit if

$$\chi^2_{.975} \leq \text{Observed chi-squared} \leq \chi^2_{.025}$$

For our example, with 24 degrees of freedom,

$$\chi^2_{.975} = 12.40$$

$$\chi^2_{.025} = 39.36$$

so we tentatively accept the null hypothesis that the Poisson distribution fits the observed data satisfactorily.

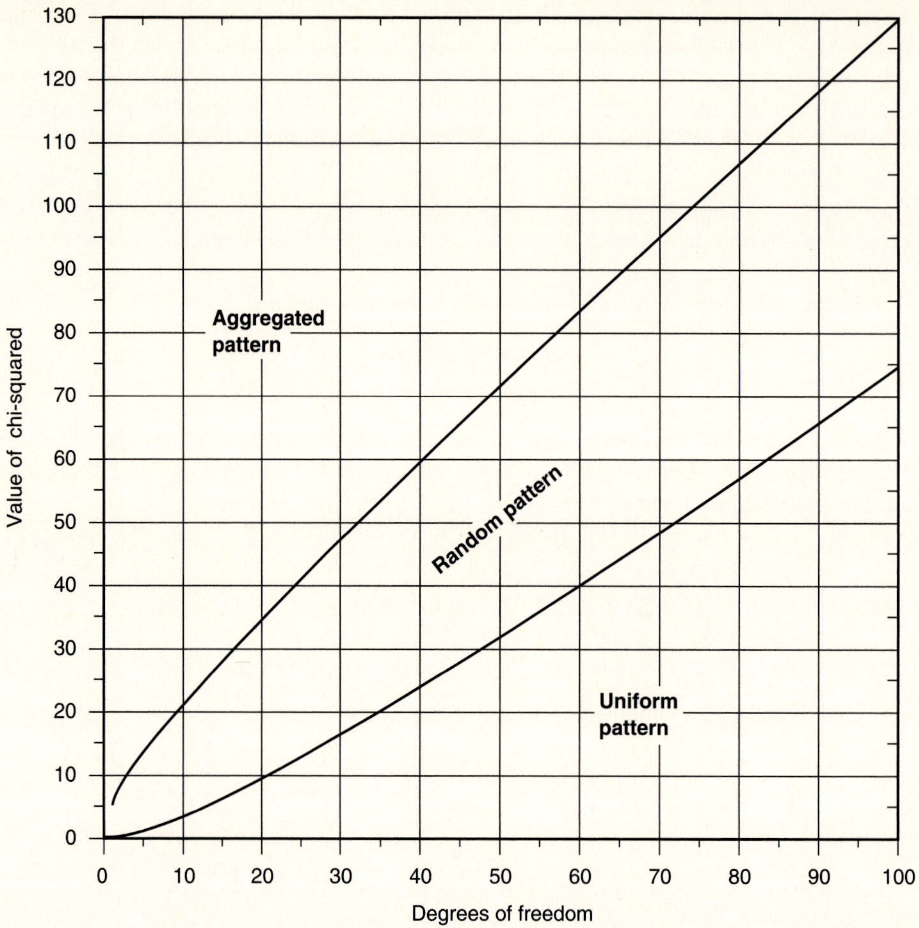

Figure 4.5 Critical values of the chi-square test of the index of dispersion (I) for $\alpha = 0.05$ and for $n < 101$. Large values of χ^2 indicate an aggregated pattern, and small values indicate a uniform pattern.

If the sample size is large ($n > 101$), Figure 4.5 cannot be used to determine the critical values of chi-squared. In this case calculate the normal approximation to the chi-squared value:

$$z = \sqrt{2\chi^2} - \sqrt{(2\nu - 1)} \tag{4.5}$$

where z = Standard normal deviate ($\mu = 0$, $\sigma = 1$)
χ^2 = Observed value of chi-square
ν = Number of degrees of freedom = $(n - 1)$

The decision rule is to accept the null hypothesis that the spatial pattern is random if z is between 1.96 and -1.96 (for $\alpha = 0.05$).

Chi-Squared Goodness-of-Fit Test. To test the null hypothesis that the Poisson distribution provides an adequate fit to the observed data, we can use the old standard:

$$\chi^2 = \sum \left[\frac{(\text{Observed frequency} - \text{expected frequency})^2}{\text{Expected frequency}} \right] \tag{4.6}$$

where we sum over each possible class (0, 1, 2 . . .) with due attention to the rule that all *expected* values should be 3 or more.* The tails of the distribution will have to be summed to satisfy this criterion.

For the earthworm data given above,

$$\chi^2 = \frac{(4 - 2.66)^2}{2.66} + \frac{(8 - 5.96)^2}{5.96} + \frac{(2 - 6.68)^2}{6.68} + \frac{(5 - 4.99)^2}{4.99} + \frac{(6 - 4.71)^2}{4.71}$$

$$= 5.00 \qquad (3 \text{ d.f.})$$

Observed and expected numbers are shown in Figure 4.4. Note that quadrat counts for 4, 5, and 6 were added together to form the last class.†

The number of degrees of freedom for this test is two less than the number of terms in the summation (because two parameters, the mean and the sample size, were obtained from the data). From the chi-squared table,

$$\chi^2_{.05} = 7.815 \qquad \text{for 3 d.f.}$$

and the decision rule is to reject the null hypothesis if the observed chi-squared is *larger* than the tabular value for $\alpha = .05$. Hence in this case we tentatively accept the null hypothesis, as we did above.

Note that the chi-squared goodness-of-fit test is not as sensitive as the index of dispersion test and thus is a more conservative test of the Poisson fit (prone to Type II errors).

Many statisticians recommend the log-likelihood ratio for goodness-of-fit tests that have typically used chi-squared in the past (Sokal and Rohlf 1995, 690). Twice the log-likelihood ratio is distributed approximately as chi-squared, so that the usual chi-squared tables may be used for hypothesis testing. For the goodness-of-fit of a Poisson distribution, we have

$$G = 2 \sum \left\{ (\text{Observed frequency}) \left[\log_e \left(\frac{\text{Observed frequency}}{\text{Expected frequency}} \right) \right] \right\} \tag{4.7}$$

where G = Test statistic for log-likelihood ratio

For the earthworm data, $G = 7.71$ with 4 degrees of freedom, and so again we tentatively accept the null hypothesis that the Poisson distribution adequately fits the observed counts. The G-test has the same restrictions as chi-squared for having all expected values above 1.

*This rule is somewhat subjective, and some authors say that all expected values should be 5 or more, and others say 1 or more. In my experience 3 is sufficient, and often 1 is adequate. Zar (1996, 466) discusses the rules for the use of χ^2 with low expected values.

†If we had grouped only the counts for 5 and 6, we would have obtained $\chi^2 = 7.13$ with 4 d.f. and made the same decision.

When sample sizes are small ($n < 200$), it is useful to apply Williams' (1976) correction to the G-statistic:

$$G_{\text{adj}} = \frac{G}{q} \tag{4.8}$$

where G_{adj} = G-statistic adjusted by Williams' correction for continuity
G = G-statistic calculated as in equation (4.7)

$$q = 1 + \frac{(a + 1)}{6n}$$

where a = Number of frequency classes used to calculate G
n = Total number of individuals in the sample

The result of Williams' correction is to reduce slightly the value of the log-likelihood ratio. These statistical tests are carried out in program-group QUADRAT SAMPLING (Appendix 2).

Confidence Limits for the Mean Confidence limits for the mean of counts that fit a Poisson distribution are exceptionally easy to obtain. The reliability of a Poisson mean is completely determined by the number of items counted *in total*. Table 2.1 gives the resulting 95% and 99% confidence limits, which can also be calculated in program-group EXTRAS (Appendix 2). For our earthworm example, with a total of 56 worms counted, Table 2.1 gives 95% confidence limits of 41.75 and 71.28. These limits are normally converted back to a per-quadrat basis by dividing by the number of quadrats counted. For the earthworm example,

$$\text{Mean} = \frac{56}{25} = 2.24 \text{ worms per } 0.25 \text{ m}^2$$

$$\text{Lower 95\% confidence limit} = \frac{41.75}{25} = 1.67 \text{ worms per } 0.25 \text{ m}^2$$

$$\text{Upper 95\% confidence limit} = \frac{71.28}{25} = 2.85 \text{ worms per } 0.25 \text{ m}^2$$

As the mean of a theoretical Poisson distribution gets larger, the asymmetry of the distribution disappears. Figure 4.6 illustrates the changing shape of the Poisson distribution as the mean increases. Once the mean of the Poisson exceeds 10, or in ecological terms when the organism being counted is abundant, the Poisson takes on the shape of the *normal*, bell-shaped distribution. Indeed, the Poisson distribution approaches the normal distribution as a limit when the mean becomes very large (Pielou 1977).

If the Poisson distribution is an adequate representation of the counts you have made, you should be celebrating. The Poisson world, the world of randomness in spatial pattern, is a very convenient world, in which sampling is very easy, and the statistics are simple (see Chapters 5 and 6). But often the result of these tests is to reject the Poisson model, and we now consider what to do when the spatial pattern is not random but aggregated or clumped.

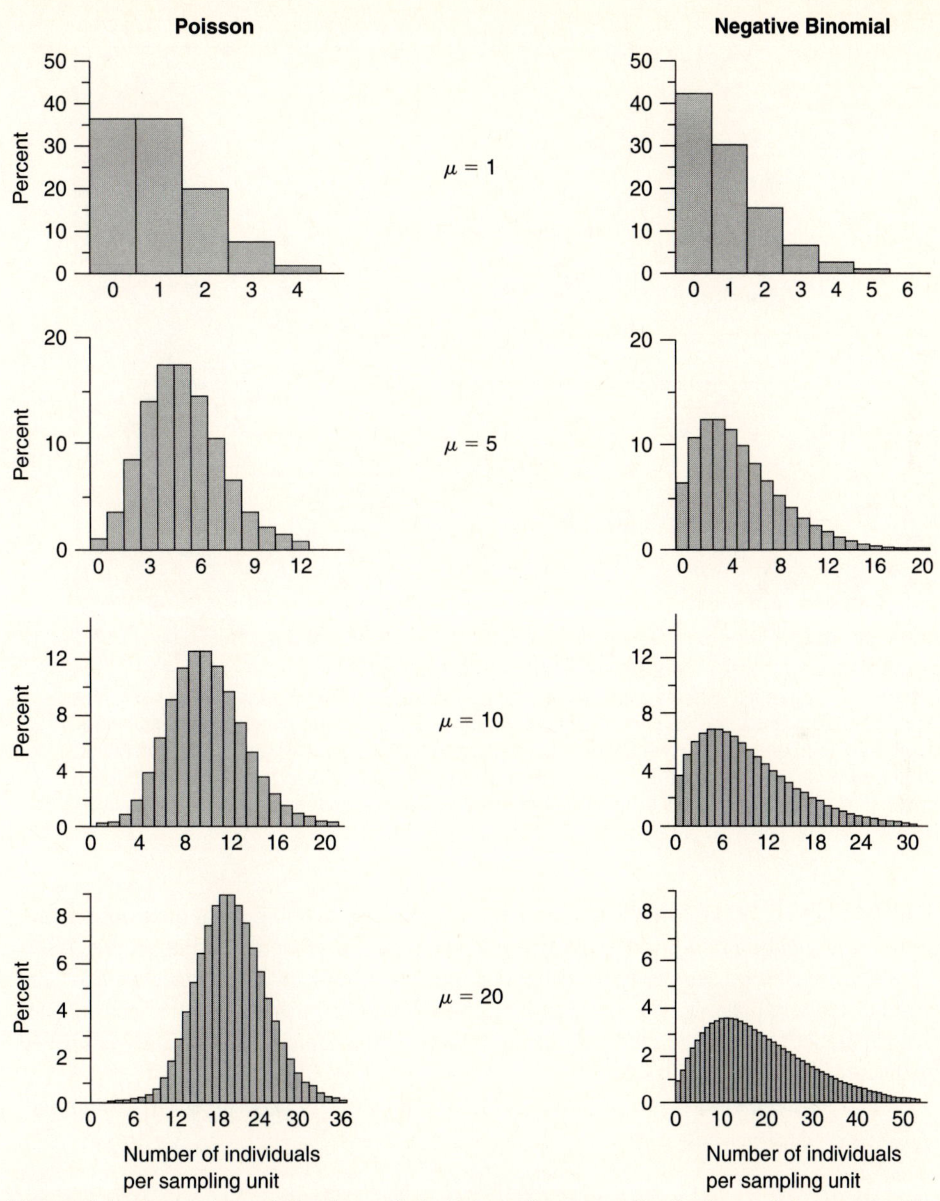

Figure 4.6 Comparison of the shape of the Poisson frequency distribution and the negative binomial distribution. A range of means from 1 to 20 is shown for each theoretical distribution. For all negative binomial distributions, $k = 2.5$.

4.2.2 Negative Binomial Distribution

Not very many species show random spatial patterns in nature, and we need to consider how to describe and study populations that show clumped patterns. A variety of statistical distributions have been used to describe aggregated spatial patterns in biological popula-

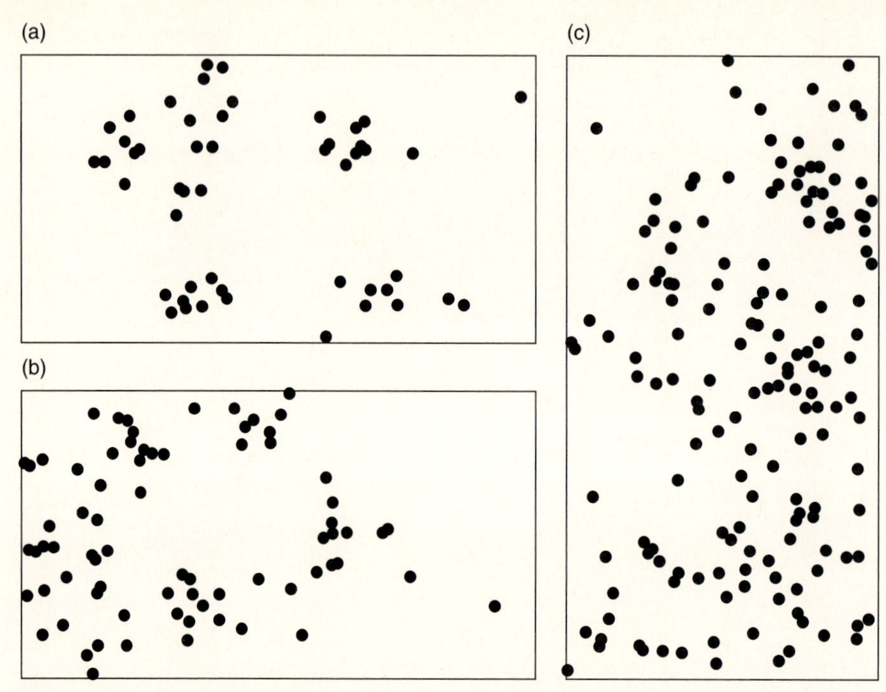

Figure 4.7 Clumped distributions can occur in an infinite variety of forms. Three different types of clumping are shown here. (a) Small clumps. (b) Large clumps with individuals randomly distributed within each clump. (c) Large clumps with individuals uniformly distributed within each clump. The ecological problem is to find the best way of obtaining confidence limits on abundance estimates from such clumped distributions and to find out the best way to sample clumped patterns.

tions (Patil et al. 1971). The most common one is the negative binomial distribution, and in the ecological literature today the two phrases *negative binomial* and *aggregated patterns* are almost used as synonyms. But it is important to remember that once we leave the world of random patterns, we open Pandora's box to find an infinite variety of possibilities (see Figure 4.7). There are many aggregated patterns that cannot be adequately described by the negative binomial distribution.

The negative binomial distribution is mathematically similar to the positive binomial, and is an expansion of the series

$$(q - p)^{-k}$$

The negative binomial is another discrete probability distribution and is governed by two parameters: the exponent k (often called *negative-binomial* k) and p ($= q - 1$) which is related to the mean of the negative binomial as follows:

$$\text{Mean} = \mu = kp$$

From our point of view, it is simplest to consider the individual terms of the negative binomial distribution:

$$\begin{Bmatrix}\text{Probability of observing zero}\\ \text{individuals in a quadrat}\end{Bmatrix} = \left(1 + \frac{\bar{x}}{k}\right)^{-k}$$

$$\begin{Bmatrix}\text{Probability of observing one}\\ \text{individual in a quadrat}\end{Bmatrix} = \left(\frac{k}{1}\right)\left(\frac{\bar{x}}{\bar{x} + k}\right)\left(1 + \frac{\bar{x}}{k}\right)^{-k}$$

$$\begin{Bmatrix}\text{Probability of observing two}\\ \text{individuals in a quadrat}\end{Bmatrix} = \left(\frac{k}{1}\right)\left(\frac{k+1}{2}\right)\left(\frac{\bar{x}}{\bar{x} + k}\right)^2\left(1 + \frac{\bar{x}}{k}\right)^{-k}$$

$$\begin{Bmatrix}\text{Probability of observing three}\\ \text{individuals in a quadrat}\end{Bmatrix} = \left(\frac{k}{1}\right)\left(\frac{k+1}{2}\right)\left(\frac{k+2}{3}\right)\left(\frac{\bar{x}}{\bar{x} + k}\right)^3\left(1 + \frac{\bar{x}}{k}\right)^{-k}$$

$$\vdots$$

The general term for the negative binomial distribution is:

$$P_x = \left[\frac{\Gamma(k + x)}{x!\,\Gamma(k)}\right]\left(\frac{\mu}{\mu + k}\right)^x\left(\frac{k}{k + \mu}\right)^k \tag{4.9}$$

where P_x = Probability of a quadrat containing x individuals
x = A counter (0, 1, 2, 3 . . .)
μ = Mean of distribution
k = negative-binomial exponent
Γ = Gamma function (see Appendix 1)

The negative binomial is a unimodal frequency distribution like the Poisson. Figure 4.6 shows how the shape of the negative binomial changes as the mean increases and k is held constant. Note that as k becomes large and the mean is above 10, the negative binomial approaches the normal distribution in shape, as shown in Figure 4.8. For this reason, k can be thought of as an inverse index of aggregation; the larger the k, the less the aggregation; and conversely the smaller the k, the greater the clumping.

The theoretical variance of the negative binomial is given by

$$\text{Variance of the negative binomial distribution} = \mu + \frac{\mu^2}{k} \tag{4.10}$$

Thus the variance of the negative binomial is *always greater than the mean*, and this is one simple hallmark of aggregation in field data.

To fit a theoretical negative binomial distribution to some actual data, two parameters must be estimated. The *mean* is the easy one, and we use the usual expectation:

$$\text{Estimated mean} = \mu = \bar{x}$$

The exponent k is more difficult to estimate because the best way to estimate it depends on the mean and k (Bliss and Fisher 1953). First, calculate an approximate estimate of k:

$$\text{Approximate } \hat{k} = \frac{\bar{x}^2}{s^2 - \bar{x}} \tag{4.11}$$

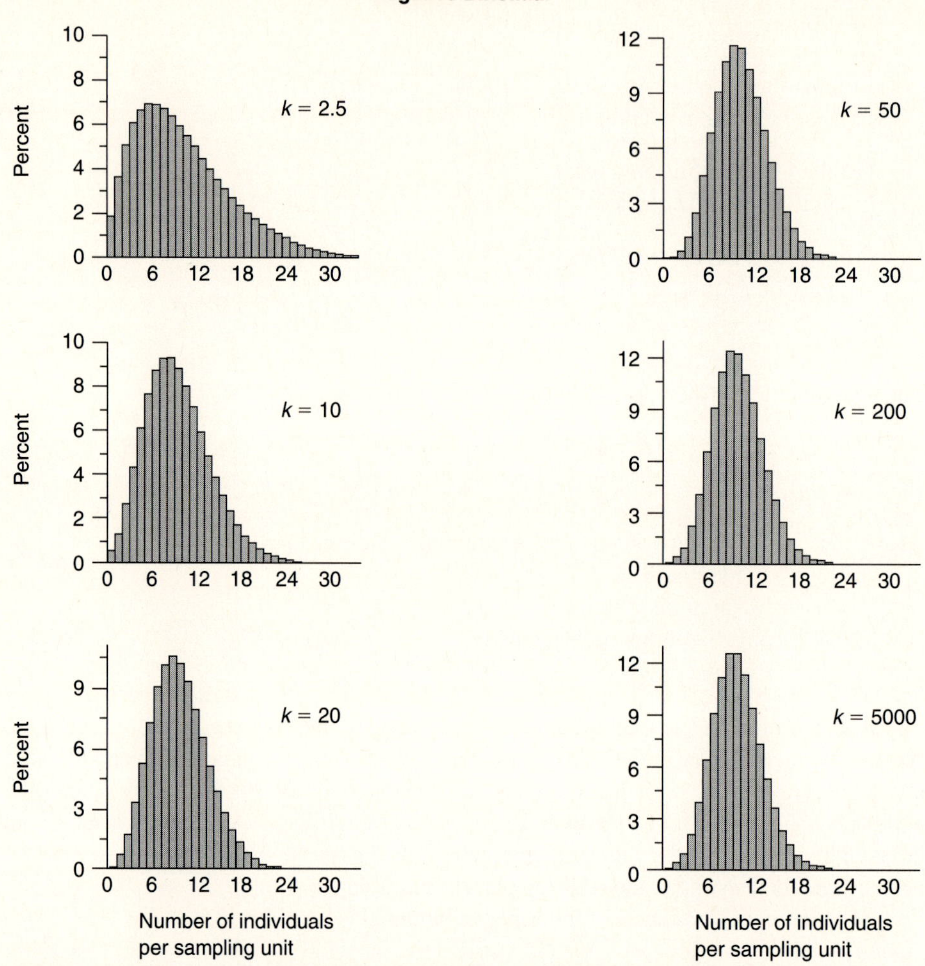

Figure 4.8 Effect of k on the shape of the negative binomial distribution. The mean of all six frequency distributions is 10, and various values of k are shown from 2.5 to infinity. As k approaches infinity, the negative binomial approaches the Poisson distribution. (After Elliott 1977.)

which is just a rearrangement of the theoretical variance formula given above. I will call this approximate method (equation [4.11]) *Method 1* for estimating k, following Anscombe (1950). After you have used Method 1, follow this key, which is summarized in Figure 4.9:

(1) *Small sample size*: Number of quadrats < 20, and counts are too few to arrange in a frequency distribution.

(a) *More than 1/3 of the quadrats are empty*: Calculate an estimate of k by solving equation (4.12) iteratively (i.e., by trial and error):

$$\log_e\left(\frac{N}{n_0}\right) = \hat{k}\,\log_e\left(1 + \frac{\bar{x}}{\hat{k}}\right) \tag{4.12}$$

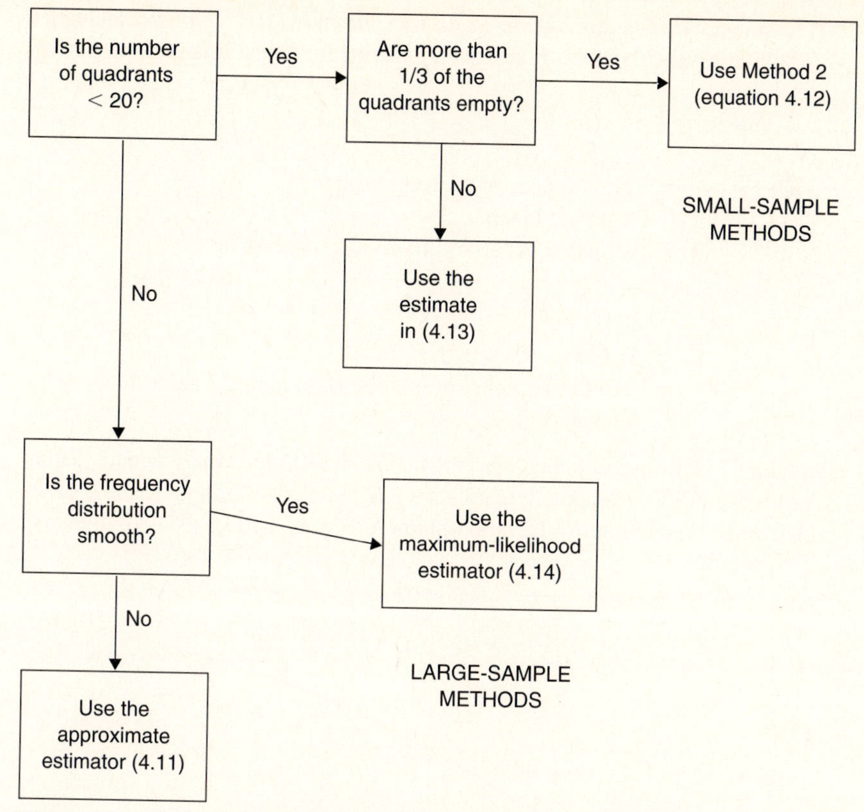

Figure 4.9 Decision tree for estimating the negative binomial exponent k.

where N = Total number of quadrats counted
n_0 = Number of quadrats containing zero individuals
$\bar{x}$ = Observed mean
$\hat{k}$ = Estimate of the negative binomial exponent

Begin with the approximate value of k calculated above and raise or lower it to make the two sides of this equation balance. Following Anscombe (1950) I will call this method of estimating k using equation (4.12) *Method 2.*

(b) *Less than 1/3 of the quadrats are empty*:

$$\hat{k} = \frac{\bar{x}^2 - (s^2/n)}{s^2 - \bar{x}} \tag{4.13}$$

I will call this *Method 3*. When the mean is above 4.0 and less than 1/3 of the quadrats are empty, Elliott (1977, 56) gives another method for estimating k using transformations, but it is usually not worth the extra effort to calculate.

(2) *Large sample size*: Number of quadrats > 20, and counts can be arranged in a frequency distribution.

(a) *Frequency distribution smooth with no extremely large counts*: Calculate a maximum-likelihood estimate for k by solving this equation by trial and error:

$$(N)\ \log_e\left(1 + \frac{\bar{x}}{\hat{k}}\right) = \sum_{i=0}^{\infty}\left(\frac{A_x}{\hat{k} + x}\right) \tag{4.14}$$

where N = Total number of quadrats counted
$\bar{x}$ = Observed mean
$\hat{k}$ = Estimated negative-binomial exponent

$$A_x = \sum_{j=x+1}^{\infty}(f_j) = f_{x+1} + f_{x+2} + f_{x+3} + f_{x+4} + \cdots$$

i = A counter (0, 1, 2, 3 . . .)
f_x = Observed number of quadrats containing x individuals
j = A counter (1, 2, 3, 4 . . .)

The best estimate of k is always obtained by this *maximum-likelihood method* using equation (4.14). Anscombe (1950) calculated the relative efficiencies of Methods 1 and 2 of estimating k, relative to the maximum-likelihood method. Figure 4.10 gives a contour map

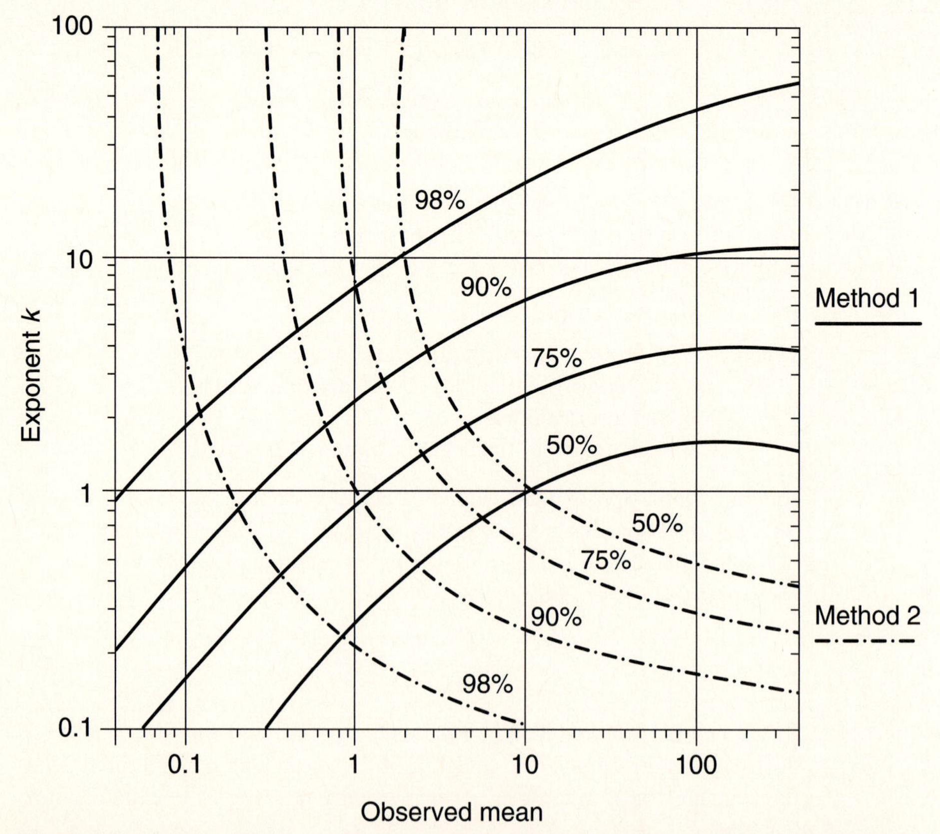

Figure 4.10 Relative efficiency of the two methods of estimating the negative binomial exponent k from field data. If you cannot use the maximum-likelihood estimator (eq. [4.14]), locate your sample on this graph using the observed mean and the approximate estimate of the exponent k from equation (4.11) and use the method with the highest relative efficiency. (After Anscombe 1950.)

of these relative frequencies and can be used to determine the best method to use (if the maximum-likelihood estimator will not converge to a solution) as follows:

1. Estimate approximate k by Method 1 above (equation [4.11]).
2. Try to use the maximum-likelihood estimator (eq. [4.14]). If it will not work on your data continue on to (3).
3. Locate your sample on Figure 4.10, given the observed mean and this approximate estimate of k.
4. Estimate from the contour lines the relative efficiency for Methods 1 and 2.
5. Use the method with the higher efficiency to recalculate k to achieve a better estimate.

Figure 4.10 may also be used to decide on the best statistical test for goodness-of-fit to the negative binomial (see below).

As an example, use the aphid data in Box 4.2. With a mean of 3.46 and an approximate k of 3.07, from the graph in Figure 4.10, we read

$$\text{Relative efficiency of Method 1} \sim 83\%$$

$$\text{Relative efficiency of Method 2} \sim 50\%$$

Thus, if we could not use the maximum-likelihood method for these data, we should prefer Method 1 for estimating k.

(b) *Frequency distribution irregular with some extreme counts*: Try to use the maximum-likelihood method just described (equation [4.14]). If it does not converge to a good estimate, use the approximate estimate for large samples (equation [4.11]).

Elliott (1977, Chap. 5) has an excellent discussion of these problems of estimating k and gives several examples of the calculations involved.

Box 4.2 illustrates the use of the maximum-likelihood estimator for k. It is more tedious than difficult. Program-group QUADRAT SAMPLING (Appendix 2) does these calculations.

Tests for Goodness-of-Fit There are three approaches to testing the null hypothesis that a negative binomial distribution is an adequate description of observed counts. Note that this is *not* equivalent to asking the more general question, *Is there evidence of clumping in the data?* If the index of dispersion $(s^2/\bar{x}) > 1$ (see page 119), there is evidence of aggregation in your data. But remember there are many different types of aggregation (Figure 4.7) and the negative binomial distribution will fit only *some* of these clumped patterns.

Chi-Squared Goodness-of-Fit Test. This is another application of the standard test:

$$\chi^2 = \sum \left[\frac{(\text{Observed frequency} - \text{expected frequency})^2}{\text{Expected frequency}} \right] \tag{4.15}$$

Box 4.2 Maximum Likelihood Estimation of Negative Binomial *k* for Counts of the Black Bean Aphid (*Aphis fabae*) on Bean Stems

Number of aphids on one stem, X	Number of stems (= quadrats), f_x	Proportion of stems, P_x
0	6	.12
1	8	.16
2	9	.18
3	6	.12
4	6	.12
5	2	.04
6	5	.10
7	3	.06
8	1	.02
9	4	.08
	n = 50	1.00

Calculate the usual statistical sums, means, and variances:

$$\sum Xf_x = 173 \qquad \bar{x} = \frac{173}{50} = 3.46$$

$$\sum X^2 f_x = 959 \qquad s^2 = 7.356$$

$$\sum X^3 f_x = 6413$$

1. Calculate the approximate *k* estimate (equation [4.11]):

$$\text{Approximate } \hat{k} = \frac{\bar{x}^2}{s^2 - \bar{x}} = \frac{3.46^2}{7.356 - 3.46} = 3.07$$

Since we have a large sample and the frequency distribution is smooth, we use the maximum-likelihood estimator (equation [4.14]) to obtain a more precise estimate of *k* as follows.

2. Determine the A_x sums:

$$A_0 = f_1 + f_2 + f_3 + f_4 + f_5 + f_6 + f_7 + f_8 + f_9$$
$$= 8 + 9 + 6 + 6 + 2 + 5 + 3 + 1 + 4 = 44$$

$$A_1 = f_2 + f_3 + f_4 + \cdots + f_9 = 9 + 6 + 6 + 2 + 5 + 3 + 1 + 4 = 36$$

$$A_2 = f_3 + f_4 + \cdots + f_9 = 6 + 6 + 2 + 5 + 3 + 1 + 4 = 27$$

Similarly, $A_3 = 21$, $A_4 = 15$, $A_5 = 13$, $A_6 = 8$, $A_7 = 5$, $A_8 = 4$, and $A_9 = 0$.

3a. Calculate the two sides of the equality (equation [4.14]):

$$(N)\log_e\left(1+\frac{\bar{x}}{\hat{k}}\right)=\sum_{i=0}^{\infty}\left(\frac{A_x}{\hat{k}+x}\right)$$

$$(N)\log_e\left(1+\frac{\bar{x}}{\hat{k}}\right)=(50)\left[\log_e\left(1+\frac{3.46}{3.07}\right)\right]=37.736$$

$$\sum_{i=0}^{\infty}\left(\frac{A_x}{\hat{k}+x}\right)=\frac{A_0}{k+0}+\frac{A_1}{k+1}+\frac{A_2}{k+2}+\cdots+\frac{A_8}{k+8}$$

$$=\frac{44}{3.07}+\frac{36}{4.07}+\frac{27}{5.07}+\cdots+\frac{4}{11.07}=37.073$$

Since $37.736 \neq 37.073$ and the summation term is smaller, we reduce our provisional estimate of k to, say, 3.04 and start again at step (3).

$$(N)\log_e\left(1+\frac{\bar{x}}{\hat{k}}\right)=(50)\left[\log_e\left(1+\frac{3.46}{3.04}\right)\right]=37.997$$

$$\sum_{i=0}^{\infty}\left(\frac{A_x}{\hat{k}+x}\right)=\frac{44}{3.04}+\frac{36}{4.04}+\frac{27}{5.04}+\cdots+\frac{4}{11.04}=37.711$$

Since $37.997 \neq 37.711$ and the summation term is still smaller, we reduce our provisional estimate of k once again to, say, 3.02 and start again at step (3).

3b. We continue this trial-and-error procedure as follows:

$$\hat{k}=3.02 \qquad 38.173 \neq 37.898$$

$$\hat{k}=2.90 \qquad 39.266 \neq 39.063$$

$$\hat{k}=2.80 \qquad 40.228 \neq 40.096$$

$$\hat{k}=2.60 \qquad 42.310 \neq 42.354$$

We have now gone too far, so we increase provisional $\hat{k}$ to $\hat{k} = 2.65$ and now obtain for the two sides of the equation:

$$41.768 = 41.764$$

which is accurate enough for $n = 50$ quadrats, and we accept 2.65 as the best estimate of the negative binomial exponent k for these data.

These calculations can be carried out in program-group QUADRAT SAMPLING (Appendix 2).

This test (or its replacement, the G-test) should be used whenever the observed data can be arranged in a frequency distribution ($n > 20$). It is not a particularly sensitive test and is prone to Type II errors when there are too few classes or when the number of quadrats counted is below 50.

The expected frequencies for a theoretical negative binomial can be calculated from the individual terms given previously. To illustrate, we use the aphid data in Box 4.2:

$$n = 50 \qquad \bar{x} = 3.46 \qquad \hat{k} = 2.65$$

$$\left\{\begin{matrix}\text{Proportion of stems expected} \\ \text{to have zero aphids}\end{matrix}\right\} = \left(1 + \frac{\bar{x}}{\hat{k}}\right)^{-k}$$

$$= \left(1 + \frac{3.46}{2.65}\right)^{-2.65} = 0.1093$$

$$\left\{\begin{matrix}\text{Proportion of stems expected} \\ \text{to have one aphid}\end{matrix}\right\} = \left(\frac{\hat{k}}{1}\right)\left(\frac{\bar{x}}{\bar{x} + \hat{k}}\right)\left(1 + \frac{\bar{x}}{\hat{k}}\right)^{-k}$$

$$= \left(\frac{2.65}{1}\right)\left(\frac{3.46}{3.46 + 2.65}\right)\left(1 + \frac{3.46}{2.65}\right)^{-2.65}$$

$$= 0.1640$$

$$\left\{\begin{matrix}\text{Proportion of stems expected} \\ \text{to have two aphids}\end{matrix}\right\} = \left(\frac{\hat{k}}{1}\right)\left(\frac{k + 1}{2}\right)\left(\frac{\bar{x}}{\bar{x} + \hat{k}}\right)\left(1 + \frac{\bar{x}}{\hat{k}}\right)^{-k}$$

$$= \left(\frac{2.65}{1}\right)\left(\frac{2.65 + 1}{2}\right)\left(\frac{3.46}{3.46 + 2.65}\right)\left(1 + \frac{3.46}{2.65}\right)^{-2.65}$$

$$= 0.1695$$

Similarly,

Proportion of stems expected with 3 aphids = 0.1488

4 aphids = 0.1190

5 aphids = 0.0896

6 aphids = 0.0647

7 aphids = 0.0453

8 aphids = 0.0309

9 aphids = 0.0207

10 or more aphids = 0.0382

We convert these proportions to frequencies by multiplying each by the sample size ($n = 50$ in this example). We obtain the following:

No. of aphids, x	Observed no. of stems, f_x	Expected no. of stems from negative binomial
0	6	5.47
1	8	8.20
2	9	8.47
3	6	7.44
4	6	5.95
5	2	4.48
6	5	3.23
7	3	2.26
8	1	1.54
9	4	1.04
≥10	0	1.91

$$\chi^2 = \frac{(6 - 5.47)^2}{5.47} + \frac{(8 - 8.20)^2}{8.20} + \cdots + \frac{(0 - 1.91)^2}{1.91} = 13.51$$

Classes 7 and 8 and classes 9 and 10 could be lumped to keep the expected frequency from being too low, but I have not done so here. The number of degrees of freedom for this test are (number of frequency classes used $-$ 3), since three statistics—n, $\bar{x}$, and $\hat{k}$—have been used to fit the negative binomial to these data. Each frequency class used should have an expected frequency above 1. For these data, d.f. = 8 and the critical value of chi-squared is

$$\chi^2_{0.05} = 15.51$$

and since the observed χ^2 is less than this critical value, we accept the null hypothesis that the negative binomial distribution is an adequate description of the clumped pattern in these data.

U-Statistic Goodness-of-Fit Test. This test compares the observed and expected variances of the negative binomial distribution (Evans 1953) and has more statistical power than the chi-squared goodness-of-fit test just described.

$$U = \text{observed variance} - \text{expected variance} \tag{4.16}$$

$$= s^2 - \left(\bar{x} + \frac{\bar{x}^2}{k}\right)$$

The expected value of U is zero. To test if the observed value of U is different from zero, calculate the standard error of U as follows:

$$a = \frac{\bar{x}}{\hat{k}}$$

$$b = 1 + a$$

$$c = \frac{ba^4}{[b(\log_e b) - a]^2}\left[b^{(1+\bar{x}/a)} - (\bar{x} + b)\right]$$

Then

$$\text{S.E.}(\hat{U}) = \sqrt{\frac{1}{n}\left\{2\bar{x}(\bar{x} + a)(b)\left[\frac{(b^2)(\log_e b) - a(1 + 2a)}{(b)(\log_e b) - a}\right] + c\right\}} \tag{4.17}$$

If the observed value of U exceeds 2 standard errors of U, then we reject the null hypothesis that the negative binomial is an adequate fit to the observed data (at $\alpha = 0.05$). Note that this is true only for large samples and only approximately correct for small samples.

For the bean aphid data in Box 4.2, we obtain

$$U = s^2 - \left(\bar{x} + \frac{\bar{x}^2}{k}\right)$$

$$= 7.356 - \left(3.46 + \frac{3.46^2}{2.65}\right) = -0.622$$

The standard error of U is estimated as

$$a = \frac{\bar{x}}{\hat{k}} = \frac{3.46}{2.65} = 1.306$$

$$b = 1 + a = 2.306$$

$$c = \frac{ba^4}{[b(\log_e b) - a]^2}\left[b^{(1+\bar{x}/a)} - (\bar{x} + b)\right] = \frac{2.306(1.306^4)}{[2.306(\log_e 2.306) - 1.306]}\left[(2.306^{3.65}) - 5.765\right]$$

$$= 266.8957$$

$$\text{S.E.}(\hat{U}) = \sqrt{\frac{1}{50}\left\{2(3.46)(4.76)(2.306)\left[\frac{(2.306^2)(\log_e 2.306) - 1.306(3.611)}{(2.306)(\log_e 2.306) - 1.306}\right] + 266.8957\right\}}$$

$$= 2.16$$

Since the observed value of U (−0.62) is much less than twice its standard error (4.32), we accept the null hypothesis that the negative binomial fits these data.

T-Statistic Goodness-of-Fit Test. This test compares the observed and expected third moments of the negative binomial distribution. The third moment is a measure of skewness of a distribution.

$$T = \text{Observed measure of skewness} - \text{expected measure of skewness}$$

$$T = \left(\frac{\Sigma fx^3 - 3\bar{x}\,\Sigma fx^2 + 2\bar{x}^2\,\Sigma fx}{n}\right) - \hat{s}^2\left(\frac{2\hat{s}^2}{\bar{x}} - 1\right) \tag{4.18}$$

The expected value of T is zero. To determine whether the observed data deviate significantly from the amount of skewness expected, calculate the standard error of T as follows. As above,

$$a = \frac{\bar{x}}{k}$$

These confidence limits must be transformed back to the original scale of counts by reversing the original transformation (log-antilog, inverse hyperbolic sine-hyperbolic sine). Elliott (1977) gives an example of these calculations, which are also provided in program-group QUADRAT SAMPLING (Appendix 2).

Box 4.3 Line Intercept Method of Density Estimation for Willow Shrubs

Three lines were laid out at random along a 125 m baseline to estimate the willow density on an irregular area bordering a stream. The study area was 6.3 ha. The following data were recorded for the intercept distances (w_i):

Line 1 (438 m): 1.3, 3.1, 0.8, 2.2, 0.4, 1.7, 0.2, 1.5, 1.9, 0.4, 0.1 m ($n = 11$)

Line 2 (682 m): 1.1, 0.1, 1.8, 2.7, 2.4, 0.7, 0.4, 0.3, 1.4, 0.1, 2.1, 2.3 m ($n = 12$)

Line 3 (511 m): 0.3, 1.7, 2.1, 0.2, 0.2, 0.4, 1.1, 0.3 m ($n = 8$)

Line 4 (387 m): 3.3, 3.0, 1.4, 0.2, 1.7, 1.1, 0.2, 1.9, 0.9 m ($n = 9$)

Begin by calculating the sum of the reciprocals: for line 1:

$$y_i = \sum \frac{1}{w_i} = \frac{1}{1.3} + \frac{1}{3.1} + \frac{1}{0.8} + \cdots + \frac{1}{0.1} = 24.577$$

An estimate of population size can be obtained for each line from equation (4.23). For line 1:

$$\hat{N} = \left[\frac{W}{n}\right] \sum_{i=1}^{k} \left(\frac{1}{w_i}\right) = \left[\frac{125}{1}\right](24.577) = 3072 \text{ shrubs}$$

Repeating these calculations for each of the four lines gives these results:

	Line 1	Line 2	Line 3	Line 4	Totals
$y_i = \sum \frac{1}{w_i}$	24.577	31.139	21.140	14.485	91.342
$\hat{N}$	3072	3892	2642	1811	2854
$\hat{D}$	488	618	419	287	453

Since we know the area is 6.3 ha, the density estimates ($\hat{D}$) for each line are obtained by dividing the population size by this area, and the results are given in the above table as number of willows per hectare.

Combined population size and density estimates for all the four lines can be obtained from these equations applied to all the data or simply by averaging the results of each of the individual lines.

To calculate confidence limits for the overall density of willows on the study area, we use equation (4.26) to calculate the standard error of the density estimate:

$$s_{\hat{D}} = \hat{D}\sqrt{\frac{(s_{\hat{y}}/\bar{y})^2 + (S_L/\bar{L})^2 - 2(C_{yL})}{(n - 1)}}$$

From the usual statistical formulas and the data in the table above, we can calculate the following:

$$\bar{y} = \text{Observed mean value of } y_i = \sum \frac{1}{w_i} = 22.8355$$

$$s_{\hat{y}} = \text{Standard deviation of the } y_i \text{ values}$$

$$= \frac{\Sigma\,(y_i - \bar{y})^2}{n - 1} = \frac{(24.577 - 22.835)^2 + \cdots}{4 - 1} = 6.942$$

$$\bar{L} = \text{Observed mean value of the lengths of each line} = 504.5$$

$$s_L = \text{Standard deviation of the lengths of the lines}$$

$$= \frac{\Sigma\,(L_i - \bar{L})^2}{n - 1} = \frac{(438 - 504.5)^2 + \cdots}{4 - 1} = 128.82$$

We need to determine the sum of cross-products C_{yL} from equation (4.27):

$$C_{yL} = \frac{\Sigma\,(y_i - \bar{y})(L_j - \bar{L})}{(n - 1)\bar{y}\bar{L}} = \frac{[(24.577 - 22.835)(438 - 504.5)] + \cdots}{(4 - 1)(22.835)(504.5)}$$

$$= \frac{2328.113}{34560.77} = 0.06736$$

Substituting these values in equation (4.26), we obtain

$$s_{\hat{D}} = \hat{D}\sqrt{\frac{(s_{\hat{y}}/\bar{y})^2 + s_L/\bar{L})^2 - 2(C_{yL})}{(n - 1)}}$$

$$= 453.085\sqrt{\frac{(6.942/22.835)^2 + (128.81/504.5)^2 - 2(0.067)}{(4 - 1)}}$$

$$= 39.59$$

Calculate the 95% confidence limits in the usual way: from the t-table with 3 degrees of freedom, $t_\alpha = 3.182$, and we obtain

$$\hat{D} \pm t_\alpha s_{\hat{D}}$$

$$453 \pm 3.182(39.59)$$

or 327 to 579 willows per hectare for this area. Because of the small sample size in this example, the confidence limits are quite wide.

An alternative strategy for estimating confidence limits for a negative binomial distribution is to use a generalized transformation like the Box-Cox transform (Chapter 15, Section 15.1.2) to normalize the data and then to use conventional confidence limits based on the *t*-distribution. Unfortunately this strategy will not always work well with small samples and with highly skewed data with many zero counts.

Many other statistical frequency distributions have been used to describe quadrat counts of plants and animals (Evans 1953; Patil et al. 1971; Elliott 1977). You should refer to the more specialized literature or consult a professional statistician if you need to consider other frequency distributions because you have clumped patterns that do not fit the negative binomial. Johnson and Kotz (1969) give a good general summary of many discrete frequency distributions. For much ecological data, the Poisson and the negative binomial distributions are adequate descriptors.

Much of the early literature on the statistical analysis of quadrat counts was infused with the belief that general laws might be derived from count data, and that one might infer biological mechanisms of pattern formation by fitting different statistical frequency distributions. It was quickly realized, however, that this could not be done, and that these statistical distributions are only approximate descriptions and tell us little about the biological processes generating the observed spatial patterns. Even if we cannot infer ecology from statistics, the adequate description of the statistical distribution of quadrat counts is still important to an ecologist because it dictates sampling strategies, as we shall see in Chapters 7 and 8, and allows for valid tests of hypotheses about abundance data gathered from quadrat counts.

4.3 LINE INTERCEPT METHOD

Plant ecologists have for many years been interested in measuring the cover of plants in a community using line transects. This is a general procedure that can be applied to estimating the cover of tree or shrub canopies, the area covered by lakes within a geographical region, or the area covered by squirrel burrows. To estimate cover is relatively simple. Figure 4.11 illustrates the line intercept method and the measurements that need to be taken.

Estimates of abundance or density of plants can also be derived from line intercept data (Eberhardt 1978b). To set out line intercept transects, you need to establish a baseline of length W along which you randomly locate individual transects. For each plant or sample unit intercepted, measure the longest perpendicular width w. This width determines the probability that any individual plant will be bisected by the sampling line, as you can see from Figure 4.11. Eberhardt shows that the appropriate estimate of population size is

$$\hat{N} = \left[\frac{W}{n}\right] \sum_{i=1}^{k} \left(\frac{1}{w_i}\right) \tag{4.24}$$

where $\hat{N}$ = Estimate of population size
W = Width of the baseline from which the transects begin
w_i = Perpendicular width of plants intersected
n = Number of transect lines sampled
k = Total number of plants intercepted on all lines ($i = 1, 2, 3, \ldots k$)

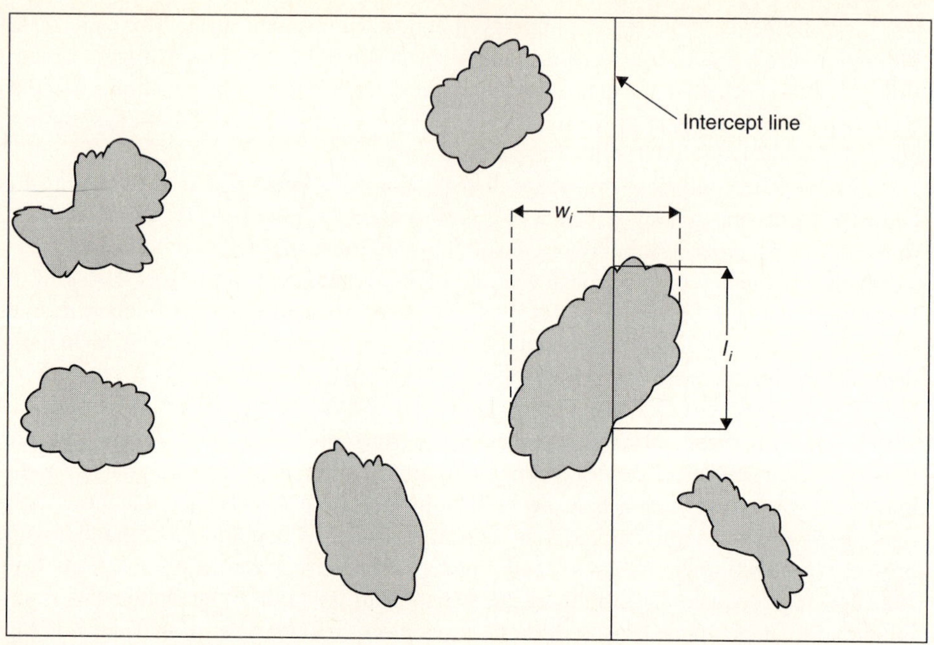

Figure 4.11 Schematic illustration of the line intercept method as used for density estimation of shrubs and trees. The shaded areas represent the canopy coverage of a shrub. The measurements l_i are for cover estimates (fraction of the line covered by the canopy of this particular species), and the intercept distances w_i are the maximum perpendicular distance coverage and are used to estimate numbers and density of the plants.

To estimate the density of organisms for any shape of area, simply divide this estimate of numbers by the area being studied. If the area being studied has not been measured, density can be estimated from randomly oriented transects by the following equation:

$$\hat{D} = \left[\frac{1}{L}\right] \sum_{i=1}^{k} \left(\frac{1}{w_i}\right) \tag{4.25}$$

where $\hat{D}$ = Estimate of population density
L = Length of all lines combined
w_i = Perpendicular width of plants intersected
k = Total number of plants intercepted on all lines ($i = 1, 2, 3, \ldots k$)

If a series of line intercepts are measured, each one can be used to generate an estimate of population size and thus an estimate of variability in order to obtain confidence limits in the usual way. But if the line intercept lengths vary, it is necessary to obtain the standard error of the mean density estimate from Eberhardt as follows:

$$s_{\hat{D}} = \hat{D}\sqrt{\frac{(s_{\hat{y}}/\bar{y})^2 + s_L/\bar{L})^2 - 2(C_{yL})}{(n - 1)}} \tag{4.26}$$

where $s_{\hat{D}}$ = Standard error of the mean density estimate

$s_{\hat{y}}$ = Standard deviation of the observed $y_i = \sum \frac{1}{w_i}$ for each line

$\bar{y}$ = Observed mean value of $y_i = \sum \frac{1}{w_i}$ for each line

s_L = Standard deviation of the observed lengths of each line

$\bar{L}$ = Observed mean value of the lengths of each line

$$C_{yL} = \frac{\Sigma\,(y_i - \bar{y})(L_j - \bar{L})}{(n - 1)\bar{y}\bar{L}} \qquad (4.27)$$

n = Number of lines in sample

Box 4.3 illustrates the use of the line intercept method to estimate the density of willow shrubs.

4.4 AERIAL SURVEYS OF WILDLIFE POPULATIONS

Many wildlife surveys are carried out by aerial census, a specialized form of quadrat sampling. Aerial surveying is discussed in detail in Norton-Griffiths (1978), and I will discuss here the additional statistical problems that this form of quadrat sampling entails. This discussion is equally applicable to ground surveys or vehicle surveys, and will lead us into a discussion of line transect sampling in the next chapter.

Most research programs being done on large animals require three types of information:

1. Total numbers
2. Size and structure of populations (age ratios, sex ratios)
3. Distribution and movements

The first important point to remember in designing a census is *decide in advance exactly what the objectives of the census are.* Many research workers come to grief trying to get all possible data in one census. It may be necessary to do 2 or even 3 censuses to achieve all your objectives.

Many factors will influence the way in which a census is done:

- *Resources*: aircraft, vehicles, manpower, quality of crew
- *Size of area*: small or very large
- *Nature of the vegetation*: open plains, thick brush
- *Nature of the country*: flat, mountainous, no roads
- *Species concerned*

Norton-Griffiths (1978) discusses these factors in detail.

An aerial census can be done either as a *total count* or as a *sample count*. Total counts are expensive and rely on the assumption that no animals are counted twice and that no animals are missed. These assumptions are difficult to evaluate in any real situation, so total counts are often of dubious reliability.

Because of money alone, most wildlife surveys are done as sample counts. Efforts have been directed along three lines in aerial sampling:

1. Raise the *precision* of the estimates by good survey design, high sampling intensity, and use of stratified sampling techniques.
2. Correct the estimates for *bias*; try to remove sources of inaccuracy in the sampling program.
3. Use aerial census as a measure of *relative density*, recognizing that it is biased and that the bias cannot be removed or estimated, but only held constant.

In any census in which an observer counts animals there are two sources of error. First, an observer may undercount on some occasions and overcount on others. Any one observation may be inaccurate, but on the average the errors will cancel out. This is called *counting error*. Counting error increases with counting rate—this is very important. *Counting becomes less precise the faster you have to count*. Second, most biologists tend to undercount, and this is called *counting bias*. The direction of this error is consistent, but unfortunately this undercounting is not a constant bias because it usually gets worse the more animals there are to count and the faster they have to be counted. This can be complicated by the difficulties of spotting animals in thick brush. In addition bias may vary among observers.

Counting bias can be measured only on populations of known size that are censused by aerial survey. This has been done only a few times. LeResche and Rausch (1974) tried total counts on moose (*Alces alces*) populations enclosed in 2.6 sq. km fenced areas located in a 25-year-old burn in Alaska. They knew from ground counts exactly how many moose were present in each enclosure. They used 49 observers, and allowed each to fly 15 minutes over each square mile enclosure. The pilot was always the same, and he did not sight animals for the counting. Two variables were categorized: snow conditions (poor, good, excellent) and the experience level of the observer. They got the results given in Table 4.4.

TABLE 4.4 COUNTING BIAS IN AERIAL SURVEYS

Observer's experience	Observer's history	Snow conditions	Proportion of total moose seen
Inexperienced	None	Excellent	0.43
Inexperienced	None	Good	0.44
Inexperienced	None	Poor	0.19
Experienced	Not current		0.46
Experienced	Current[a]	Excellent	0.68
Experienced	Current	Good	0.61
Experienced	Current	Poor	0.40
Experienced with pilot counting	Current	Excellent	0.70

Source: LeResche and Rausch 1974.

Note: Total counts were made of moose (*Alces alces*) in 2.6 sq. km fenced areas in Alaska. The number of moose in each enclosure was known. Even the best observers saw only about 70% of the animals.

[a] *Current* means having been involved in an active counting program within the previous few weeks.

Three experienced individuals were tested several times and on replicate counts the 95% confidence limits were from ±10% to ±18% of their means (and the means are only 71% of the true population value). LeResche and Rausch concluded that aerial counts were not valid estimates of absolute moose numbers because of the large counting bias. Caughley (1974) listed 17 analyses of the accuracy of aerial censusing for large mammals, and the percent counted of the known population ranged from 23% to 89%. Undercounting is thus the rule in aerial census.

Undercounting bias results from two different factors (Marsh and Sinclair 1989). *Availability* bias results from animals being concealed when the counting is being done. For example, a moose may be under a tree and not visible, or a whale may be diving and not near the surface. *Perception* bias results from observers missing animals that are in view.

A number of computer simulation programs are now available to illustrate counting bias in the classroom or in training sessions. WILDLIFE COUNTS* is one simulation program for the IBM PC that illustrates well the problems of counting organisms quickly. Let us consider first how we might correct for counting bias in an aerial survey.

4.4.1 Correcting for Bias in Aerial Surveys

Four approaches have been used. The simplest approach is to photograph groups of animals as they are counted visually and get the bias directly

$$\text{Counting bias} = \frac{\text{Number visually counted}}{\text{Number actually present on photographs}} \tag{4.28}$$

In actual field work two complications may intervene. First we may decide not to photograph groups smaller than, say, 10 because the counting bias is negligible in that range. Second, we may miss taking some photographs, and this complicates the bias correction.

We can define

$\sum X_1$ = Number of animals in small groups (<10) not photographed

$\sum X_2$ = Number of animals in large groups that were not photographed

$\sum X_3$ = Number of animals counted *visually* (field count) and photographed

$\sum X_4$ = Number of animals in the X_3 groups that are counted *photographically* (photograph count)

Y = Total number of animals on the transect, corrected for counting bias

If all animals are photographed in the larger groups, then

$$Y = \sum X_1 + \sum X_4 \tag{4.29}$$

*WILDLIFE COUNTS, 318 Coleman Street, Juneau, Alaska 99801.

If some animals cannot be photographed, we correct these counts by the observed bias in the other counts, and hence

$$Y = \sum X_1 + \frac{\Sigma X_2}{B} + \sum X_4 \tag{4.30}$$

where B = Counting bias $= \dfrac{\Sigma X_3}{\Sigma X_4}$

This analysis presumes that the photographs provide a count without error. This is an important assumption and needs to be checked. Harris and Lloyd (1977) showed that photographic counts of seabird colonies could also vary greatly among observers (Table 4.5) and that undercounting was the rule. Even experienced counters varied $\pm 10\%$ in their repeated counts of birds in photographs. Both photo quality and observer carefulness are important factors.

A variant of this approach is to use complete ground counts as a form of double sampling for at least part of the study zone (Jolly 1969b). The ground count is assumed to be precise, and this is probably correct only for highly visible species in open habitats. For most species, ground counts cannot be complete, and we must rely on other methods for bias correction.

A second approach to bias correction in aerial survey is to use double counts by ground and air observers (Pollock and Kendall 1987). If individual animals can be recognized, the ground count does not need to be complete. One can use a variant of the Petersen method (Section 2.1) in which M individuals are seen in the ground survey, and C individuals are seen in the second (aerial) survey, of which R individuals are recaptures in the sense of being seen by both ground and air observers. Clearly this approach demands individual recognition, which may not be possible for many species. It has been used suc-

TABLE 4.5 COUNTS OF NESTS OF GANNETS (*Sula bassana*) FROM AN AERIAL PHOTOGRAPH OF GRASSHOLM, DYFED, WALES

Observer	Number of counts	Mean count	Range	Standard error
A[a]	10	3222	3077–3323	28
B[a]	8	3051	2852–3192	38
C	3	2949	2823–3014	63
D[a]	3	3359	3358–3362	13
E[a]	3	3301	3274–3315	14
F	2	3173	3138–3209	
G[a]	1	3092		
H	1	3324		
I	1	3228		
J[a]	1	3000		

Source: Harris and Lloyd, 1977.

Note: The mean count of all observers was 3170 nests.

[a]Observers with previous experience in counting seabirds from photographs.

cessfully for bald eagle nests by Grier (1982), and for emus in Australia by Caughley and Grice (1982).

A third approach to bias correction is to mark and release a subset of animals in the study area. The proportion of the tagged animals seen from the air is then used to correct the aerial counts. One variant of this approach is to use radio-collared animals as marked animals. Bear et al. (1989) used colored tags to evaluate this method for an elk population. Packard et al. (1985) used radio-marked manatees to estimate the visibility bias for aerial censuses of this aquatic mammal, and found that there was high variability in the estimated bias. One must assume that the visibility bias is the same for the radio-tagged animals as it is for the population at large to use this approach.

A fourth approach to avoid bias in aerial counts is to use the line transect method described in the next chapter (Section 5.1). This approach assumes that all the individuals along the flight line are counted without error. With the use of helicopters for counting this assumption would be more likely to be correct. White et al. (1989) evaluated line transect methodology for mule deer populations using a helicopter survey and found that population estimates were negatively biased, possibly because of a failure to see all deer along the flight lines.

A critical factor in all aerial census is whether or not animals are seen. *Sightability* is the probability that an animal within the field of search will be seen by an observer. Sightability is related to many variables, but Caughley (1974) suggests that three are most important:

- *Transect width*: As strip width is increased, the mean distance between an animal and its observer is increased, and the time available to locate an animal decreases. The human eye has to move more to cover the strip, and more animals may be hidden by the vegetation.
- *Altitude*: As altitude is increased, the mean distance between the observer and the animal increases, and this might be expected to increase the counting bias; but on the positive side, the required eye movements decrease, and the amount of obscuring vegetation decreases because vision is more vertical.
- *Speed*: As speed is increased, the time available to locate and count an animal decreases, and the rate of eye movement must increase.

Transect width is probably the most complex factor influencing the chances of counting an animal. As transect width changes, sightability changes but not in a simple way. There is a growing list of factors that affect sightability. Caughley et al. (1976) examined the effects of 7 factors on aerial counts of kangaroos and sheep in Australia:

Speed Altitude above ground Transect width Observer	Significant effects
Time of day Fatigue of observers Length of survey	Unimportant effects

This list is clearly not exhaustive. Bayliss and Giles (1985) showed that ambient temperature affected kangaroo counts very strongly, and Broome (1985) showed that different bird species did not react the same to aerial counts. Gasaway et al. (1986) showed that moose in Alaska were more visible in May than in June. All these results show clearly that there will be no universal method for correcting biases in visibility from aerial counts. In some cases the biases may remain of unknown magnitude, and aerial counts should then *not* be used as absolute population estimates.

The general principles of determining optimal quadrat size and shape can also be applied to aerial censusing. In practice the best procedure is to try to check for counting bias using one of the four methods just described (Seber 1992) to see if bias is significant in your particular sampling problem. Then it is important to standardize speed, altitude, transect width, time of day, and any other variables that affect your counts.

4.4.2 Sampling in Aerial Surveys

We define a *census zone* as the whole area in which the number of animals is to be estimated. The sample zone is that part of the census zone that is searched and counted. Many of the problems that have to be overcome in sampling wildlife populations stem from the simple fact that animals are not distributed evenly. In most cases a clumped distribution occurs, and unless we take account of this fact, we cannot census very accurately.

We can sample the census zone in one of two basic ways: simple random sampling or stratified random sampling. *Sampling methods* are discussed in detail in Chapter 8, and I give here only a brief summary of the methods as applied in aerial surveys. The census zone is divided into a number of discrete units known as *sample units* (equal or unequal size), and a number of these are chosen to sample. Figure 4.12 illustrates several ways the sample units may be arranged geographically. There are three basic types of aerial surveys.

Aerial Transect Sampling The most common type of sampling is aerial transect sampling. The aircraft flies in a straight line from one side of the census zone to the other at a fixed height above the ground. Streamers are attached to the wing struts of the plane so that the observer sees a strip demarcated on the ground. The width of the strip is decided in advance, and the observer counts all the animals within the streamers.

Sample units are located by drawing a baseline on a map of the census zone (see Figure 4.12). The baseline is divided like a ruler into pieces the width of the transect strip. The numbers of transects to be sampled are located by a random number table (once n has been set), and the transects are run at right angles to the baseline. The transects can all be of different lengths if necessary and may be split if the census zone has a very irregular shape.

Aerial Quadrat Sampling The sample units are square or rectangular quadrats located at random within the census zone. The whole census zone can be set up as a checkerboard, and the quadrats to be searched can be determined by random numbers. The aircraft can spend as long as necessary searching each quadrat.

Aerial Block Sampling Block sampling is similar to quadrat sampling, except that the sample units are blocks of land demarcated by physical features, such as rivers. A sample

These estimates are then converted to total population size for the whole census zone:

$$\text{Total numbers} = \hat{X} = N\bar{x} \tag{4.31}$$

The variance of this estimate of total numbers depends on the type of sampling used:

Sampling with replacement Each quadrat has the possibility of being selected more than once. Of course the aerial count would be done only once, but the sample could be included 2, 3, or more times in the statistical calculations. For sampling *with replacement*,

$$\text{Variance of total numbers} = (Ns_{\bar{x}})^2 = \frac{N^2}{n}s^2$$

$$\text{Standard error of total numbers} = \sqrt{\text{variance of total numbers}} \tag{4.32}$$

$$\text{95\% confidence limits for total numbers} = \hat{X} \pm t_{.025}\begin{pmatrix}\text{standard error} \\ \text{of total numbers}\end{pmatrix}$$

where $t_{.025}$ is Student's t-value for $(n - 1)$ degrees of freedom.

Norton-Griffiths (1978) works out an example of these calculations in detail.

Sampling without replacement Each quadrat is struck off a list as it is selected in the random sample, so it can appear only once in the statistical calculations. In this case the variance formula above overestimates the true variance and must be corrected as follows: for sampling *without replacement*,

$$\text{Variance of total numbers} = (Ns_{\bar{x}})^2 = \frac{N^2}{n}s^2\left(1 - \frac{n}{N}\right) \tag{4.33}$$

where $\left(1 - \frac{n}{N}\right)$ = Finite population correction (see Chapter 8)

The standard error and the confidence limits are calculated as above. This method was first used on aerial surveys by Siniff and Skoog (1964) on caribou (*Rangifer tarandus*).

Method 2: Unequal-Size Units (The Ratio Method) In a census with aerial transects of differing lengths, the approach is to calculate *density* for each transect and extrapolate this to the total census zone (Jolly 1969a).

The first step is to calculate average density for the whole area:

$$\text{Average density} = \hat{R} = \frac{\text{Total animals counted}}{\text{Total area searched}} = \frac{\Sigma x_i}{\Sigma z_i} \tag{4.34}$$

where x_i = Total animals counted in transect i
z_i = Area of transect i
i = Sample number $(1, 2, 3 \ldots n)$
n = Total number of transects counted

The estimate of the total population is therefore

$$\hat{X} = \hat{R}Z$$

where Z = Area of total census zone
$\hat{R}$ = Average density per unit area

The variance of this estimate is more difficult to determine because both the density and the area of the transects vary. Jolly gives these formulas for the variance:

Sampling with replacement

$$\text{Variance of total numbers} = \frac{N^2}{n(n-1)}\left[\sum x^2 + R^2 \sum z^2 - 2R \sum (xz)\right] \quad (4.35)$$

where N = Total number of possible transects, and all the other terms are as defined above.

Sampling without replacement

$$\text{Variance of total numbers} = \frac{N(N-n)}{n(n-1)}\left[\sum x^2 + R^2 \sum z^2 - 2R \sum (xz)\right] \quad (4.36)$$

Note that these are large-sample estimates of the variance, and they are only approximate when sample size is small ($n < 30$). Confidence intervals are obtained in the same way as in Method 1 above. Box 4.4 gives an example of these calculations.

Box 4.4 Population Estimates from Aerial Census Using Jolly's (1969) Methods 2 and 3

Method 2: The Ratio Method

Topi were counted by Norton-Griffiths (1978) on 12 transects of unequal length in a census area like that in Figure 4.12a. The area of the census zone was 2829 km^2. There were 126 possible transects, of which 12 were selected at random without replacement.

Transect no., i	Area of transect, z_i (km^2)	No. of topi counted, x_i
1	8.2	2
2	13.7	26
3	25.8	110
4	25.2	82
5	21.9	89
6	20.8	75
7	23.0	42
8	19.2	50
9	21.4	47
10	17.5	23
11	19.2	30
12	20.8	54

$$\sum z = 8.2 + 13.7 + \cdots = 236.7$$

$$\sum z^2 = 8.2^2 + 13.7^2 + \cdots = 4930.99$$

$$\sum x = 2 + 26 + 110 + \cdots = 630$$

$$\sum x^2 = 2^2 + 26^2 + 110^2 + \cdots = 43{,}868$$

$$\sum xz = (8.2)(2) + (13.7)(36) + \cdots = 13{,}819.6$$

$$n = 12$$

$$N = 126$$

$$Z = 2829 \text{ km}^2$$

1. *Average density* (equation [4.34]):

$$\hat{R} = \frac{\text{Total animals counted}}{\text{Total area searched}} = \frac{\Sigma\, x_i}{\Sigma\, z_i} = \frac{630}{236.7} = 2.661597 \text{ topi/km}^2$$

2. *Total population of topi:*

$$\hat{X} = \hat{R}Z = (2.66)(2829) = 7530 \text{ topi}$$

3. *Variance of total population* (sampling without replacement, equation [4.36]):

$$\text{Variance of total numbers} = \frac{N(N-n)}{n(n-1)}\left[\sum x^2 + R^2 \sum z^2 - 2R \sum (xz)\right]$$

$$= \frac{126(114)}{12(11)}[43{,}868 + (2.66)^2(4931) - (2)(2.66)(13{,}819.6)]$$

$$= 569{,}686.1$$

4. *Standard error of total population:*

$$\text{S.E.}(\hat{X}) = \sqrt{\text{Variance}(\hat{X})} = \sqrt{569{,}686.1} = 754.775$$

5. *95% confidence limits on total population size* ($t_{.025}$ for 11 d.f. is 2.201):

$$\hat{X} \pm t_{.025}[\text{S.E.}(\hat{X})]$$

$$7530 \pm (2.201)(754.775) = 7530 \pm 1661 \text{ topi}$$

Method 3: Probability-Proportional-to-Size Sampling

Moose were counted on irregular blocks in the southern Yukon (see Figure 4.12c). The total census zone was 5165 km^2 subdivided by topography into 23 blocks. Random points were placed on the map ($n = 12$), and two blocks (C and G) received two random points, while all the others got only one. Thus blocks C and G are counted twice in all the statistical calculations that follow.

Block	Area of block, z (km²)	No. of moose counted, x	Density of moose, $d = x/z$
A	225	63	0.2800
B	340	52	0.1529
C	590	110	0.1864
D	110	15	0.1364
E	63	26	0.4127
F	290	30	0.1034
G	170	42	0.2471
H	410	79	0.1927
I	97	60	0.6186
J	198	51	0.2576

Only 10 blocks were counted, but the sample size is 12 because 2 blocks are used twice (equation [4.38]).

1. *Mean density* $= \hat{\bar{d}} = \dfrac{\Sigma d}{n}$

$$= \frac{0.2800 + 0.1529 + 0.1864 + 0.1864 + 0.1364 + \cdots}{12}$$

$$= \frac{3.0213}{12} = 0.25177 \text{ moose/km}^2$$

2. *Variance of density* (equation [4.41]):

$$\hat{s}_d^2 = \frac{\Sigma d^2 - (\Sigma d)^2/n}{n - 1}$$

$$= \frac{0.97916 - (3.021276)^2/12}{11} = 0.0198622$$

3. *Total population of moose* (equation [4.39]):

$$\hat{X} = \bar{d}Z = (0.25177)(5165) = 1300 \text{ moose}$$

4. *Variance of total population* (equation [4.40]):

$$\text{Var}(\hat{Y}) = \frac{Z^2}{n}\hat{s}_d^2 = \frac{(5165)^2}{12} = (0.0198622) = 44{,}155.7$$

5. *Standard error of total population size:*

$$\text{S.E.}(\hat{X}) = \sqrt{\text{Var}(\hat{X})} = \sqrt{44{,}155.7} = 210.13$$

6. *95% confidence limits on total population size:*

$$\hat{X} \pm t_{.025}[\text{S.E.}(\hat{X})]$$

$$1300 \pm (2.201)(210.13) = 1300 \pm 462 \text{ moose}$$

These calculations can be done in the aerial census subprogram of program-group QUADRAT SAMPLING (Appendix 2).

Method 3: Sampling with Probability Proportional to Size

Method 3 can be used on equal-size sampling units or unequal-size units. Each stratum is divided into sampling units of any size or shape—transects, quadrats, or blocks.

Instead of selecting the units in a stratum from a single set of random numbers, the traditional way, we use a pair of random numbers to define the coordinates of a point on a map of the study area. We thus locate n_i random points on the map within the area, and our sample thus becomes the sample of units containing one or more of these points. If a sample unit happens to contain 2 or 3 points, it is counted 2 or 3 times in the subsequent calculations. We need to make the aerial count of it only once of course. This is another example of *sampling with replacement*, and method 3 does not utilize sampling without replacement (Caughley 1977b). The chance of a sample unit being included in a calculation is thus *proportional to its size*, and there is no problem with sample units of irregular shape. This type of sampling is referred to as PPS sampling (probability proportional to size).

The calculations are as follows:

Calculate for each sampling unit the density:

$$\hat{d} = \frac{x}{z} \tag{4.37}$$

where d = Density in a sample unit
x = Number of animals counted in this sample unit
z = Area of sample unit

Determine the average density for the sample units, noting that each d may be included several times in the totals if it was selected more than once:

$$\text{Average density} = \bar{d} = \frac{\Sigma\, \hat{d}}{n} \tag{4.38}$$

where n = Number of units sampled (random points).

Determine the total population size from

$$\text{Total population} = \hat{X} = \bar{d}Z \tag{4.39}$$

where $\bar{d}$ = Average density per unit area
Z = Total area of census zone

Calculate the variance of total population size:

$$\text{Variance of total numbers} = \frac{Z^2}{n}\,\hat{s}_d^2 \tag{4.40}$$

$$\text{where}\quad \hat{s}_d^2 = \frac{\Sigma\, d^2 - (\Sigma\, d)^2/n}{n-1} \tag{4.41}$$

where d = Density in each sample unit selected
n = Number of random points

This method is preferred by Jolly (1969a) for all aerial census situations except those in which a nearly complete count is required. Box 4.4 illustrates these calculations. Program-group QUADRAT SAMPLING (Appendix 2) calculates total population estimates for these three methods of aerial census.

4.5 SUMMARY

One way to estimate density is to count the number of individuals in a series of quadrats. Quadrats may be small or large, circular or rectangular, and the first question you need to decide is the optimal shape and size of quadrat to use. The best quadrat is that which gives the highest precision for the lowest cost.

To calculate confidence limits on estimates of abundance from quadrat data, you need to know the spatial pattern of the individuals. If organisms are spread randomly in space, counts from quadrats will fit a *Poisson* frequency distribution in which the variance and the mean are equal. More often, animals and plants are aggregated or clumped in space, and the variance of quadrat counts exceeds the mean count. Some types of clumped spatial patterns can be described by the *negative binomial* distribution. Techniques of fitting these distributions to observed quadrat counts are described, and ways of testing goodness-of-fit are available. Confidence intervals for means can be computed once the underlying statistical distribution is determined.

Line intercept sampling is used in plant ecology to estimate plant cover, but it can also provide estimates of population abundance and population density. Measurements of plant size must be made perpendicular to the intercept lines to estimate density and parallel to the intercept line to estimate cover.

Aerial counts are a particularly graphic type of quadrat sampling in which the quadrats are often long, thin strips. Not all animals are seen in aerial counts, and a serious undercounting bias is almost always present. Four methods are available for estimating the counting bias so that aerial counts can estimate true density. The sampling problems associated with aerial counts are easily seen and bring into focus problems that will be discussed in detail in Chapter 8.

SELECTED READING

Caughley, G. 1977. Sampling in aerial survey. *Journal of Wildlife Management* 41: 605–615.

Eberhardt, L. L. 1978. Transect methods for population studies. *Journal of Wildlife Management* 42: 1–31.

Elliott, J. M. 1977. Some methods for the statistical analysis of samples of benthic invertebrates. *Freshwater Biological Association, Scientific Publication No. 25*: 1–142.

Evans, D. A. 1953. Experimental evidence concerning contagious distributions in ecology. *Biometrika* 40: 186–211.

Pringle, J. D. 1984. Efficiency estimates for various quadrat sizes used in benthic sampling. *Canadian Journal of Fisheries and Aquatic Sciences* 41: 1485–1489.

Seber, G. A. F., and J. R. Pemberton. 1979. The line intercept methods for studying plant cuticles from rumen and fecal samples. *Journal of Wildlife Management* 43: 916–925.

Wiegert, R. G. 1962. The selection of an optimum quadrat size for sampling the standing crop of grasses and forbs. *Ecology* 43: 125–129.

QUESTIONS AND PROBLEMS

4.1. Review the arguments in Box 4.1 and read the conclusions in Pringle (1984; see "Selected Reading"). Pringle recommends the use of 0.5 × 0.5 m quadrats from these same data. Why do the recommendations about optimal quadrat size not agree? How would the recommendation differ if (1) relative costs were equal for all sizes of quadrats; (2) relative costs were directly proportional to quadrat area?

4.2. A field plot was divided into 16 quadrats of 1 m^2, and the numbers of the herb *Clintonia borealis* were counted in each quadrat with these results:

3	0	5	1
5	7	2	0
3	2	0	7
3	3	3	4

Calculate the precision of sampling this universe with two possible shapes of 4 m^2 quadrats: 2×2 m and 4×1 m.

4.3. McNeil (1967) recorded the digging associated with spawning in female pink salmon in a 3×66 m section of stream in Alaska. He obtained these results:

No. of times digging occurred in a given quadrat	Observed occurrences
0	116
1	59
2	29
3	10
4	0
5	2
>5	0

Fit a Poisson distribution to these data and test the goodness-of-fit in the two ways given in the text. Discuss the general biological interpretation of your results. State the statistical null hypotheses and their ecological interpretation for these data.

4.4. Rice (1967) recommends that quadrat size be selected so that the resulting measures of plant density have a reasonably normal distribution for the various species being measured. Why might this recommendation be impossible to implement?

4.5. Beall (1940) counted the number of European corn borer (*Pyrausta nubilalis*) larvae on four study areas, using 120 quadrats on each area. He obtained these data:

Number of individuals per plot	Study areas			
	1a	2a	3a	4a
0	19	24	43	47
1	12	16	35	23
2	18	16	17	27
3	18	18	11	9
4	11	15	5	7
5	12	9	4	3
6	7	6	1	1
7	8	5	2	1

Number of individuals per plot	Study areas 1a	2a	3a	4a
8	4	3	2	0
9	4	4		0
10	1	3		1
11	0	0		1
12	1	1		
13	1			
14	0			
15	1			
16	0			
17	1			
18	0			
19	1			
26	1			
n	120	120	120	120

Determine if these data show a random or a clumped pattern. Fit the appropriate distribution and test the goodness-of-fit in the best manner.

4.6. Sinclair (1972) counted wildebeest in the Serengeti by aerial transect sampling of a census zone of 3245 km^2. He selected by sampling with replacement 15 transects as his sample, and he weighted the probability of selecting a given transect by the area of the transect. One transect (7) was selected twice, all others once. There were 96 possible transects in the total census zone, of unequal length. He got these results:

Transect number	Length (km)	Width (km)	No. of wildebeest counted
1	32	0.55	134
2	35	0.55	157
3	46	0.55	174
4	47	0.55	198
5	38	0.55	46
6	31	0.55	98
7	25	0.55	73
8	37	0.55	128
9	32	0.55	156
10	15	0.55	83
11	23	0.55	67
12	28	0.55	130
13	19	0.55	86
14	30	0.55	170

Estimate the total population of wildebeest in the census zone and calculate the 90% confidence interval for this estimate. What statistical and biological assumptions must you make to do these calculations?

4.7. The following data were gathered by students on three line intercepts randomly placed along a 760 m baseline. The data are the perpendicular width of oak trees intercepted by each line in southern Missouri.

Observation no.	Line A (350 m)	Line B (295 m)	Line C (375 m)
1	7.6 m	4.6 m	0.8 m
2	5.2	2.7	4.9
3	1.2	1.8	3.6
4	4.7	3.1	3.2
5	4.4	4.2	1.7
6	3.0	2.8	5.4
7	3.1	2.3	7.6
8	6.9	1.8	3.1
9	3.5	4.8	2.0
10	5.3	2.9	4.6
11	1.9	3.6	3.7
12	2.7	3.1	4.9
13	6.7	5.6	
14	0.9	6.8	
15	4.1	0.5	
16	1.8	4.6	
17	4.6		
18	2.3		
19	6.2		
20	5.2		

Estimate the population density of oak trees in this stand and compute the 90% confidence limits for this estimate.

4.8. Calculate 95% confidence limits for the mean density of black bean aphids from the data in Box 4.2. Imagine that you were under the mistaken belief that counts of the black bean aphid were well described by the Poisson distribution. Compute confidence limits for these data under this mistaken assumption, and compare them with the correct limits. Is there a general relationship between confidence limits based on these two distributions such that one is always wider or narrower than the other for the same data?

4.9. Why are the data in Table 4.2 not adequate for deciding that the best plot size is 4 × 4 m for these trees?

CHAPTER 5

Estimating Abundance: Line Transects and Distance Methods

Sampling plants or animals with quadrats is not the only alternative to mark-recapture estimation of abundance. Quadrats are not natural sampling units, and one must always decide what size and shape of quadrat to use. One alternative is to use "plotless" sampling procedures. These techniques have been developed by plant ecologists and have been applied recently by animal ecologists. They are useful for plants or animals that move little or can be located before they move. Plotless methods provide a third general class of methods for estimating abundance in plant and animal populations, and along with mark-recapture and quadrat counts, they are important tools for the field ecologist.

5.1 LINE TRANSECTS

The line intercept method discussed in Chapter 4 is one example of a family of methods for estimating abundance from transects. Another important method for estimating populations with transect lines is *line transect* sampling. Much of the material on line transect sampling has been brought together in Buckland et al. (1993) which provides a detailed

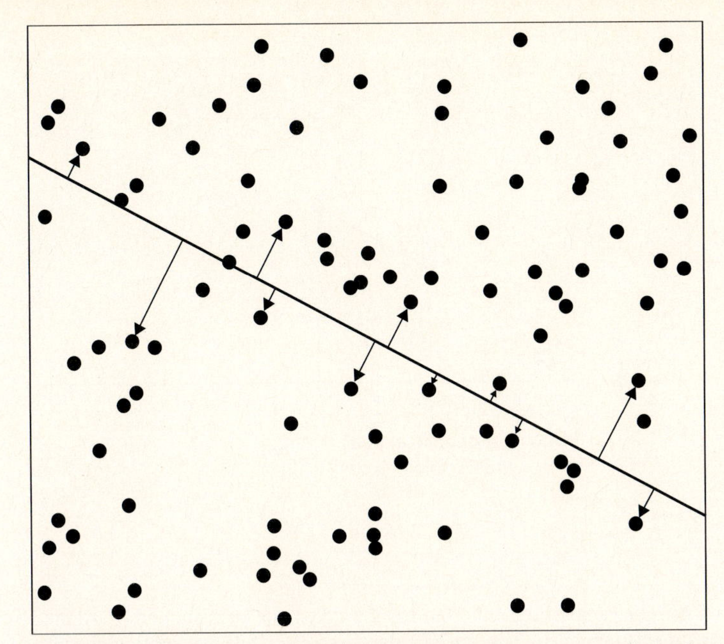

Figure 5.1 Schematic view of the method of line transect sampling. The census zone is the whole area of the square. Only one transect is shown for illustration. The observer moves along the transect line and the distances indicated by the arrows are measured. In this example 13 animals were seen (including two right on the transect line). Note that many individuals were not seen, and that detection falls off with distance from the transect line.

reference for these methods. Here I will summarize the general procedures of line transect sampling and highlight the assumptions you must make to use these methods to estimate abundance.

Figure 5.1 illustrates how line transect sampling is done. A transect line is searched, and each animal seen provides one measurement of the perpendicular distance to the transect line. Since in practice animals are often seen along the line, three measurements can be taken for each individual sighted, as shown in Figure 5.2:

1. Sighting distance (r_i)
2. Sighting angle (θ_i)
3. Perpendicular distance (x_i)*

Transect lines may be traversed on foot, on horseback, in a vehicle, or in a helicopter or airplane.

If a fixed width of strip is counted, and if all organisms in the strip are seen, estimates of population size are simple, because strips are just long, thin quadrats. All the principles

*The perpendicular distance can be calculated from the other two by $x = r \sin \theta$.

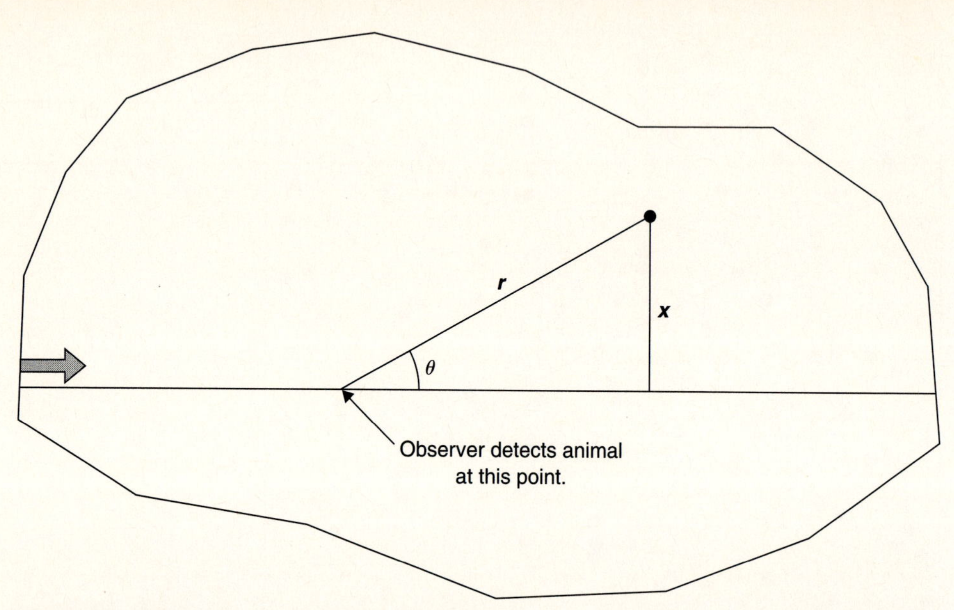

Figure 5.2 Illustration of the basic measurements that can be taken for each individual sighted along a line transect. The key measurement is the perpendicular distance (x_i). If the sighting distance (r_i) is easier to record in the field, the sighting angle (θ) must also be measured. The perpendicular distance $x = r \sin(\theta)$.

of quadrat sampling discussed in Chapter 4 apply to this situation. Plant ecologists sometimes use *line transects* to mean these long, thin quadrats that are completely censused.

In practice some organisms are undetected as one moves along a transect, and in these cases it is best not to limit observations to a fixed strip width. Because individuals are missed, an undercounting bias occurs. In these cases estimation of population density is more difficult because we need to estimate the *detection function*, as shown in Figure 5.3. The figure shows that in general the detectability will fall off with distance from the centerline of the transect. If we can make four assumptions, we can estimate population density from the detection function:

1. Animals directly on the transect line will never be missed (i.e., their detection probability = 1).
2. Animals are fixed at the initial sighting position; they do not move before being detected, and none are counted twice.
3. Distances and angles are measured exactly, with no measurement error and no rounding errors.
4. Sightings of individual animals are independent events.

If these assumptions are valid, we can estimate the density of the population by

$$\hat{D} = \frac{n}{2La} \tag{5.1}$$

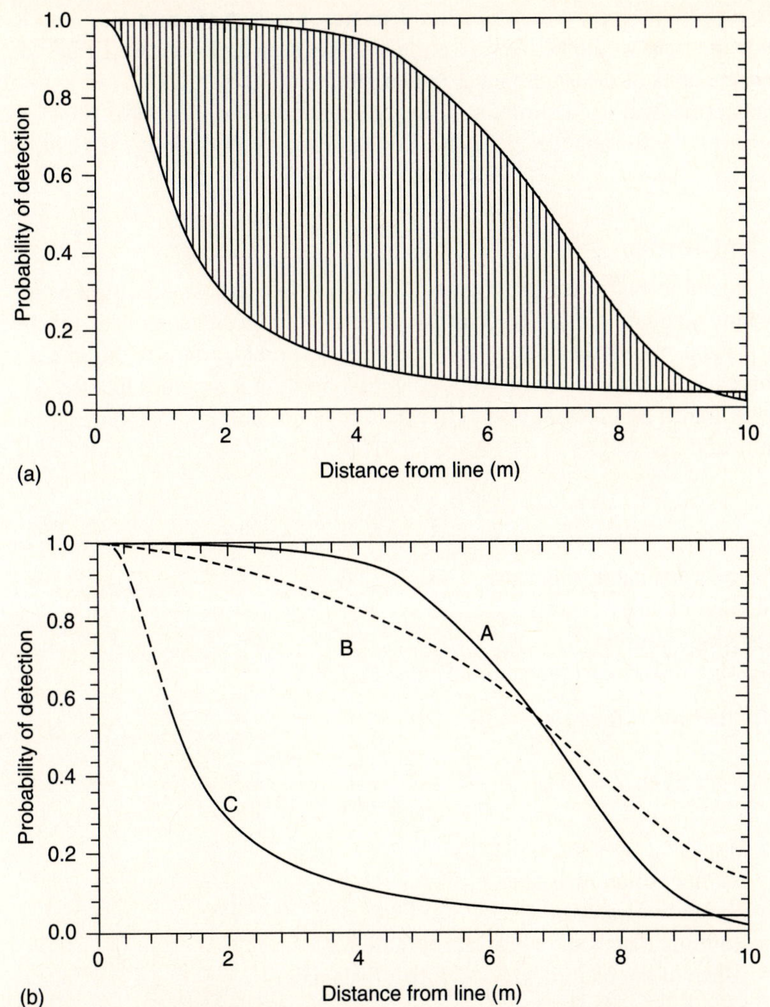

Figure 5.3 Detection function of a line transect survey. The basic idea of these models is that the probability of detection falls off the farther an animal is from the line transect baseline. (a) The shaded area encloses the general zone for detection functions for wildlife populations. (b) The detection function for any particular set of data may take a variety of shapes, and the statistical problem is to decide what mathematical function to use and what values of its parameters fit best. The generalized exponential (A), the half-normal (B), and the Hayes and Buckland (1983) function (C) are illustrated here. (Modified from Burnham et al. 1980 and Routledge and Fyfe 1992a.)

where D = Density of animals per unit area
n = Number of animals seen on transect
L = Total length of transect
a = Half the effective strip width (a constant that must be estimated)

The constant a is simply the total area under the detection function (Figure 5.3), and it estimates how wide the strip would be if every organism was seen and none were missed. It is scaled in the same units of measurement as the lengths.

There are numerous ways of estimating a in the literature, and they have been reviewed comprehensively by Burnham et al. (1980, Table 24) and by Buckland et al. (1993). We discuss three here.

5.1.1 Hayne Estimator

Hayne (1949) developed a way to estimate densities of birds like grouse that flush as an observer comes within a certain radius. The basic assumption of this estimator is that there is a fixed flushing distance r such that if an observer comes closer than r units to the animal, it flushes and is observed. This is a restrictive assumption because it assumes that the detection function of Figure 5.3 is rectangular. If this assumption is correct, then population density can be estimated by

$$\hat{D}_H = \frac{n}{2L}\left(\frac{1}{n}\sum\frac{1}{r_i}\right) \tag{5.2}$$

where $\hat{D}_H$ = Hayne's estimator of density
n = Number of animals seen
L = Length of transect
r_i = Sighting distance to each animal i (see Figure 5.2)

The variance of this density estimate is:

$$\text{Var}(\hat{D}_H) = D_H^2\left[\frac{\text{var}(n)}{n^2} + \frac{\Sigma\,(1/r_i - R)^2}{R^2 n(n-1)}\right] \tag{5.3}$$

where $\hat{D}_H$ = Hayne's estimator of density
n = Number of animals seen
$\text{var}(n)$ = Variance of $n \cong n$
r_i = Sighting distance for animal i (Figure 5.2)
R = Mean of the reciprocals of sighting distances i

The standard error of the mean density is estimated by the square root of this variance.

The one critical assumption of the Hayne estimator is that the sines of the angles (θ) of the observed sightings are a sample from a uniform random variable ranging from 0 to 1. This assumption implies that the average sighting angle is 32.7°.* This can be tested by the statistic:

$$z = \frac{\sqrt{n}(\bar{\theta} - 32.7)}{21.56} \tag{5.4}$$

where z = Standard normal deviate
n = Observed number of sightings
$\bar{\theta}$ = Observed mean sighting angle (Figure 5.2)

*If the sine of θ ranges uniformly from 0 to 1, the mean value of θ is $(D/2) - 1$ radians, or 32.7°. See Hayne 1949, 157.

The decision rule is to reject the null hypothesis that the average angle is 32.7° if z is greater than 1.96 or less than -1.96 for $\alpha = .05$. If this null hypothesis is rejected, the Hayne estimator of density should not be used.

If the Hayne model is not applicable because the angle of sightings does not average 32.7°, you may be able to use a modification of the Hayne model. Burnham and Anderson (1976) found that the average sighting angle was usually in the range 32° to 45°, and that in these cases a reasonable estimator is

$$\hat{D}_{MH} = c\hat{D}_H \tag{5.5}$$

where $\hat{D}_{MH}$ = Modified Hayne estimator
$\hat{D}_H$ = Hayne estimator (formula above)
c = Correction factor = $1.9661 - 0.02954\ \bar{\theta}$
$\bar{\theta}$ = Mean sighting angle for all n observations

The variance of this estimator is

$$\text{Var}(\hat{D}_{MH}) = \hat{D}_{MH}^2 \left\{ \frac{\text{var}(n)}{n^2} + \frac{0.000876}{c^2} \left[\frac{\Sigma\ (\theta_i - \bar{\theta})^2}{n(n-1)} \right] + \frac{\Sigma\ (1 - r_i) - R^2}{R^2(n)(n-1)} \right\} \tag{5.6}$$

Box 5.1 illustrates these calculations.

Box 5.1 Line Transect Method of Density Estimation Using the Hayne Estimator

The following data are part of the data from a line transect to estimate the abundance of white-eared kob in the Sudan. They are used only to illustrate the method, since in practice a larger sample size would be obtained.

Animal no.	Perpendicular distance, y_i (m)	Sighting distance, r_i (m)	Measured angle (θ)
1	92.35	150	38°
2	163.80	200	55
3	22.27	160	8
4	58.47	200	17
5	157.30	250	39
6	86.99	130	42
7	26.05	150	10
8	50.80	130	23
9	163.80	200	55
10	71.93	100	46
11	72.11	140	31
12	84.52	200	25

Transect length was 10 km. For calculating convenience, express all the distances in kilometers (not in meters).

1. To calculate population density from equation (5.2), we have

$$\hat{D}_H = \frac{n}{2L}\left(\frac{1}{n}\sum\frac{1}{r_i}\right)$$

$$= \frac{12}{2(10)}\left[\frac{1}{12}\left(\frac{1}{0.150} + \frac{1}{0.200} + \frac{1}{0.160} + \frac{1}{0.200} + \cdots + \frac{1}{0.200}\right)\right]$$

$$= \frac{12}{20}(6.342) = 3.8055 \text{ animals/km}^2$$

The variance of this density estimate is, from equation (5.3),

$$\text{Var}(\hat{D}_H) = D_H^2\left[\frac{\text{var}(n)}{n^2} + \frac{\Sigma\,(1/r_i - R)^2}{R^2 n(n-1)}\right]$$

$$= (3.8055)^2\left[\frac{12}{12^2} + \left(\frac{1/0.150 - 6.34)^2 + (1/0.200 - 6.34)^2 + \cdots}{6.34^2(12)(11)}\right)\right]$$

$$= 1.2902$$

$$\text{S.E.}(\hat{D}_H) = \sqrt{1.2902} = 1.13589$$

2. Calculate the 95% confidence interval in the usual manner: with 11 d.f. the t-value is 2.20, and

$$\hat{D}_H \pm t_{.025}[\text{S.E.}(\hat{D}_H)]$$

$$3.8055 \pm 2.20(1.13589)$$

$$3.8055 \pm 2.4990 \text{ kob/km}^2$$

The wide confidence interval is due to the small sample size; thus the recommendation that n should be 40 or more.

3. To test the critical assumption of the Hayne method that the average angle of detection is 32.7° for these data (equation [5.4]):

$$z = \frac{\sqrt{n}(\bar{\theta} - 32.7)}{21.56}$$

$$z = \frac{\sqrt{12}(32.42 - 32.7)}{21.56} = -0.05$$

so the null hypothesis that the sighting angle is 32.7° is tentatively accepted for these data.

Program-group LINE TRANSECTS (Appendix 2) can do these calculations.

5.1.2 Fourier Series Estimator

This is a robust estimator that is sufficiently general to fit almost any type of line transect data. It is a good, general-purpose estimator and is strongly recommended by Burnham et al. (1980) as the best model available. The estimate of density is the usual one:

$$\hat{D} = \frac{n}{2La} \tag{5.7}$$

where $\hat{D}$ = Estimated population density
n = Number of animals seen on the transect
L = Length of transect
a = Half the effective strip width (a constant)

The critical parameter a is estimated as

$$\frac{1}{\hat{a}} = \frac{1}{w^*} + \sum_{r=1}^{m} a_k \tag{5.8}$$

where w* = Transect width (largest perpendicular distance observed)

$$\hat{a}_k = \frac{2}{nw^*}\left[\sum_{i=1}^{n} \cos\left(\frac{K\pi y_i}{w^*}\right)\right] \tag{5.9}$$

where n = Number of animals seen
π = 3.14159
x_i = Perpendicular distance of animal i (Fig. 5.2)
k = Number of term in equation (1, 2, 3, 4 . . .)
m = Maximum number of cosine terms in the summation (<6)

Burnham et al. (1980) give a stopping rule for m as follows: Choose m to be the smallest integer at which this inequality holds true:

$$\frac{1}{w^*}\left(\frac{2}{n+1}\right)^{1/2} \geq |a_{m+1}| \tag{5.10}$$

where w^* = Transect width (largest perpendicular distance observed)
n = Number of animals counted
$|a_{m+1}|$ = Absolute value of a_{m+1} (defined above)

The variance of the Fourier series estimator is complex and should probably be done with the computer using Program TRANSECT described by Burnham et al. (1980). The procedure in capsule form is as follows:

Estimate the variances of the a_k:

$$\text{var}(\hat{a}_k) = \frac{1}{n-1}\left[\frac{1}{w^*}\left(a_{2k} + \frac{2}{w^*}\right) - a_k^2\right] \quad \text{for } k \geq 1 \tag{5.11}$$

Estimate all the covariances of the a_k:

$$\text{cov}(\hat{a}_k, \hat{a}_j) = \frac{1}{n-1}\left[\frac{1}{w^*}(a_{k+j} + a_{k-j}) - a_k a_j\right] \quad \text{for } k \geq j \geq 1 \tag{5.12}$$

Having determined m above, and given that $\text{cov}(a_k, a_k)$ is the same as $\text{var}(a_k)$, we obtain

$$\text{var}\left(\frac{1}{\hat{a}}\right) = \sum_{j=1}^{m} \sum_{k=1}^{m} \text{cov}(\hat{a}_j, \hat{a}_k) \tag{5.13}$$

Estimate the variance of $\hat{D}$, the estimated population density, by

$$\text{var}(\hat{D}) = \hat{D}^2\left[\frac{\text{var}(n)}{n} + \frac{\text{var}(1/\hat{a})}{(1/\hat{a})^2}\right]$$

$$\text{S.E.}(\hat{D}) = \sqrt{\text{var}(\hat{D})} \tag{5.14}$$

$$95\% \text{ confidence interval} = \hat{D} \pm (1.96)[\text{S.E.}(\hat{D})]$$

One problem in these estimates for the variance of line transect estimators is determining the variance of n, the number of animals counted. It is usually assumed that n is a Poisson variable, so the variance = the mean = n, but this assumption could be in error (Burnham et al. 1980). A better way to use the line transect statistically is to count a series of replicate lines, ensuring that each line is long enough to have an adequate sample size ($n > 20$ or 25). Then the variance of the density estimate D can be obtained directly. We define, for R replicate lines,

$$\hat{D}_i = \frac{n_i}{2l_i\hat{a}_i} \tag{5.15}$$

where $\hat{D}_i$ = Density estimate for line i
n_i = Number of animals counted on line i
l_i = Length of transect line i
$\hat{a}_i$ = Estimated parameter for line i
i = 1, 2, 3 . . . R lines (replicate number)

For each line we estimate a using one of the techniques just described.

The overall density estimate is the weighted average:

$$\hat{D} = \frac{\sum_{i=1}^{R} l_i D_i}{\sum_{i=1}^{R} l_i} \tag{5.16}$$

The empirical estimate of the variance of $\hat{D}$ is

$$\text{var}(\hat{D}) = \frac{\sum_{i=1}^{R} [l_i(\hat{D}_i - \hat{D})^2]}{L(R - 1)} \tag{5.17}$$

where $L = \Sigma\, l_i$
$\hat{D}$ = Overall density estimate from equation (5.16)
R = Number of replicate line transects

The standard error of the overall density estimate is

$$\text{S.E.}(\hat{D}) = \sqrt{\text{var}(\hat{D})} \tag{5.18}$$

and the 95% confidence interval is

$$\hat{D} \pm t_{.025}[\text{S.E.}(\hat{D})] \tag{5.19}$$

where $t_{.025}$ has $(R - 1)$ degrees of freedom.

5.1.3 Shape-Restricted Estimator

An efficient estimator of population density can be constructed for line transect data by placing two restrictions on the shape of the detection curve (Johnson and Routledge 1985). The first restriction is that the detection function must be a continuously decreasing function, as shown in Figure 5.3. The second restriction is more powerful and requires that the curve must have a concave shoulder followed by a convex tail with an inflection point between these two regions. Given these restrictions, Johnson and Routledge (1985) utilize a flexible least-squares procedure to specify the detection function. The procedure is too complex to summarize here, and is coded in a computer program TRANSAN described by Routledge and Fyfe (1992b). This procedure permits one to specify the length of the horizontal shoulder (i.e., the zone in which detectability is near 1), the inflection point, or the height of the right tail of the detection function. By altering assumptions about the shoulder width, the biologist can explore the impact on the resulting density estimates. For example, a biologist may from experience be able to state that the horizontal shoulder extends 10 meters on either side of the transect line. Figure 5.4 illustrates data obtained on a line transect study of starfish (*Pisaster brevispinus*). The data suggest a rise in sightability

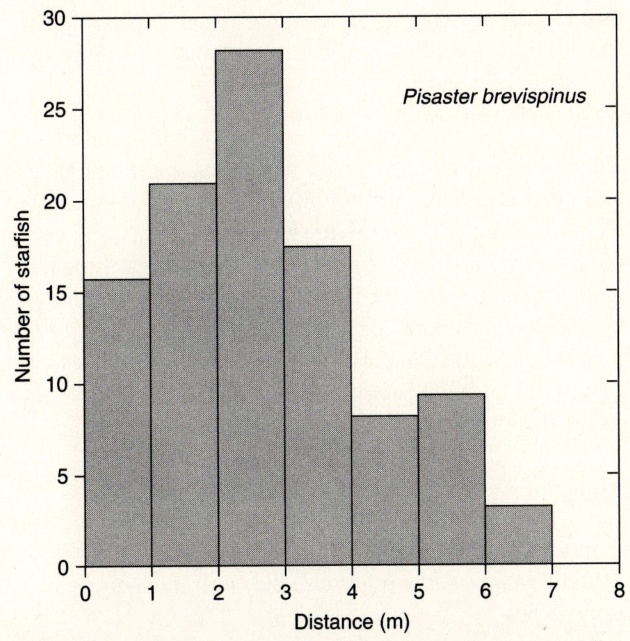

Figure 5.4 Observed perpendicular distances for a set of line transects of starfish (*Pisaster brevispinus*) near Vancouver, B.C. The shape of the smoothed detection function for these data will have a broad shoulder out to about 3 m, and will fall off to zero around 8–10 m. (Data from Routledge and Fyfe 1992a.)

over the first 3 meters from the transect line, but this is apparently only chance variation, and the first shape-restriction on these data would be that the shoulder width be at least 3 meters. A second restriction is the upper limit to the sighting distance. If starfish are not visible beyond 8 meters from the transect line, the upper limit can be constrained to be 8 m. Clearly the more you know about the natural history of the study animal, the more closely you can constrain the detection function. The detection function for these data is thus more like line A in Figure 5.3 than line B or C. Johnson and Routledge (1985) showed in simulations that the shape-restricted estimator of detection functions was more robust than any of the more commonly used methods like the Fourier series estimator or the half-normal function.

Anderson et al. (1979) provide a useful set of guidelines for anyone using the line transect technique for population estimation. They recommend nine cautions in data collection:

1. The centerline of the transect must be straight and well marked.
2. Care must be taken that all animals on the centerline are seen with certainty.
3. Transect width should be effectively unbounded, and all animals seen should be recorded.
4. All measurements of distances and angles should be accurately done with a tape measure and compass.
5. All measurements should be taken in the field: perpendicular distance, sighting angle, sighting distance (Figure 5.2).
6. Measurements should be recorded separately for convenient lengths of transect; they can be combined later as needed.
7. Sample size (n) should be at least 40, and 60–80 would be better, if possible.
8. Transects along roads or ridgetops should be avoided; a randomization procedure is essential for proper statistical inference.
9. Only competent, interested personnel should be used.

Extensive discussion of the practical and statistical problems of line transect sampling appears in Pollock (1978), Gates (1979), Burnham et al. (1980), Burnham and Anderson (1984), Johnson and Routledge (1985), Routledge and Fyfe (1992a), and Buckland et al. (1993). Several computer programs are available to do line transect calculations. Program-group LINE TRANSECTS (Appendix 2) computes the Hayne estimator and the modified Hayne estimator for line transect sampling. Program DISTANCE of Buckland et al. (1993) is much larger and more comprehensive and will compute the Fourier series estimator as well as a variety of other parametric functions like the half-normal. Program TRANSAN of Routledge and Fyfe (1992b) computes the shape-restricted estimator for line transect data.

5.2 DISTANCE METHODS

Plant ecologists have developed a variety of plotless sampling methods that work well on trees and shrubs. These are all called *distance methods* because they utilize distances measured by means of two general approaches:

1. Select random organisms and measure the distance to their nearest neighbors.
2. Select random points and measure the distance from the point to the nearest organisms.

Plotless sampling is usually applied to a single species of plant or animal, and is popular in ecology because it is a two-edged sword. We can use plotless sampling to estimate the *density* of the population. On the other hand, if we know the density of a population, we can use plotless sampling to determine whether the *spatial pattern* is random, aggregated, or uniform. We are concerned here with the first approach, estimating density. We discuss the investigation of spatial pattern in the next chapter.

The general procedure in plotless sampling is illustrated in Figure 5.5. The *census zone* is the area of interest, but in practice a boundary strip must be added so that measurements made on individuals at the edge of the census zone are not biased. The data of interest are the measurements from a series of random points to the nearest organism, or from organisms to their nearest neighbors. In principle, one could extend this approach to measure, in addition, the distance to the second-nearest neighbor, the distance to the third-nearest neighbor, and so on. But, to begin, let us consider the simple case shown in Figure 5.5.

If the entire population can be mapped as shown in Figure 5.5, we would know population density exactly, and our estimation problem would be simple. But in most cases we must sample individuals scattered over a large area. For example, we may need to estimate the density of a tree species in a large region of 10 km^2 of forest. Statisticians call this "sparse sampling" and state that the primary requirement of sparse sampling schemes

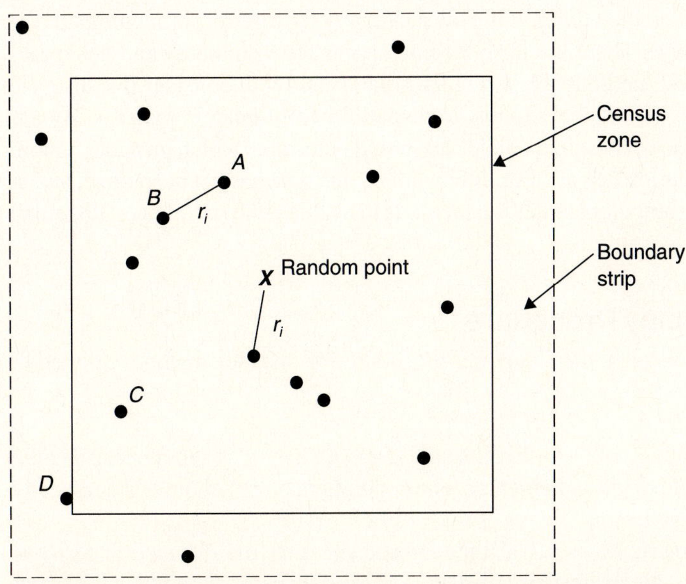

Figure 5.5 Schematic illustration of distance measures. The census zone is surrounded by a boundary strip, and in some cases the nearest organism is located in the boundary strip. Two types of measurements can be taken: (1) A random point X is located in the census zone, and the distance r_i to the nearest organism is measured. (2) A random organism A is selected, and the distance r_i to the nearest neighbor B is measured. The nearest neighbor might be in the boundary strip outside the census zone (e.g., C to D).

is that the sample points should be well separated so that observations can be assumed to be independent. Usually this requires that we do not sample more than 5–10% of the total population, so that we assume sampling without replacement. Using the same general approach shown in Figure 5.5, we can make two kinds of measurements:

1. from random *points* to the nearest organism
2. from a random *organism* to its nearest neighbor.

There is in principle no problem with setting out random points in a large area, although it may be very time-consuming and less easy to achieve in the field than statisticians tend to assume (Pollard 1971). But there is definitely a problem with selecting a random individual. In principle it is easy: mark each individual with a number and select n of these at random. But of course this is impossible in the field because, we would have to enumerate the entire population! The next idea was to select a random *point* and use the individual closest to that random point as the equivalent of a random organism. But Pielou (1977, 154) showed that this procedure is biased in favor of isolated individuals and *thus should not be used to select a random individual.*

There appears to be no easy way out of this dilemma. The best compromises all involve some type of systematic sampling (see Chapter 8, Section 8.4 for more discussion of systematic sampling). A variety of plotless sampling methods have been proposed, and while most of them perform adequately for organisms that are randomly spread in the landscape, there is considerable variation in how accurately they perform for clumped patterns (Engeman et al. 1994). There is a need for interaction between theory and field practice in developing the best methods for particular organisms. I will consider five distance techniques that appear to be relatively robust to deviations from random patterns. By applying them for equal time periods in the field, one could determine which provides a more precise estimate of population density, but little work of this type has yet been done, and at present we must rely on computer simulation methods to evaluate the techniques (Engeman et al. 1994).

5.2.1 Byth and Ripley Procedure

Byth and Ripley (1980) suggest the following procedure for distance methods applied to large areas:

1. Set out $2n$ sampling points within the study zone (where n = sample size desired for nearest-neighbor measurements). These points may be set out systematically for ease of location.
2. Select half of these $2n$ points at random and measure the distance from these random points to the nearest organism ($x_1, x_2 \dots x_n$).
3. Around the remaining half of the $2n$ sampling points, lay out a small plot of a size large enough to contain about 5 individuals on average. Number all these individuals on all of the small plots and select n of these enumerated organisms at random.
4. Measure the distance between the selected organism and its nearest neighbor ($r_1, r_2, r_3 \dots r_n$).

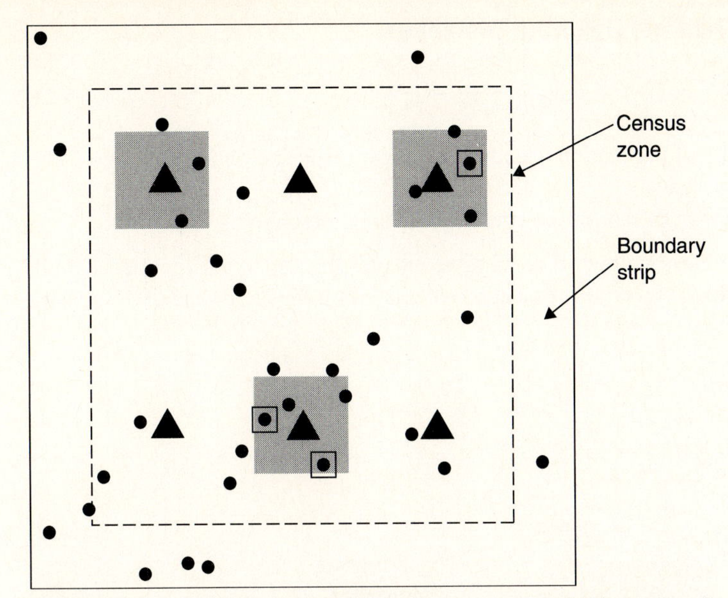

Figure 5.6 Semisystematic sampling scheme recommended by Byth and Ripley (1980) for distance sampling of large areas in which a complete enumeration is not possible. In this example the desired sample size is 3 (n), so 6 ($2n$) points (triangles) are laid out in a systematic manner within the study area (boundary is dashed line). Three points are randomly selected for point-to-organism measurements, and the three points remaining (shaded areas) are used for setting out small plots for a random selection of individuals for nearest-neighbor distances. The three trees selected at random for nearest-neighbor measurements are shown as boxed squares. (After Byth and Ripley 1980.)

Figure 5.6 illustrates this sampling approach, which is still time-consuming to apply in the field.

The reason for developing distance measures was to estimate population density without the need to lay out quadrats. But in doing this we run into a fundamental problem: density estimates from all distance methods are sensitive to the spatial pattern. If the animals or plants have a random pattern, all distance measures should provide an unbiased estimate of population density. If the pattern is aggregated, estimates of density are biased, and we need to look into how large this bias can be.

Either of the distance measurements suggested by Byth and Ripley (1980) will provide an unbiased estimate of population density if the spatial pattern is random:

For Point-to-Organism Distances

$$\hat{N}_1 = \frac{n}{\pi \Sigma (x_i^2)} \qquad (5.20)$$

where $\hat{N}_1$ = Estimate of population density from point-to-organism data
n = Sample size
x_i = Distance from random point i to nearest organism

For Organisms-to-Nearest Neighbor Distances

$$\hat{N}_2 = \frac{n}{\pi \, \Sigma \, (r_i^2)} \tag{5.21}$$

where $\hat{N}_2$ = Estimate of population density from organism-to-neighbor data
n = Sample size
r_i = Distance from random organism i to nearest-neighbor

The variances of these estimators are similar and may be used to put confidence limits on the density estimates. The variances must be calculated for the reciprocal of the density. Define

$$\hat{y} = \frac{1}{\hat{N}}$$

Then

$$\text{Variance } (\hat{y}) = \frac{\hat{y}^2}{n} \tag{5.22}$$

$$\text{Standard error } (\hat{y}) = \sqrt{\frac{\text{Variance}(\hat{y})}{n}} \tag{5.23}$$

where $\hat{N}$ can be either $\hat{N}_1$ or $\hat{N}_2$, and n is sample size.

In most natural populations, however, spatial patterns are not random but clumped. How can we estimate population density in these situations? The general strategy that has developed for constructing unbiased density estimators from distance data is based on the observation that distances from random points to the nearest organism are *increased* if the spatial pattern is clumped, while distances from random organisms to their nearest neighbors are *decreased* (Diggle 1975). Thus some kind of average density estimator that combines N_1 and N_2 should be better and less biased than either of these simple estimates by itself.

Diggle (1975) suggested that the best compound estimator for many nonrandom patterns was the geometric mean of $\hat{N}_1$ and $\hat{N}_2$:

$$\hat{N}_3 = \sqrt{\hat{N}_1 \hat{N}_2} \tag{5.24}$$

where $\hat{N}_3$ = Diggle's estimator of population density for nonrandom patterns
$\hat{N}_1$ = Point-to-organism density estimator (equation [5.20])
$\hat{N}_2$ = Organism-to-nearest-neighbor density estimator (equation [5.21])

Diggle showed that this estimator has a low bias over a wide range of clumped-to-uniform spatial patterns, and is the best estimator available at present for data gathered by the Byth and Ripley procedure.

The variance of Diggle's estimator is calculated on the reciprocal of density:

$$\text{Variance}\left(\frac{1}{\hat{N}_3}\right) = \frac{(1/\hat{N}_3)^2}{n} \tag{5.25}$$

and the standard error is as usual:

$$\text{Standard error}\left(\frac{1}{\hat{N}_3}\right) = \sqrt{\frac{\text{Variance}(1/\hat{N}_3)}{n}} \tag{5.26}$$

Box 5.2 gives an example of these calculations, and program-group DISTANCE METHODS does these calculations (Appendix 2).

5.2.2 T-Square Sampling Procedure

An alternative sampling scheme to the Byth and Ripley procedure was first described by Besag and Gleaves (1973) and is called T-square sampling. It is simpler to implement in the field than the Byth and Ripley procedure, and thus might be preferred by field workers.

Figure 5.7 illustrates the principles of T-square sampling. Random points are located in the study region, and at each random point two distances are measured:

1. The distance (x_i) from the random point (O) to the nearest organism (P)
2. The distance (z_i) from the organism (P) to its nearest neighbor (Q) with the restriction that the angle OPQ must be more than 90° (the T-square distance). If the closest neighbor is not within this angle, use the next closest organism until you find one that satisfies this angle.

The point-to-organism distances obtained in T-square sampling are identical to those obtained previously (see equation [5.20]). The T-square distances however are constrained by the 90° rule, and consequently the density estimator that utilizes T-square distances (z_i) has a different formula:

$$\hat{N}_4 = \frac{2n}{\pi \sum (z_i^2)} \tag{5.27}$$

where $\hat{N}_4$ = T-square estimate of population density (analogous to $\hat{N}_2$)
n = Number of samples
z_i = T-square distance associated with random point i

This estimator should not be used unless it is known that the organisms being sampled have a random pattern.

Byth (1982) showed that the most robust estimator of population density for use with T-square sampling was the following compound measure of $\hat{N}_1$ and $\hat{N}_4$:

$$\hat{N}_T = \frac{n^2}{2 \sum (x_i)[\sqrt{2} \sum (z_i)]} \tag{5.28}$$

The standard error of $\hat{N}_T$ is calculated on the reciprocal of the density and is given by Diggle (1983):

$$\text{Standard error}\left(\frac{1}{\hat{N}_T}\right) = \sqrt{\frac{8(\bar{z}^2 s_x^2 + 2\bar{x}\bar{z}s_{xz} + \bar{x}^2 s_z^2)}{n}} \tag{5.29}$$

where $\bar{x}$ = Mean value of point-to-organism distances
$\bar{z}$ = Mean value of T-square organism-to-neighbor distances
n = Sample size
s_x^2 = Variance of point-to-organism distances
s_z^2 = Variance of T-square organism-to-neighbor distances
s_{xz} = Covariance of x and z distances

Box 5.2 Estimating Population Density from Distance Measurements of Random Points-to-Organism and Random Organisms-to-Nearest-Neighbor

Using the Byth and Ripley procedure (Figure 5.6), ecology students measured the following distances on subalpine fir trees in the Coast Mountains of British Columbia ($n = 20$).

Sample no.	Point-to-tree distances, x_i (m)	Tree-to-nearest-neighbor distances, r_i (m)
1	8.65	3.60
2	12.20	8.55
3	6.95	2.15
4	3.05	6.80
5	9.65	5.05
6	4.35	10.60
7	7.10	4.35
8	15.20	2.85
9	6.35	7.95
10	12.00	3.15
11	2.80	6.90
12	5.55	3.95
13	8.10	8.10
14	11.45	4.50
15	13.80	7.65
16	7.35	1.10
17	6.30	3.40
18	9.60	4.80
19	10.35	6.25
20	3.15	2.90

For the point-to-tree data:

$$\sum (x_i^2) = 8.65^2 + 12.20^2 + 6.95^2 + \cdots = 1587.798$$

For the tree-to-neighbor data:

$$\sum (r_i^2) = 3.60^2 + 8.55^2 + 2.15^2 + \cdots = 665.835$$

Three density estimates are available from equations (5.20), (5.21), and (5.24):

$$\hat{N}_1 = \frac{n}{\pi \Sigma (x_i^2)} = \frac{20}{(3.14)(1587.798)} = 401 \times 10^{-3} \text{ trees/m}^2$$

or 40 trees/hectare.

$$\hat{N}_2 = \frac{n}{\pi \Sigma (r_i^2)} = \frac{20}{(3.14)(665.835)} = 9.56 \times 10^{-3} \text{ trees/m}^2$$

or 96 trees/hectare.

$$\hat{N}_3 = \sqrt{\hat{N}_1 \hat{N}_2} = \sqrt{(4.01)(9.56 \times 10^{-6})} = 6.19 \times 10^{-3} \text{ trees/m}^2$$

or 62 trees/hectare.

These trees show a *clumped* spatial pattern and are not randomly distributed (see Chapter 6 for a test of this spatial pattern). Consequently neither $\hat{N}_1$ nor $\hat{N}_2$ is a good unbiased estimator of density. The best estimator is $\hat{N}_3$, Diggle's estimator (equation [5.24]), which has the following variance (equation [5.25]):

$$\text{Variance of}\left(\frac{1}{\hat{N}_3}\right) = \frac{(1/\hat{N}_3)^2}{n} = \frac{(161.5)^2}{20} = 1304.27$$

and (equation [5.26]):

$$\text{Standard error of}\left(\frac{1}{\hat{N}_3}\right) = \sqrt{\frac{\text{Variance}(1/\hat{N}_3)}{n}}$$

$$= \sqrt{\frac{1304.27}{20}} = 8.075$$

Hence the 95% confidence limits on $1/\hat{N}_3$ will be

$$\frac{1}{\hat{N}_3} \pm t_\alpha[\text{S.E.}(1/\hat{N}_3)] \qquad [\text{where } t_\alpha \text{ has } (n-1) \text{ or 19 d.f.}]$$

$$161.55 \pm (2.09)(8.075) \quad \text{or} \quad 144.673 \text{ to } 178.429$$

or, taking reciprocals, from 0.00560 to 0.00691 trees/m^2, or 56 to 69 trees/ha.

Program-group DISTANCE METHODS (Appendix 2) can do these calculations.

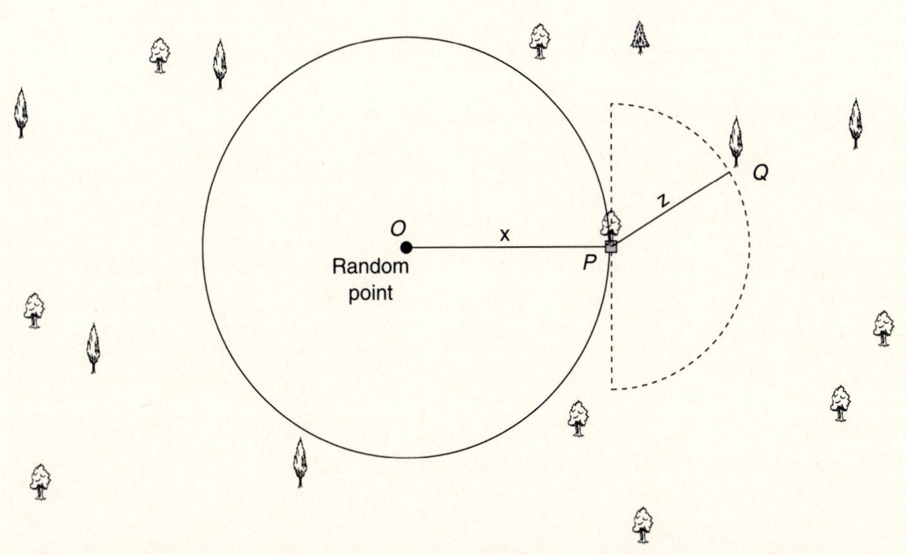

Figure 5.7 Schematic view of T-square sampling. A random point O is located within the study area, and the distance x is measured from this point to the nearest organism P. A second distance z is measured from organism P to its nearest neighbor constrained to be in the hemisphere to the right of the dashed line. The angle OPQ must be more than 90°. The study is sampled with a series of n random points like this. Trees symbolize individual organisms.

Box 5.3 Estimating Population Density from T-Square Sampling

Using the T-square sampling procedure illustrated in Figure 5.7, an ecology class measured the following distances on white spruce trees in the southwestern Yukon ($n = 16$):

Sample point no.	Point-to-tree distance, x_i (m)	T-square distance from tree to neighbor, z_i (m)
1	12.6	8.7
2	9.3	16.4
3	7.5	9.3
4	16.2	12.6
5	8.8	3.5
6	10.1	11.2
7	6.2	13.6
8	1.5	9.1
9	14.3	2.7
10	9.6	8.6
11	11.3	7.9
12	8.9	12.1
13	6.3	15.6
14	13.9	9.9
15	10.8	13.7
16	7.6	8.4
Sum	154.9	163.3
Sum of items squared	1694.93	1885.05
Mean	9.681	10.206

From the usual statistical formulas, we calculate

$$\text{Variance of } (x) = \frac{\Sigma\, x^2 - (\Sigma\, x)^2/n}{n - 1} = \frac{1694.93 - (154.9)^2/16}{15} = 13.020$$

$$\text{Variance of } (z) = \frac{\Sigma\, z^2 - (\Sigma\, z)^2/n}{n - 1} = \frac{1885.05 - (163.3)^2/16}{15} = 14.558$$

$$\text{Covariance of } x \text{ and } z = \frac{\Sigma\, xz - (\Sigma\, x)(\Sigma\, z)/n}{n - 1} = \frac{1543.72 - (154.9)(163.3)/16}{15}$$

$$= -2.4819$$

The density of trees is estimated from equation (5.28) as

$$\hat{N}_T = \frac{n^2}{2\, \Sigma\, (x_i)[\sqrt{2}\, \Sigma\, (z_i)]}$$

$$= \frac{16^2}{2\, (154.9)[\sqrt{2}(163.3)]} = 0.003578 \text{ trees per m}^2$$

Calculate the standard error of the reciprocal of this density estimate from equation (5.29):

$$\text{Standard error}\left(\frac{1}{\hat{N}_T}\right)$$

$$= \sqrt{\frac{8(\bar{z}^2 s_x^2 + 2\bar{x}\bar{z}s_{xz} + \bar{x}^2 s_z^2)}{n}}$$

$$= \sqrt{\frac{8[(10.206)^2(13.02) + 2(9.681)(10.206)(-2.4819) + (9.681)^2(14.558)]}{16}}$$

$$= 33.3927$$

The 95% confidence interval for the reciprocal of this density estimate is thus

$$1/\hat{N}_T \pm t_\alpha[\text{S.E.}(1/\hat{N}_T)] \qquad (t_\alpha = 2.113 \text{ for 15 d.f.})$$

$$279.49 \pm (2.113)(33.3927) \qquad \text{or} \qquad 208.93 \text{ to } 350.05$$

Taking reciprocals, we obtain confidence limits of 2.9×10^{-3} to 4.8×10^{-3} trees/m^2, or 29 to 48 trees per hectare for this small sample.

Program-group DISTANCE METHODS (Appendix 2) does these calculations.

Box 5.3 gives an example of the use of these estimators. Program-group DISTANCE METHODS (Appendix 2) does these calculations.

5.2.3 Ordered Distance Method

The ordered distance method was first suggested by Morisita (1957) and further developed by Pollard (1971). The method involves measuring the distance from a random sampling point to the *n*th closest individual. Pollard (1971) was the first to recognize that the variance of the density estimate falls as *n* increases, so that measurements to the second nearest individual have more precise density estimates than those to the nearest individual. In practice it is difficult in the field to locate individuals beyond the third-nearest, and this is the method we will discuss here. Simulation studies suggest that the third-nearest individual gives satisfactory results for clumped patterns, although the method may also be used for the nearest or second-nearest organism.

The general formula for estimating population density for the ordered distance method for the third-nearest individual is as follows:

$$\hat{D} = \frac{3n - 1}{\pi \Sigma (R_i^2)} \tag{5.30}$$

where $\hat{D}$ = Population density estimated by the ordered distance method
n = Number of random points sampled
π = 3.14159
R_i = Distance from random point to third-nearest organism

The variance of this density estimate is given by

$$\text{Variance}(\hat{D}) = \frac{(\hat{D})^2}{3n - 2} \tag{5.31}$$

and the standard error of the density estimate is the square root of this variance.

Confidence intervals for D can be obtained in the following way when $4n > 30$. The 95% confidence limits are given by Seber (1982, 42) as

$$\text{Lower confidence limit for } \sqrt{\hat{D}} \text{ is } \frac{\sqrt{12n - 1} - 1.96}{\sqrt{4\pi \Sigma (R_i^2)}} \tag{5.32}$$

$$\text{Upper confidence limit for } \sqrt{\hat{D}} \text{ is } \frac{\sqrt{12n - 1} + 1.96}{\sqrt{4\pi \Sigma (R_i^2)}} \tag{5.33}$$

and these limits are then squared to convert them to population densities.

To apply this sampling method, one proceeds as follows:

1. Locate a random point in the study zone.
2. Determine the nearest individual, the second-nearest individual, and the third-nearest individual to the random point. You will need to use exact measurements to determine these if spacing of individuals is close.
3. Measure the distance from the random point to the third individual (R_i).
4. Repeat the entire procedure for the next random point. Try to obtain $n = 30$ to 50.

Note that you do *not* need to measure the distances to the nearest and second-nearest individuals. Only one distance is measured per random point. If you are measuring trees, measure to the center of the tree. If you measure in meters, your density estimate will be in numbers per square meter.

The procedure is illustrated schematically in Figure 5.8 and Box 5.4 gives some sample calculations for the ordered distance method of density estimation.

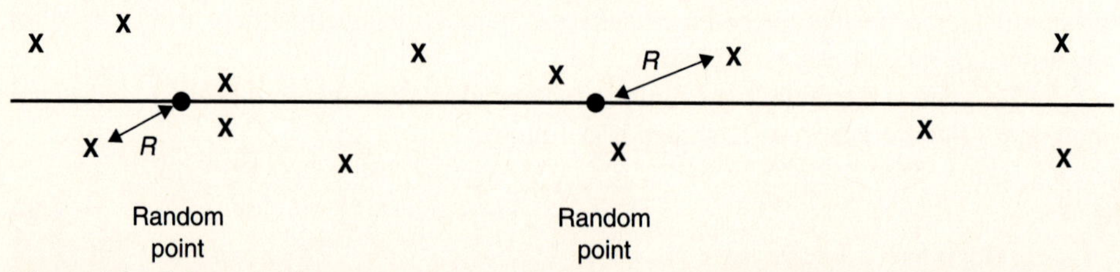

Figure 5.8 Schematic illustration of the ordered distance method of Morisita (1957) utilizing the third-nearest organism. Only one distance is recorded for each random point (•), the distance to the third-nearest organism (R_i). Note that one does not need to measure the distances to the first and second-nearest organisms (unless you need to do so to decide which is the third-nearest organism). Each x represents one organism.

Box 5.4 Estimating Population Density from Ordered Distance Sampling

A forest ecologist used the ordered distance sampling procedure illustrated in Figure 5.8 to determine the regeneration success of loblolly pine trees in Georgia. She measured the following distances from 24 random points to the third-nearest pine tree:

Sample point no.	Point-to-third-nearest tree distance, x_i (m)
1	18.5
2	4.1
3	7.2
4	6.2
5	8.3
6	10.1
7	16.2
8	3.5
9	13.2
10	6.6
11	10.9
12	5.9
13	6.8
14	3.9
15	2.8
16	7.2
17	8.1
18	5.2
19	9.0
20	4.4
21	3.2
22	7.5
23	9.9
24	7.6
Sum	186.3
Sum of items squared	1802.75
Mean	7.7625

Estimate the density of trees from equation (5.30):

$$\hat{D} = \frac{3n - 1}{\pi \sum (R_i^2)} = \frac{(3)(24) - 1}{3.14159(1802.75)} = 0.012536 \text{ trees per m}^2$$

or, expressed per hectare, 125 trees per ha.

The variance of this density estimate is from equation (5.31):

$$\text{Variance}(\hat{D}) = \frac{(\hat{D})^2}{3n - 2} = \frac{0.012536^2}{3(24) - 2} = 2.2451 \times 10^{-7}$$

which gives the standard error of the density as

$$\text{Standard error of } (\hat{D}) = \sqrt{\text{Variance of } (\hat{D})}$$

$$= \sqrt{2.24516 \times 10^{-7}} = 0.001498$$

The 90% confidence interval for this density estimate is obtained from equations (5.32) and (5.33), with a change of the z-value to $z_{.10} = 1.645$ from $z_{.05} = 1.96$:

$$\text{Lower confidence limit for } \sqrt{\hat{D}} = \frac{\sqrt{12n - 1} - 1.645}{\sqrt{4\pi \Sigma (R_i^2)}}$$

$$= \frac{\sqrt{12(24) - 1} - 1.645}{\sqrt{4(3.14159)(1802.75)}} = 0.10162$$

so the lower 90% confidence limit is 0.10161^2, or 0.01033 trees per m^2, or 103 trees per hectare.

$$\text{Upper confidence limit for } \sqrt{\hat{D}} = \frac{\sqrt{12n - 1} + 1.645}{\sqrt{4\pi \Sigma (R_i^2)}}$$

$$= \frac{\sqrt{12(24) - 1} + 1.645}{\sqrt{4(3.14159)(1802.75)}} = 0.12348$$

so the upper 90% confidence limit is 0.12348^2, or 0.015249 trees per m^2, or 152 trees per hectare.

In practice one would like a larger sample size to reduce the width of the confidence band.

Program-group DISTANCE METHODS (Appendix 2) can do these calculations.

5.2.4 Variable-Area Transect Method

This method is a combination of distance and quadrat methods and was first suggested by Parker (1979). A fixed-width strip is searched from a random point until the nth individual is located in the strip. A field worker needs to search in only one direction from the random point, and once the nth individual is found, the length of transect is measured from the random point to the point at which the nth individual occurred. This method can be used for any number of individuals, and we shall use $n = 3$ as a convenient number for field workers.

The formula for population density estimated by the variable area transect methods was derived by Parker (1979) as follows:

$$\hat{D}_v = \frac{3n - 1}{w \, \Sigma (l_i)} \tag{5.34}$$

where $\hat{D}$ = Estimate of population density for the variable-area transect method
n = Number of random points
w = Width of transect searched (fixed)
l_i = Length of transect i searched until the third organism was found

The variance of this estimate of population density is given by Parker as

$$\text{Variance}(\hat{D}_v) = \frac{(\hat{D}_v)^2}{3n - 2} \tag{5.35}$$

Note that this is the same variance formula used in the ordered distance estimator. Confidence limits for population density are given by Parker as

$$\text{Lower 95\% confidence limit} = \frac{C_1}{2wn\bar{l}}$$
$$\text{Upper 95\% confidence limit} = \frac{C_2}{2wn\bar{l}} \tag{5.36}$$

where C_1 = Value from chi-squared distribution at $\alpha = 0.025$ for $6n$ degrees of freedom
C_2 = Value from chi-squared distribution at $\alpha = 0.975$ for $6n$ d.f.
w = Transect width
n = Sample size
$\bar{l}$ = Mean length of transect searched until third organism is located

The variable-area transect method is illustrated in Figure 5.9.

To apply the variable-area transect method, one proceeds as follows:

1. Locate a random point on a transect line in your study area.
2. Move along the transect from the random point until you have found three individuals of the species being studied. Project a perpendicular to the line of travel (marked by a tape measure usually).
3. Measure the distance l along the tape from the random point to the perpendicular line that projects to the third individual.
4. Repeat the procedure until you have a sample of at least 30–50 distances.

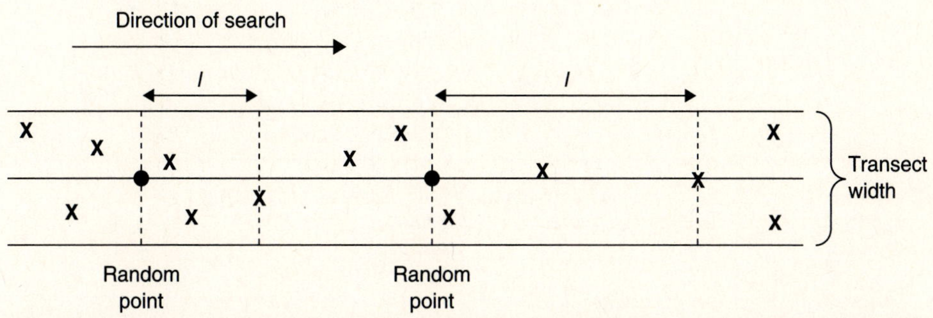

Figure 5.9 Schematic illustration of the variable-area transect method of population estimation. This is a combination of quadrat and distance methods. One moves along a transect line from a random point, counting the organisms within a specified distance from the line, and marks the distance at which the third individual is located. These distances (l_i) are the data used in equation (5.32). Each x indicates one organism.

Density estimates from these plotless sampling methods can be obtained from program-group DISTANCE METHODS (Appendix 2), which includes the confidence limits for the density estimates.

5.2.5 Point-Quarter Method

The classic distance method is the point-quarter method developed by the first land surveyors in the United States in the nineteenth century. The four trees nearest to the corner of each section of land (1 sq. mile) were recorded in the first land surveys and they form a valuable database on the composition of the forests in the eastern United States before much land had been converted to agriculture. The point-quarter technique has been a commonly used distance method in forestry. It was first used in plant ecology by Cottam et al. (1953) and Cottam and Curtis (1956). Figure 5.10 illustrates the technique. A series of random points is selected, often along a transect line, with the constraint that points should not be so close that the same individual is measured at two successive points. The area around each random point is divided into four 90° quadrants, and the distance to the nearest tree is measured in each of the four quadrants. Thus four point-to-organism distances are generated at each random point, and this method is similar to measuring the distances from a random point to the first-, second-, third-, and fourth-nearest neighbors.

The appropriate unbiased estimate of population density for the point-quarter method is from Pollard (1971):

$$\hat{N}_p = \frac{4(4n - 1)}{\pi \sum (r_{ij}^2)} \tag{5.37}$$

where $\hat{N}_p$ = Point-quarter estimate of population density
n = Number of random points
π = 3.14159
r_{ij} = Distance from random point i to the nearest organism in quadrant j ($j = 1, 2, 3, 4$; $i = 1, \ldots n$)

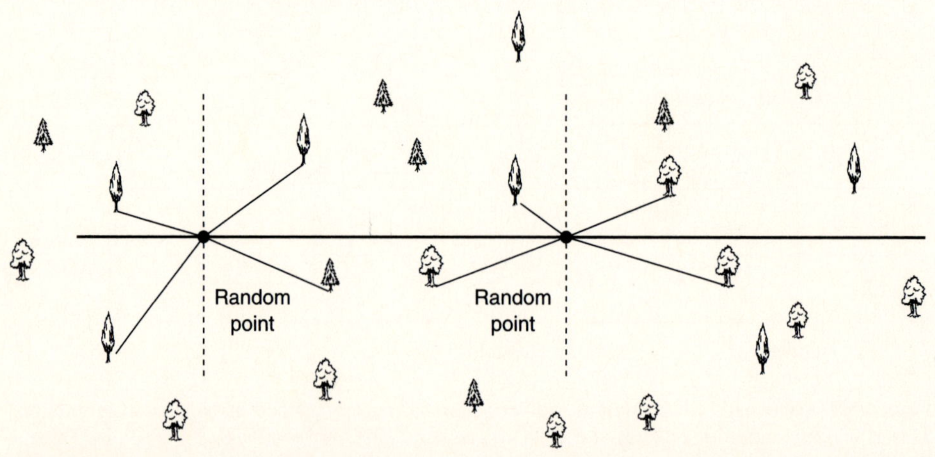

Figure 5.10 Point-quarter method of density estimation. The area around each random point is subdivided into four 90° quadrats, and the nearest organism to the random point is located in each quadrat. Thus four point-to-organism distances are obtained at each random point. This method is commonly used on forest trees. Trees illustrate individual organisms.

The variance of this density estimate is given by Pollard as

$$\text{Variance}(\hat{N}_p) = \frac{\hat{N}_p^2}{4n - 2} \tag{5.38}$$

and the standard error is

$$\text{Standard error of } \hat{N}_p = \sqrt{\frac{\text{Variance of } N_p}{4n}} \tag{5.39}$$

Confidence intervals for N_p can be obtained in the following way when $4n > 30$. The 95% confidence limits are given by Seber (1982, 42) as

$$\text{Lower confidence limit for } \sqrt{\hat{N}_p} = \frac{\sqrt{16n - 1} - 1.96}{\sqrt{\pi \sum (r_{ij}^2)}} \tag{5.40}$$

$$\text{Upper confidence limit for } \sqrt{\hat{N}_p} = \frac{\sqrt{16n - 1} + 1.96}{\sqrt{\pi \sum (r_{ij}^2)}} \tag{5.41}$$

These limits are then squared to convert them to population densities.

Program-group DISTANCE METHODS in Appendix 2 can do these calculations for point-quarter data.

The point-quarter method is very efficient when it is easy to divide the area around the random points into four quadrants accurately, and when random points take a long time to locate in the field. There have been two general criticisms of the point-quarter method. First, since 4 trees are measured at each point, the number of points sampled is often too low to be representative of a large population spread over a large area. Second, the density estimates obtained from the point-quarter method are susceptible to bias if the spatial pattern is not random (Pollard 1971). But recent simulation studies by Engeman et al. (1994) suggest that in general the bias of the point-quarter estimator for clumped patterns is as low as that of the ordered distance estimator and the variable-area transect estimator that they recommend. Table 5.1 summarizes the size of the bias of these estimators in the simulations run by Engeman et al. (1994). It may be better to use the more statistically robust ordered distance sampling procedure (Figure 5.8) or variable-area transect sampling

TABLE 5.1 RELATIVE BIAS OF DENSITY ESTIMATES FOR FIVE DISTANCE METHODS. Simulation studies of five spatial patterns ranging from uniform to random to three types of clumped distributions were run 5000 times for four different sample sizes to obtain these estimates. Relative bias is expressed as a proportion of the true density. Negative bias means that the estimates were too low relative to the true values, and positive bias means that the estimates were too high. (Data from Engeman et al. 1994.)

Estimator	Random pattern	Uniform pattern	Clumped-50 pattern[a]	Clumped-15 pattern[a]	Double clumping[a]
Byth and Ripley	0.12	0.13	0.07	1.05	1.99
T-square	0.28	0.26	0.21	0.52	0.50
Ordered distance	0.02	0.13	−0.09	−0.25	−0.48
Variable-area transect	0.02	0.13	−0.06	−0.21	−0.41
Point-quarter	0.01	0.17	0.06	−0.22	−0.29

[a] The clumped-50 pattern is moderately clumped, the clumped-15 pattern more severely clumped, and the double-clumping pattern is extremely clumped.

(Figure 5.9) in preference to point-quarter sampling, but more field data are needed to test these potential biases when organisms are clumped and not spread in a random pattern. If point-quarter sampling is used, it is important to measure the 90° quadrant boundaries accurately. The difficulty of doing this in the field has been one of the arguments used against the application of this method in routine field surveys (Engeman et al. 1994).

Other possible distance measures are reviewed in Diggle (1983) and Seber (1982). Batcheler (1971, 1975) has developed techniques for estimating population density from distance measurements that utilize empirically derived correction factors. Batcheler (1975) used this approach to estimate deer density from fecal pellet groups in New Zealand. Byth (1982) suggested that these kinds of empirically derived estimates are not robust to many types of nonrandom spatial patterns in natural populations, and thus she recommends avoiding them in favor of more robust estimators like those discussed above.

5.3 SUMMARY

Plotless sampling methods provide another method for estimating the abundance of plants and animals. This chapter discusses two approaches to estimating density that do not involve setting out quadrats or marking individuals. The line transect method utilizes the distances at which organisms are sighted from transect lines to estimate density. Distance methods use distances from random points to the nearest organism and distances from random organisms to their nearest neighbors to estimate density. For all these plotless estimators, sample sizes of the order of 40–60 are needed to achieve good precision.

Line transects can be done on foot, or in land vehicles, or from airplanes, and the critical assumption is that all the organisms directly on the transect line are seen. Detection falls with distance away from the transect line, and the rate at which it falls can be used to estimate the effective width of strip counted along the transect. The critical problem is specifying the form of the detection function, and many different mathematical functions can be used. If ecological constraints can be set on the shape of the detection function, more precise estimates of density can be obtained.

Distance measures have been used extensively in plant ecology. Five methods are described that utilize distances from random points to the nearest organism and distances from random individuals to their nearest neighbors to estimate density. While random points are easy to locate, the practical problems associated with sampling random individuals are considerable. When organisms are randomly spread in space, there are few problems. But when the spatial pattern is clumped, most distance measures are negatively biased. Three sampling procedures are available that successfully overcome many of these problems. The ordered distance method, the variable-area plot method, and the point-quarter sampling procedure are relatively easy to apply in the field and statistically robust. When patchiness occurs, distance methods typically underestimate the true density, but for the better methods this bias is relatively small.

SELECTED READING

Bonnell, M. L., and Ford, R. G. 1987. California sea lion distribution: A statistical analysis of aerial transect data. *Journal of Wildlife Management* 51: 13–19.

Buckland, S. T., Anderson, D. R., Burnham, K. P., and Laake, J. L. 1993. *Distance Sampling*. London: Chapman and Hall, Chaps. 1, 7.

Burnham, K. P., and Anderson, D. R. 1984. The need for distance data in transect counts. *Journal of Wildlife Management* 48: 1248–1254.

Engeman, R. M., R. T. Sugihara, L. F. Pank, and W. E. Dusenberry. 1994. A comparison of plotless density estimators using Monte Carlo simulation. *Ecology* 75: 1769–1779.

Greig-Smith, P. 1979. Pattern in vegetation. *Journal of Ecology* 67: 755–779.

Healy, W. M., and Welsh, C. J. E. 1992. Evaluating line transects to monitor gray squirrel populations. *Wildlife Society Bulletin* 20: 83–90.

Ripley, B. D. 1981. *Spatial Statistics*. New York: John Wiley and Sons.

Seber, G. A. F. 1982. *The Estimation of Animal Abundance and Related Parameters*. 2d ed. London: Griffin, Chaps. 7, 9.

Wywialowski, A. P., and Stoddart, L. C. 1988. Estimation of jackrabbit density: methodology makes a difference. *Journal of Wildlife Management* 52: 57–59.

QUESTIONS AND PROBLEMS

5.1. Duck nests were searched for on foot by walking transect lines on the Monte Vista National Wildlife Refuge in Colorado. The strip width was limited to 3.66 m (12 ft). In 1986, 102 duck nests were found with the following distribution of perpendicular distances to the transect lines:

Perpendicular distance class (ft.)	Number of nests found
0–1	10
1–2	4
2–3	14
3–4	14
4–5	10
5–6	10
6–7	11
7–8	6
8–9	8
9–10	4
10–11	7
11–12	4

Discuss the shape of the detection function for these data and suggest possible improvements in how these surveys might be carried out.

5.2. Discuss the possibility of applying line transect methods of analysis to data collected by aerial survey. What factors might limit the use of line transect methods from aircraft? What advantages would accrue if you could use line transects for aerial surveys? Read Burnham and Anderson (1984) and discuss the application of their recommendations.

5.3. The following data were gathered from a 160 m line transect for pheasants in southern England. Strip width (w) was 65 m.

Observation no.	Sighting distance, x_i	Sighting angle, θ
1	27.6 m	46°
2	25.2	27
3	16.2	8
4	24.7	31
5	44.4	42
6	48.0	28
7	13.1	2
8	6.9	18
9	23.5	48
10	5.3	29
11	14.9	36
12	23.7	31
13	36.7	56
14	10.9	68
15	24.1	0
16	61.8	46
17	27.6	22
18	8.3	18
19	16.2	27
20	25.2	34

Calculate the Hayne estimate of population density for these data and compare the resulting confidence limits with those obtained using the Modified Hayne estimator and the Fourier series estimator.

5.4. Calculate an estimate of density from the variable-area transect method for ground squirrel burrows counted along a transect of 7 m total width in which the following distances were measured from the start of the random quadrat to the perpendicular at which the third burrow was located: (n = 30)

6.7, 19.6, 32.2, 7.2, 12.4, 11.1, 27.8, 12.4, 16.4, 8.9, 19.0, 12.1, 19.5, 23.3, 17.9, 12.3, 18.0, 24.7, 21.8, 28.0, 12.2, 8.4, 19.2, 13.0, 26.2, 21.8, 13.9, 14.7, 37.7, and 24.5 m.

Estimate 90% confidence limits for the density of burrows.

5.5. Point-quarter sampling was done on Amabilis fir in the Coast Range of British Columbia by an ecology class, with the following results: (n = 10 random points)

	Distance from point to tree (m)			
Point	Quadrat 1	Quadrat 2	Quadrat 3	Quadrat 4
1	3.05	4.68	9.15	7.88
2	2.61	12.44	19.21	3.87
3	9.83	5.41	7.55	11.16
4	7.41	9.66	1.07	3.93
5	1.42	7.75	3.48	1.88
6	8.86	11.81	6.95	7.32
7	12.35	9.00	8.41	3.16
8	10.18	3.16	7.14	2.73
9	3.49	5.70	9.12	8.37
10	5.88	4.15	13.95	7.10

Estimate population density for these trees and calculate a 95% confidence interval for your estimate. What assumptions must you make to do these estimates?

5.6 Plan a study to estimate the density of grizzly bears in a national park. Consider mark-recapture, quadrat-based, and plotless-type methods of estimating density, and list their potential advantages and disadvantages.

PART TWO

Spatial Pattern in Animal and Plant Populations

Organisms are not spread at random across the landscape, and one important question of landscape ecology concerns the pattern of individuals in space. This question is important in its own right, since for zoologists spacing behavior has been a central problem in behavioral ecology, and for botanists the investigation of pattern has played a prominent role in the methods for the study of plants as individuals.

The investigation of spatial pattern has become a specialized area of study in plant ecology, and an imposing array of methods has been developed over the last 50 years (Dale 1998). In this section we introduce only a few of the methods that can be used, and interested students are encouraged to go to the more specialized literature. Similarly, animal ecologists have developed a large literature on home range estimation and territoriality in animals that we can only touch on here.

Our primary focus is on spatial pattern from two perspectives. First, we wish to determine what spatial pattern a population shows in the field, because that pattern has consequences for the estimation of abundance and the construction of confidence intervals for these estimates. In addition, if we are to census a population efficiently, the design of our sampling program will be affected by the spatial pattern of the organisms under study. These are practical matters that address statistical problems of estimation. Second, we need to develop a set of metrics that will provide a measure of spatial pattern. If one population of shrubs has a more clumped pattern than another population, how can we measure the degree of clumping? The metrics of spatial pattern are usually referred to as *indices of dispersion*, and there is a large literature on how to measure quantitatively spatial pattern or, more directly, which index of dispersion is best. We will approach this problem with a quantitative outlook in this section, but remember that the important part of spatial pattern is the ecology behind it, the mechanisms that generate and maintain spatial patterns. We will not deal with these mechanisms in this book, but only with the foundation of the metrics of measuring spatial pattern efficiently and accurately.

CHAPTER 6

Spatial Pattern and Indices of Dispersion

Organisms may form random, uniform, or clumped spatial patterns in nature, and the first question we must ask is how we can decide statistically which of these patterns is a good description for our particular population. We have already addressed this question for the case of quadrat sampling in Chapter 4, and in this chapter we carry this discussion forward with more sophisticated measures for spatial pattern.

Two situations separate the methods for pattern detection in plant and animal populations in the field. First, we may have a complete spatial map of the population we are studying. This is an ideal situation, because we can apply an array of sophisticated mathematical methods to such spatial maps (Diggle 1983). When we have a spatial map, we know the density of the population because it is completely enumerated. Spatial maps can be constructed for plants in quadrats, for bird territories in forests or grasslands, or for nest sites of geese. Many hypotheses about ecological processes can be tested with spatial maps, and much effort has been put into obtaining data to construct them for many populations of plants and animals.

Second, we may be interested in broad-scale sampling of populations that cannot be completely mapped and enumerated in order to quantify pattern. Forest ecologists typically deal with extensive stands of trees, and yet need information on spatial pattern both to test ecological hypotheses and to estimate abundance efficiently. Many of the methods discussed in the last three chapters can be used to quantify spatial patterns, and in addition an array of new methods based on contiguous quadrats can be utilized to measure spatial pattern.

6.1 METHODS FOR SPATIAL MAPS

6.1.1 Nearest-Neighbor Methods

In some situations an ecologist may have an exact map of the geographical location of each individual organism. Figure 6.1 illustrates this type of spatial data for redwood seedlings in California. Such data represent a complete enumeration of the population being studied; thus population density is *known* and not estimated.* Spatial maps like Figure 6.1 can be used to measure the pattern of a population.

Clark and Evans (1954) were the first to suggest a method for analyzing pattern for spatial maps. They used the distance from an individual to its nearest neighbor (see Figure 5.5: A→B) as the relevant measure. These distances are measured for *all* individuals in the spatial map, and thus there are no problems with random sampling because it is a complete enumeration of this local population. The nearest neighbor of an individual within the study area can be an individual located outside of the study area (see Figure 5.1). For this simple situation we have

$$\bar{r}_A = \text{Mean distance to the nearest neighbor} = \frac{\sum r_i}{n} \tag{6.1}$$

where r_i = Distance to nearest neighbor for individual i
n = Number of individuals in study area

Clark and Evans (1954) showed that the expected distance to the nearest neighbor could be calculated very simply for a large population that has a random spatial pattern. Define:

$$\rho = \text{Density of organisms} = \frac{\text{Number in study area}}{\text{Size of study area}}$$

Then

$$\bar{r}_E = \text{Expected distance to nearest neighbor} = \frac{1}{2\sqrt{\rho}} \tag{6.2}$$

*If, on the other hand, there are several quadrats like that shown in Figure 6.1, one is dealing with quadrat counts; see Chapter 4.

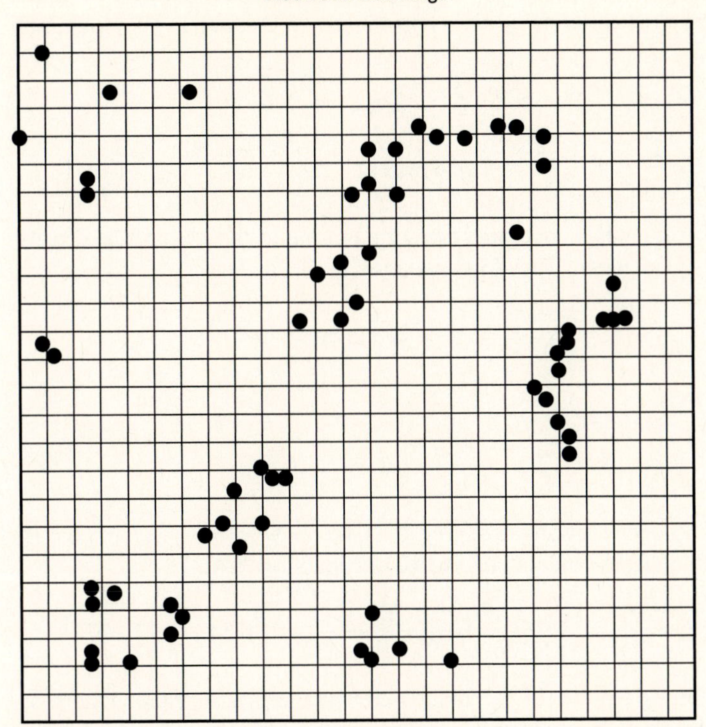

Figure 6.1 Spatial map of the locations of 62 redwood tree seedlings (*Sequoia sempervirens*) in a 23 × 23 m quadrat in California. (Data from Strauss 1975.)

We can measure the deviation of the observed pattern from the expected random pattern by the following ratio:

$$R = \frac{\bar{r}_A}{\bar{r}_E} = \text{Index of aggregation} \tag{6.3}$$

If the spatial pattern is random, $R = 1$. When clumping occurs, R approaches zero; in a regular pattern, R approaches an upper limit around 2.15.

A simple test of significance for deviation from randomness is available, because the standard error of the expected distance is known exactly from plane geometry:

$$z = \frac{\bar{r}_A - \bar{r}_E}{s_r} \tag{6.4}$$

where z = Standard normal deviate

s_r = Standard error of the expected distance to nearest neighbor

$= \dfrac{0.26136}{\sqrt{n\rho}}$

n = Number of individuals in study area

ρ = Density of individuals in study area

Consider an example. Campbell and Clarke (1971) measured nearest-neighbor distances for 39 singing crickets (*Teleogryllus commodus*) in Australia. They obtained these data:

$$n = 39 \qquad \Sigma\, r_i = 63.4 \qquad \Sigma (r_i)^2 = 136.78$$

$$\bar{r}_A = \frac{\Sigma\, \bar{r}_i}{n} = \frac{63.4}{39} = 1.626 \text{ meters}$$

For this population of 39 animals in a circular area of 12 m radius (452.4 m^2) the density is 0.08621 crickets per square meter. Thus from equation (6.2),

$$\bar{r}_E = \frac{1}{2\sqrt{0.08621}} = 1.703 \text{ meters}$$

The index of aggregation is

$$R = \frac{\bar{r}_A}{\bar{r}_E} = \frac{1.626}{1.703} = 0.95$$

which indicates a slight tendency toward an aggregated pattern away from randomness. To test for significant deviation from random pattern, we use equation (6.4):

$$z = \frac{\bar{r}_A - \bar{r}_E}{s_r} = \frac{1.626 - 1.703}{0.26136/\sqrt{(39)(0.08621)}} = -0.54$$

Since $|z|$ is less than 1.96, we tentatively accept the null hypothesis at $\alpha = 0.05$ that these crickets are randomly spaced in the study area.

The Clark and Evans test (equation [6.4]) is unbiased only when a boundary strip is included in the study, but ecologists often do not use boundary strips because they wish to maximize sample size (e.g., Figure 6.1). If the Clark and Evans test is applied in this situation without using a boundary strip, the test is biased. Organisms close to the boundary will tend to have larger nearest-neighbor distances than those well inside (e.g., individual C in Figure 5.5). Sinclair (1985) showed that the uncorrected Clark and Evans test is biased in favor of regular patterns, so that many aggregated patterns are judged to be random, and many random patterns are judged to be uniform. This bias is enormous with small sample sizes ($n < 100$).

If a boundary strip is *not* included in the study, you should use the Donnelly (1978) modification of the Clark and Evans test as follows:

$$\bar{r}_c = \text{Expected distance to nearest neighbor corrected for lack of a boundary strip} \qquad (6.5)$$

$$= \bar{r}_E + \left[\left(0.051 + \frac{0.041}{\sqrt{n}}\right)\left(\frac{L}{n}\right)\right]$$

where L = Length of the boundary of the whole study area

The standard error of this expected distance is given by Donnelly as

$$s_r = \frac{\sqrt{0.07A + \left(0.037L\sqrt{\frac{A}{n}}\right)}}{n} \qquad (6.6)$$

where A = Area of study zone
L = Length of boundary of study zone
n = Number of individuals in study zone

These values are used in the z-test (equation [6.4]) in the same manner as for the Clark and Evans test. Donnelly (1978) suggests that this z-test is unbiased if $n > 7$ and the study area has a smooth boundary like a square or circle. It is not recommended if the study zone is a long, thin rectangle because then the edge effects are overwhelming.

Box 6.1 illustrates the use of the Donnelly modification of the Clark and Evans test for spatial pattern of tropical ant nests.

6.1.2 Distances to Second–*n*th Nearest Neighbors

If distances to nearest neighbors are useful in revealing spatial patterns in populations, additional information might be gained by measuring the distance to the second, third, fourth and nth nearest neighbors. In practice it is difficult in the field to determine easily which is the third, fourth, and fifth nearest neighbor, and no one seems to have gone beyond the fifth nearest neighbor. Thompson (1956) showed that the Clark and Evans approach could be easily generalized to the second, third, . . . nth nearest neighbors. Table 6.1 gives the expected distances and their standard errors for the first to fifth nearest neighbors. Two additional tests are available for testing spatial pattern in nearest-neighbor data.

Thompson's Test Thompson (1956) suggested a chi-square test for evaluating the null hypothesis of random spatial patterning. This test can be applied to nearest-neighbor data, as discussed above for the Clark and Evans test, or for more general studies of second

Box 6.1 Donnelly Modification of Clark and Evans Test for Spatial Pattern

Ant nests are typically thought by ecologists to be uniformly spaced in both tropical and temperate ant communities. Levings and Franks (1982) reported the following data from ground ants (*Ectatomma ruidum*) from one study area on Barro Colorado Island in Panama:

Study area = 100 m^2 n = 30 nests L = 40 m boundary length

Mean distance to nearest nest = 1.0863 m

They did not include a boundary strip around their plots, and we will use these data to illustrate the errors this introduces into the test of the null hypothesis of random spatial pattern of these ant nests.

1. Clark and Evans test: from equation (6.2)

$\bar{r}_E$ = Expected distance to nearest neighbor if random pattern

$$= \frac{1}{2\sqrt{\rho}} = \frac{1}{2\sqrt{.30}} = 0.91287 \text{ meters}$$

The index of aggregation is thus, from equation (6.3),

$$R = \frac{\bar{r}_A}{\bar{r}_E} = \frac{1.0863}{0.91287} = 1.19$$

which suggests a tendency toward a regular pattern of nest spacing. To test for a significant deviation from the expected random pattern, we use equation (6.4):

$$z = \frac{\bar{r}_A - \bar{r}_E}{s_r} = \frac{1.0863 - 0.91287}{(0.26136/\sqrt{(30)(0.30)})} = 1.99$$

The value of z has probability 0.046, and consequently (at $\alpha = 0.05$) we reject the null hypothesis of a random spatial pattern for these ant nests in favor of a uniform pattern. This is an error because a boundary strip was not included in the study, and the correct statistical test is given next.

2. Donnelly modification of Clark and Evans test: from equation (6.5),

$\bar{r}_c$ = Expected distance to nearest neighbor corrected for lack of a boundary strip

$$= \bar{r}_E + \left[\left(0.051 + \frac{0.041}{\sqrt{n}}\right)\left(\frac{L}{n}\right)\right]$$

$$= 0.91287 + \left[\left(0.051 + \frac{0.041}{\sqrt{30}}\right)\left(\frac{40}{30}\right)\right] = 0.99085$$

The standard error of this expected distance is given by equation (6.6):

$$s_r = \frac{\sqrt{0.07A + \left(0.037L\sqrt{\frac{A}{n}}\right)}}{n}$$

$$= \frac{\sqrt{0.07(100) + \left(0.037(40)\sqrt{\frac{100}{30}}\right)}}{30} = 0.10383$$

We use the same test as above (equation [6.4]):

$$z = \frac{\bar{r}_A - \bar{r}_E}{s_r} = \frac{1.0863 - 0.99085}{0.10383} = 0.92$$

This value of z has probability 0.36 under the null hypothesis of a random pattern, and we tentatively accept the null hypothesis.

The Clarke and Evans test, if used without a boundary strip, biases the test for random patterns in the direction of uniform or regular patterns, as this example illustrates. In this case we have no evidence that these ant nests are spaced in a uniform pattern, a pattern expected by most ant ecologists.

TABLE 6.1 EXPECTED DISTANCES TO FIRST, SECOND, . . . , FIFTH NEAREST NEIGHBORS AND ASSOCIATED STANDARD ERRORS FOR A LARGE POPULATION WITH A RANDOM PATTERN[a]

Parameter	Nearest neighbor				
	First	Second	Third	Fourth	Fifth
Expected mean distance	$\frac{0.5000}{\sqrt{\rho}}$[b]	$\frac{0.7500}{\sqrt{\rho}}$	$\frac{0.9375}{\sqrt{\rho}}$	$\frac{1.0937}{\sqrt{\rho}}$	$\frac{1.2305}{\sqrt{\rho}}$
Standard error of expected distance	$\frac{0.2614}{\sqrt{n\rho}}$	$\frac{0.2723}{\sqrt{n\rho}}$	$\frac{0.2757}{\sqrt{n\rho}}$	$\frac{0.2774}{\sqrt{n\rho}}$	$\frac{0.2821}{\sqrt{n\rho}}$

Source: After Thompson 1956.

[a] A boundary strip as in Figure 5.5 is assumed.

[b] ρ = Population density = Number of individuals/Area of study = n/A

nearest neighbor, third nearest neighbor and so on, and every one of these tests is independent. It is possible for nearest neighbors to be uniformly spaced, while second nearest neighbors may be clumped in their distribution. Thompson's test is given by

$$\chi^2 = 2\pi\rho \sum_{i=1}^{n} (r_i^2) \qquad (\text{d.f.} = 2nk) \tag{6.7}$$

where ρ = Population density on study area = $\frac{n}{A}$

r_i = Distance to kth nearest neighbor for individual i

k = Rank of neighbors being measured ($k = 1$ for nearest neighbor; $k = 2$ for second nearest neighbor, etc.)

n = Number of individuals measured

π = 3.14159

For example, for the cricket data given on page 194 (with $\Sigma\, r_i^2 = 136.78$),

$$\chi^2 = 2(3.14159)(0.08621)(136.78)$$
$$= 74.09 \qquad [\text{d.f.} = 2(39)(1) = 78]$$

This value of χ^2 must be tested against two alternatives: a significantly *small* value of χ^2 indicates a *clumped* pattern, and a significantly *large* value of χ^2 indicates a *uniform* pattern. For this example at $\alpha = .05$, we have two decision rules:

1. If observed χ^2 is less than $\chi^2_{.975}$ (59.0 in this example), we have evidence of a clumped pattern.
2. If observed χ^2 is greater than $\chi^2_{.025}$ (99.6 in this example), we have evidence of a uniform pattern.

In this example neither is true, and the cricket data are consistent with random patterning.

For large numbers of degrees of freedom, the usual chi-square test tables are not adequate, and we must use the normal approximation,

$$z = \sqrt{2\chi^2} - \sqrt{4nk - 1} \tag{6.8}$$

where n and k are as defined above, and z is the standard normal deviate. Negative values of z indicate a tendency to aggregation, and positive values indicate a tendency toward regularity of pattern. For this cricket example,

$$z = \sqrt{2(74.09)} - \sqrt{4(39)(1) - 1}$$

$$= -0.28$$

which is not significant ($p > .78$), and thus we tentatively accept the null hypothesis of a random pattern.

If distances to the second, third, . . . nth nearest neighbor are available, Thompson's test is applied n times. It is possible that nearest neighbors are uniformly distributed but that fifth nearest neighbors are clumped, so each test must be done independently.

Thompson's test is not as powerful as the following test if sample size is large ($n > 50$).

Goodness-of-Fit Test Campbell and Clark (1971) proposed a chi-square goodness-of-fit test to compare the observed and expected distribution of nearest-neighbor distances. To use this test, you must first group the observed measurements of nearest-neighbor distances into a frequency distribution (see Box 6.2 for an illustration). The class limits for the frequency distribution must be chosen in advance, and this decision is arbitrary. Since the chi-square test requires an expected frequency in each cell of 3, the class limits picked should not be too narrow. A useful range would be to have about 5–10 classes in your frequency distribution, and this will require a large sample size of 50–100 measurements of nearest-neighbor distances. The expected frequency in each class of distances can be estimated by a two-step procedure:

Step 1: Estimate the cumulative probability for distances from 0 to r_i (where r_i is the upper class limit for each class):

$$F_x = 1 - e^{-x/2} \tag{6.9}$$

where $x = 2\pi\rho r_i^2$
ρ = Population density
F_x = Probability of obtaining a distance in the range $0 - r_i$

For example, if the first class of nearest-neighbor distances ranges from 0 to 0.85 m and the second class from 0.85 to 1.65 m, then for the cricket data used above where density is 0.08621,

For class 1: $x = 2(3.14159)(0.08621)(0.85^2) = 0.39136$

$$F_x = 1 - e^{-(0.39/2)} = 0.178$$

For class 2: $x = 2(3.14159)(0.08621)(1.65^2) = 1.4747$

$$F_x = 1 - e^{-(1.47/2)} = 0.522$$

Calculate F_x values for all the classes in the frequency distribution.

Box 6.2 Campbell and Clarke Test of Goodness-of-Fit for Detecting Spatial Patterning in Known Populations

Nearest-neighbor distances were measured in a low-density cricket population in Australia by Campbell and Clarke (1971). They obtained these data on March 15, 1968, on a totally enumerated population: $n = 51$, $\Sigma r^2 = 4401.05$, $\bar{r}_A = 7.32$ m. Their distances ranged from 0.02 m to 23 m, and they divided this range into 7 classes in intervals of 3.5 meters and tallied their raw data as follows:

Distance to nearest neighbor-class limits (m)	Observed frequency
0–3.55	15
3.56–7.05	16
7.06–10.55	6
10.56–14.05	9
14.06–17.55	1
17.56–21.05	1
21.06–∞	3

The study area was 17,800 m^2, and hence the known density was 51/17,800, or 0.002865 crickets/m^2.

1. Determine the expected frequency for these cricket data using equation (6.9) to fill in this table:

Class no.	Upper class limit (m)	x	$e^{-x/2}$	Cumulative probability, $F_x = 1 - e^{-x/2}$
1	3.55	0.2269	0.8928	0.1072
2	7.05	0.8948	0.6393	0.3607
3	10.55	2.0037	0.3672	0.6328
4	14.05	3.5537	0.1692	0.8308
5	17.55	5.5448	0.0625	0.9375
6	21.05	7.9769	0.0185	0.9815
7	∞	∞	0.0	1.0000

The density of crickets (ρ) was 51 individuals on 17,800 m^2 or 2.865×10^{-3} individuals/m^2.

For class 1,

$$x = 2\pi\rho r_1^2 = 2(3.14159)(0.002865)(3.55^2)$$
$$= 0.2269$$

and

$$F_x = 1 - e^{-X/2} = 1 - (2.718^{-0.1134})$$
$$= 0.1072$$

For class 2,

$$x = 2(3.14159)(0.002865)(7.05^2)$$
$$= 0.8948$$

and

$$F_x = 1 - (2.718^{-0.4474})$$
$$= 0.3607$$

We proceed similarly for classes 3–7, with the results shown in the above table.

2. We calculate the expected frequencies from the sample size ($n = 51$) and the formula

$$\left\{\begin{matrix}\text{Expected relative}\\ \text{frequency in}\\ \text{class } x+1\end{matrix}\right\} = \left(\begin{matrix}\text{Cumulative}\\ \text{probability in}\\ \text{class } x+1\end{matrix}\right) - \left(\begin{matrix}\text{Cumulative}\\ \text{probability in}\\ \text{class } x\end{matrix}\right)$$

with the expected relative frequency in class 1 being equal to F_x for class 1.

Class no.	Cumulative probability F_x	Expected relative frequency	Expected no. of distances
1	0.1072	0.1072	5.47
2	0.3607	0.2535	12.93
3	0.6328	0.2721	13.88
4	0.8308	0.1980	10.10
5	0.9375	0.1067	5.44
6	0.9815	0.0440	2.24
7	1.000	0.0185	0.95
	Total	1.0000	51.00

The expected relative frequencies must sum to 1.00 if you have done the calculations correctly.

To get the final expected number of distances, use the following formula:

$$\left\{\begin{matrix}\text{Expected number}\\ \text{of distances}\end{matrix}\right\} = \left(\begin{matrix}\text{Expected relative}\\ \text{frequency}\end{matrix}\right)\left(\begin{matrix}\text{Sample}\\ \text{size}\end{matrix}\right)$$

For example, with $n = 51$,

$$\left\{\begin{matrix}\text{Expected number of distances}\\ \text{in class 1}\end{matrix}\right\} = (0.1072)(51) = 5.47$$

$$\left\{\begin{matrix}\text{Expected number of distances}\\ \text{in class 2}\end{matrix}\right\} = (0.2535)(51) = 12.93$$

and so on for the other classes.

We can now compute the chi-square test statistic:

$$\chi^2 = \sum \frac{(\text{Observed} - \text{expected})^2}{\text{Expected}}$$

taking care to combine the last two classes so that the expected value is above 3.

Class (m)	Observed no. of crickets	Expected no.	$\frac{(O - E)^2}{E}$
0–3.55	15	5.47	16.60
3.55–7.05	16	12.93	0.73
7.05–10.55	6	13.88	4.47
10.55–14.05	9	10.10	0.12
14.05–17.55	1	5.44	3.62
17.55–21.05	1	2.24	0.21
21.05–∞	3	0.95	
		Total	$\chi^2 = 25.75$

with (6 − 1) degrees of freedom. Since the critical value of χ^2 is 15.09 for 5 d.f., this value is highly significant, and we reject the null hypothesis of random spacing. Since there are more crickets at short nearest-neighbor distances than expected, the data suggest a clumped distribution.

The nearest-neighbor subprogram in program-group SPATIAL PATTERN (Appendix 2) will do these calculations from the raw data of nearest-neighbor distances.

Step 2: Estimate the expected frequency in each class. These expected values can be obtained directly by subtraction as follows. Thus class 1 should contain 17.8% of the observed distances and class 2 should contain (52.2 − 17.8%), or 34.4% of the observed distances. With a sample size of 39 in this example, we should expect (39)(17.8%) or 6.94 distances to fall in the 0–0.85 m category and (39)(34.4%) or 13.42 distances to fall in the 0.85–1.65 m category. These expected frequencies should be calculated for all classes in the frequency distribution.

The goodness-of-fit test follows directly from these calculations and is of the usual form:

$$\chi^2 = \sum \frac{(\text{Observed} - \text{expected})^2}{\text{Expected}} \quad (\text{d.f.} = \text{no. of classes} - 1) \tag{6.10}$$

The only problem is to make sure to combine classes in the tails of the frequency distribution so that the expected value in each class is at least 3. Box 6.2 illustrates these calculations and program-group SPATIAL PATTERN in Appendix 2 can do these calculations.

6.1.3 More Sophisticated Techniques for Spatial Maps

If an ecologist has a spatial map of a population (as in Figure 6.1) it is possible to achieve a much more detailed description of the map than one obtains with nearest-neighbor data. In most cases special stochastic models are fitted to map data, and such models are at a level of mathematical complexity that is beyond the scope of this book. I sketch here only the outlines of this approach; see Ripley (1981), Diggle (1983), and Perry (1995a) for further details.

A spatial map of the locations of animals or plants can be represented by points on a plane, as in Figure 6.2. We can assign to each organism a "territory" that includes all the space that is closer to this individual than it is to any other individual. These "territories" should not be confused with the home ranges or territories of vertebrates; they are perhaps better thought of in terms of plants competing for soil water or soil nutrients. This mathematical construction of territories is called the *Dirichlet tessellation* of the organisms in the study area and is illustrated in Figure 6.2.

Individuals that share a common boundary of their territories can be connected with lines that are the equivalent of the nearest-neighbor distances discussed above. These lines connect to form a series of triangles called the *Delaunay triangulation*, as shown in Figure 6.2. The computation of the Dirichlet tessellation becomes much more complicated as the number of individuals in the study area increases. We shall not attempt these cal-

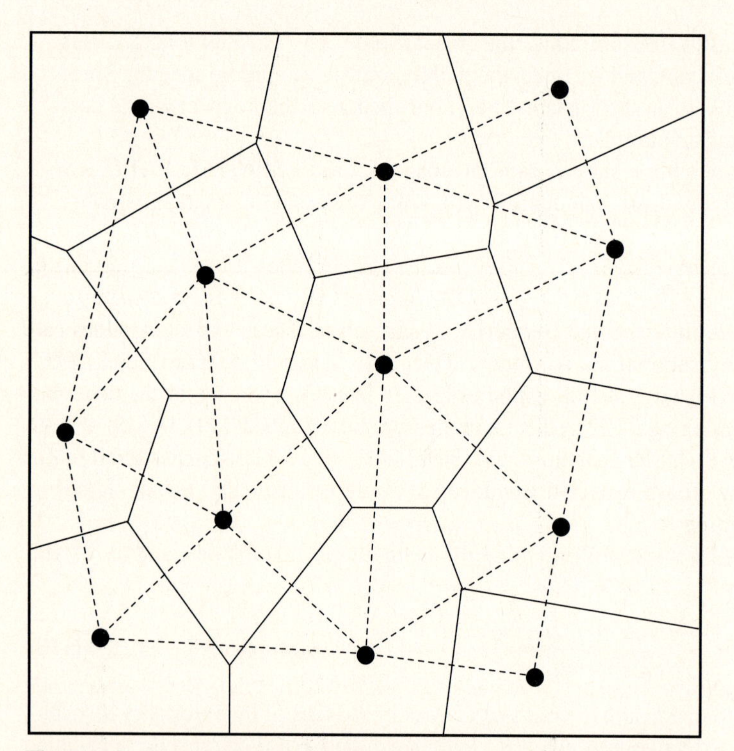

Figure 6.2 Hypothetical spatial map of a population of 12 organisms in a study area. The *Dirichlet tessellation* (solid lines) assigns a "territory" to each organism that contains all the space that is closer to that organism than it is to any other. The resulting polygons are shown. Lines that join all individuals with a common boundary to their territory define a set of triangles called the *Delaunay triangulation* (dashed lines). The sides of these triangles are a set of nearest-neighbor distances.

culations in this book, and interested students should refer to Diggle (1983) and Perry (1995a).

With a spatial map like that of Figure 6.2 we can compute *all* of the distances between each individual and every other individual in the study area. Using Monte Carlo simulations with a computer, we can compare the observed distances to those expected under complete spatial randomness (see Diggle 1983, 11–16). We can also locate random points on the map, measure point-to-nearest-organism distances, and compare these with computer-generated expected values. There is an extensive literature on the analysis of mapped patterns and the fitting of stochastic models to such data. Consult Pielou (1977), Diggle (1983), and Ripley (1981) for the details of this approach, which moves fairly rapidly into difficult mathematics.

The analysis of spatial maps becomes even more complex when the study area is not spatially homogenous. For example, birds nesting in isolated patches of forest often position their nests near the edge of the wood. If small blocks of woodland are patchily distributed, the nearest neighbor of a nest in one wood might be another nest in the next patch of forest. Edge effects in patchy environments can cause severe biases in nearest-neighbor methods, and the commonsense notion of discarding all nearest-neighbor distances that cross habitat boundaries is not a solution (Ripley 1985). One possible method of approach useful in patchy environments is outlined in Ripley (1985), but the problem awaits more theoretical analysis.

6.2 CONTIGUOUS QUADRATS

Quadrats are widely used in plant ecology to analyze the spatial patterning of vegetation, as we have already seen in Chapter 4. Here we introduce a new type of quadrat sampling based on adjacent or contiguous quadrats. In this case a series of contiguous quadrats are typically counted, as illustrated in Figure 6.3. These quadrats are clearly not independent random samples from the population, and consequently they cannot be used to estimate population density, percent cover, or any other measure of abundance. They are best viewed as a statistical universe about which questions of pattern may be asked, and the major operational question is where to start the quadrats.

There are two main methods of analyzing data from contiguous quadrats; they are discussed in detail by Dale (1998) and by Ludwig and Reynolds (1988). I will present here only the simplest method, the blocked-quadrat variance method developed by Hill (1973). The specific methods developed by Hill (1973) have been given the acronym TTLQV (two-term local quadrat variance) and are applied to data gathered from a series of contiguous quadrats like those shown in Figure 6.3. Blocks are defined by adding together adjacent quadrat data as follows:

Block size	Quadrats
1	(1) (2) (3) (4) (5) (6) (7) (8)
2	(1, 2) (3, 4) (5, 6) (7, 8)
3	(1, 2, 3) (4, 5, 6)
4	(1, 2, 3, 4) (5, 6, 7, 8)

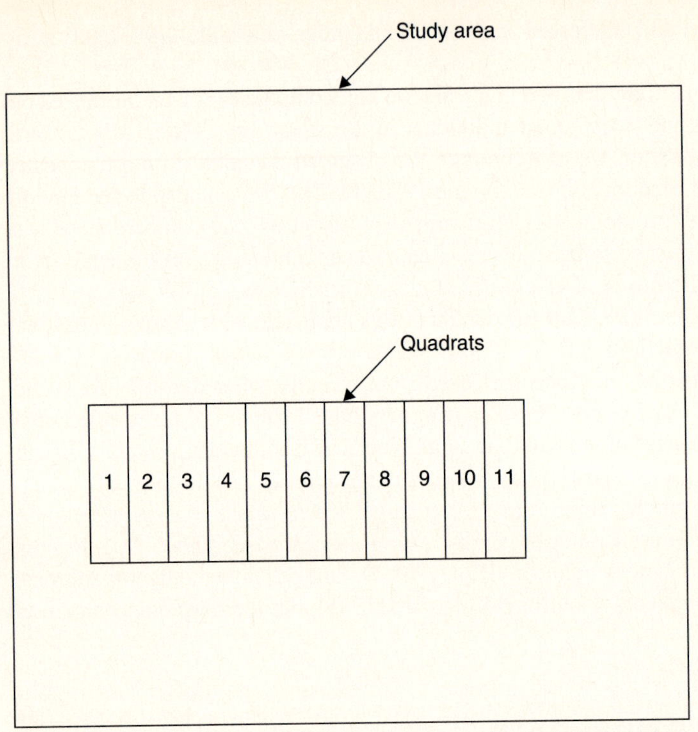

Figure 6.3 Schematic illustration of a series of 11 contiguous quadrats for the analysis of statistical pattern in plant communities. Such quadrats are not independent and cannot be used to estimate abundance or cover, but they can be used to determine spatial pattern.

Given this grouping of data into block sizes, we proceed to calculate the variances for each block size, as follows:

$$\text{Var}_1(X) = \left(\frac{1}{n-1}\right)\left[\left(\frac{(x_1 - x_2)^2}{2}\right) + \left(\frac{(x_2 - x_3)^2}{2}\right) + \cdots + \left(\frac{(x_{n-1} - x_n)^2}{2}\right)\right] \tag{6.11}$$

$$\text{Var}_2(X) = \left(\frac{1}{n-3}\right)\left[\left(\frac{(x_1 + x_2 - x_3 - x_4)^2}{4}\right) + \left(\frac{(x_2 + x_3 - x_4 - x_5)^2}{4}\right) + \cdots \right. \\ \left. \cdots + \left(\frac{(x_{n-3} + x_{n-2} - x_{n-1} - x_n)^2}{4}\right)\right] \tag{6.12}$$

where $\text{Var}_1(X)$ = Variance of counts at block size 1
$\text{Var}_2(X)$ = Variance of counts at block size 2
n = Number of quadrats
x_1 = Number of organisms counted in quadrat 1, and so on

and these calculations are carried forward in a similar manner for block sizes 3, 4, 5 and so on to the upper limits of the possible pooling of quadrats. The upper limit of block size is $n/2$, but the recommended upper limit is $n/10$. For example, if you counted 150 contiguous quadrats, you could calculate block sizes for up to two blocks of 75 quadrats each, but it is recommended that you stop at block size 15. A plot of these variances against the block size can be used to determine the spatial pattern shown in the species being analyzed.

6.2.1 Testing for Spatial Pattern

After computing the variances from grouped sets of adjacent quadrats (TTLQV method), we can ask how to use these estimates to determine spatial pattern. A plot of the variances (equation [6.11]) against block size for the TTLQV method will show in general one of three patterns corresponding to the spatial patterns of the organisms (Figure 6.4). If the individuals are dispersed at random over the study zone, the plot of variance against block

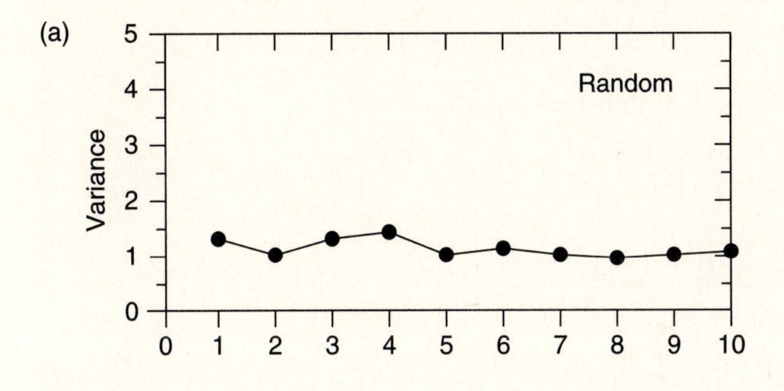

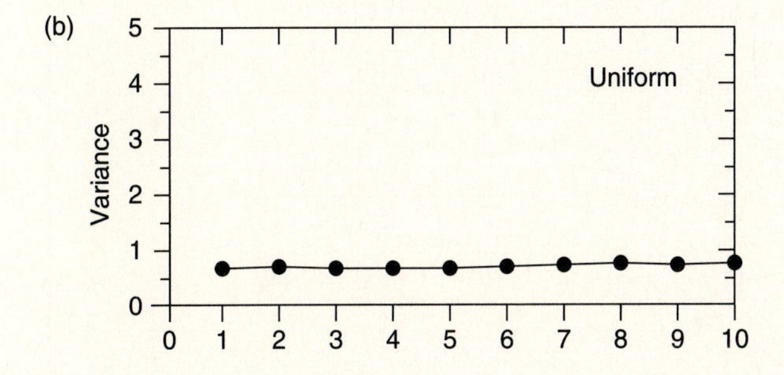

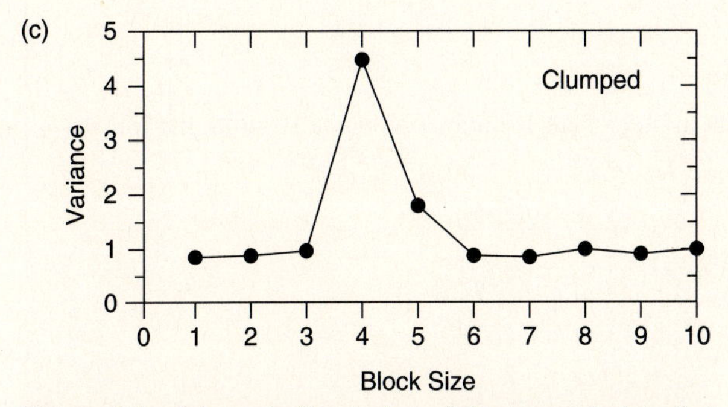

Figure 6.4 Schematic illustration of the types of plots that are found for Hill's TTLQV method for contiguous quadrats. A plot of the block variances against the size of the blocks will give different patterns for (a) random, (b) uniform, and (c) clumped distributions.

Box 6.3 Calculation of Variances for Contiguous Quadrats Using Hill's TTLQV Method

Amabilis fir seedlings were counted in a series of 104 contiguous quadrats, each 10 by 10 cm. The seedling counts obtained were as follows:

0 0 1 0 5 7 2 1 0 0 0 1 0 4 6 3 0 0 1 1 0 0 0 2 5 8

0 2 4 6 5 2 1 0 0 1 2 4 7 3 0 0 1 0 1 0 3 6 5 2 0 0

0 1 3 8 4 1 0 0 1 1 0 5 6 3 1 0 0 0 1 0 1 4 7 4 2 0

1 2 0 3 4 6 4 0 1 0 1 0 3 6 7 5 2 0 1 0 0 0 2 3 7 4

1. Calculate the variance for block size 1 from equation (6.11):

$$\text{Var}_1(X) = \left(\frac{1}{n-1}\right)\left[\left(\frac{(x_1 - x_2)^2}{2}\right) + \left(\frac{(x_2 - x_3)^2}{2}\right) + \cdots + \left(\frac{(x_{n-1} - x_n)^2}{2}\right)\right]$$

$$\text{Var}_1(X) = \left(\frac{1}{104-1}\right)\left[\left(\frac{(0 - 0)^2}{2}\right) + \left(\frac{(0 - 1)^2}{2}\right) + \cdots + \left(\frac{(7 - 4)^2}{2}\right)\right]$$

$$= 2.6505$$

2. Calculate the variance for block size 2 from equation (6.12):

$$\text{Var}_2(X) = \left(\frac{1}{n-3}\right)\left[\left(\frac{(x_1 + x_2 - x_3 - x_4)^2}{4}\right) + \left(\frac{(x_2 + x_3 - x_4 - x_5)^2}{4}\right) + \cdots + \left(\frac{(x_{n-3} + x_{n-2} - x_{n-1} - x_n)^2}{4}\right)\right]$$

$$\text{Var}_2(X) = \left(\frac{1}{104-3}\right)\left[\left(\frac{(0 + 0 - 1 - 0)^2}{4}\right) + \left(\frac{(0 + 1 - 0 - 5)^2}{4}\right) + \cdots + \left(\frac{(2 + 3 - 7 - 4)^2}{4}\right)\right]$$

$$= 8.3243$$

3. Continue these calculations to block size 10 (approximate $n/10$) with the following results:

$$\text{Var}_3(X) = 12.4226$$

$$\text{Var}_4(X) = 13.7281$$

$$\text{Var}_5(X) = 11.4558$$

$$\text{Var}_6(X) = 7.6541$$

$$\text{Var}_7(X) = 4.7724$$

$$\text{Var}_8(X) = 3.2008$$

$$\text{Var}_9(X) = 2.7497$$

$$\text{Var}_{10}(X) = 2.8006$$

4. These variances can be plotted as a function of block size, as in Figure 6.4c. The resulting pattern shows a strong peak of variance in the range of block sizes 3 to 5. The ecological interpretation is that *amabilis* fir seedlings in this community are strongly clumped, and on average the clumps are 60–100 cm apart (twice the block-size peak multiplied by the quadrat size of 10 cm).

Program-group SPATIAL PATTERN (Appendix 2) can do these calculations.

size will fluctuate irregularly with no pattern. If the spacing pattern is uniform, the variances estimated will all be low and will not tend to fluctuate with block size. Finally, if the individuals are clumped in their distribution, the variances will tend to peak at a block size equivalent to the radius of clump size (the average area occupied by a clump). The average distance between the clumps will be twice this block size. If the variance peak is high and sharp, the clumps are tight, and the clumped pattern is said to be of high intensity, with distinct clumps and large open spaces between the clumps. If the variance peak is low, the pattern is of low intensity, and the clumps are not well defined. Since many organisms are clumped in natural systems, the TTLQV approach can be used effectively to define and quantify the type of clumping. Box 6.3 provides an illustration of this approach.

6.3 SPATIAL PATTERN FROM DISTANCE METHODS

In many cases the entire population cannot be mapped as it is in Figure 6.1, and we must sample individuals scattered over a large area. We have discussed the use of distance methods to estimate population density. We now wish to use these same methods to determine spatial pattern. Using the same general approach shown in Figure 5.5, we can make two kinds of measurements:

1. From random *points* to the nearest organism
2. From a random *organism* to its nearest neighbor

6.3.1 Byth and Ripley Procedure

Given these data collected with the Byth and Ripley procedure (Figure 5.6), we can test whether or not the individuals in the population sampled have a random pattern by means of the tests designed by Hopkins (1954).

$$h = \frac{\Sigma\,(x_i^2)}{\Sigma\,(r_i^2)} \tag{6.13}$$

where h = Hopkins' test statistic for randomness
x_i = Distance from random point i to the nearest organism
r_i = Distance from random organism i to its nearest neighbor

Hopkins showed that h is distributed as F with $2n$ degrees of freedom in the numerator and the same in the denominator. The intuitive justification of the Hopkins test is that if the organisms are clumped, point-to-organism distances will be large, relative to organism-to-organism distances. The opposite will occur if the spacing pattern is uniform. Thus the F-test for h is a two-tailed F-test in which h will be significantly small if there is uniformity.

An index of pattern ranging from 0 to 1 can be estimated by

$$I_H = \frac{h}{1 + h} = \frac{\Sigma\,(x_i^2)}{\Sigma\,(x_i^2) + \Sigma\,(r_i^2)} \tag{6.14}$$

This index will approach 1 when clumping increases and approach 0 when uniformity is maximal. Under the null hypothesis of randomness, the index of pattern should be 0.5.

Box 6.4 illustrates the use of Hopkins' test to determine spatial pattern.

6.3.2 T-Square Sampling Procedure

Given the T-square distance data illustrated in Figure 5.7, we can test the hypothesis of a random spatial pattern in the population. The most powerful test statistic, recommended by Hines and Hines (1979) is

$$h_T = \frac{2n[2\,\Sigma\,(x_i^2) + \Sigma\,(z_i^2)]}{[(\sqrt{2}\,\Sigma\,x_i) + \Sigma\,z_i]^2} \tag{6.15}$$

Box 6.4 Hopkins Test for Spatial Pattern with Point-to-Organism Data and Organism-to-Nearest-Neighbor Distances

To illustrate this test, I have selected a small sample of measurements taken on red alder trees by an ecology class. A sample of 24 points on a 1 ha field was used to obtain 12 measurements, as illustrated in Figure 5.6 following the Byth and Ripley procedure for distance sampling, as follows:

Sample no.	Point-to-nearest-tree distance, x_i (m)	Tree-to-nearest-neighbor distance, r_i (m)
1	6.2	3.2
2	9.8	2.8
3	3.4	1.1
4	1.2	4.6
5	5.7	4.2
6	6.1	1.3
7	3.4	5.9
8	5.7	0.4
9	7.2	4.1
10	4.1	3.8
11	6.9	6.9
12	2.8	1.8

$$\sum (x_i^2) = 6.2^2 + 9.8^2 + 3.4^2 + \cdots = 385.330$$
$$\sum (r_i^2) = 3.2^2 + 2.8^2 + 1.1^2 + \cdots = 176.850$$

To test the null hypothesis of a random spatial pattern, we use equation (6.13):

$$h = \frac{\Sigma (x_i^2)}{\Sigma (r_i^2)} = \frac{385.330}{176.850} = 2.18$$

Under the null hypothesis of a random spatial pattern, h is distributed as F with degrees of freedom $2n$ (= 24) in both numerator and denominator. From a table of F-values,

$$F_{.025} = 0.44 \qquad \text{(with 24, 24 d.f.)}$$
$$F_{.975} = 2.27$$

Thus the decision rules are, for $\alpha = 0.05$ and a two-tailed test,

1. If observed h is less than 0.44, reject the null hypothesis of randomness in favor of a *uniform* pattern.
2. If observed h is greater than 2.27, reject the null hypothesis of randomness in favor of a *clumped* pattern.

Since our observed F value is within this range, we cannot reject the null hypothesis of randomness. This may be because the sample size is so small, and it would be advisable to have 20–40 random points rather than 12 as in this example.

Note that when using a table of F values with n_1 degrees of freedom in the numerator and n_2 in the denominator,

$$F_{\alpha[n_1,\, n_2]} = \frac{1}{F_{1-\alpha[n_2,\, n_1]}}$$

For this particular case with $n_1 = n_2$,

$$F_{.025} = \frac{1}{F_{.975}} \quad \text{and} \quad F_{.05} = \frac{1}{F_{.95}}$$

Consequently, if the F-table contains only values of $F_{.05}$ or $F_{.025}$, you can calculate the critical value for the lower tail of the distribution.

For these data, the index of pattern (equation [6.14]) is,

$$I_H = \frac{h}{1 + h} = \frac{2.18}{3.18} = 0.69$$

This suggests a tendency away from randomness (0.5) toward aggregation (1.0).

Program-group SPATIAL PATTERN (Appendix 2) can do these calculations.

where h_T = Hines's test statistic for randomness of T-square data
n = Sample size (no. of random points)
x_i = Point-to-organism distances
z_i = T-square organism-to-neighbor distances

This test statistic is evaluated by referring to critical values in Table 6.2. Low values of h_T indicate a uniform pattern, and high values indicate aggregation. In a random world, h_T is 1.27; smaller values indicate a uniform pattern, and larger values indicate clumping.

Diggle (1983) and Hines and Hines (1979) consider other test statistics for T-square data, but none of them seems superior to h_T in statistical power.

TABLE 6.2 CRITICAL VALUES FOR THE HINES TEST STATISTIC h_T (EQ. [6.15]), WHICH TESTS THE NULL HYPOTHESIS THAT SPATIAL PATTERN IS RANDOM IN A POPULATION SAMPLED WITH THE T-SQUARE SAMPLING PROCEDURE ILLUSTRATED IN FIGURE 5.7[a]

	Regular alternative				Aggregated alternative			
n/α	0.005	0.01	0.025	0.05	0.05	0.025	0.01	0.005
5	1.0340	1.0488	1.0719	1.0932	1.4593	1.5211	1.6054	1.6727
6	1.0501	1.0644	1.0865	1.1069	1.4472	1.5025	1.5769	1.6354
7	1.0632	1.0769	1.0983	1.1178	1.4368	1.4872	1.5540	1.6060
8	1.0740	1.0873	1.1080	1.1268	1.4280	1.4743	1.4743	1.5821
9	1.0832	1.0962	1.1162	1.1344	1.4203	1.4633	1.4539	1.5623
10	1.0912	1.1038	1.1232	1.1409	1.4136	1.4539	1.4456	1.5456
11	1.0982	1.1105	1.1293	1.1465	1.4078	1.4456	1.4384	1.5313
12	1.1044	1.1164	1.1348	1.1515	1.4025	1.4384	1.4319	1.5189
13	1.1099	1.1216	1.1396	1.1559	1.3978	1.4319	1.4261	1.5080
14	1.1149	1.1264	1.1439	1.1598	1.3936	1.4261	1.4209	1.4983
15	1.1195	1.1307	1.1479	1.1634	1.3898	1.4209	1.4098	1.4897
17	1.1292	1.1399	1.1563	1.1710	1.3815	1.4098	1.4008	1.4715
20	1.1372	1.1475	1.1631	1.1772	1.3748	1.4008	1.3870	1.4571
25	1.1498	1.1593	1.1738	1.1868	1.3644	1.3870	1.3768	1.4354
30	1.1593	1.1682	1.1818	1.1940	1.3565	1.3768	1.3689	1.4197
35	1.1668	1.1753	1.1882	1.1996	1.3504	1.3689	1.3625	1.4077
40	1.1730	1.1811	1.1933	1.2042	1.3455	1.3625	1.3572	1.3981
45	1.1782	1.1859	1.1976	1.2080	1.3414	1.3572	1.3528	1.3903
50	1.1826	1.1900	1.2013	1.2112	1.3379	1.3528	1.3377	1.3837
75	1.1979	1.2043	1.2139	1.2223	1.3260	1.3377	1.3260	1.3619
100	1.2073	1.2130	1.2215	1.2290	1.3189	1.3289	1.3189	1.3492
150	1.2187	1.2235	1.2307	1.2369	1.3105	1.3184	1.3105	1.3344
200	1.2257	1.2299	1.2362	1.2417	1.3055	1.3122	1.3055	1.3258
300	1.2341	1.2376	1.2429	1.2474	1.2995	1.3049	1.2995	1.3158
400	1.2391	1.2422	1.2468	1.2509	1.2960	1.3006	1.2960	1.3099
500	1.2426	1.2454	1.2496	1.2532	1.2936	1.2977	1.2936	1.3059

Source: Hines and Hines 1979.

[a] n = Sample size. Low values of h_T indicate a clumped pattern.

6.3.3 Eberhardt's Test

One test for spatial pattern and the associated index of dispersion that can be used on random-point-to-nearest-organism distances was suggested by Eberhardt (1967) and analyzed further by Hines and Hines (1979):

$$I_E = \left(\frac{s}{\bar{x}}\right)^2 + 1 \tag{6.16}$$

where I_E = Eberhardt's index of dispersion for point-to-organism distances
s = Observed standard deviation of distances
$\bar{x}$ = Mean of point-to-organism distances

This index does not depend on the density of the population, and it is easy to calculate. For example, using the spruce tree data in Box 5.3,

$$n = 16 \qquad \bar{x} = 9.681 \qquad s^2 = 13.020$$

we get

$$I_E = \left(\frac{\sqrt{13.020}}{9.681}\right)^2 + 1 = 1.14$$

Critical values of I_E have been computed by Hines and Hines (1979) and are given in Table 6.2, which is entered with sample size $2n$. The expected value of I_E in a random population is 1.27. Values below this suggest a regular pattern, and larger values indicate clumping.

Eberhardt's index can be applied to point-quarter data by using only the nearest of the four distances measured at each point in equation (5.37). This is very inefficient compared with the previous methods.

6.3.4 Variable-Area Transects

The variable-area transect method described in Chapter 5 can also provide data for a test of the null hypothesis of a random spatial pattern. Parker (1979) suggests a statistical test by calculating two values for each observation:

$$P_i = \frac{i}{n} \tag{6.17}$$

$$S_k = \frac{\sum_{i=1}^{k} l_i}{\sum_{i=1}^{n} l_i}$$

where P_i = Proportion of sample counted up to quadrat i
n = Total numbers of quadrats sampled
S_k = Proportion of total lengths measured up to quadrat k
l_i = Length to third organism for quadrat number i ($i = 1, 2, 3, 4, \ldots n$)

Under the null hypothesis of randomness, the two proportions P_i and S_k will rise together. To test the null hypothesis, search for the largest deviation between the two proportions

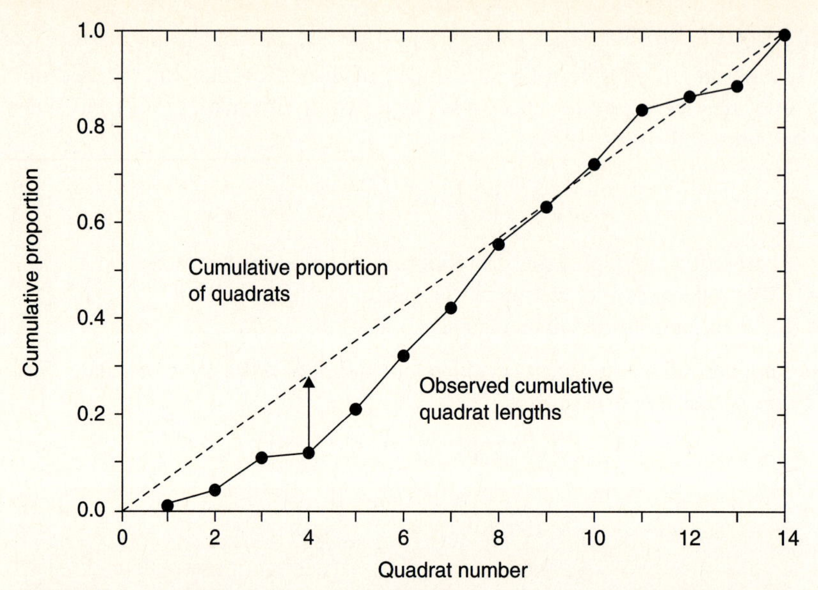

Figure 6.5 Test of the null hypothesis of random spatial pattern for the variable-area transect method of Parker (1979). Two cumulative distributions are plotted for the sample data in Box 6.5 on pismo clams (*Tivela stultorum*) in California. The dashed line gives the cumulative proportion of quadrats searched and is a straight line. The data points are the cumulative proportion of the total quadrat lengths searched up to each quadrat number. The maximum deviation between these two distributions (arrow) is the Kolmogorov-Smirnov statistic D tabled in Zar (1996, Table B.9).

and refer this maximum deviation to the table for the Kolmolgorov-Smirnov goodness-of-fit test (Zar 1996, 474 and Table B.9; Sokal and Rohlf 1995, 708). Figure 6.5 illustrates this test for randomness, and Box 6.5 provides an example of the calculations.

6.4 INDICES OF DISPERSION FOR QUADRAT COUNTS

Many populations of animals and plants are aggregated in nature, and a few are spaced out in a regular pattern. One reason for determining these patterns is that they affect decisions about what method to use for estimating population density, as we have seen. A second reason is to describe these patterns objectively and to try to explain them biologically. In this section we describe various measures that have been proposed for quadrat counts to quantify pattern in natural populations.

A series of counts of the numbers of individual animals or plants are taken in n quadrats of defined size and shape, as discussed in Chapter 4. If we wish to use these counts to construct an index of dispersion, the idealized index should have three properties (Elliott 1977):

1. It should change in a smooth manner as one moves from maximum uniformity to randomness to maximum aggregation.
2. It should not be affected by sample size (n), population density ($\bar{x}$), or by variation in the size and shape of the sampling quadrat.

Box 6.5 Parker's Test for Random Spatial Pattern with Variable Area Transect Data

Parker (1979) reported the following data on the abundance of pismo clams (*Tivela stultorum*) at Seal Beach, California. The transect width was 0.23 m, and the following data was obtained, following the procedures illustrated in Figure 5.9:

Quadrat no.	Length of quadrat searched to reach the third clam, l_i (m)	Cumulative proportion of quadrats searched, P_i	Cumulative proportion of lengths measured, S_k
1	5.4	0.071	0.011
2	15.3	0.143	0.041
3	33.2	0.214	0.106
4	7.8	0.286	0.121
5	43.0	0.357	0.205
6	60.5	0.429	0.323
7	49.8	0.500	0.421
8	63.4	0.571	0.545
9	42.3	0.643	0.628
10	35.8	0.714	0.698
11	65.8	0.786	0.827
12	15.3	0.857	0.857
13	13.9	0.929	0.884
14	59.3	1.000	1.000

1. To calculate the cumulative proportion of quadrats searched, use equation (6.17):

$$P_i = \frac{i}{n}$$

$$P_1 = \frac{1}{14} = 0.071 \qquad P_2 = \frac{2}{14} = 0.143 \qquad \cdots$$

2. To calculate the cumulative proportion of lengths measured, we need to sum the 14 lengths:

$$\sum l_i = 5.4 + 15.3 + 33.2 + \cdots = 510.8 \text{ meters}$$

From this, using equation (6.17), we obtain

$$S_k = \frac{\sum_{i=1}^{k} l_i}{\sum_{i=1}^{n} l_i}$$

$$S_1 = \frac{5.4}{510.8} = 0.011$$

$$S_2 = \frac{5.4 + 15.3}{510.8} = 0.041 \qquad \cdots$$

These calculations fill in the third and fourth columns of the above table.

3. Calculate the difference between these two cumulative frequency distributions:

$$d_i = |P_i - S_i|$$

where d_i = Differences between the two frequency distributions for quadrat i

For these data, it is clear that the largest difference between the cumulative proportions is at sample 4, where $d_4 = |0.286 - 0.121| = 0.165$.

From Table B.9 in Zar (1996), the critical value of the Kolmogorov-Smirnov test statistic at $\alpha = 0.05$ is 0.349 for $n = 14$, and since our observed value is less than the tabled value, we tentatively accept the null hypothesis of a random pattern.

In practice, of course, this sample size is very small, and we should try to measure 30–40 variable area quadrats to have a more powerful test.

3. It should be statistically tractable, so that a confidence belt can be specified and comparisons between samples can be tested for significance.

All of these properties are important, but perhaps the second is most important. Unfortunately there is no perfect index of dispersion that fulfills all these criteria. All indices have their defects, but some are better than others. I will discuss five possible indices of dispersion here. Elliott (1977) and Myers (1978) discuss several other indices, but none seems very useful in practice.

6.4.1 Variance-to-Mean Ratio

The variance-to-mean ratio is one of the oldest and one of the simplest measures of dispersion. The ratio ($s^2/\bar{x}$) is usually called the *index of dispersion* (I; see equation [4.3]) and is based on the observation that in a random pattern, described by the Poisson distribution, the variance equals the mean, so $I = 1$. Table 6.3 gives the expected values for the index of dispersion when there is a uniform or a clumped pattern. There is a clear problem with the index of dispersion when it is applied to clumped populations: the upper limit is a function of sample size, which violates property 2 above. Nevertheless, Myers (1978) showed in a simulation analysis that the variance-to-mean ratio is only weakly affected by population density and is thus a good (if not best) measure of dispersion. This conclusion was criticized by Hurlbert (1990) and by Perry (1995b). The essence of these criticisms is that certain nonrandom patterns may produce variance-to-mean ratios of 1. Two examples from quadrat counts follow:

(a)

2	6	6	$n = 9$
2	6	6	$\bar{x} = 4$
2	2	4	$s^2 = 4$

Variance/mean ratio = 1.0

(b)

8	6	3	$n = 9$
5	4	2	$\bar{x} = 4$
3	3	2	$s^2 = 4$

Variance/mean ratio = 1.0

TABLE 6.3 LIMITS OF FIVE COMMONLY USED INDICES OF SPATIAL PATTERNING IN POPULATIONS SAMPLED BY QUADRAT COUNTS[a]

	Value expected under		
	Maximum uniformity	Randomness	Maximum aggregation
Variance/mean ratio	0	1	ΣX
Reciprocal of k (negative binomial exponent)	$\frac{-1}{x}$	0	$n - \frac{1}{\bar{x}}$
Green's coefficient	$\frac{-1}{\Sigma (x) - 1}$	0	1
Morisita coefficient	$1 - \left(\frac{n - 1}{\Sigma (x) - 1}\right)$	0	n
Standardized Morisita coefficient	-1	0	$+1$

Note: n = Sample size, $\bar{x}$ = Mean population density.

[a] Values of these indices are given for conditions of maximum uniformity, randomness, and maximum aggregation.

Clearly both these patterns are nonrandom, (a) being aggregated and bimodal, and (b) showing a strong gradient. This nonrandomness is not captured by the variance-to-mean ratio. Other measures of dispersion need to be considered.

6.4.2 *k* of the Negative Binomial

The exponent k is frequently used as an index of dispersion for sets of data that fit a clumped pattern and are adequately described by the negative binomial distribution (see Chapter 4, section 4.2.2). Thus the first requirement for the use of this index is that you must check your data for agreement with the negative binomial. Since small values of k indicate maximum clumping, it is customary to use the reciprocal ($1/k$) as the index of pattern, rather than k itself. Table 6.3 shows that $1/k$ is possibly affected by sample size and by quadrat size. Myers (1978) showed that all indices based on k were strongly correlated with population density, violating property 2 above. You should not use ($1/k$) as an index of pattern unless all the data sets being compared have the same sample size and similar population densities. Since these conditions rarely occur, this index of dispersion is not recommended.

6.4.3 Green's Coefficient

Green (1966) set out to devise a coefficient of dispersion that would have all the desirable properties described above. He based his coefficient on the variance-to-mean ratio, and thus it is simple to compute:

$$\text{Green's coefficient of dispersion} = \frac{(s^2/\bar{x}) - 1}{\Sigma (X) - 1} \tag{6.18}$$

Negative values of Green's coefficient indicate a uniform pattern, and positive values indicate a clumped pattern. Myers (1978) found that Green's coefficient was one of the best available for describing pattern because in simulation tests it was nearly independent of population density and sample size. Unfortunately the sampling distribution of Green's coefficient has not been worked out, and it is difficult to assign confidence limits to it.

6.4.4 Morisita's Index of Dispersion

Morisita (1962) developed an index of dispersion that has some of the desirable attributes of an index:

$$I_d = n\left[\frac{\Sigma\, x^2 - \Sigma\, x}{(\Sigma\, x)^2 - \Sigma\, x}\right] \tag{6.19}$$

where I_d = Morisita's index of dispersion
n = Sample size
$\Sigma\, x$ = Sum of the quadrat counts $= x_1 + x_2 + x_3 \cdots$
$\Sigma\, x^2$ = Sum of quadrat counts squared $= x_1^2 + x_2^2 + x_3^2 \cdots$

This coefficient is relatively independent of population density but is affected by sample size, so it is not as good as Green's coefficient. It does have the desirable statistical property of having a known sampling distribution. Morisita (1962) showed that one could test the null hypothesis of randomness by

$$\chi^2 = I_d(\Sigma\, x - 1) + n - \Sigma\, x \qquad (\text{d.f.} = n - 1) \tag{6.20}$$

where χ^2 = Test statistic for Morisita's index of dispersion (chi-square distribution), and other terms are as defined above.

Most of the desirable attributes of the Morisita index are used in the next index which is the best currently available.

6.4.5 Standardized Morisita Index

Smith-Gill (1975) set out to improve Morisita's index by putting it on an absolute scale from −1 to +1. The resulting calculations are more tedious than difficult. Proceed as follows. Given a set of counts (x_i) of organisms in a set of quadrats (n = no. quadrats):

Calculate the Morisita index of dispersion using equation 6.19. Then calculate two critical values for the Morisita index from the following formulas:

$$\text{Uniform index} = M_u = \frac{\chi^2_{.975} - n + \Sigma\, x_i}{(\Sigma\, x_i) - 1} \tag{6.21}$$

where $\chi^2_{.975}$ = Value of chi-squared from table with $(n - 1)$ degrees of freedom that has 97.5% of the area to the right
x_i = Number of organisms in quadrat $i\,(i = 1, \ldots n)$
n = Number of quadrats

$$\text{Clumped index} = M_c = \frac{\chi^2_{.025} - n + \Sigma\, x_i}{(\Sigma\, x_i) - 1} \tag{6.22}$$

where $\chi^2_{.025}$ = Value of chi-squared from table with $(n - 1)$ d.f. that has 2.5% of the area to right

Then calculate the standardized Morisita index by one of the following four formulas:

When $I_d \geq M_c > 1.0$,

$$I_p = 0.5 + 0.5\left(\frac{I_d - M_c}{n - M_c}\right) \tag{6.23}$$

When $M_c > I_d \geq 1.0$,

$$I_p = 0.5\left(\frac{I_d - 1}{M_u - 1}\right) \tag{6.24}$$

When $1.0 > I_d > M_u$,

$$I_p = -0.5\left(\frac{I_d - 1}{M_u - 1}\right) \tag{6.25}$$

When $1.0 > M_u > I_d$,

$$I_p = -0.5 + 0.5\left(\frac{I_d - M_u}{M_u}\right) \tag{6.26}$$

The standardized Morisita index of dispersion (I_p) ranges from -1.0 to $+1.0$, with 95% confidence limits at $+0.5$ and -0.5. Random patterns give an I_p of zero, clumped patterns above zero, uniform patterns below zero.

In a simulation study Myers (1978) found the standardized Morisita index to be one of the best measures of dispersion because it was independent of population density and sample size. Box 6.6 illustrates the calculations for the standardized Morisita index.

The sample size necessary to obtain a reliable index of dispersion is difficult to estimate because estimates of confidence limits are not available for most coefficients. Green (1966) recommends that a minimum sample size should be 50 quadrats, and that when the pattern is highly clumped, at least 200 quadrats are required.

6.4.6 Distance-to-Regularity Indices

Perry and Hewitt (1991) criticized the conventional approaches to measuring aggregation and uniformity because all the existing methods ignore the movements of organisms. Perry (1995a, 1995b) developed two indices of dispersion based on the concept of distance to regularity. Given a set of quadrat counts, how many individuals would have to be moved about to produce a set of counts that showed maximum regularity? In a simple world, maximum regularity would be shown by equal counts in all quadrats. This approach computes the number of organisms that have to be moved to produce regularity and compares it with the expected distance to regularity calculated by randomly and independently assigning individuals to quadrats in a computer, with an equal probability of assignment to each quadrat. The computer reclassification is typically done 1000 times to generate the expected distance to regularity. Perry (1995a) proposed the following index:

$$I_r = \frac{D}{E_r} \tag{6.27}$$

where I_r = Index of regularity
D = Observed distance to regularity
E_r = Expected distance to regularity for the permuted data

Box 6.6 Calculation of the Standardized Morisita Index

A. R. E. Sinclair counted elephants on 26 ten-hectare plots and obtained the following data: one plot had 20 elephants, one plot had 30 elephants, one plot had 10, and the remaining 23 plots had no elephants.

$$\sum x_i = 20 + 30 + 10 + 0 + 0 + \cdots = 60$$

1. Calculate Morisita's index of dispersion from equation (6.19):

$$I_d = n\left[\frac{\Sigma x_i^2 - \Sigma x_i}{(\Sigma x_i)^2 - \Sigma x_i}\right]$$

$$I_d = 26\left[\frac{(30^2 + 20^2 + 10^2) - 60}{60^2 - 60}\right] = 26\left(\frac{1340}{3540}\right) = (0.3785)(26)$$

$$= 9.842$$

2. Calculate the two critical points from equations (6.21) and (6.22): for $(n - 1)$ or 25 degrees of freedom from the chi-squared table, we read

$$\chi^2_{.975} = 13.1 \qquad \chi^2_{.025} = 40.6$$

$$\text{Uniform index} = M_u = \frac{\chi^2_{.975} - n + \Sigma x_i}{(\Sigma x_i) - 1}$$

$$= \frac{13.1 - 26 + 60}{60 - 1} = 0.7983$$

$$\text{Clumped index} = M_c = \frac{\chi^2_{.025} - n + \Sigma x_i}{(\Sigma x_i) - 1}$$

$$= \frac{40.6 - 26 + 60}{60 - 1} = 1.2644$$

3. Calculate standardized Morisita index: since I_d (9.84) is greater than M_c (1.26) and greater than 1.0, we use equation (6.23) to calculate I_P:

$$I_P = 0.5 + 0.5\left(\frac{I_d - M_c}{n - M_c}\right)$$

$$= 0.5 + 0.5\left(\frac{9.842 - 1.2644}{26 - 1.2644}\right) = 0.6734$$

4. Since the standardized Morisita index is above 0.5, we can be 95% confident that we have a clumped distribution of elephants in our sample data.

The negative binomial subprogram in program-group QUADRAT SAMPLING (Appendix 2) calculates the standardized Morisita index.

This index is 1.0 when the sample is random, less than 1 when uniform, and greater than 1 when aggregated. This index can be calculated for any set of quadrats for which the spatial framework is known. Box 6.7 gives an illustration of this method.

A second index proposed by Perry (1995a) is based on a similar approach, but instead of reassigning the individuals at random to quadrats, you reassign the counts to different quadrats. This approach tests the spatial arrangement of the counts, and, for example, if

Box 6.7 Calculation of Perry's Index of Regularity and Index of Aggregation

Monte Lloyd counted centipedes (*Lithobius crassipes*) on 6 quadrats equally spaced at 2 ft intervals and obtained these data:

2 1 2 4 5 10 with a total of 24 individuals in 6 quadrats

Quadrat A	B	C	D	E	F
2	1	2	4	5	10

We can compute two indices from these data.

1. Calculate the observed distance to regularity for the data.

In this simple example, a regular or uniform distribution would be:

4 4 4 4 4 4

To achieve this uniform pattern, we can hypothetically move centipede individuals one at a time from the original observed pattern. In this simple case,

Move 2 individuals from quadrat F to quadrat A: 2×10 ft = 20 ft

Move 3 individuals from quadrat F to quadrat B: 3×8 ft = 24 ft

Move 1 individual from quadrat F to quadrat C: 1×6 ft = 6 ft

Move 1 individual from quadrat E to quadrat C: 1×4 ft = 4 ft

All quadrats in this hypothetical universe now have 4 centipedes. The total distance we had to move centipedes to achieve a regular pattern is

$$D = 20 + 24 + 6 + 4 = 54 \text{ ft}$$

2. We can reassign the 24 individuals to the 6 quadrats at random, with an equal probability of being put in any one of the 6 quadrats. For each random rearrangement, we can compute the moves we would have to make to achieve a regular pattern. For example, one random toss for the individuals might be:

Quadrat A	B	C	D	E	F
2	3	7	3	4	5

By moving these hypothetical individuals in the direction of regularity, we can verify that for this arrangement we need to move:

$$4 + 2 + 4 = 10 \text{ ft}$$

By repeating this 400 times you can determine that the expected value of the distance moved to obtain a regular pattern is 21.9 ft. From equation (6.27):

$$I_r = \frac{D}{E_r} = \frac{54}{21.9} = 2.47$$

From the 400 random selections, only 0.7% had values above the observed 54 ft, and this suggests a significantly aggregated pattern in these centipedes.

This index of regularity is 1.0 for random counts and above 1 for aggregated patterns. Note that this index does not take into account the position of the quadrats.

3. Perry (1995b) developed a second index to measure aggregation when the position of the quadrat is important. For example there may be a gradient in the abundances. To test for these types of nonrandom patterns, proceed as follows.

To calculate the expected distance to regularity, we need to randomly assign these counts to randomly selected quadrats. For example, one of the random rearrangements might be:

Quadrat A	B	C	D	E	F
4	2	10	1	2	5

Note that in this case we do not rearrange the individuals but rearrange the counts for the quadrats. This is equivalent to moving the quadrats in space and thus directly addresses the question of whether or not the geographic positions of the quadrat are effectively random. By moving these hypothetical individuals in the same way shown above, you can verify that for this arrangement you need to move:

$$4 + 6 + 4 + 2 = 16 \text{ ft}$$

If you repeat these permutations many times and average them you will find the expected distance to regularity is 29.3 ft. From equation (6.28):

$$I_a = \frac{D}{E_a} = \frac{54}{29.3} = 1.84$$

From 400 simulations on the computer, only 1.5% of the values exceeded the observed 54 ft, and we can thus conclude that the spatial pattern of these centipedes is significantly aggregated. An index of aggregation of 1 indicates random patterns, and greater than 1, clumped patterns.

This is a simple case for illustration. A much larger sample size of 30–50 quadrats would be desirable in a field study of spatial pattern.

These calculations are clearly most easily done in a computer program, and Perry (1995) has developed Program SADIE to do these permutations.

there is a gradient among the quadrats (as shown above, page 214), it will be detected. The second index is

$$I_a = \frac{D}{E_a} \qquad (6.28)$$

where I_a = Index of aggregation
D = Observed distance to regularity
E_a = Average distance to regularity for the permuted counts

Both these indices should be utilized for quadrat count data since they detect different forms of aggregation. Box 6.7 illustrates this approach.

Perry (1995b) has developed another method called SADIE (Spatial Analysis by Distance Indices) for the analysis of mapped data in which the exact coordinates of individuals are known. Again the principle of moving individuals is utilized, but now since the spatial map is known, it is possible to move them directionally and stepwise toward complete regularity (Figure 6.6). The observed number and direction of moves to regularity in the

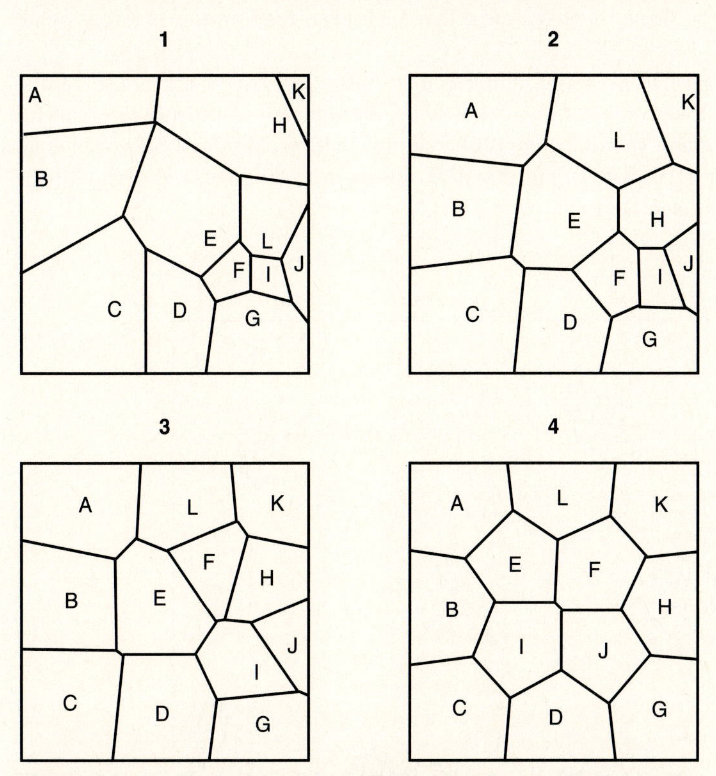

Figure 6.6 An illustration of the SADIE approach to measuring the distance to regularity for a spatial map of 12 sycamore aphids. (1) The starting arrangement of the 12 aphids with the Dirichlet tessellation outlined. (2) The first movement of the SADIE algorithm, in which each aphid has been moved by the computer in the direction of a uniform pattern. (3) The position of the aphids after 4 moves. (4) The final position of the 12 aphids after 325 moves on the computer, showing a regular pattern. The sum of the moves between the initial and final positions of each aphid gives the distance-to-regularity measurement. (Modified from Perry 1995a.)

data are then compared to the simulated number and direction of moves needed to change a randomly generated map into a regular pattern. An index of pattern identical to those given above can be generated with this approach. For example, the redwood seedling map in Figure 6.1 provides an index of pattern of 1.28, and shows significant aggregation. The mathematical details are given in Perry (1995b), and the procedures are performed by Program SADIEM as described in Perry (1995b).

It is important to remember when sampling with quadrats that *the spatial pattern obtained, and the resulting index of dispersion, depends on quadrat size and shape*. Elliott (1977) shows this graphically for a population with a clumped pattern in which the clumps themselves are uniformly spaced (Figure 6.7). The only resolution to this problem is to sample the population with a series of quadrats of varying size and to plot empirically how the index of dispersion changes with quadrat area.

If a population is completely random in its spatial pattern, there will be no change in the index of dispersion with quadrat size, contrary to the results shown in Figure 6.7. In this special case, quadrat counts are always described by the Poisson distribution, no matter what the quadrat size or shape. It seems clear that such ideal randomness is rarely found in nature.

There is little information on the sampling distributions of any of the indices of dispersion, so it is impossible to suggest any simple way of producing confidence intervals for these indices. In principle it should be possible to do this with jackknife or bootstrap techniques (Sokal and Rohlf 1995, 820) but this does not seem to be discussed in any of the literature on indices of dispersion.

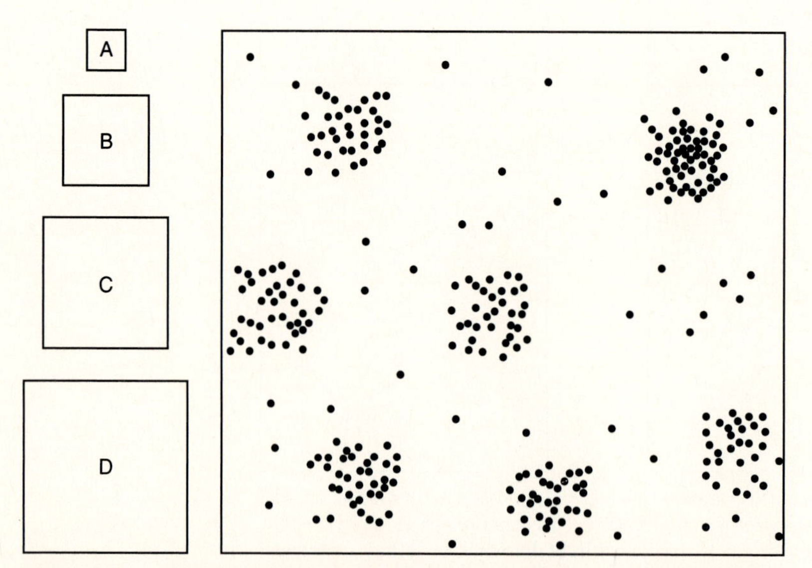

Figure 6.7 A hypothetical clumped population with regularly distributed clumps. If four quadrat sizes are used to sample this population, the index of dispersion obtained will show apparent randomness with quadrat size A, a clumped pattern with quadrat size B, apparent randomness with quadrat size C, and finally a uniform pattern with quadrat size D. (*Source:* Elliott 1977.)

6.5 SUMMARY

Spatial pattern is of interest in itself because we would like to understand the ecological processes determining the locations of individuals, which are rarely spread at random over the landscape. When a spatial map is available for a particular population, nearest-neighbor distances can be measured for the entire population to assess spatial pattern. Distances to second and third nearest neighbors can provide additional information about the spatial positioning of individuals. A series of techniques of increasing sophistication are available for analyzing patterns in spatial maps.

Plant ecologists use contiguous quadrats to analyze the spatial patterning of plant populations. The variance among quadrats changes as adjoining quadrats are grouped into larger and larger units, and this change can help identify the scale of clump sizes in aggregated distributions.

For large areas sampled with distance measures, spatial patterns can be analyzed from data on distances from random points to nearest organisms. The degree of nonrandomness can be measured with two different measures of dispersion for distance methods.

Many indices of dispersion for quadrat sampling have been proposed. For quadrat sampling the best one seems to be the standardized Morisita coefficient, which is relatively unaffected by population density and sample size. New, computer-intensive methods (SADIE) have also been suggested for measuring aggregation in organisms sampled with quadrats. With quadrat sampling, it is critical to remember that the spatial pattern obtained usually depends on quadrat size and shape, and a series of quadrat sizes should be used to measure the scale of the patterns shown by the population.

SELECTED READING

Clark, P. J., and Evans, F. C. 1954. Distance to nearest neighbor as a measure of spatial relationships in populations. *Ecology* 35: 445–453.

Dale, M. 1998. *Spatial Pattern Analysis in Plant Ecology*. Cambridge University Press, Cambridge.

Diggle, P. J. 1983. *Statistical Analysis of Spatial Point Patterns*. Academic Press, London.

Engeman, R. M., Sugihara, R. T., Pank, L. F. and Dusenberry, W. E. 1994. A comparison of plotless density estimators using Monte Carlo simulation. *Ecology* 75: 1769–1779.

Hurlbert, S. H. 1990. Spatial distribution of the montane unicorn. *Oikos* 58: 257–271.

Mumme, R. L., Koenig, W. D., and Pitelka, F. A. 1983. Are acorn woodpecker territories aggregated? *Ecology* 64: 1305–1307.

Myers, J. H. 1978. Selecting a measure of dispersion. *Environmental Entomology* 7: 619–621.

Perry, J. N. 1995. Spatial analysis by distance indices. *Journal of Animal Ecology* 64: 303–314.

Ripley, B. D. 1981. *Spatial Statistics*. John Wiley and Sons, New York.

QUESTIONS AND PROBLEMS

6.1. A series of 64 contiguous quadrats of 1 cm^3 of forest soil was counted for collembolans with the following results:

4	3	10	5	8	3	2	8	0	0	5	4	13	17	7	4
1	7	11	6	5	9	8	2	3	2	6	19	6	4	5	2
2	5	5	7	16	8	9	9	3	3	2	9	11	5	2	0
0	2	0	3	5	4	2	1	0	4	11	2	3	10	3	2

What spatial pattern do these collembolans show in this forest soil?

6.2. For a goldenrod (*Solidago* spp.) plot of 136 by 176 ft. (23,936 sq. ft.), Clark and Evans (1954) counted 89 plants and measured nearest-neighbor distances for each. They obtained

$$n = 89 \qquad \sum r = 530.24 \qquad \sum r^2 = 4751.5652$$

(a) Calculate the Clark and Evans test for a random pattern for these data, and compare the results with that of Thompson's test.

(b) Assume that no boundary strip was included in these data and calculate the Donnelly modification of the Clark and Evans test. How do your conclusions differ?

6.3. Burgess et al. (1982) tested the spatial patterning of acorn woodpecker territories in California by measuring the distances to the first through the fifth nearest neighbors in their study area. They used the woodpeckers' storage trees (granaries) as the point from which distances were measured, and got these results (Study area = 1313.5 sq. distance units):

	Nearest-neighbor distance[a]				
	First	Second	Third	Fourth	Fifth
	2.9	4.7	7.8	6.9	7.0
	3.2	3.6	3.7	4.7	5.2
	2.7	3.6	4.4	4.9	6.7
	3.6	3.8	4.4	4.6	4.9
	2.7	3.6	5.1	5.8	6.9
	4.9	6.7	8.2	8.4	8.9
	0.6	3.2	4.1	4.8	5.5
	3.7	5.6	7.1	8.6	9.8
	4.7	4.8	6.9	7.1	8.2
	4.9	5.0	6.1	6.9	7.2
	0.6	3.7	3.9	4.8	4.9
	5.3	5.7	6.7	6.9	9.1
	1.1	3.7	4.8	4.8	6.0
	4.4	5.8	7.7	9.0	10.8
	2.9	4.1	4.8	5.0	5.1
	4.8	5.1	5.3	5.7	8.0
	1.1	3.9	4.1	4.5	5.1
	3.9	5.6	8.2	[b]	[b]
	5.7	8.9	9.8	[b]	[b]
	3.7	3.9	[b]	[b]	[b]
Mean distance =	3.37	4.75	5.79	6.08	7.02
Sum of squares =	273.42	487.26	694.39	666.52	892.61
Density =	0.0152	0.0152	0.0145	0.0129	0.0129

[a]Each sample distance unit = 48 m.
[b]These neighbors fell outside the sample perimeter.

Use Thompson's test to evaluate the null hypothesis of a random pattern for each of the first to fifth nearest neighbors. Review the controversy arising from this study (Mumme et al. 1983, Burgess 1983) and suggest techniques for further analysis of the problems discussed.

6.4. Neil Gilbert sampled two species on randomly selected cabbage leaves and obtained the following data:

Leaf number	Cabbage aphid	Predatory beetle
1	5	0
2	4	0
3	5	0
4	1	0
5	2	1
6	1	0
7	0	2
8	1	1
9	2	0
10	4	1
11	4	0
12	0	0
13	1	1
14	1	1
15	0	1
16	2	0
17	1	0
18	3	0
19	4	0
20	0	1
21	1	0
22	1	1
23	2	1
24	1	2

Calculate an index of dispersion for each of these two species and estimate the confidence limits for these indices.

6.5 Discuss the statistical and ecological reasons why the hypothesis of a uniform spatial pattern might not be accepted for data like that given for ants in Box 6.1.

6.6 Calculate an index of dispersion for the data on subalpine fir trees in Box 5.2.

PART THREE

Sampling and Experimental Design

Ecologists are more and more confronted with the need to do their work in the most efficient manner. To achieve this practical goal, an ecologist must learn something about sampling theory and experimental design. These two subjects are well covered in many statistical books such as Cochran (1977), Cox (1958), and Winer et al. (1991), but the terminology is unfortunately foreign to ecologists, and some translation is needed. In the next four chapters I discuss sampling and experimental design from an ecological viewpoint. I emphasize methods and designs that seem particularly needed in ecological research. All of this should be viewed as a preliminary discussion that will direct students to more comprehensive texts in sampling and experimental design.

Sampling and experimental design are statistical jargon for the three most obvious questions that can occur to a field ecologist: *Where* should I take my samples, *how* should I collect the data in space and time, and *how many* samples should I try to take? Over the past 90 years statisticians have provided a great deal of practical advice on how to answer these questions and—for those that plan ahead—how *not* to answer these questions. Let us begin with the simplest question of the three: How many samples should I take?

CHAPTER 7

Sample Size Determination and Statistical Power

A recurrent practical question in ecological research projects is *How much work do I need to do?* The statistician translates this question into the more easily answered question, *How large a sample should I take?* In spite of the apparent similarity of these two questions, there is a great gap between them. The detailed objectives of the research program need to be specified for a complete answer to the first question, and a statistician is not very helpful at this stage. For example, you may wish to estimate the density of ring-necked pheasants on your study area. Given this overall objective, you must specify much more ecological detail before you see your statistician. You must decide, for example, how often each year estimation is required, the type of sampling gear you will use, and whether your study area contains a closed population or is subject to extensive immigration and emigration. All of this information and more must be used to help decide on the method of population estimation you should apply. Do not expect a statistician to help you with these difficult decisions.—But when you have decided, for example, to use the Petersen method of mark-recapture estimation, it is useful to ask a statistician, *How large a sample should I take?*

Throughout the last five chapters we have periodically come across this question and tried to provide a partial answer. This chapter draws together a series of general methods that you can apply to any situation, and will guide you toward more complex methods when necessary. There is an excellent review of the sample-size problem in population ecology by Eberhardt (1978) and general surveys by Kraemer and Thiemann (1987), Mace (1964) and Cochran (1977).

We will start here with the simplest question of statistical *estimation*, and see how to determine the sample size necessary to obtain a specified level of precision. For example, we might ask how many fish we need to weigh to obtain a mean weight with 95% confidence limits of ±2.5 grams. From this simple question we move on to discuss the more complex questions of the sample sizes required for statistical *inference*, in which, for example, we wish to compare the size of fish in two or more different lakes. To test hypotheses about statistical populations, we must clearly understand the concept of statistical power.

7.1 SAMPLE SIZE FOR CONTINUOUS VARIABLES

We will begin by considering variables measured on the interval or ratio scale, which can show a normal distribution. How can we decide how big a sample to take?

7.1.1 Means from a Normal Distribution

If you need to measure a sample to estimate the average length of whitefish in a particular lake, you can begin with a very simple statistical assumption that lengths will have a normal, bell-shaped frequency distribution. The two-sided confidence interval for a normal variable is

$$\text{Probability}[(\bar{x} - t_\alpha s_{\bar{x}}) < \mu < (\bar{x} + t_\alpha s_{\bar{x}})] = 1 - \alpha$$

where t_α = Student's t-value for $n - 1$ degrees of freedom for $\frac{\alpha}{2}$

$s_{\bar{x}}$ = Standard error of mean = $\sqrt{\frac{\text{Variance of mean}}{\text{Sample size}}}$

n = Sample size

This confidence interval formulation is usually written in statistical shorthand as

$$\bar{x} \pm t_\alpha s_{\bar{x}}$$

The width of the confidence interval depends on the t-value and the standard error. To proceed further, some information is required, and since the steps involved are general, I will describe them in some detail. Three general steps must be taken to estimate sample size:

Step 1. *Decide what level of precision you require.* Do you want your estimate of the mean to be accurate within ±10% of the true population mean? Within ±1% of the true mean? You must decide what the desired error limits will be for your sample. This is not a statistical question but an ecological one, and it will depend on what theory you are testing or what comparisons you will make with your data.

The desired level of precision can be stated in two different ways; these can be very confusing if they are mixed up. You can define the *absolute* level of precision desired. For example, you may wish the 95% confidence interval to be ±2.8 mm or less for your whitefish. Alternatively, you can define the *relative* level of precision desired. For example, you may wish the 95% confidence interval to be ±6% of the mean or less. These two are related simply:

$$\text{Percent relative error desired} = \left(\frac{\text{Absolute error desired}}{\bar{x}}\right)100$$

I will use the absolute level of precision to begin, but for some purposes as we proceed, we will find it easier to use relative error as a measure of precision.

A second potential source of confusion involves how to specify the level of precision you require. Some authors define the level of precision in terms of ±*1 standard error* of the mean. Others define the level of precision in terms of the *width of the confidence interval*, which for 95% confidence gives approximately ±2 standard errors of the mean. In this chapter I will always define the desired level of precision in terms of the width* of the confidence interval. If you use other statistical texts, be sure to check which definition of "desired precision" they adopt.

Step 2. *Find some equation that connects sample size (n) with the desired precision of the mean.* This equation will depend on the kind of sampling being done and the type of variable being measured. The purpose of this chapter is to give you these equations. For the case of the mean from a normal distribution, we have the following equation:

$$\text{Desired absolute error} = d = t_\alpha s_{\bar{x}}$$

or, expanding:

$$d = \frac{t_\alpha s}{\sqrt{n}}$$

Rearranging, we have

$$n = \left(\frac{t_\alpha s}{d}\right)^2 \tag{7.1}$$

where n = Sample size needed to estimate the mean
t_α = Student's t-value for $n - 1$ degrees of freedom for the $1 - \alpha$ level of confidence
s = Standard deviation of variable
d = Desired absolute error

This is the formula to estimate sample size for a mean from a normal distribution.

*Technically speaking, I should say the "half-width" of the confidence interval; but this cumbersome wording is omitted. I use the term to indicate that in general we think of confidence intervals as a parameter ± width of the confidence interval.

Step 3. *Estimate or guess the unknown parameters of the population that you need to solve the equation.* In this simple case we need to have an estimate of s, the standard deviation of the variable measured. How can we get an estimate of the standard deviation? There are four ways you may get this for the variable you are studying:

1. *By previous sampling of a similar population*: You may know, for example, from work done last year on the whitefish in another lake that the standard deviation was about 6.5 mm.
2. *By the results of a pilot study*: You may wish to spend a day or two sampling to get an estimate of the standard deviation in your population.
3. *By guesswork*: Often an experienced person may have implicit knowledge of the amount of variability in a variable. From this type of information on the *range* of measurements to be expected (maximum value − minimum value), you can estimate the standard deviation of a measure that has a normal distribution by the formulas given in Table 7.1. For example, if you know that in a sample of about 50 fish, length will vary from 15–70 mm, then from Table 7.1,

$$\text{Estimated standard deviation} \approx (0.222)(\text{range})$$

$$\approx (0.222)(55) = 12.2 \text{ mm}$$

4. *By two-stage sampling*: If it is feasible for you to sample in two steps, you can first take a sample of size n_1 and calculate from these measurements a preliminary estimate of the standard deviation. Cochran (1977, 79) shows that in the

TABLE 7.1 ESTIMATION OF THE STANDARD DEVIATION (s) OF A VARIABLE FROM KNOWLEDGE OF THE RANGE (w) FOR SAMPLES OF VARIOUS SIZES[a]

Sample size	Conversion factor	Sample size	Conversion factor
2	0.886	19	0.271
3	0.591	20	0.268
4	0.486	25	0.254
5	0.430	30	0.245
6	0.395	40	0.231
7	0.370	50	0.222
8	0.351	60	0.216
9	0.337	70	0.210
10	0.325	80	0.206
11	0.315	90	0.202
12	0.307	100	0.199
13	0.300	150	0.189
14	0.294	200	0.182
15	0.288	300	0.174
16	0.283	500	0.165
17	0.279	1000	0.154
18	0.275		

Source: Dixon and Massey 1983.

[a]Range = maximum value − minimum value. Multiply the observed range by the tabled values to obtain an unbiased estimate of the standard deviation. A normal distribution is assumed.

second sample you need to take additional measurements to make a total sample size of:

$$n = \frac{(t_\alpha s_1)^2}{d^2}\left(1 + \frac{2}{n_1}\right) \tag{7.2}$$

where n = Final total sample size needed
n_1 = Size of first sample taken
s_1 = Standard deviation estimated from first sample taken
t_α = Student's t with $n - 1$ d.f. for $1 - \alpha$ level of confidence
d = Desired absolute error

Since you are using data from the first sample to estimate total sample size, there is an element of statistical circularity here, but the important point is to remember that these procedures are approximations to help you to decide the appropriate sample size needed.

An example will illustrate the application of these three steps with equation (7.1). Assume you wish to measure whitefish to obtain a 95% confidence interval of ±2.8 mm. From previous sampling, you know that the standard deviation of lengths is about 9.4 mm. Thus, from equation (7.1),

$$n = \left(\frac{t_\alpha s}{d}\right)^2 = \left[\frac{t_\alpha(9.4)}{2.8}\right]^2$$

We immediately have another problem: to look up t_α-value we need to know the sample size n, so we are trapped in a statistical catch-22. For those who desire precision, equation (7.1) can be solved iteratively by trial and error. In practice it is never worth the effort because t-values for 95% confidence limits are almost always around 2 (unless n is very small), and we will use in equation (7.1) the approximation $t_\alpha = 2$ when we need 95% confidence limits.* Thus,

$$n \approx \left[\frac{2(9.4)}{2.8}\right]^2 \approx 45.1$$

The recommendation is to measure about 45 fish.

This technique for estimating sample size can also be used with relative measures of precision (Eberhardt 1978a). The most convenient measure of relative variability is the *coefficient of variation*:

$$\text{CV} = \frac{s}{\bar{x}} \tag{7.3}$$

where s = Standard deviation
$\bar{x}$ = Observed mean

If you know the coefficient of variation for the variable you are measuring, you can estimate the sample size needed in the following way. From the formulas given above,

$$\text{Desired relative error} = r = \left(\frac{t_\alpha s_{\bar{x}}}{\bar{x}}\right)100$$

*If you need to work in 90% confidence intervals, you can use $t_\alpha \sim 1.7$. If you use 99% confidence intervals, use $t_\alpha \sim 2.7$. These are approximations.

or, expanding,

$$r = \left(\frac{t_\alpha s}{\bar{x}\sqrt{n}}\right)100$$

Rearranging,

$$n = \left(\frac{s}{\bar{x}}\right)^2\left(\frac{t_\alpha^2}{r^2}\right)100^2$$

or

$$n = \left(\frac{100\ \text{CV}\ t_\alpha}{r}\right)^2 \tag{7.4}$$

We can simplify this equation by assuming for 95% confidence limits that $t_\alpha = 2$, so this reduces to

$$n \cong \left(\frac{200\ \text{CV}}{r}\right)^2 \tag{7.5}$$

where r = Desired relative error (width of confidence interval as percentage)
CV = Coefficient of variation (equation [7.3])

As an example, suppose you are counting plankton samples and know that the coefficient of variation is about 0.70 for such data. If you wish to have ±25% in the relative precision of your mean estimate, then

$$n \approx \left[\frac{(200)(0.70)}{25}\right]^2 = 31.4$$

and you require a sample size of about 31 samples to get a confidence interval of approximately ±25% of the mean.

Eberhardt (1978a) has drawn together estimates of the coefficient of variation from published ecological data (Table 7.2). Clearly it is an oversimplification to assume that all different sampling regions will have a constant coefficient of variation for any ecological measurement. But the point is that as an *approximate* guide and in the absence of more detailed background data, Table 7.2 and equation (7.5) can supply a rough guide to the sampling intensity you should provide.

All of the above formulas assume that the population is very large relative to the number sampled. If you are measuring more than 5–10% of the whole population, you do not need to take so large a sample. The procedure is to calculate the sample size from equation (7.1) or (7.5) above and then correct it with the following *finite population correction*:

$$n^* \cong \frac{n}{1 + (n/N)} \tag{7.6}$$

where n^* = Estimated sample size required for finite population of size N
n = Estimated sample size from (7.1) and (7.5) for infinite population
N = Total size of finite population

TABLE 7.2 COEFFICIENTS OF VARIATION OBSERVED IN A VARIETY OF POPULATION SAMPLING TECHNIQUES TO ESTIMATE POPULATION SIZE[a]

Group of organisms	Coefficient of variation
Aquatic organisms	
Plankton	0.70
Benthic organisms	
Surber sampler, counts	0.60
Surber sampler, biomass or volume	0.80
Grab samples or cores	0.40
Shellfish	0.40
Fish	0.50–2.00
Terrestrial organisms	
Roadside counts	0.80
Call counts	0.70
Transects (on foot)	0.50–2.00
Fecal pellet counts	1.00

[a] Average values compiled by Eberhardt (1978a).

For example, if we know that the whitefish population described above totals only 250 individuals, to get a 95% confidence interval of ±2.8 mm, we first estimate $n = 45.1$ as above, and then

$$n^* \approx \frac{n}{1 + n/N} = \frac{45.1}{1 + (45.1/250)} = 38.2 \text{ fish}$$

Note that the required sample size is always *less* when we use the finite population correction, so that less effort is required to sample a finite population.

These equations to estimate sample size are derived from the normal distribution. What happens if the variable we are measuring does not have a normal, bell-shaped distribution? Fortunately, it does not matter much because of the Central Limit Theorem: *As sample size increases, the means of samples drawn from a population with any shape of distribution will approach the normal distribution.** In practice this theorem means that with large sample sizes ($n > 30$) we do not have to worry about the assumption of a normal distribution. This theorem is among the most important practical findings of theoretical statistics because it means that you do not need to worry if you are measuring a variable that has a skewed distribution—you can still use the approach outlined above (Sokal and Rohlf 1995). But if the distribution is *strongly* skewed, you should be cautious and apply these formulas conservatively.

7.1.2 Comparison of Two Means

Planning experiments that involve comparisons of two or more means is more complicated than deciding on a single sample size. You may, for example, wish to compare whitefish lengths in two different lakes. After the data are collected, you will apply a *t*-test to the

* Assuming the variance is finite for the distribution.

means. How can you decide beforehand how big a sample to use? To do this, you must first decide on the smallest difference you wish to be able to detect:

$$d = |\mu_A - \mu_B| \tag{7.7}$$

where d = Smallest difference you wish to detect
μ_A = Mean value for population A
μ_B = Mean value for population B

For example, you may decide you wish to pick up a difference of 8 mm or more in fish lengths between lakes A and B. This difference must be expressed in units of the standard deviation of the variable being measured:

$$D = \frac{d}{s} \tag{7.8}$$

where D = Standardized smallest difference you wish to detect
d = Smallest difference you wish to detect
s = Standard deviation of variable measured

The standard deviation is assumed to be the same for both populations A and B, and is estimated from previous knowledge or guesswork.

The second decision you must make is the probability you will tolerate for making a Type I or a Type II error; these are usually called α and β:

α = Probability of rejecting the null hypothesis of no difference when in fact it is true (Type I error)

β = Probability of accepting the null hypothesis when in fact it is false and the means really do differ (Type II error)

By convention α is often set to 0.05, but of course it can be set to any other value depending on the consequences of making a Type I mistake. The probability β is less well known to ecologists and yet is critical. It is related to the power of the statistical test:

$$\text{Power} = 1 - \beta \tag{7.9}$$

When β is very small, the test is said to be very powerful, which means that you will not make Type II errors very often. Ecologists are less familiar with β because it is fixed in any typical study after you have specified α, measured a particular sample size, and specified the alternative hypothesis. But if you have some choice in deciding on what sample size to take, you can specify β independently of α. Sokal and Rohlf (1995, 167) and Zar (1996, 81) have a good discussion of the problem of specifying α and β.

Given that you have made these decisions, you can now estimate the sample size needed either by solving the equations given in Mace (1964, 39) or by the use of Table 7.3 from Davies (1956). In most cases the table is more convenient. For example, suppose for your comparison of fish lengths you have

$$d = |\mu_A - \mu_B| = 8 \text{ mm}$$

$$s = 9.4 \text{ mm} \quad \text{(from previous studies)}$$

You decide you wish $\alpha = 0.01$ and $\beta = 0.05$ and are conducting a two-tailed test. Calculate the standardized difference D from equation (7.8):

$$D = \frac{d}{s} = \frac{8}{9.4} = 0.85$$

From Table 7.3 in the second column, we read

$$n = 51$$

This means that you should measure 51 fish from lake A *and* 51 fish from lake B to achieve your stated goal.

Note from Table 7.3 that the smaller the α, and the smaller the β, and the smaller the standardized distance D that you wish to detect, the *larger* the sample size required. This table thus restates the general principle in statistics that if you wish to gain in precision you must increase sample size.

The approach used in Table 7.3 can be generalized for the comparison of 3, 4, 5 . . . means from different populations; thus, this table can be useful for planning relatively complex experiments. Kastenbaum et al. (1970) give a more detailed discussion of sample sizes required when several means are to be compared.

If Table 7.3 does not cover your particular needs, you can use the approximate equation suggested by Snedecor and Cochran (1967, 113) for a comparison of two means:

$$n \cong \frac{2(z_\alpha + z_\beta)^2 s^2}{d^2} \tag{7.10}$$

where n = Sample size required from *each* of the two populations
z_α = Standard normal deviate for α level of probability ($z_{.05} = 1.96$; $z_{.01} = 2.576$)
z_β = Standard normal deviate for the probability of a Type II error (see table below)
s^2 = Variance of measurements (known or guessed)
$d = |\mu_A - \mu_B|$ = Smallest difference between means you wish to detect with probability $1 - \beta$

Equation (7.10) is only approximate because we are using z-values in place of the more proper t-values, but this is a minor error when $n > 50$. The z_β values are obtained from tables of the standard normal deviate (z). A few examples follow (Eberhardt 1978a):

Type II error (β)	Power ($1 - \beta$)	Two-tailed z_β
0.40	0.60	0.25
0.20	0.80	0.84
0.10	0.90	1.28
0.05	0.95	1.64
0.01	0.99	2.33
0.001	0.999	2.58

TABLE 7.3 NUMBER OF SAMPLES REQUIRED FOR A *t*-TEST OF DIFFERENCE BETWEEN THE MEANS OF TWO POPULATIONS[a]

	Level of *t*-test																				
	0.01					0.02					0.05					0.1					
Single-sided test	$\alpha = 0.005$					$\alpha = 0.01$					$\alpha = 0.025$					$\alpha = 0.05$					
Double-sided test	$\alpha = 0.01$					$\alpha = 0.02$					$\alpha = 0.05$					$\alpha = 0.1$					
$\beta =$	0.01	0.05	0.1	0.2	0.5	0.01	0.05	0.1	0.2	0.5	0.01	0.05	0.1	0.2	0.5	0.01	0.05	0.1	0.2	0.5	
0.05																					0.05
0.10																					0.10
0.15																					0.15
0.20																				137	0.20
0.25															124					88	0.25
0.30										123					87					61	0.30
VALUE OF *D* 0.35					110					90					64				102	45	0.35
0.40					85					70				100	50			108	78	45	0.40
0.45				118	68				101	55			105	79	39		108	86	62	28	0.45
0.50				96	55			106	82	45		106	86	64	32		88	70	51	23	0.50
0.55			101	79	46		106	88	68	38		87	71	53	27	112	73	58	42	19	0.55
0.60		101	85	67	39		90	74	58	32	104	74	60	45	23	89	61	49	36	16	0.60
0.65		87	73	57	34	104	77	64	49	27	88	63	51	39	20	76	52	42	30	14	0.65
0.70	100	75	63	50	29	90	66	55	43	24	76	55	44	34	17	66	45	36	26	12	0.70
0.75	88	66	55	44	26	79	58	48	38	21	67	48	39	29	15	57	40	32	23	11	0.75
0.80	77	58	49	39	23	70	51	43	33	19	59	42	34	26	14	50	35	28	21	10	0.80
0.85	69	51	43	35	21	62	46	38	30	17	52	37	31	23	12	45	31	25	18	9	0.85
0.90	62	46	39	31	19	55	41	34	27	15	47	34	27	21	11	40	28	22	16	8	0.90
0.95	55	42	35	28	17	50	37	31	24	14	42	30	25	19	10	36	25	20	15	7	0.95
1.00	50	38	32	26	15	45	33	28	22	13	38	27	23	17	9	33	23	18	14	7	1.00

1.1	42[a]	32	27	22	13	38	28	23	19	11	32	23	19	14	8	27	19	15	12	6	1.1
1.2	36	27	23	18	11	32	24	20	16	9	27	20	16	12	7	23	16	13	10	5	1.2
1.3	31	23	20	16	10	28	21	17	14	8	23	17	14	11	6	20	14	11	9	5	1.3
1.4	27	20	17	14	9	24	18	15	12	8	20	15	12	10	6	17	12	10	8	4	1.4
1.5	24	18	15	13	8	21	16	14	11	7	18	13	11	9	5	15	11	9	7	4	1.5
1.6	21	16	14	11	7	19	14	12	10	6	16	12	10	8	5	14	10	8	6	4	1.6
1.7	19	15	13	10	7	17	13	11	9	6	14	11	9	7	4	12	9	7	6	3	1.7
1.8	17	13	11	10	6	15	12	10	8	5	13	10	8	6	4	11	8	7	5		1.8
1.9	16	12	11	9	6	14	11	9	8	5	12	9	7	6	4	10	7	6	5		1.9
2.0	14	11	10	8	6	13	10	9	7	5	11	8	7	6	4	9	7	6	4		2.0
2.1	13	10	9	8	5	12	9	8	7	5	10	8	6	5	3	8	6	5	4		2.1
2.2	12	10	8	7	5	11	9	7	6	4	9	7	6	5		8	6	5	4		2.2
2.3	11	9	8	7	5	10	8	7	6	4	9	7	6	5		7	5	5	4		2.3
2.4	11	9	8	6	5	10	8	7	6	4	8	6	5	4		7	5	4	4		2.4
2.5	10	8	7	6	4	9	7	6	5	4	8	6	5	4		6	5	4	3		2.5
3.0	8	6	6	5	4	7	6	5	4	3	6	5	4	4		5	4	3			3.0
3.5	6	5	5	4	3	6	5	4	4		5	4	4	3		4	3				3.5
4.0	6	5	4	4		5	4	4	3		4	4	3			4					4.0

Source: Davies 1956, Table E.1.

[a] The entries in the table show the number of observations needed to test for differences of specified size (D) between two means at fixed levels of α and β. Entries in the table show the sample size needed in each of two samples of equal size.

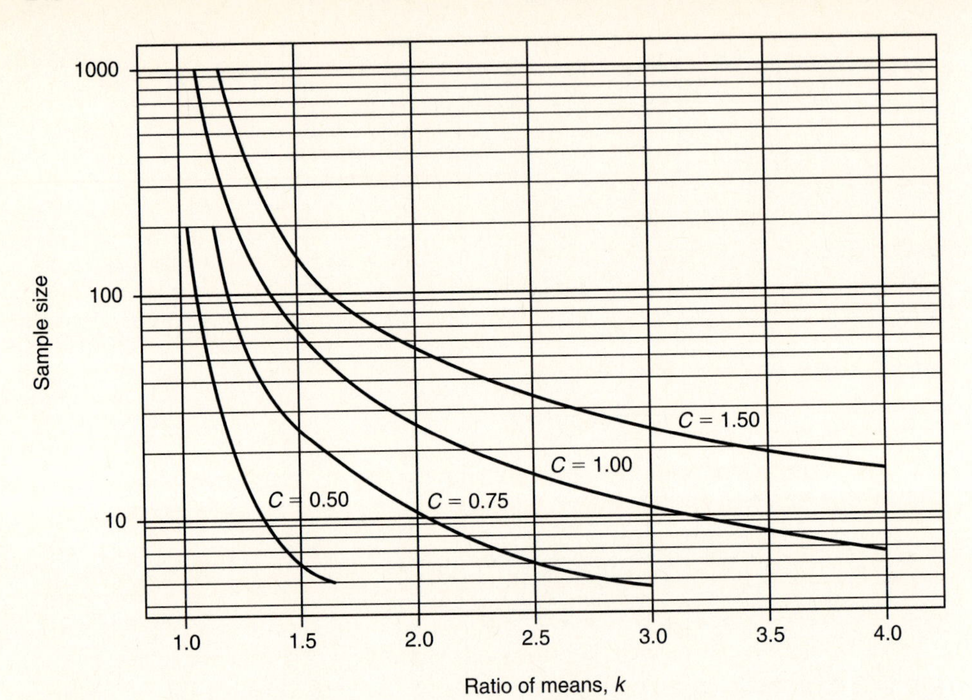

Figure 7.1 Sample sizes required to detect a significant difference between two means expressed as a ratio (k) with $\alpha = 0.05$ and $\beta = 0.20$, plotted for different coefficients of variation (C) that are typical of many population techniques. (From Eberhardt 1978a.)

For example, if you wished to detect a difference in fish lengths in the above example of 4 mm, then

$$d = |\mu_A - \mu_B| = 4 \text{ mm}$$

$$s^2 = (9.4)^2 \text{ mm} \quad \text{(from previous studies)}$$

and $\alpha = 0.01$ and $\beta = 0.05$. Thus from equation (7.10):

$$n \cong \frac{2(2.576 + 1.64)^2(9.4)^2}{4^2} = 196.3 \text{ fish}$$

Note that Table 7.3 does not cover this particular combination of d and s^2.

An alternative approach is to use the tables in Kastenbaum et al. (1970) to estimate the required sample size. Figure 7.1 is derived from these tables. In this case the means are expressed as a ratio you wish to detect:

$$k = \frac{\mu_A}{\mu_B} \tag{7.11}$$

where μ_A = Larger mean
μ_B = Smaller mean
k = Ratio you wish to detect between the means

If you have a rough idea of the coefficient of variation in your data (see Table 7.2), Figure 7.1 can be used to plan a study involving the comparison of two or more means.

Note that the sample size given in Figure 7.1 is required for *each* of the two populations being compared.

7.1.3 Variances from a Normal Distribution

In some cases an ecologist may wish to estimate the variance of a given measurement to a specified level of precision. How can you estimate how large a sample you should take? First, you must specify the allowable limits for the confidence limits on the variance. This is most easily expressed as a percentage (±25%) or as a proportion (±0.25). Second, you must choose the probability (α) of the confidence interval not including the true variance. Then, if sample sizes are not too small ($n > 30$), from Mace (1964, 57),

$$n \cong \frac{3}{2} + z_\alpha^2\left[\frac{1}{\nu}\left(\frac{1}{\nu} + \sqrt{\frac{1}{\nu^2} - 1}\right) - \frac{1}{2}\right] \tag{7.12}$$

where n = Approximate sample size required to estimate the variance
ν = Allowable limits of error (expressed as a proportion) of the variance

In equation (7.12) substitute $z_\alpha = 1.96$ for $\alpha = 0.05$; if you wish to have $\alpha = 0.01$, use $z_\alpha = 2.58$. Other constants may be chosen for other α values from tables of the standard normal distribution.

For example, suppose that you wish to estimate the variance of a measurement with allowable limits for the confidence limits of ±35% and you wish to use $\alpha = 0.05$. Then from equation (7.12),

$$n \cong \frac{3}{2} + 1.96^2\left[\frac{1}{0.35}\left(\frac{1}{0.35} + \sqrt{\frac{1}{0.35} - 1}\right) - \frac{1}{2}\right]$$
$$= 60.3$$

so you would need to take about 60 samples.

Unfortunately, this procedure is applicable only to variances from a *normal* distribution and is quite sensitive to departures from normality.

7.2 SAMPLE SIZE FOR DISCRETE VARIABLES

Counts of the numbers of plants in a quadrat or the numbers of eggs in a nest differ from continuous variables in their statistical properties. The frequency distribution of counts will often be described by either the binomial distribution, the Poisson distribution, or the negative binomial distribution (Elliott 1977). The sampling properties of these distributions differ, so we require a different approach to estimating sample sizes needed for counts.

7.2.1 Proportions and Percentages

Proportions like the sex ratio or the fraction of juveniles in a population are described statistically by the binomial distribution. All the organisms are classified into two classes, and the distribution has only two parameters:

p = Proportion of x types in the population

$q = 1 - p$ = Proportion of y types in the population

We must specify a margin of error (d) that is acceptable in our estimate of p, and the probability (α) of not achieving this margin of error. If sample size is above 20, we can use the normal approximation to the confidence interval:

$$\hat{p} \pm t_\alpha s_{\hat{p}} \tag{7.13}$$

where $\hat{p}$ = Observed proportion
t_α = Value of Student's t-distribution for $n - 1$ degrees of freedom
$s_{\hat{p}}$ = Standard error of $\hat{p} = \sqrt{\hat{p}\hat{q}/n}$

Thus the desired margin of error is

$$d = t_\alpha s_{\hat{p}} = t_\alpha \sqrt{\frac{\hat{p}\hat{q}}{n}}$$

Solving for n, the sample size required is

$$n = \frac{t_\alpha^2 \hat{p}\hat{q}}{d^2} \tag{7.14}$$

where n = Sample size needed for estimating the proportion p
d = Desired margin of error in our estimate

and the other parameters are as defined above.

As a first approximation for $\alpha = 0.05$ we can use $t_\alpha = 2.0$. We need to have an approximate value of p to use in this equation. Prior information, or a guess, should be used; the only rule-of-thumb is that when in doubt, pick a value of p closer to 0.5 than you guess. This will make your answer conservative.

As an example, suppose you wish to estimate the sex ratio of a deer population. You expect p to be about 0.40, and you would like to estimate p within an error limit of ± 0.02 with $\alpha = 0.05$. From equation (7.14),

$$n \cong \frac{(2.0^2)(0.40)(1 - 0.40)}{(0.02)^2} = 2400 \text{ deer}$$

Given this estimate of n, we can recycle to equation (7.14) with a more precise value of $t_\alpha = 1.96$ to get a better estimate of $n = 2305$. So you must classify 2305 deer to achieve this level of precision. If you wish to have $\alpha = 0.01$ ($t_\alpha = 2.576$) you must classify 3981 deer, while if you will permit $\alpha = 0.10$ ($t_\alpha = 1.645$) you must classify only 1624 deer.

You can also use two-stage sampling (Section 8.5) to estimate the proportion p (Cochran 1977, 79). Take a first sample of size n and calculate a preliminary estimate of $\hat{p}$. In the second sample take additional samples to make a total sample size of:

$$n = \left(\frac{\hat{p}_1\hat{q}_1}{\nu}\right) + \left(\frac{3 - 8\hat{p}_1\hat{q}_1}{\hat{p}_1\hat{q}_1}\right) + \left(\frac{1 - 3\hat{p}_1\hat{q}_1}{\nu n_1}\right) \tag{7.15}$$

where n = Total final sample size needed to estimate proportion
$\hat{p}_1$ = Proportion of x types in first sample
$\hat{q}_1 = 1 - p_1$
$\nu = \dfrac{d^2}{t_\alpha^2} \left(\text{for } \alpha = .05,\ \nu \approx \dfrac{d^2}{3.48};\ \text{for } \alpha = .01,\ \nu \approx \dfrac{d^2}{6.64}\right)$
d = Desired margin of error in $\hat{p}$

If you are sampling a finite population of size N, you may then correct your estimated sample size n by equation (7.6), using the finite population correction to reduce the actual

Precision corresponding to curve	Population ratio				
	20:100	40:100	50:100	70:100	100:100
A	±4:100	±6:100	±7:100	±9:100	±13:100
B	±3:100	±4:100	±5:100	±7:100	±10:100
C	±2:100	±3:100	±4:100	±5:100	±7:100

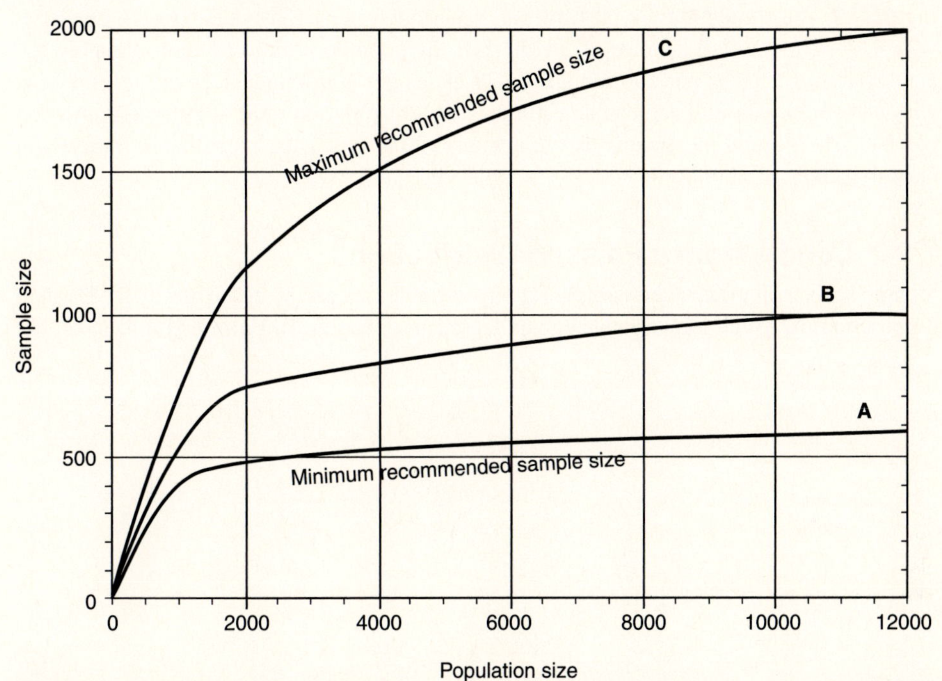

Figure 7.2 Recommended sample sizes for estimating a "population ratio" of two sex or age classes in populations of finite size. Three curves are shown for the three levels of precision given at the top of the graph. Precision is specified here as a 90% confidence interval. A finite population correction is included in these curves, and sampling without replacement is assumed. (From Czaplewski et al. 1983.)

sample size needed. For example, suppose the deer population is known to be only about 1500 animals. From equation (7.6),

$$n^* = \frac{n}{1 + n/N} = \frac{2305}{1 + 2305/1500} = 909 \text{ deer}$$

so a much smaller sample is required in this example if there is a finite population.

Wildlife managers often measure attributes of wildlife populations in a slightly different way with "population ratios." These are expressed, for example, as fawns/100 does, or males/100 females.* For ratios of this type, Czaplewski et al. (1983) have presented a useful series of charts and graphs to determine how large a sample size you need to attain a certain level of precision. In all cases they have assumed sampling without replacement.

*The more conventional notation is as a proportion or percentage; e.g., if 70 males per 100 females, then 70/(100 + 70) or 0.412, or more usually 41% males in the population.

Figure 7.2 shows their recommended sample sizes for a range of total population sizes and for 90% confidence intervals. Note that if the total population is below 2000–4000 animals, the finite population correction (equation [7.6]) reduces the required sample size considerably. If population size exceeds 12,000, for all practical purposes it can be considered infinite.

One example will illustrate the use of Figure 7.2. Assume you have a deer population of about 3000 animals, and you expect a sex ratio of 70 males per 100 females. From Figure 7.2, you should classify at least 520 deer and at most 1350 deer.* In the first case, with $n = 520$, you will achieve an estimate of the population ratio of approximately ± 9 per 100 (i.e., 61 males per 100 females to 79 males per 100 females). In the second case, with $n = 1350$, you will achieve an estimate of the population ratio of approximately ± 5 per 100. The general message is that to achieve a high precision in estimating ratios, you need to take large samples.

7.2.2 Counts from a Poisson Distribution

Sample size estimation is very simple for any variable that can be described by the Poisson distribution, in which the variance equals the mean (Chapter 4). From this it follows that

$$\mathrm{CV} = \frac{s}{\bar{x}} = \frac{\sqrt{s^2}}{\bar{x}} = \frac{\sqrt{\bar{x}}}{\bar{x}}$$

or

$$\mathrm{CV} = \frac{1}{\sqrt{\bar{x}}}$$

Thus from equation (7.5), assuming $\alpha = 0.05$:

$$n \approx \left(\frac{200\mathrm{CV}}{r}\right)^2 = \left(\frac{200}{r}\right)^2 \frac{1}{\bar{x}} \tag{7.16}$$

where n = Sample size required for a Poisson variable
r = Desired relative error (as percentage)
CV = Coefficient of variation $= \frac{1}{\sqrt{\bar{x}}}$

For example, if you are counting eggs in starling nests and know that these counts fit a Poisson distribution and that the mean is about 6.0, then if you wish to estimate this mean with precision of $\pm 5\%$ (width of confidence interval), you have:

$$n \cong \left(\frac{200}{5}\right)^2 \left(\frac{1}{6.0}\right) = 266.7 \text{ nests}$$

Equation (7.16) can be simplified for the normal range of relative errors as follows:

$$\text{For } \pm 10\% \text{ precision} \quad n \cong \frac{400}{\bar{x}}$$

*Note that you can get the same result using equation (7.14).

$$\text{For } \pm 25\% \text{ precision} \quad n \cong \frac{64}{\bar{x}}$$

$$\text{For } \pm 50\% \text{ precision} \quad n \cong \frac{16}{\bar{x}}$$

Note that these are all expressed in terms of the width of the confidence interval, and all are appropriate only for variables that fit a Poisson distribution.

7.2.3 Counts from a Negative Binomial Distribution

Since many animals and plants have a clumped pattern, quadrat counts will often be described by the negative binomial distribution instead of the Poisson (Chapter 4). To estimate the sample size for a series of counts that fit a negative binomial, you must know two variables and decide two more:

1. Mean value you expect in the data ($\bar{x}$)
2. Negative binomial exponent (k)
3. Desired level of error (r) as a percentage
4. Probability (α) of not achieving your desired level of error

The level of error is set as an expected confidence interval, for example, as ±20% of the mean. Since the variance of the negative binomial is given by

$$s^2 = \bar{x} + \frac{\bar{x}^2}{k}$$

we can substitute this expected variance into equation (7.1) and do some algebra to derive an estimated sample size:

$$n = \frac{(100t_\alpha)^2}{r^2}\left(\frac{1}{\bar{x}} + \frac{1}{k}\right) \tag{7.17}$$

where n = Sample size required for a negative binomial variable
t_α = Student's t-value for $n - 1$ degrees of freedom for α probability
$\bar{x}$ = Estimated mean of counts
k = Estimated negative binomial exponent
r = Desired level of error (percent)

For most practical purposes we can assume $t_\alpha = 2.0$ (for 95% confidence limits). As an example, using the data from Box 4.2 on the black-bean aphid with $\bar{x} = 3.46$ and $\hat{k} = 2.65$, assume we would like to have confidence limits of ±15% of the mean:

$$n \cong \frac{200^2}{15^2}\left(\frac{1}{3.46} + \frac{1}{2.65}\right) = 118.5 \text{ quadrats (stems)}$$

As with the Poisson distribution, we can simplify equation (7.17) for the normal range of relative errors used in ecological research:

$$\text{For } \pm 10\% \text{ precision} \quad n \cong 400\left(\frac{1}{\bar{x}} + \frac{1}{k}\right)$$

$$\text{For } \pm 25\% \text{ precision} \quad n \cong 64\left(\frac{1}{\bar{x}} + \frac{1}{k}\right)$$

$$\text{For } \pm 50\% \text{ precision} \quad n \cong 16\left(\frac{1}{\bar{x}} + \frac{1}{k}\right)$$

By comparing these formulas with those developed above for the Poisson, you can see the extra sampling required to attain a given confidence interval with the negative binomial. For example, if you had erroneously assumed that the black-bean aphids had a Poisson distribution with a mean of 3.46, you would estimate for 95% confidence limits of ±15% that you would require a sample size of 51 quadrats, rather than the correct value of 118 quadrats calculated above. This is a vivid illustration of why the tests outlined in Chapter 4 are so critical in ecological sampling and the design of experiments.

7.3 SAMPLE SIZE FOR SPECIALIZED ECOLOGICAL VARIABLES

Some variables that ecologists estimate, like population density, are basically continuous variables that are estimated in indirect ways. For example, we might use a line transect technique (Section 5.1) to estimate population density. Such derived variables are more difficult to analyze in order to decide in advance how big a sample you need, but some approximate techniques are available to help you plan a field study.

7.3.1 Mark-Recapture Estimates

In Chapter 2 we analyzed the simplest mark-recapture method (the Petersen method) in considerable detail, and we included the Robson and Regier (1964) charts (Figures 2.3 and 2.4) for estimating the number of animals that need to be marked and recaptured to achieve a given level of accuracy in the estimate of population size.

An alternative procedure is to use the coefficient of variation of the estimated population size to estimate the required sample size directly. Seber (1982, 60) gives the coefficient of variation of the Petersen population estimate as

$$\text{CV}(\hat{N}) \cong \frac{1}{\sqrt{R}} = \frac{1}{\sqrt{MC/N}} \tag{7.18}$$

where R = Expected number of marked animals to be caught in the second Petersen sample
M = Number of animals marked and released in the first Peterson sample
C = Total number of animals caught in the second Petersen sample
$\hat{N}$ = Estimated population size

This formula can be used in two ways. First, if you have started a Petersen sample and thus know the value of M, and you can guess the approximate population size, you can determine C for any desired level of the coefficient of variation. For example, if you have marked and released 200 fish (M) and you think the population is about 3000 (N), and you wish the coefficient of variation of the estimated population size to be about ±25% (corresponding to a level of precision of ±50%),* we have from equation (7.18)

$$\text{CV}(\hat{N}) = 0.25 = \frac{1}{\sqrt{MC/N}} = \frac{1}{\sqrt{200C/3000}}$$

Solving for C, we obtain

$$C = \frac{3000}{(200)(0.25)^2} = 240 \text{ fish}$$

so we should capture about 240 fish in our second Petersen sample, and we would expect in these 240 fish to get about 16 marked individuals (R). This is only a crude approximation because equation (7.18) assumes large sample sizes for R, and to be conservative you should probably take a somewhat larger sample than 240 fish.

A second use for equation (7.18) is to get a rough estimate of the number of marked individuals (R) that should be obtained in the second Petersen sample. In this case sampling might continue until a prearranged number of recaptures is obtained (Eberhardt 1978a). For example, suppose in the fish example above you wish to have a coefficient of variation of ($\hat{N}$) of 0.05 (5%). From equation (7.18),

$$\text{CV}(\hat{N}) = 0.05 = \frac{1}{\sqrt{R}}$$

Solving for R,

$$R = \frac{1}{(0.05)^2} = 400 \text{ fish}$$

If you wish to obtain such a high degree of accuracy in a Petersen estimate you must recapture 400 *marked* fish, so clearly you must mark and release *more* than 400 in the first Petersen sample. These examples assume a large population, so that the finite population correction is negligible.

The Schnabel method of population estimation utilizes a series of samples (e.g., Table 2.2) and can be readily adapted to a predetermined level of precision. The coefficient of variation for a Schnabel estimate of population size is, from Seber (1982, 190),

$$\text{CV}(\hat{N}) \cong \frac{1}{\sqrt{\Sigma R_t}} \tag{7.19}$$

where $\text{CV}(\hat{N})$ = Expected coefficient of variation for Schnabel estimate of population size

R_t = Number of marked individuals caught in sample t

This formula is similar to equation (7.18). It can be used as follows to decide when to stop sampling in a Schnabel experiment. Consider the data in Table 2.2. After June 11

*Note that the coefficient of variation for population estimates is equal to approximately *one-half* the relative level of precision (r) defined above for $\alpha = 0.05$ (p. 233).

(sample 10) there was a total of 13 recaptures (ΣR_t). Thus

$$CV(\hat{N}) = \frac{1}{\sqrt{13}} = 0.277$$

and thus the 95% confidence limits would be about twice 0.277, or ±55% of the estimated population size $\hat{N}$. By the end of this experiment on June 15 there were 24 recaptures, and thus

$$CV(\hat{N}) \cong \frac{1}{\sqrt{24}} = 0.204$$

so the 95% confidence limits would be approximately ±40% of $\hat{N}$. If you wanted to reduce the confidence interval to approximately ±20% of $\hat{N}$, you would have to continue sampling until $\Sigma R_t = 100$. These estimates of sample size are only a rough guide because they ignore finite population corrections, and equation (7.19) assumes a normal approximation that is not entirely appropriate when sample sizes are small. The important point is that they provide some guidance in planning a marking experiment using the Schnabel method.

There are no simple methods available to estimate sample sizes needed to attain a given level of precision with the Jolly-Seber model. At present the only approach possible is to use a simulation model of the sampling process to help plan experiments. Arnason and Baniuk (1980) have written a computer program named POPAN that will allow you to specify certain sampling rates in a Jolly-Seber model and will simulate the estimation so you can see the levels of precision obtained. Roff (1973) and Pollock et al. (1990) have emphasized that sampling intensity must be very high to attain good levels of precision with Jolly-Seber estimation.

7.3.2 Line Transect Estimates

If you wish to design a line transect survey, you need to determine in advance what length of transect (L) or what sample size (n) you need to attain a specified level of precision. We discussed in Chapter 5 the general methods of estimation used in line transect surveys and in aerial surveys. How can we estimate the sample size we require in line transect surveys?

If we measure the radial distances (r_i) to each animal seen (see Figure 5.2), Eberhardt (1978a) has shown that the coefficient of variation of the density estimate is given by

$$CV(\hat{D}) \approx \sqrt{\frac{1}{n}\left[1 + CV^2\left(\frac{1}{r_i}\right)\right]} \tag{7.20}$$

where $CV(\hat{D})$ = Coefficient of variation of the line transect density estimate D of equation (5.1)

n = Sample size

$CV^2\left(\frac{1}{r_i}\right)$ = (Coefficient of variation of the reciprocals of the radial distances)^2

There are two possibilities for using this equation. The coefficient of variation of the reciprocals of the radial distances can be estimated empirically in a pilot study, and this value used to solve equation (7.20). Alternatively, Eberhardt (1978b) and Seber (1982, 31) suggest that the coefficient of variation of the reciprocals of radial distances is often in the range of 1–3, so that equation (7.20) can be simplified to

$$\mathrm{CV}(\hat{D}) = \sqrt{\frac{b}{n}} \tag{7.21}$$

where b = A constant (typically 1–4)

For any particular study, you could assume b to be in the range of 1–4 and be conservative by estimating it to be relatively high. For example, suppose that you wish to have a coefficient of variation of density of about ±10% so that the 95% confidence interval is about ±20% of the population size. A pilot study has determined that $b = 2$. From equation (7.21),

$$\mathrm{CV}(\hat{D}) = \sqrt{\frac{2}{n}} \quad \text{(assumes } b = 2\text{)}$$

$$0.10 = \sqrt{\frac{2}{n}} \quad \text{or} \quad n = 200$$

Alternatively, if the perpendicular distances (x_i) are used in a line transect estimator (such as equation [5.7]), Eberhardt (1978b) suggests that the approximate coefficient of variation is

$$\mathrm{CV}(\hat{D}) = \sqrt{\frac{4}{n}} \quad \text{(assumes } b = 4\text{)} \tag{7.22}$$

which is exactly double that given above for radial distances.

One alternative strategy to use with line transect data is two-step sampling, in which the first sample is used as a pilot survey. Burnham et al. (1980, 31–37) discuss the details of planning a line transect study by two-stage sampling. They provide a rule-of-thumb equation to predict the length of a line transect you would need to achieve a given precision:

$$\hat{L} = \frac{b}{[\mathrm{CV}(\hat{D})]^2}\left(\frac{L_1}{n_1}\right) \tag{7.23}$$

where
$\hat{L}$ = Length of total survey line required
b = A constant between 1.5 and 4 (recommend $b = 3$)
$\mathrm{CV}(\hat{D})$ = Desired coefficient of variation in the final density estimate
L_1 = Length of pilot survey line
n_1 = Number of animals seen on pilot survey line

There is some disagreement about the value of b in ecological line transect studies (Eberhardt 1978b; Burnham et al. 1980). It may be between 1.5 and 4 but is more likely to be about 3, so Burnham et al. (1980, 36) recommend using $b = 3$ to be conservative or $b = 2.5$ to be more liberal. The ratio (L_1/n_1) could be known from earlier studies and will vary greatly from study to study. For example, in surveys of duck nests it could be 10 nests/km. If the ratio (L_1/n_1) is not known already, you can estimate it from a pilot study.

Equation (7.23) can also be used backwards to estimate what the coefficient of variation in density will be if you can sample a line of length L (due to financial constraints or time). To illustrate, consider the data in Box 5.1 as data from a pilot line transect survey. From equation (7.23), assuming you wish to have a coefficient of variation in the final density of ±10% (so that a 95% confidence interval would be about ±20%), and given that you have seen 12 animals in a line transect of 10 km and wish to be conservative, assume that $b = 3$:

$$\hat{L} = \frac{3}{(0.10)^2}\left(\frac{10}{12}\right) = 250 \text{ km}$$

Consequently you should plan a total transect of 250 km if you wish to achieve this level of precision. But suppose that you only have money to sample 60 km of transect. From equation (7.23), with a conservative estimate of $b = 3$:

$$\hat{L} = 60 = \frac{3}{[\text{CV}(\hat{D})]^2}\left(\frac{10}{12}\right)$$

Solving for $\text{CV}(\hat{D})$,

$$\text{CV}(\hat{D}) = 0.204$$

Thus you could achieve ±20% precision (or a 95% confidence interval of ±40%) in your final density estimate if you could do only 60 km of transect in this particular example.

7.3.3 Distance Methods

There has been relatively little analysis done of the sample size requirements for population density estimates that use plotless sampling methods. Seber (1982, 43) suggests that the coefficient of variation of density estimates from distance sampling will be approximately

$$\text{CV}(\hat{N}) = \frac{1}{\sqrt{sr - 2}} \tag{7.24}$$

where $\text{CV}(\hat{N})$ = Coefficient of variation for plotless sampling estimate of population density
s = Number of random points or random organisms from which distances are measured
r = Number of measurements made from each point (i.e., if only nearest neighbor, $r = 1$)

Thus in sampling with plotless methods you can achieve the same level of precision with (say) 100 random points at which you measure only nearest neighbors as with 50 random points at which you measure the distance to the first and second nearest neighbors.

For example, if you wish to achieve a coefficient of variation of ±20% for population density estimated from distance sampling (which will give a 95% confidence interval of about ±40%), and you are measuring only nearest neighbors ($r = 1$), you have, from equation (7.24),

$$\text{CV}(\hat{N}) = 0.20 = \frac{1}{\sqrt{s - 2}}$$

Solving for s,

$$s = 27 \text{ random points}$$

As with all estimates of population size, this should be taken as an approximate guideline to achieve the desired precision. If you are lucky, this sample size will give you better precision than you expect, but if you are unlucky, you may not achieve the precision specified.

7.3.4 Change-in-Ratio Methods

To estimate population density from the change-in-ratio estimator discussed previously in Chapter 3 (Section 3.1.1), you should consult Paulik and Robson (1969) and Figures 3.1 and 3.2. These graphs will allow you to select the required sample size directly without the need for any computations.

7.4 STATISTICAL POWER ANALYSIS

Much of the material presented in this chapter has been discussed by statisticians under the rubric of power analysis. Statistical power can be viewed most clearly in the classical decision tree involved in statistical inference (hypothesis testing):

	Decision	
State of real world	Do not reject null hypothesis	Reject the null hypothesis
Null hypothesis is actually true	Correct decision (probability = $1 - \alpha$)	Type I error (probability = α)
Null hypothesis is actually false	Type II error (probability = β)	Correct decision (probability = $(1 - \beta)$ = power)

Most ecologists worry about α, the probability of a Type I error, but there is abundant evidence now that we should worry just as much or more about β, the probability of a Type II error (Peterman 1990; Fairweather 1991).

Power analysis can be carried out before you begin your study (*a priori*, or prospective power analysis) or after you have finished (retrospective power analysis). Here we discuss *a priori* power analysis as it is used for the planning of experiments. Thomas (1997) discussed retrospective power analysis.

The key point you should remember is that there are four variables affecting any statistical inference:

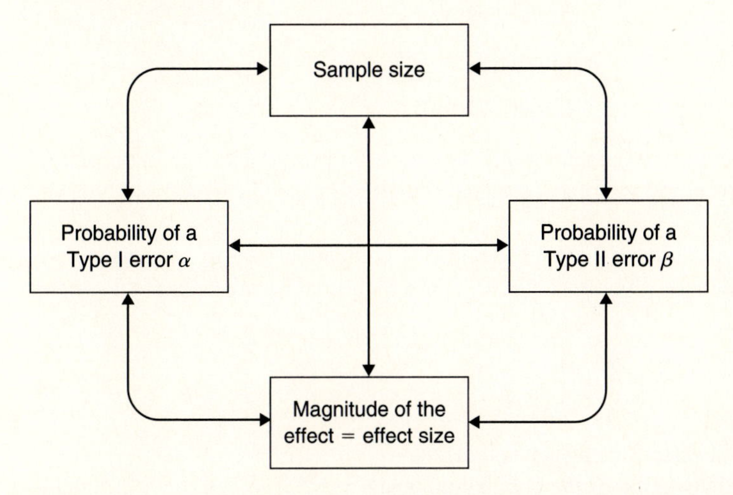

These four variables are interconnected, and once any three of them are fixed, the fourth is automatically determined. Looked at from another perspective, given any three of these, you can determine the fourth. We used this relationship previously when we discussed how to determine sample size (page 230). We can now broaden the discussion and discuss effect size in particular.

7.4.1 Estimates of Effect Size for Continuous Variables

Effect size is a general term that is interpreted in different ways by ecologists and statisticians. To an ecologist, the effect size is typically the difference in means between two populations: Fish in lake A are 2.5 kg heavier than fish in lake B. But statisticians prefer to convert these biological effect sizes to standardized effect sizes so they can discuss the general case. Thus an ecologist must first learn how to translate biological effect sizes into statistical effect sizes.

The most common situation involves the means of two populations that we would like to test for equality. The statistical effect size used here is the standardized difference in the means, defined on page 236 as:

$$D = \frac{\bar{x}_A - \bar{x}_B}{s} \tag{7.25}$$

where D = Standardized effect size or standardized difference in means
$\bar{x}_A, \bar{x}_B$ = Postulated mean values for group A and group B
s = Standard deviation of the variable being measured

D is a dimensionless number, so it can be used for any continuous variable that has an approximately normal distribution. Note that by scaling our means in relation to the standard deviation, we implicitly bring the variability of the data into our power analysis.

The effect size that is ecologically significant will depend on the specific situation, and is part of the overall problem of defining biological significance in contrast to statistical significance. Cohen (1988) has provided very general guidelines for effect sizes for comparison of two means. He defines statistical effect sizes from equation (7.25) as

Small $D = 0.2$

Medium $D = 0.5$

Large $D = 0.8$

These must be considered only general guidelines for effect sizes, and it is important to measure effects with respect to specific ecological situations or specific hypotheses you are testing.

For the analysis of variance in which we need to compare more than two means, we need to define a more general effect size. Cohen (1988) defines the statistical effect size for several means as

$$f = \frac{s_m}{s} \tag{7.26}$$

where f = Statistical effect size for several means
s_m = Standard deviation of the population means
s = Standard deviation within the populations

In the usual analysis of variance, the standard deviation s is assumed to be the same in all groups being compared and is computed as the square root of the "error" variance. The standard deviation of the population means is easy to determine when all groups have equal sample sizes (a balanced design). From Cohen (1988, 275),

$$\hat{s}_m = \sqrt{\frac{\Sigma_k (\bar{x}_i - \bar{\bar{x}})^2}{k}} \tag{7.27}$$

where k = Number of means being compared
$\bar{x}_i$ = Mean value for group i
$\bar{\bar{x}}$ = Grand mean value for all groups being compared

When several means are being compared, the spread of the means can be highly variable. For example, means may be equally spaced from high to low, or several means may be equal and only one divergent. This variation is measured by s_m. When sample sizes are not equal in all groups, weighted means must be used in equation (7.27) as described in Cohen (1988, 359). This general approach can be applied to all analysis of variance problems, including interaction effects. Cohen (1988) gives the details.

Cohen (1988) provides general guidelines for statistical effect sizes for analysis of variance problems in the same way discussed above:

Small $f = 0.10$

Medium $f = 0.25$

Large $f = 0.40$

These are only general guidelines for researchers and not measures of biological significance.

7.4.2 Effect Size for Categorical Variables

In the simplest case we wish to compare two proportions, for example the proportion of males in two bird species. The simplest measure of effect size for a biologist would be the expected difference in the proportions—for example, 0.45 vs. 0.55 for the two populations. But there is a statistical problem in using this intuitive difference as a measure of effect size, because the variance of the binomial distribution changes with the p-values, becoming maximal at $p = 0.5$. The statistical solution is to transform the proportions to φ values:

$$\varphi = 2 \arcsin \sqrt{p} \tag{7.28}$$

We can then define the effect size as

$$\begin{aligned} h &= \varphi_1 - \varphi_2 \quad \text{(for one-tailed alternatives)} \\ h &= |\varphi_1 - \varphi_2| \quad \text{(for two-tailed alternatives)} \end{aligned} \tag{7.29}$$

Cohen (1988) defines a range of effect sizes for proportions, as follows:

Small $h = 0.20$

Medium $h = 0.50$

Large $h = 0.80$

For proportions between about 0.2 and 0.8, these three levels correspond to differences of 0.09 to 0.10 for small, 0.23 to 0.25 for medium, and 0.35 to 0.39 for large effect sizes.

7.4.3 Power Analysis Calculations

The calculations involved in power analysis are complex and are not presented in detail here. Figure 7.3 illustrates graphically how power analysis works in principle. In this case we have a fixed effect size and a fixed sample size, and the power of the test is the area of nonoverlap of the expected statistical distributions. Like any statistical test, power calculations involve probabilities, and we cannot be certain for any particular set of observations that we will in fact achieve the intended result.

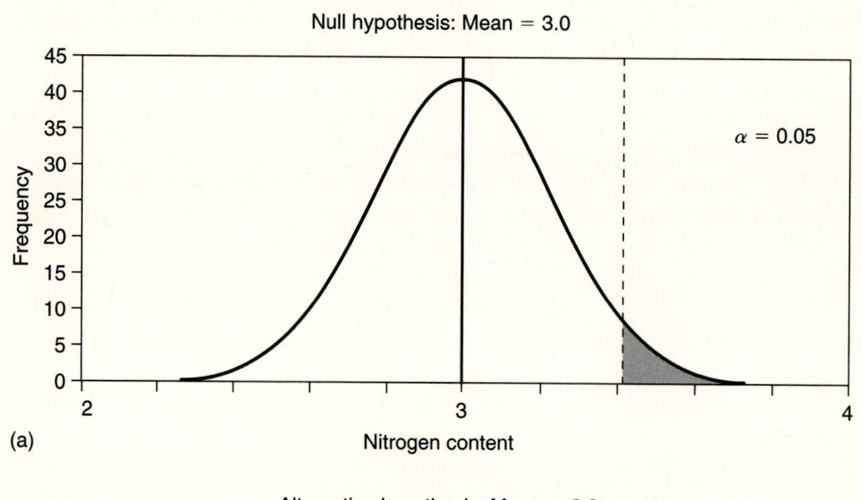

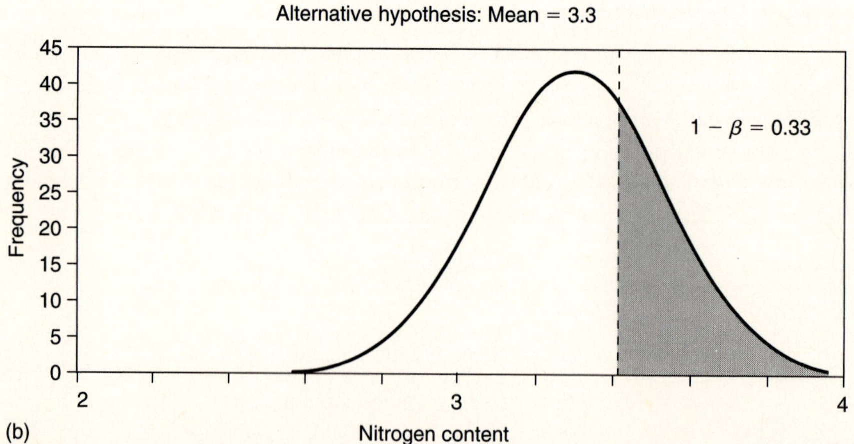

Figure 7.3 An example of how power calculations can be visualized. In this simple example, a t-test is to be carried out to determine if the plant nitrogen level has changed from the base level of 3.0% (the null hypothesis) to the improved level of 3.3% (the alternative hypothesis). Given $n = 100$, $s^2 = 2.5$, and $\alpha = 0.05$, we can see that under the alternative hypothesis the distribution of possible means will follow a t-distribution that is shifted to the right. The shaded area is the power $(1 - \beta)$ that can be expected in this study, and consequently the likelihood of a Type II error is relatively high.

A power analysis consists of fixing three of the four variables α, β, sample size, and effect size, and then calculating the fourth. It is often desirable to investigate a range of values rather than just one. For example, we may wish to know for a particular system the relationship between sample size and power for a given α and a specified effect size. Figure 7.4 illustrates one example for a study of parasitism rates.

Power analysis is a relatively recent form of analysis in ecological studies, but it has been discussed by statisticians for over 50 years. Only recently have ecologists begun to worry about Type II errors arising from studies in which the power to reject the null hypothesis is very low. Weak studies, studies with low power, may be difficult to avoid if money or time is limited, but we should be aware of the relative precision of our decisions. Failing to reject the null hypothesis should not be confused with accepting the truth of the null hypothesis. Power analysis is useful precisely because it forces ecologists to provide specific alternative hypotheses in order to estimate power, and these alternative hypotheses must reflect ecological theory or ecologically significant effects.

Power is relatively easy to compute now, even for complex experimental designs, using the personal computer. Power calculations can be obtained from a variety of specific programs that are commercially available (nQuery, PASS, or Stat-Power) or freeware that is available from a variety of sites (Goldstein 1989; Thomas and Krebs 1997). These programs can facilitate the planning of ecological studies and should be a part of every ecologist's repertoire.

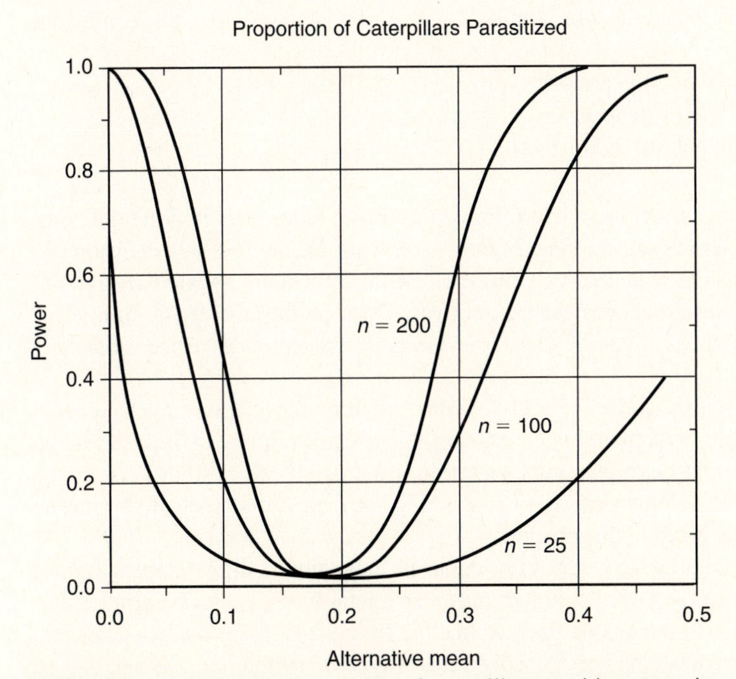

Figure 7.4 Power curves for a study of caterpillar parasitism rates, in which the expected average parasitism rate is 18%, and $\alpha = 0.01$. Power curves for sample sizes of $n = 25$, $n = 100$, and $n = 200$ are shown for a spectrum of alternative rates. Clearly, for this ecological situation, sample sizes of $n = 25$ cannot achieve high statistical power. Calculations were carried out with program PASS 6.0 (see Thomas and Krebs 1997).

7.5 WHAT TO DO IF NOTHING WORKS

Some situations will arise in any ecological study in which none of the methods outlined above seem to work. This may be because you have a complex sampling problem or because you do not have any of the background information needed to use the techniques given in this chapter. In this case you can do one of two things. First, consult a professional statistician. Statisticians have developed many clever techniques for estimating sample size in complex experimental designs that are beyond the scope of this book. In many cases it is possible to use computer-intensive randomization procedures like bootstrapping to calculate the statistical power of particular experimental designs. This is a rapidly developing area in statistics and particularly useful for ecologists.

Alternatively, you can adopt a completely empirical approach and decide the question in a stepwise manner. Every variable you can measure has an appropriate variance estimate attached to it, so that you can compute 95% confidence limits after each set of samples. This crude, empirical approach is more feasible now that computers can be taken directly into the field complete with programs to compute tedious formulas. The general procedure is as follows:

1. Collect samples for 1 day or 1 week or whatever time unit is relevant to study.
2. Compute at the end of that period the mean estimate and its 95% confidence interval for all the data gathered to date.
3. Decide on the basis of your desired level of precision whether the confidence interval is
   (a) sufficiently small—go to **4**;
   (b) still too large—go to **1**.
4. Stop and compute final estimates.

Figure 7.5 illustrates this simple procedure for counts of red alder trees in 4 m^2 quadrats. Clearly as sample size grows, the confidence interval will shrink, and the only decision you must make is when to stop. Note that you must *not* use this procedure to calculate a test of significance (like the *t*-test) and continue sampling until you get the answer you want! This approach can be considered a type of *sequential sampling*, which is described more fully in Chapter 9.

The approach used in Figure 7.5 will not work when it takes a long time to process samples, as, for example, with plankton counts taken on oceanic cruises. In these cases you must obtain prior information to plan your sampling, either from earlier studies, or from a pilot experiment whose sole purpose is to give initial estimates of statistical parameters needed in the equations listed in this chapter.

In some ecological areas enough wisdom has accumulated to allow empirical rules to be formulated for sampling even though it may take a long time to process samples. For example, Downing (1979) suggested that for benthic organisms in freshwater lakes and large rivers—no matter what the species, substrate, or type of sampling machinery—you can predict the standard error of a series of samples from the multiple regression equation:

$$\text{S.E.} = \frac{\text{antilog}(0.581 + 0.696 \log \bar{x} - 2.82 \times 10^{-4} A)}{\sqrt{n}} \tag{7.30}$$

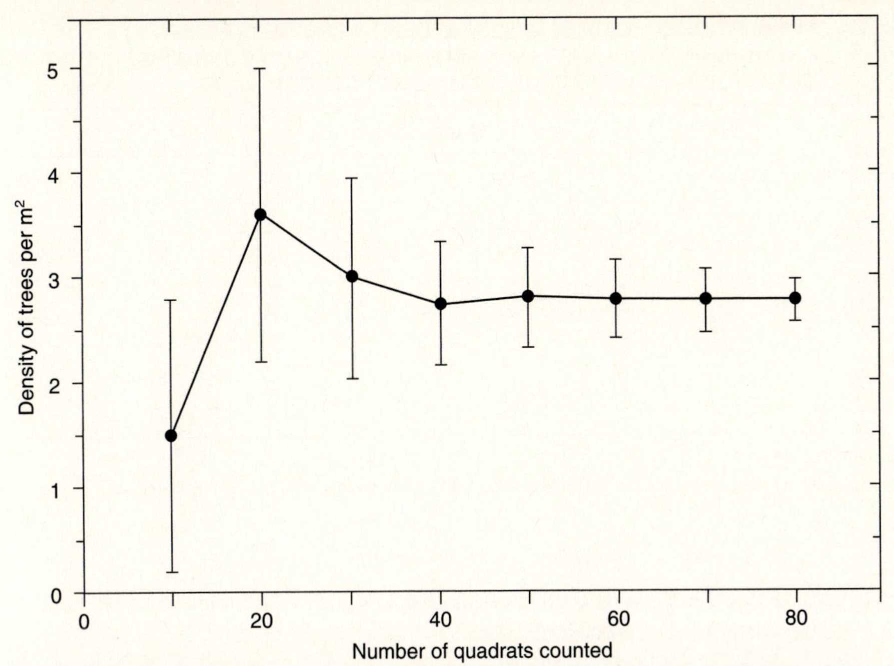

Figure 7.5 An empirical approach to determining how large a sample to take. An ecology class counted red alder trees on an area undergoing secondary succession. After each 10 quadrats were counted, the mean and 95% confidence interval were plotted. Sampling was continued until the confidence interval was judged to be sufficiently small.

where S.E. = Standard error of mean density estimate
$\bar{x}$ = Mean density in numbers per m^2
A = Area removed by benthic sampler in cm^2 (sample area)
n = Sample size

You can solve this equation for sample size n for any particular application to benthic sampling, or consult Table 7.4, which gives sample sizes needed to achieve a confidence interval of about ±40% of the mean. This type of approximate rule-of-thumb might be useful for other ecological sampling methods in which some prior information is useful in planning field studies (Taylor 1980).

Similarly, for marine and freshwater zooplankton samples, Downing et al. (1987) found that data from over a thousand sets of replicate samples could all be described by the following general relationship:

$$s^2 = 0.745m^{1.622}v^{-0.267} \quad (7.31)$$

where s^2 = Estimated variance among replicate zooplankton samples
m = Mean number of organisms per liter
v = Volume of the sampler you are using (in liters)

TABLE 7.4 NUMBER OF REPLICATE SAMPLES NEEDED TO SAMPLE THE BENTHIC INVERTEBRATES OF LAKES AND LARGE RIVERS IN RELATION TO SIZE OF SAMPLER AND DENSITY OF ORGANISMS STUDIED[a]

Density (number/m²)	Size of sampler (cm²)						
	20	50	100	250	500	750	1000
30	45	43	40	33	24	17	12
50	33	32	30	24	18	13	9
100	22	21	19	16	12	8	6
300	11	11	10	8	6	4	3
500	8	8	7	6	4	3	2
1,000	5	5	5	4	3	2	<2
5,000	<2	<2	<2	<2	<2	<2	<2
10,000	<2	<2	<2	<2	<2	<2	<2

Source: Downing 1979.

[a] These sample sizes are designed to achieve a precision of about ±40% of the mean for the 95% confidence interval.

This empirical relationship can be used to estimate the number of samples you need to take to achieve a specified level of precision:

$$\hat{n} = 0.754m^{-0.378}\nu^{-0.267}p^{-2} \tag{7.32}$$

where $\hat{n}$ = Estimated sample size needed for zooplankton sampling

m = Estimated number of organisms per liter

ν = Volume of the sampler you are using (in liters)

$p = \dfrac{\text{S.E.}}{m}$ = Relative level of precision desired (as a proportion)

S.E. = Estimated standard error of the zooplankton counts, from equation (7.31)

Once you have decided on the level of precision you need (p), you must guess the density of zooplankton you wish to sample (m) to obtain an estimate of the sample size you will need. Downing et al. (1987) discuss how much time and money can be saved by adopting this approach to zooplankton sampling in both marine and in freshwater environments.

There is probably more money wasted in ecological research by the use of poorly designed studies than anyone cares to guess. In some studies too much work is done, and estimates are accurate to ±5% when ±20% will do. Much more frequently, however, too little work is done, and the data have such wide confidence limits as to be useless. If you are trying to test a hypothesis that predicts a 10% change in metabolism association with water quality, and your sample sizes are fixed so that 95% confidence limits of ±40% are expected, you might as well not do the experiment. So the message is simple: *Plan ahead.*

To assist in planning, program-group SAMPLE SIZE (Appendix 2) will compute the sample sizes required for a specified level of precision for all the formulas given in this chapter.

7.6 SUMMARY

The most common question in ecological research is, *How large a sample should I take*? This chapter attempts to give a general answer to this question by providing a series of equations from which sample size may be calculated. It is always necessary to know something about the population you wish to analyze and this leads into a statistical catch-22 unless you use guesswork or prior observations. You must also make some explicit decision about how much error you will allow in your estimates (or how small a confidence interval you wish to have).

For continuous variables like weight or length, we can assume a normal distribution and calculate the required sample sizes for means and for variances quite precisely. For counts, we need to know the underlying statistical distribution—binomial, Poisson, or negative binomial—before we can specify sample sizes needed. Patchy patterns described by the negative binomial require much larger sample sizes to describe precisely than do random patterns.

Estimates of population size and density from mark-recapture methods can be achieved to a specified level of precision by the use of graphs and a few simple, rule-of-thumb approximations. Similar approximate methods for deciding on sample sizes are available for line transect studies, plotless sampling, and change-in-ratio estimators of population size.

Power analysis explores the relationships between the four interconnected variables α (probability of Type I error), β (probability of Type II error), effect size, and sample size. Fixing three of these automatically fixes the fourth, and ecologists should explore these relationships before they begin their experiments. Significant effect sizes should be specified on ecological grounds before a study is begun.

If all else fails, a completely empirical trial-and-error method can be used to decide on sample size by taking samples in steps and computing the confidence limits after each step. Computer-intensive randomization methods can be used to investigate the statistical power of simple to complex experimental designs. The important message is always to *plan ahead*.

SELECTED READING

Cochran, W. G. 1977. *Sampling Techniques*. Chapter 4. pp. 72–88, John Wiley, New York.

Cohen, J. 1992. A power primer. *Psychological Bulletin* 112: 155–159.

Downing, J. A. 1979. Aggregation, transformation, and the design of benthos sampling programs. *Journal of the Fisheries Research Board of Canada* 36: 1454–1463.

Eberhardt, L. L. 1978. Appraising variability in population studies. *Journal of Wildlife Management* 42: 207–238.

Kastenbaum, M. A., Hoel, D. G., and Bowman, K. O. 1970. Sample size requirements: One-way analysis of variance. *Biometrika* 57: 421–430.

Kraemer, H. C., and S. Thiemann. 1987. *How many subjects?* Sage Publications, Newbury Park, California.

Peterman, R. M. 1990. Statistical power analysis can improve fisheries research and management. *Canadian Journal of Fisheries and Aquatic Sciences* 47: 2–15.

Robson, D. S., and Regier, H. A. 1964. Sample size in Petersen mark-recapture experiments. *Transactions of the American Fisheries Society* 93: 215–226.

QUESTIONS AND PROBLEMS

7.1. Estimate the sample size required for quadrat counts of goldenrod (*Solidago* spp.) in an old field in which a random spatial pattern occurs, and the expected density on 1 m^2 quadrats is 3.32 plants/m^2. Assume that you wish to have a 95% confidence interval of $\pm 1\%$ of the mean.
(a) What changes in sampling must occur if you count 4 m^2 quadrats instead of 1 m^2 quadrats?
(b) If you count 16 m^2 quadrats?

7.2. A mammal ecologist wishes to compare the mean litter size of ground squirrels on a control area with that of squirrels living in a radiation-disposal site. She specifies the probability of a Type I error as 0.01 and the probability of a Type II error as 0.05. The standard deviation of litter sizes from previous ground squirrel studies has been 0.90 to 1.65. She wishes to pick up a difference between her two populations of 1.0 embryos or more. What sizes of samples should she take?
(a) What recommendation can you make to her if it is much more difficult to sample from site B than from site A, so that 3/4 of the total samples will come from site A?

7.3. A limnologist wishes to estimate the size of a *Daphnia magna* population in a lake. He knows that replicate samples fit a negative binomial distribution, and from past experience, k is approximately 3.8. He wishes to know the mean density within a 99% confidence interval of $\pm 40\%$, and he expects the mean density to be about 18.0 animals/sample. How many samples should he take?
(a) If he can accept a 99% confidence interval of $\pm 50\%$, how much work can he save?

7.4. A marine ecologist is doing an aerial census of porpoises off California. He expects to see about 0.04 porpoises per km of travel, and he needs to know population density within an 80% confidence belt of $\pm 30\%$. How long a transect should he plan?

7.5. A wildlife ecologist wishes to estimate the population ratio of fawns/100 does in an antelope population of northern Utah. She thinks there are between 8000 and 10,000 antelope on her study area. She expects the ratio to be approximately 80 fawns/100 does, and she would like to achieve a 95% confidence belt of ± 5 fawns/100 does. How many animals should she count and classify?
(a) How would the sample size be affected if there were really only 4000 antelope in the area?

7.6. A marine ecologist is planning a mark-recapture program using the Schnabel method to estimate the size of a beluga whale population. How many recaptures are needed to achieve a coefficient of variation of the population estimate of $\pm 20\%$? Discuss several possible ways of achieving this goal of getting the number of recaptures required.

7.7. Compare the relative efficiency of using quadrats to estimate population density and using plotless sampling. What ecological factors and what statistical factors will influence your decision about which technique to use?

7.8. Construct a power curve relating effect size (x-axis) to sample size (y-axis) to answer the following question: What sample size is required to recognize a specific effect of fertilizer on plant growth at $\alpha = 0.05$ and $\beta = 0.10$, when the expected mean is 4.8 g with expected standard deviation 0.5? Express the effect size in biological units (g). Table 7.3 can be utilized, or more specific power software packages.

7.9. Discuss in general the question of how large a benthic sampling device you should use, given the recommendations in Table 7.4. Refer to Chapter 3 and discuss exactly how you would make this decision statistically. What practical matters might influence your decision about size of sampler?

CHAPTER 8

Sampling Designs: Random, Adaptive, and Systematic Sampling

Ecologists sample whenever they cannot do a complete enumeration of the population. Very few plant and animal populations can be completely enumerated, so most of our ecological information comes from samples. Good sampling methods are critically important in ecology because we want our samples to be representative of the population under study. *How do we sample representatively*? This chapter attempts to answer this question by summarizing the most common sampling designs that statisticians have developed over the last 80 years. Sampling is a practical business, and there are two parts of its practicality. First, the gear used to gather the samples must be designed to work well under field conditions. In all areas of ecology there has been tremendous progress in the past 40 years to improve sampling techniques. These improvements are the subject of many detailed handbooks, so if you need to know what plankton sampler is best for oligotrophic lakes, or what light trap is best for nocturnal moths, you should consult the specialist literature in your subject area. Second, the method of placement and the number of samples must be decided, and this is what statisticians call *sampling design*. Should samplers be placed randomly or systematically? Should different habitats be sampled separately or all together? These are the general statistical questions I will address in this and the next chapter. I will develop a series of guidelines that will be useful in sampling plankton with nets, moths with light traps, and trees with distance methods. The methods discussed here are addressed in more detail by Cochran (1977), Jessen (1978), and Thompson (1992).

8.1 SIMPLE RANDOM SAMPLING

The most convenient starting point for discussing sampling designs is simple random sampling. Like many statistical concepts, random sampling is easier to explain on paper than it is to apply in the field. Some background is essential before we can discuss random sampling. First, you must specify very clearly what the *statistical population* is that you are trying to study. The statistical population may or may not be a biological population, and the two ideas should not be confused. In many cases the statistical population is clearly specified: the white-tailed deer population of the Savannah River Ecological Area, or the whitefish population of Brooks Lake, or the black oaks of Warren Dunes State Park. But in other cases the statistical population has poorly defined boundaries: the mice that will enter live traps, the haddock population of George's Bank, the aerial aphid population over southern England, the seed bank of *Erigeron canadensis*. Part of this vagueness in ecology depends on spatial scale and has no easy resolution. Part of this vagueness also flows from the fact that biological populations also change over time (Van Valen 1982). One strategy for dealing with this vagueness is to define the statistical population very sharply on a local scale and then draw statistical inferences about it. But the biological population of interest is usually much larger than a local population, and one must then extrapolate to draw some general conclusions. You must think carefully about this problem. If you wish to draw statistical inferences about a widespread biological population, you should sample the widespread population. Only this way can you avoid extrapolations of unknown validity. The statistical population you wish to study is a function of the question you are asking. This problem of defining the statistical population and then relating it to the biological population of interest is enormous in field ecology, and almost no one discusses it. It is the *first* thing you should think about when designing your sampling scheme.

Second, you must decide what the *sampling unit*, or experimental unit, is in your population. The sampling unit could be simple, like an individual oak tree or an individual deer, or it can be more complex like a plankton sample, or a 4 m^2 quadrat, or a branch of an apple tree. The sample units must cover the whole of the population, and they must not overlap. In most areas of ecology there is considerable practical experience available to help decide on what sampling unit to use.

The third step is to select a sample, and a variety of sampling plans can be adopted. The aim of the sampling plan is to maximize efficiency—to provide the best statistical estimates with the smallest possible confidence limits at the lowest cost. To achieve this aim, we need some help from theoretical statistics, so that we can estimate the precision and the cost of the particular sampling design we adopt. Statisticians always assume that a sample is taken according to the principles of *probability sampling*, as follows:

1. Define a set of distinct samples S_1, S_2, S_3 . . . in which certain specific sampling units are assigned to S_1, some to S_2, and so on.
2. Assign each possible sample a probability of selection.
3. Select one of the S_i samples by the appropriate probability and a random number table.

If you collect a sample according to these principles of probability sampling, a statistician can determine the appropriate sampling theory to apply to the data you gather.

Some types of probability sampling are more convenient than others, and simple random sampling is one. *Simple random sampling* is defined as follows:

1. A statistical population is defined that consists of N sampling units;
2. We select n units from the possible samples in such a way that every unit has an *equal* chance of being chosen.

The usual way of achieving simple random sampling is to number each possible sample unit from 1 to N. A series of random numbers between 1 and N is then drawn either from a table of random numbers or from a set of numbers in a hat. The sample units that happen to have these random numbers are measured and constitute the sample to be analyzed. It is hard in ecology to follow these simple rules of random sampling if you wish the statistical population to be much larger than the local area you are actually studying.

Usually, once a number is drawn in simple random sampling, it is not replaced, so we have *sampling without replacement*. Thus, if you are using a table of random numbers and you get the same number twice, you ignore it the second time. It is possible to replace each unit after measuring so that you can sample *with replacement*, but this is less often used in ecology. Sampling without replacement is more precise than sampling with replacement (Caughley 1977b).

Simple random sampling is sometimes confused with other types of sampling that are not based on probability sampling. Examples abound in ecology:

1. *Accessibility sampling*: The sample is restricted to those units that are readily accessible. Samples of forest stands may be taken only along roads, or deer may be counted only along trails.
2. *Haphazard sampling*: The sample is selected haphazardly. A bottom sample may be collected whenever the investigator is ready, or ten dead fish may be picked up for chemical analysis from a large fish kill.
3. *Judgmental sampling*: The investigator selects on the basis of experience a series of "typical" sample units. A botanist may select "climax" stands of grassland to measure.
4. *Volunteer sampling*: The sample is self-selected by volunteers who will complete some questionnaire or be used in some physiological test. Hunters may complete survey forms to obtain data on kill statistics.

The important point to remember is that all of these methods of sampling *may* give the correct results under the right conditions. Statisticians, however, reject all of these types of sampling because they can not be evaluated by the theorems of probability theory. Thus the universal recommendation: *Random sample*! But in the real world it is not always possible to use random sampling, and ecologists are often forced to use nonprobability sampling if they wish to get any information at all. In some cases it is possible to compare the results obtained with these methods to those obtained with simple random sampling (or with known parameters), so that you could decide empirically if the results were representative and accurate. But remember that you are always on shaky ground if you must use nonprobability sampling; the means and standard deviations you calculate may not be close to the true values. Whenever possible use some conventional form of random sampling.

8.1.1 Estimation of Parameters

In simple random sampling, one or more characteristics are measured on each experimental unit. For example, in quadrat sampling you might count the number of individuals of *Solidago* spp. and the number of individuals of *Aster* spp. In sampling deer, you might record for each individual its sex, age, weight, and fat index. In sampling starling nests you might count the number of eggs and measure their length. In all these cases and in many more, ecological interest is focused on four characteristics of the population:*

1. Total $= X$; for example, the total number of *Solidago* individuals in the entire 100 ha study field.
2. Mean $= \bar{x}$; for example, the average number of *Solidago* per m^2.
3. Ratio of two totals $= R = x/y$; for example, the number of *Solidago* per *Aster* in the study area.
4. Proportion of units in some defined class; for example, the proportion of male deer in the population.

We have seen numerous examples of characteristics of these types in the previous seven chapters, and their estimation is covered in most introductory statistics books. We cover them again here briefly because we need to add to them an idea that is not usually considered in introductory books: the *finite population correction*. For any statistical population consisting of N units, we define the finite population correction (*fpc*) as

$$fpc = \frac{N - n}{N} = 1 - \frac{n}{N} \qquad (8.1)$$

where fpc = Finite population correction
N = Total population size
n = Sample size

and the fraction of the population sampled (n/N) is sometimes referred to as f. In a very large population the finite population correction will be 1.0, and when the whole population is measured, the *fpc* will be 0.0.

Cochran (1977) demonstrates that unbiased estimates of the population mean and total are given by the following formulas. For the mean,

$$\bar{x} = \frac{\Sigma x_i}{n} \qquad (8.2)$$

*We may also be interested in the *variance* of the population, and the same general procedures apply.

where $\bar{x}$ = Population mean
x_i = Observed value of x in sample i
n = Sample size

For the variance of the measurements, we have the usual formula:*

$$s^2 = \frac{\Sigma (x_i - \bar{x})^2}{n - 1} \tag{8.3}$$

and the standard error of the population mean $\bar{x}$ is given by:

$$s_{\bar{x}} = \sqrt{\frac{s^2}{n}(\sqrt{1 - f})} \tag{8.4}$$

where $s_{\bar{x}}$ = Standard error of the mean $\bar{x}$
s^2 = Variance of the measurements as defined above (8.3)
n = Sample size
f = Sampling fraction = n/N

These formulas are similar to those you have always used, except for the introduction of the finite population correction. Note that when the sampling fraction (n/N) is low, the finite population correction is nearly 1.0, so the size of the population has no effect on the size of the standard error. For example, if you take a sample of 500 measurements from two populations with the same variance ($s^2 = 484.0$) and population A is small ($N = 10{,}000$) and population B is a thousand times larger ($N = 10{,}000{,}000$), the standard errors of the mean differ by only 2.5% because of the finite population correction. For this reason, the finite population correction is usually ignored whenever the sampling fraction (n/N) is less than 5% (Cochran 1977).

Estimates of the population total are closely related to these formulas for the population mean. For the population total,

$$\hat{X} = N\bar{x} \tag{8.5}$$

where $\hat{X}$ = Estimated population total
N = Total size of population
$\bar{x}$ = Mean value of population

The standard error of this estimate is given by

$$s_x = N s_{\bar{x}} \tag{8.6}$$

where s_x = Standard error of the population total
N = Total size of the population
$s_{\bar{x}}$ = Standard error of the mean from equation (8.4)

Confidence limits for both the population mean and the population total are usually derived by the normal approximation:

$$\bar{x} \pm t_\alpha s_{\bar{x}} \tag{8.7}$$

where t_α = Student's t value for $n - 1$ degrees of freedom for the $(1 - \alpha)$ level of confidence

*If the data are compiled in a frequency distribution, the appropriate formulas are supplied in Appendix 1.

This formula is used all the time in statistics books with the warning not to use it except with "large" sample sizes. Cochran (1977, 41) gives a rule-of-thumb that is useful for data that has a strong positive skew (like the negative binomial curves in Figure 4.8). First, define Fisher's measure of skewness:

$$\begin{aligned} g_1 &= \text{Fisher's measure of skewness} \\ &= \frac{1}{ns^3} \sum (x - \bar{x})^3 \end{aligned} \quad (8.8)$$

Cochran's rule is that you have a large enough sample to use the normal approximation (equation [8.7]) if

$$n > 25g_1^2 \quad (8.9)$$

where n = Sample size
s = Standard deviation

Sokal and Rohlf (1995, 116) show how to calculate g_1 from sample data, and many statistical packages provide computer programs to do these calculations. Box 8.1 illustrates the use of these formulas for simple random sampling.

8.1.2 Estimation of a Ratio

Ratios are not as commonly used in ecological work as they are in taxonomy, but sometimes ecologists wish to estimate from a simple random sample a ratio of two variables, both of which vary from sampling unit to sampling unit. For example, wildlife managers

Box 8.1 Estimation of Population Mean and Population Total from Simple Random Sampling of a Finite Population

A biologist obtained body weights of male reindeer calves from a herd during the seasonal roundup. He obtained weights on 315 calves out of a total of 1262 in the herd, and summarized them as follows:

Body weight class (kg)	Midpoint, x	Observed frequency, f_x
29.5–34.5	32	4
34.5–39.5	37	13
39.5–44.5	42	20
44.5–49.5	47	49
49.5–54.5	52	61
54.5–59.5	57	72
59.5–64.5	62	57
64.5–69.5	67	25
69.5–74.5	72	12
74.5–79.5	77	2

The observed mean is, from the grouped version of equation (8.2),

$$\bar{x} = \frac{\Sigma f_x x}{n} = \frac{(4)(32) + (13)(37) + (20)(42) + \cdots}{315}$$

$$= \frac{17{,}255}{315} = 54.78 \text{ kg}$$

The observed variance is, from the grouped version of equation (8.3),

$$s^2 = \frac{\Sigma f_x x^2 - (\Sigma f_x x)^2/n}{n - 1}$$

$$= \frac{969{,}685 - (17{,}255)^2/315}{314} = 78.008$$

The standard error of the mean weight, from equation (8.4), is

$$s_{\bar{x}} = \sqrt{\frac{s^2}{n}}\sqrt{1 - f}$$

$$= \sqrt{\frac{78.008}{315}}\sqrt{1 - \frac{315}{1262}} = 0.4311$$

From equation (8.8) we calculate $g_1 = (1/ns^3)\,\Sigma\,(x - \bar{x})^3 = -0.164$. Hence, applying equation (8.9),

$$n > 25g_1^2$$

$$315 >> 25(-0.164)^2 = 0.67$$

so that we have a "large" sample according to Cochran's rule-of-thumb. Hence we can compute normal confidence limits from equation (6.7): for $\alpha = 0.05$, the t-value for 315 d.f. is 1.97:

$$\bar{x} \pm t_\alpha s_{\bar{x}}$$

$$54.78 \pm 1.97(0.431)$$

or 95% confidence limits of 53.93 to 55.63 kg.

To calculate the population total biomass, we have from equation (8.5),

$$\hat{X} = N\bar{x} = 1262(54.78) = 69{,}129.6 \text{ kg}$$

and the standard error of this total biomass estimate for the herd is, from equation (8.6),

$$s_{\hat{X}} = Ns_{\bar{x}} = 1262(0.4311) = 544.05$$

and normal 95% confidence limits can be calculated as above to give for the population total

$$69{,}130 \pm 1072 \text{ kg}$$

for $\alpha = 0.05$

may wish to estimate the wolf/moose ratio for several game-management zones, or behavioral ecologists may wish to measure the ratio of breeding females to breeding males in a bird population. Ratios are peculiar statistical variables with strange properties that few biologists appreciate (Atchley et al. 1976). Ratios of two variables are *not* just like ordinary measurements and to estimate means, standard errors, and confidence intervals for ecological ratios, you should use the following formulas from Cochran (1977):

For the mean ratio:

$$\hat{R} = \frac{\bar{x}}{\bar{y}} \tag{8.10}$$

where $\hat{R}$ = Estimated mean ratio of x to y
$\bar{x}$ = Observed mean value of x
$\bar{y}$ = Observed mean value of y

The standard error of this estimated ratio is

$$s_{\hat{R}} = \frac{\sqrt{1-f}}{\sqrt{n}\,\bar{y}} \sqrt{\frac{\Sigma x^2 - 2\hat{R}\,\Sigma xy + \hat{R}^2\,\Sigma y^2}{n-1}} \tag{8.11}$$

where $s_{\hat{R}}$ = Estimated standard error of the ratio R
f = Sampling fraction = n/N
n = Sample size
$\bar{y}$ = Observed mean of Y measurement (denominator of ratio)

and the summation terms are the usual ones defined in Appendix 1.

The estimation of confidence intervals for ratios from the usual normal approximation (eq. [8.7]) is not valid unless sample size is large as defined in (equation [8.9]) (Sukhatme and Sukhatme 1970). Ratio variables are often skewed to the right and not normally distributed, particularly when the coefficient of variation of the denominator is relatively high (Atchley et al. 1976). The message is that you should treat the computed confidence intervals of a ratio as only an approximation unless sample size is large. Box 8.2 illustrates the use of these formulas for calculating a ratio estimate.

8.1.3 Proportions and Percentages

The use of proportions and percentages is common in ecological work. Estimates of the sex ratio in a population, the percentage of successful nests, the incidence of disease, and a variety of other measures are all examples of proportions. In all these cases we assume there are two classes in the population, and all individuals fall into one class or the other. We may be interested in either the *number* or the *proportion* of type X individuals from a simple random sample:

	Population	Sample
No. of total individuals	N	n
No. of individuals of type X	A	a
Proportion of type X individuals	$P = \frac{A}{N}$	$\hat{p} = \frac{a}{n}$

Box 8.2 Estimation of a Ratio of Two Variables from Simple Random Sampling of a Finite Population

Wildlife ecologists interested in measuring the impact of wolf predation on moose populations in British Columbia obtained estimates by aerial counting of the population size of wolves and moose in 11 subregions, which constituted 45% of the total game management zone.

Subregion	No. of wolves	No. of moose	Wolves/moose
A	8	190	0.0421
B	15	370	0.0405
C	9	460	0.0196
D	27	725	0.0372
E	14	265	0.0528
F	3	87	0.0345
G	12	410	0.0293
H	19	675	0.0281
I	7	290	0.0241
J	10	370	0.0270
K	16	510	0.0314
			$\bar{x}$ = 0.03333
			SE = 0.00284
			95% CL = 0.02701 to 0.03965

The mean numbers of wolves and moose are

$$\bar{x} = \frac{\Sigma x}{n} = \frac{8 + 15 + 9 + 27 + \cdots}{11} = \frac{140}{11} = 12.727 \text{ wolves}$$

$$\bar{y} = \frac{\Sigma y}{n} = \frac{190 + 370 + 460 + \cdots}{11} = \frac{4352}{11} = 395.64 \text{ moose}$$

The mean ratio of wolves to moose is estimated from equation (8.10):

$$\hat{R} = \frac{\bar{x}}{\bar{y}} = \frac{12.727}{395.64} = 0.03217 \text{ wolves/moose}$$

The standard error of this estimate (equation [8.11]) requires three sums of the data:

$$\sum x^2 = 8^2 + 15^2 + 9^2 + \cdots = 2214$$

$$\sum y^2 = 190^2 + 370^2 + 460^2 + \cdots = 2{,}092{,}844$$

$$\sum xy = (8)(190) + (15)(370) + (9)(460) + \cdots = 66{,}391$$

From equation (8.11),

$$s_{\hat{R}} = \frac{\sqrt{1 - f}}{\sqrt{n}\,\bar{y}} \sqrt{\frac{\Sigma x^2 - 2\hat{R}\,\Sigma xy + \hat{R}^2\,\Sigma y^2}{n - 1}}$$

$$= \frac{\sqrt{1 - 0.45}}{\sqrt{11}\ (395.64)} \sqrt{\frac{2214 - 2(0.032)(66{,}391) + (0.032^2)(2{,}092{,}844)}{11 - 1}}$$

$$= \frac{0.7416}{1312.19} \sqrt{\frac{108.31}{10}} = 0.00186$$

The 95% confidence limits for this ratio estimate are thus (t_α for 10 d.f. for $\alpha = 0.05$ is 2.228):

$$\hat{R} \pm t_\alpha s_R \quad \text{or} \quad 0.03217 \pm 2.228(0.00186)$$

or 0.02803 to 0.03631 wolves per moose.

In statistical work the *binomial* distribution is usually applied to samples of this type, but when the population is finite, the more proper distribution to use is the *hypergeometric* distribution (Cochran 1977).* Fortunately, the binomial distribution is an adequate approximation for the hypergeometric except when sample size is very small.

For proportions, the sample estimate of the proportion P is simply

$$\hat{p} = \frac{a}{n} \tag{8.12}$$

where $\hat{p}$ = Proportion of type X individuals
a = Number of type X individuals in sample
n = Sample size

The standard error of the estimated proportion $\hat{p}$ is from Cochran (1977),

$$s_{\hat{p}} = \sqrt{1 - f} \sqrt{\frac{\hat{p}\hat{q}}{n - 1}} \tag{8.13}$$

where $s_{\hat{p}}$ = Standard error of the estimated population p
f = Sampling fraction = n/N
$\hat{p}$ = Estimated proportion of X types
$\hat{q} = 1 - \hat{p}$
n = Sample size

For example, if in a population of 3500 deer, you observe a sample of 850, of which 400 are males,

$$\hat{p} = 400/850 = 0.4706$$

$$s_{\hat{p}} = \sqrt{1 - \frac{850}{3500}} \sqrt{\frac{(0.4706)(1 - 0.4706)}{849}} = 0.0149$$

*Zar (1996, 520) has a good, brief description of the hypergeometric distribution.

To obtain confidence limits for the proportion of x-types in the population, several methods are available (as we have already seen in Chapter 2, Section 2.1.1). Confidence limits can be read directly from graphs such as Figure 2.2, or obtained more accurately from tables such as Burnstein (1971) or from program-group EXTRAS (Appendix 2). For small sample sizes, the exact confidence limits can be read from tables of the hypergeometric distribution in Lieberman and Owen (1961).

If sample size is large, confidence limits can be approximated from the normal distribution. Table 8.1 lists sample sizes that qualify as large. The normal approximation to the binomial gives confidence limits of

$$\hat{p} \pm \left[\left(z_\alpha \sqrt{1-f} \sqrt{\frac{\hat{p}\hat{q}}{n-1}}\right) + \frac{1}{2n}\right]$$

or

$$\hat{p} \pm \left(z_\alpha s_{\hat{p}} + \frac{1}{2n}\right) \tag{8.14}$$

where $\hat{p}$ = Estimated proportion of X types
z_α = Standard normal deviate (1.96 for 95% confidence intervals, 2.576 for 99% confidence intervals)
$s_{\hat{p}}$ = Standard error of the estimated proportion (equation [8.13])
f = Sampling fraction = n/N
$\hat{q} = 1 - \hat{p}$ = proportion of Y types in sample
n = Sample size

The fraction ($1/2n$) is a correction for continuity, which attempts to correct partly for the fact that individuals come in units of one, so it is possible, for example, to observe 216 male deer or 217, but not 216.5. Without this correction the normal approximation usually gives a confidence belt that is too narrow.

TABLE 8.1 SAMPLE SIZES NEEDED TO USE THE NORMAL APPROXIMATION (EQUATION [8.14]) FOR CALCULATING CONFIDENCE INTERVALS FOR PROPORTIONS[a]

Proportion, p	Number of individuals in the smaller class, np	Total sample size, n
0.5	15	30
0.4	20	50
0.3	24	80
0.2	40	200
0.1	60	600
0.05	70	1400

Source: Cochran 1977.

[a] For a given value of p do not use the normal approximation unless you have a sample size this large *or larger.*

TABLE 8.2 WIDTH OF 90% CONFIDENCE INTERVALS FOR DISEASE INCIDENCE[a]

No. of groups, *n*	Percent disease incidence															
	1.0%				2.0%				5.0%				10%			
	Group size, *k*				Group size, *k*				Group size, *k*				Group size, *k*			
	1	5	10	159	1	5	10	79	1	5	10	31	1	5	10	15
12	4.7	2.1	1.5	—	6.6	3.0	2.2	—	10.3	4.9	3.7	—	14.2	7.1	—	—
20	3.6	1.6	1.2	—	5.1	2.3	1.7	—	8.0	3.8	2.8	—	11.0	5.5	4.5	—
30	3.0	1.3	1.0	0.4	4.2	1.9	1.4	0.7	6.5	3.1	2.3	1.8	9.0	4.5	3.7	3.5
60	2.1	1.0	0.7	0.3	3.0	1.4	1.0	0.5	4.6	2.2	1.6	1.3	6.4	3.2	2.6	2.5
90	1.7	0.8	0.6	0.2	2.4	1.1	0.8	0.4	3.8	1.8	1.3	1.0	5.2	2.6	2.1	2.0
120	1.5	0.7	0.5	0.2	2.1	1.0	0.7	0.4	3.3	1.5	1.2	0.8	4.5	2.2	1.8	1.8
150	1.3	0.6	0.4	0.2	1.9	0.8	0.6	0.3	2.9	1.4	1.0	0.8	4.0	2.0	1.6	1.6
250	1.0	0.5	0.3	0.1	1.4	0.7	0.5	0.2	2.3	1.1	0.8	0.6	3.1	1.6	1.3	1.2
350	0.9	0.4	0.3	0.1	1.2	0.6	0.4	0.2	1.9	0.9	0.7	0.5	2.6	1.3	1.1	1.0
450	0.8	0.3	0.2	0.1	1.1	0.5	0.4	0.2	1.7	0.8	0.6	0.5	2.3	1.2	1.0	0.9

Source: Worlund and Taylor 1983.

[a] A number of groups (n) of group size k are tested for disease. If one individual in a group has a disease, the whole group is diagnosed as disease-positive. The number in the table should be read as "$\pm d\%$," that is, as one-half of the width of the confidence interval.

For the example above, the 95% confidence interval would be

$$0.4706 \pm \left[1.96(0.0149) + \frac{1}{2(850)}\right]$$

$$\text{or } 0.4706 \pm 0.0298 \text{ (0.44 to 0.50 males)}$$

Note that the correction for continuity in this case is very small and if ignored would not change the confidence limits except in the fourth decimal place.

Not all biological attributes come as two classes like males and females, of course, and we may wish to estimate the proportion of organisms in three or four classes (instar I, II, III and IV in insects for example). These data can be treated most simply by collapsing them into two classes, instar II vs. all other instars for example, and using the methods described above. A better approach is described in Section 13.4.1 for multinomial data.

One practical illustration of the problem of estimating proportions comes from studies of disease incidence (Ossiander and Wedemeyer 1973). In many hatchery populations of fish, samples need to be taken periodically and analyzed for disease. Because of the cost and time associated with disease analysis, individual fish are not always the sampling unit. Instead, groups of 5, 10 or more fish may be pooled, and the resulting pool analyzed for disease. One diseased fish in a group of 10 will cause that whole group to be assessed as disease-positive. Worlund and Taylor (1983) developed a method for estimating disease incidence in populations when samples are pooled. The sampling problem is acute here because disease incidence will often be only 1–2%, and at low incidences of disease, larger group sizes are more efficient in estimating the proportion diseased. Table 8.2 gives the confidence intervals expected for various sizes of groups and number of groups when the expected disease incidence varies from 1–10%. For group size = 1, these limits are the same as those derived above (equation [8.14]). But Table 8.2 shows clearly that, at low incidence, larger group sizes are much more precise than smaller group sizes. Worlund and Taylor (1983) provide more details on optimal sampling design for such disease studies. One problem with disease studies is that diseased animals might be much easier to catch than healthy animals, and one must be particularly concerned with obtaining a random sample of the population.

8.2 STRATIFIED RANDOM SAMPLING

One of the most powerful tools you can use in sampling design is to *stratify* your population. Ecologists do this all the time intuitively. Figure 8.1 gives a simple example. Population density is one of the most common bases of stratification in ecological work. An ecologist who recognizes good and poor habitats is implicitly stratifying the study area.

In stratified sampling the statistical population of N units is divided into subpopulations that do not overlap; together they comprise the entire population. Thus,

$$N = N_1 + N_2 + N_3 + \cdots + N_L$$

where L = total number of subpopulations

The subpopulations are called *strata* by statisticians. Clearly if there is only one stratum, we are back to the kind of sampling we have discussed earlier in this chapter. To obtain the full benefit from stratification you must know the sizes of all the strata (N_1, N_2, . . .). In

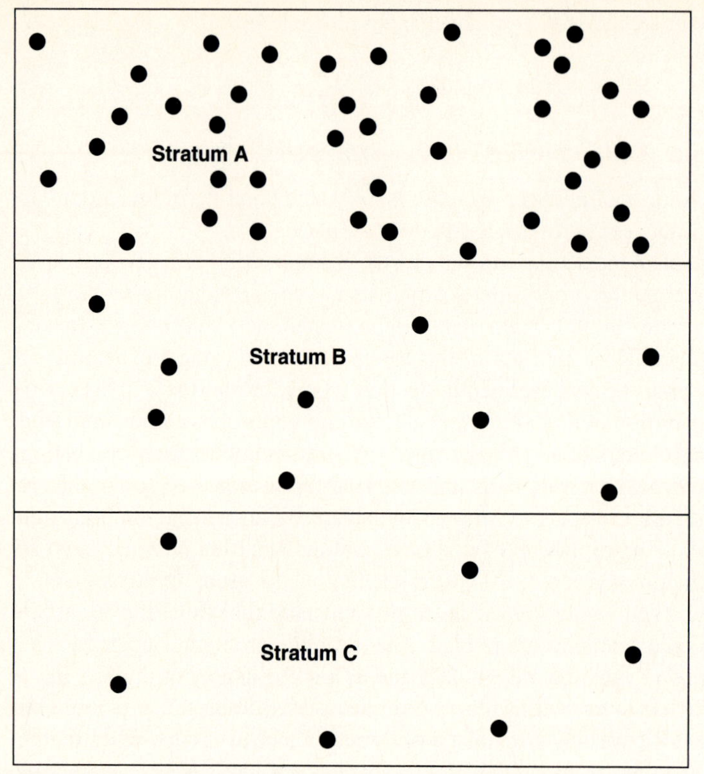

Figure 8.1 The idea of stratification in estimating the size of a plant or animal population. Stratification is made on the basis of population density. Stratum A has about ten times the density of Stratum C.

many ecological examples, stratification is done on the basis of geographical area, and the sizes of the strata are easily found in m^2 or km^2, for example. There is no need for the strata to be of the same size.

Once you have determined what the strata are, you sample each stratum *separately*. The sample sizes for each stratum are denoted by subscripts:

$$n_1 = \text{sample size in stratum 1}$$
$$n_2 = \text{sample size in stratum 2}$$

and so on. If within each stratum you sample using the principles of simple random sampling outlined above (Section 8.1), the whole procedure is called *stratified random sampling*. It is not necessary to sample each stratum randomly, and you could, for example, sample systematically within a stratum. But the problems outlined above would then mean that it would be difficult to estimate how reliable such sampling is, so we recommend sampling randomly within each stratum.

Ecologists have many different reasons for wishing to stratify their sampling. Four general reasons are common (Cochran 1977):

1. Estimates of means and confidence intervals may be required separately for each subpopulation.

2. Sampling problems may differ greatly in different areas. Animals may be easier or harder to count in some habitats than they are in others. Offshore samples may require larger boats and be more expensive to get than nearshore samples.
3. Stratification may result in a gain in precision in the estimates of the parameters of the whole population. Confidence intervals can be narrowed appreciably when strata are chosen well.
4. Administrative convenience may require stratification if different field laboratories are doing different parts of the sampling.

Point 3 is perhaps the most critical one on this list, and I will now discuss how estimates are made from stratified sampling and illustrate the gains one can achieve.

8.2.1 Estimation of Parameters

For each of the subpopulations ($N_1, N_2, \ldots$), all of the principles and procedures of estimation outlined above can be used. Thus, for example, the mean for stratum 1 can be estimated from equation (8.2) and the variance from equation (8.3). New formulas are required, however, to estimate the mean for the whole population N. It will be convenient, before I present these formulas, to outline one example of stratified sampling, so that the equations can be related more easily to the ecological framework.

Table 8.3 gives information on a stratified random sample taken on a caribou herd in central Alaska by Siniff and Skoog (1964). They used as their sampling unit a quadrat of 4 square miles, and they stratified the whole study zone into six strata, based on a pilot survey of caribou densities in different regions. Table 8.3 shows that the 699 total sampling units were divided very unequally into the six strata, so that the largest stratum (A) was 22 times the size of the smallest stratum (D).

We define the following notation for use with stratified sampling:

$$\text{Stratum weight} = W_h = \frac{N_h}{N} \tag{8.15}$$

TABLE 8.3 STRATIFIED RANDOM SAMPLING OF THE NELCHINA CARIBOU HERD IN ALASKA BY SINIFF AND SKOOG (1964)[a]

Stratum	Stratum size, N_h	Stratum weight, W_h	Sample size, n_h	Mean no. of caribou counted per sampling unit, $\bar{x}_h$	Variance of caribou counts, s_h^2
A	400	0.572	98	24.1	5575
B	30	0.043	10	25.6	4064
C	61	0.087	37	267.6	347,556
D	18	0.026	6	179.0	22,798
E	70	0.100	39	293.7	123,578
F	120	0.172	21	33.2	9795
Total	699	1.000	211		

Source: Siniff and Skoog 1964.

[a] Six strata were delimited in preliminary surveys based on the relative caribou density. Each sampling unit was 4 square miles. A random sample was selected in each stratum, and counts were made from airplanes.

where N_h = Size of stratum h (number of possible sample units in stratum h)
N = Size of entire statistical population

The stratum weights are proportions and must add up to 1.0 (Table 8.3). Note that the N_h must be expressed in sample units. If the sample unit is 0.25 m^2, the sizes of the strata must be expressed in units of 0.25 m^2 (not as hectares, or km^2).

Simple random sampling is now applied to each stratum separately and the means and variances calculated for each stratum from equations (8.2) and (8.3). We will defer until the next section a discussion of how to decide sample size in each stratum. Table 8.3 gives sample data for a caribou population.

The overall mean per sampling unit for the entire population is estimated as follows (Cochran 1977):

$$\bar{x}_{ST} = \frac{\sum_{h=1}^{L} N_h \bar{x}_h}{N} \tag{8.16}$$

where $\bar{x}_{ST}$ = Stratified population mean per sampling unit
N_h = Size of stratum h
h = Stratum number (1, 2, 3, . . . , L)
$\bar{x}_h$ = Observed mean for stratum h
N = Total number of sampling units = ΣN_h

Note that $\bar{x}_{ST}$ is a *weighted* mean in which the stratum sizes are used as weights.

For the data in Table 8.3, we have

$$\bar{x}_{ST} = \frac{(400)(24.1) + (30)(25.6) + (61)(267.6) + \cdots}{699}$$

$$= 77.96 \text{ caribou/sample unit}$$

Given the density of caribou per sampling unit, we can calculate the size of the entire caribou population from the equation

$$\hat{X}_{ST} = N\bar{x}_{ST} \tag{8.17}$$

where $\hat{X}_{ST}$ = Population total
N = Number of sample units in entire poulation
$\bar{x}_{ST}$ = Stratified mean per sampling unit (equation [8.16])

For the caribou example,

$$\hat{X}_{ST} = 699(77.96) = 54{,}597 \text{ caribou}$$

so the entire caribou herd is estimated to be around 55 thousand animals at the time of the study.

The variance of the stratified mean is given by Cochran (1977, 92) as

$$\text{Variance of } (\bar{x}_{ST}) = \sum_{h=1}^{L} \left[\frac{w_h^2 s_h^2}{n_h} (1 - f_h) \right] \tag{8.18}$$

where W_h = Stratum weight (equation [8.15])
s_h^2 = Observed variance of stratum h (equation [8.3])
n_h = Sample size in stratum h
f_h = Sampling fraction in stratum $h = n_h/N_h$

The last term in this summation is the finite population correction, and it can be ignored if you are sampling less that 5% of the sample units in each stratum. Note that the variance of the stratified means depends only on the size of the variances *within* each stratum. If you could divide a highly variable population into homogeneous strata such that all measurements *within* a stratum were equal, the variance of the stratified mean would be zero, which means that the stratified mean would be without any error! In practice, of course, you cannot achieve this, but the general principle still pertains: *Pick homogeneous strata and you gain precision.*

For the caribou data in Table 8.3 we have

$$\text{Variance of } (\bar{x}_{ST}) = \left[\frac{(0.572)^2(5575)}{98}\right]\left(1 - \frac{98}{400}\right) + \left[\frac{(0.043)^2(4064)}{10}\right]\left(1 - \frac{10}{30}\right) + \cdots$$

$$= 69.803$$

The standard error of the stratified mean is the square root of its variance:

$$\text{Standard error of } (\bar{x}_{ST}) = \sqrt{\text{Variance of } \bar{x}_{ST}} = \sqrt{69.803} = 8.355$$

Note that the variance of the stratified mean cannot be calculated unless there are at least two samples in each stratum. The variance of the population total is simply

$$\text{Variance of } (\hat{X}_{ST}) = (N)^2(\text{variance of } \bar{x}_{ST}) \tag{8.19}$$

For the caribou, the variance of the total population estimate is

$$\text{Variance of } (\hat{X}_{ST}) = 699^2(69.803) = 34{,}105{,}734$$

and the standard error of the total is the square root of this variance, or 5840.

The confidence limits for the stratified mean and the stratified population total are obtained in the usual way:

$$\bar{x}_{ST} \pm t_\alpha(\text{standard error of } \bar{x}_{ST}) \tag{8.20}$$

$$\hat{X}_{ST} \pm t_\alpha(\text{standard error of } \hat{X}_{ST}) \tag{8.21}$$

The only problem is what value of Student's t to use. The appropriate number of degrees of freedom lies somewhere between the lowest of the values $(n_h - 1)$ and the total sample size $(\Sigma\, n_h - 1)$. Cochran (1977, 95) recommends calculating an effective number of degrees of freedom from the approximate formula:

$$\text{d.f.} \approx \frac{\left(\Sigma_{h=1}^{L}\, g_h s_h^2\right)^2}{\Sigma_{h=1}^{L}\left[g_h^2 s_h^4/(n_h - 1)\right]} \tag{8.22}$$

where d.f. = Effective number of degrees of freedom for the confidence limits in equations (8.20) and (8.21)
$g_h = N_h(N_h - n_h)/n_h$
s_h = Observed variance in stratum h
n_h = Sample size in stratum h
N_h = Size of stratum h

For example, from the data in Table 8.3, we obtain

Stratum	g_h
A	1232.65
B	60.00
C	39.57
D	36.00
E	55.64
F	565.71

and from equation (8.22),

$$\text{d.f.} = \frac{(43{,}106{,}434)^2}{8.6614 \times 10^2} = 134.3$$

Thus for 95% confidence intervals for this example, $t_\alpha = 1.98$. Thus for the population mean from equation (8.20), the 95% confidence limits are

$$77.96 \pm 1.98(8.35)$$

or from 61.4 to 94.5 caribou per 4 sq. miles. For the population total from equation (8.21) the 95% confidence limits are

$$54{,}497 \pm 1.98(5840)$$

or from 42,933 to 66,060 caribou in the entire herd.

8.2.2 Allocation of Sample Size

In planning a stratified sampling program you need to decide how many sample units you should measure in each stratum. Two alternate strategies are available for allocating samples to strata: proportional allocation or optimal allocation.

Proportional Allocation The simplest approach to stratified sampling is to allocate samples to strata on the basis of a constant sampling fraction in each stratum. For example, you might decide to sample 10% of all the sample units in each stratum. In the terminology defined above,

$$\frac{n_h}{n} = \frac{N_h}{N} \tag{8.23}$$

For example, in the caribou population of Table 8.3, if you wished to sample 10% of the units, you would count 40 units in stratum A, 3 in stratum B, 6 in stratum C, 2 in stratum D, 7 in E, and 12 in F. Note that you should always constrain this rule so that at least 2 units are sampled in each stratum to allow the estimation of variances.

Equation (8.23) tells us what *fraction* of samples to assign to each stratum, but we still do not know how many samples we need to take in total (n). In some situations this is fixed and beyond control. But if you are able to plan ahead, you can determine the sample size you require as follows.

- Decide on the absolute size of the confidence interval you require in the final estimate. For example, in the caribou case you may wish to know the density to ±10 caribou/4 sq. miles with 95% confidence.
- Calculate the estimated total number of samples needed for an infinite population from the approximate formula (Cochran 1977, 104):

$$n \approx \frac{4 \sum W_h s_h^2}{d^2} \tag{8.24}$$

where n = Total sample size required (for large population)
W_h = Stratum weight
s_h^2 = Observed variance of stratum h
d = Desired absolute precision of stratified mean (width of confidence interval is $\pm d$

This formula is used when 95% confidence intervals are specified in d. If 99% confidence intervals are specified, replace 4 in equation (8.24) with 7.08, and for 90% confidence intervals, use 2.79 instead of 4. For a finite population, correct this estimated n by equation (7.6):

$$n^* = \frac{n}{1 + n/N}$$

where n^* = Total sample size needed in finite population of size N

For the caribou data in Table 8.3, if an absolute precision of ±10 caribou/4 square miles is needed:

$$n \approx \frac{4[(0.572)(5575) + (0.043)(4064) + \cdots + 1]}{10^2} = 1933.6$$

Note that this recommended sample size is *more* than the total sample units available! For a finite population of 699 sample units,

$$n^* \approx \frac{1933.6}{1 + 1933.6/699} = 513.4 \text{ sample units}$$

These 514 sample units would then be distributed to the six strata in proportion to the stratum weight. Thus, for example, stratum A would be given (0.572)(514), or 294 sample units. Note again the message that if you wish to have high precision in your estimates, you will have to take a large sample size.

Optimal Allocation When deciding on sample sizes to be obtained in each stratum, you will find that proportional allocation is the simplest procedure. But it is not the most efficient, and if you have prior information on the sampling methods, more powerful allocation plans can be specified. In particular, you can minimize the cost of sampling with the following general approach developed by Cochran (1977).

Assume that you can specify the cost of sampling according to a simple cost function:

$$C = c_0 + \sum c_h n_h \tag{8.25}$$

where C = Total cost of sampling
c_O = Overhead cost
c_h = Cost of taking one sample in stratum h
n_h = Number of samples taken in stratum h

Of course the cost of taking one sample might be equal in all strata, but this is not always true. Costs can be expressed in money or in time units. Economists have developed much more complex cost models, but we shall continue to use this simple model here.

Cochran (1977, 95) demonstrates that, given the cost function above, the standard error of the stratified mean is at a minimum when

$$n_h \text{ is proportional to } \frac{N_h s_h}{\sqrt{c_h}}$$

This means that we should apportion samples among the strata by the following ratio:

$$\frac{n_h}{n} = \frac{N_h s_h / \sqrt{c_h}}{\Sigma\ [N_h s_h / \sqrt{c_h}]} \tag{8.26}$$

This formula leads to three useful rules-of-thumb in stratified sampling: In a given stratum, take a larger sample if

1. The stratum is larger.
2. The stratum is more variable internally.
3. Sampling is cheaper in the stratum.

Once we have done this, we can go in one of two ways:

Step 1. *Minimize the standard error of the stratified mean for a fixed total cost.* If the cost is fixed, the total sample size is dictated by

$$n = \frac{(C - c_O)\ \Sigma\ (N_h s_h / \sqrt{c_h})}{\Sigma\ (N_h s_h \sqrt{c_h})} \tag{8.27}$$

where n = Total sample size to be used in stratified sampling for all strata combined
C = Total cost (fixed in advance)
c_O = Overhead cost
N_h = Size of stratum h
s_h = Standard deviation of stratum h
c_h = Cost to take one sample in stratum h

Box 8.3 illustrates the use of these equations for optimal allocation.

Step 2. *Minimize the total cost for a specified value of the standard error* of the stratified mean. If you specify in advance the level of precision you need in the stratified mean, you can estimate the total sample size by the formula

$$n = \frac{(\Sigma\ W_h s_h \sqrt{c_h})(\Sigma\ W_h s_h / \sqrt{c_h})}{V + (1/N)(\Sigma\ W_h s_h^2)} \tag{8.28}$$

Box 8.3 Optimal and Proportional Allocation in Stratified Random Sampling

Russell (1972) sampled a clam population using stratified random sampling and obtained the following data:

Stratum	Size of stratum, N_h	Stratum weight, W_h	Sample size, n_h	Mean (bushels), x_h	Variance, s_h^2
A	5703.9	0.4281	4	0.44	0.068
B	1270.0	0.0953	6	1.17	0.042
C	1286.4	0.0965	3	3.92	2.146
D	5063.9	0.3800	5	1.80	0.794
	N = 13,324.2	1.0000	18		

Stratum weights are calculated as in equation (8.15). I use these data to illustrate hypothetically how to design proportional and optimal allocation sampling plans.

Proportional Allocation

If you were planning this sampling program based on proportional allocation, you would allocate the samples in proportion to stratum weight (equation [8.23]):

Stratum	Fraction of samples to be allocated to this stratum
A	0.43
B	0.10
C	0.10
D	0.38

Thus, if sampling was constrained to take only 18 samples (as in the actual data), you would allocate these as 7, 2, 2, and 7 to the four strata. Note that proportional allocation can never be exact in the real world because you must always have two samples in each stratum, and you must round off the sample sizes.

If you wish to specify a level of precision to be attained by proportional allocation, you proceed as follows. For example, assume you desire an absolute precision of the stratified mean of $d = \pm 0.1$ bushels at 95% confidence. From equation (8.24),

$$n \approx \frac{4 \sum W_h s_h^2}{d^2} = \frac{4[(0.4281)(0.068) + (0.0953)(0.042) + \cdots]}{(0.1)^2}$$

$$\approx 217 \text{ samples}$$

(assuming the sampling fraction is negligible in all strata). These 217 samples would be distributed to the four strata according to the fractions given above: 43% to stratum A, 10% to stratum B, and so on.

Optimal Allocation

In this example proportional allocation is very inefficient because the variances are very different in the four strata, as well as the means. Optimal allocation is thus to be preferred. To illustrate the calculations, we consider a hypothetical case in which the cost per sample varies in the different strata. Assume that the overhead cost in equation (8.25) is \$100, and the costs per sample are

$$c_1 = \$10$$
$$c_2 = \$20$$
$$c_3 = \$30$$
$$c_4 = \$40$$

Apply equation (8.26) to determine the fraction of samples in each stratum:

$$\frac{n_h}{n} = \frac{N_h s_h / \sqrt{c_h}}{\Sigma\, N_h s_h / \sqrt{c_h}}$$

These fractions are calculated as follows:

Stratum	N_h	s_h	$\sqrt{c_h}$	$N s_h/\sqrt{c_h}$	Estimated fraction, n_h/n
A	5703.9	0.2608	3.162	470.35	0.2966
B	1270.0	0.2049	4.472	58.20	0.0367
C	1286.4	1.4649	5.477	344.06	0.2169
D	5063.9	0.8911	6.325	713.45	0.4498
				Total = 1586.06	1.0000

We can now proceed to calculate the total sample size needed for optimal allocation under two possible assumptions.

Minimize the Standard Error of the Stratified Mean

In this case cost is fixed. Assume for this example that \$200 is available. Then, from equation (8.27),

$$n = \frac{(C - c_o)(\Sigma\, N_h s_h / \sqrt{c_h})}{\Sigma\, (N_h s_h \sqrt{c_h})}$$

$$= \frac{(2000 - 100)(1586.06)}{44{,}729.745} = 70.9 \text{ (rounded to 71 samples)}$$

Note that only the denominator needs to be calculated, since we have already computed the numerator sum.

We allocate these 71 samples according to the fractions just established:

Stratum	Fraction of samples	Total no. samples allocated of 68 total
A	0.2966	21.1 (21)
B	0.0367	2.6 (3)
C	0.2169	15.4 (15)
D	0.4498	31.9 (32)

Minimize the Total Cost for a Specified Standard Error

In this case you must decide in advance what level of precision you require. In this hypothetical calculation, use the same value as above, $d = 0.1$ bushels (95% confidence limit). In this case the desired variance (V) of the stratified mean is

$$V = \left(\frac{d}{t}\right)^2 = \left(\frac{0.1}{2}\right)^2 = 0.0025$$

Applying formula (8.28),

$$n = \frac{(\Sigma\ W_h s_h \sqrt{c_h})(\Sigma\ W_h s_h / \sqrt{c_h})}{V + (1/N)(\Sigma\ W_h s_h^2)}$$

We need to compute three sums:

$$\sum W_h s_h \sqrt{c_h} = (0.4281)(0.2608)(3.162) + (0.0953)(0.2049)(4.472) + \cdots$$

$$= 3.3549$$

$$\sum W_h s_h / \sqrt{c_h} = \frac{(0.4281)(0.068)}{3.162} + \frac{(0.0953)(0.2049)}{4.472} + \cdots$$

$$= 0.1191$$

$$\sum W_h s_h^2 = (0.4281)(0.068) + (0.0953)(0.042) + \cdots$$

$$= 0.5419$$

Thus,

$$n = \frac{(3.3549)(0.1191)}{0.0025 + (0.5419/13{,}324.2)} = 157.3 \text{ (rounded to 157 samples)}$$

We allocate these 157 samples according to the fractions established for optimal allocation

Stratum	Fraction of samples	Total no. of samples allocated of 157 total
A	0.2966	46.6 (47)
B	0.0367	5.8 (6)
C	0.2169	34.1 (34)
D	0.4498	70.6 (71)

Note that in this hypothetical example, many fewer samples are required under *optimal* allocation ($n = 157$) than under *proportional* allocation ($n = 217$) to achieve the same confidence level ($d = \pm 0.1$ bushels).

Program-group SAMPLING (Appendix 2) does these calculations.

where n = Total sample size to be used in stratified sampling
W_h = Stratum weight
s_h = Standard deviation in stratum h
c_h = Cost to take one sample in stratum h
N = Total number of sample units in entire population
V = Desired variance of the stratified mean $= (d/t_\alpha)^2$
d = Desired absolute width of the confidence interval for $1 - \alpha$
t_α = Student's t-value for $1 - \alpha$ confidence limits ($t \approx 2$ for 95% confidence limits, $t \approx 2.66$ for 99% confidence limits, $t \approx 1.67$ for 90% confidence limits)

Box 8.3 illustrates the application of these formulas.

If you do not know anything about the cost of sampling, you can estimate the sample sizes required for optimal allocation from the two formulas:

1. To estimate the total sample size needed (n),

$$n = \frac{(\Sigma\ W_h s_h)^2}{V + (1/N)(\Sigma\ W_h s_h^2)} \tag{8.29}$$

where V = Desired variance of the stratified mean, and the other terms are as defined above.

2. To estimate the sample size in each stratum,

$$n_h = n\left(\frac{N_h s_h}{\Sigma\ N_h s_h}\right) \tag{8.30}$$

where n = Total sample size estimated in equation (8.29)

and the other terms are as defined above. These two formulas are just variations of the ones given above, in which sampling costs are presumed to be equal in all strata.

Proportional allocation can be applied to any ecological situation. Optimal allocation should always be preferred, if you have the necessary background information to estimate the costs and the relative variability of the different strata. A pilot survey can give much of this information and help to fine tune the stratification.

Stratified random sampling is almost always more precise than simple random sampling. If used intelligently, stratification can result in a large gain in precision, that is, in a smaller confidence interval for the same amount of work (Cochran 1977). The critical factor is always to *chose strata that are relatively homogeneous*. Cochran (1977, 98) has shown that with optimal allocation, the theoretical expectation is that

$$\text{S.E.(optimal)} \leq \text{S.E.(proportional)} \leq \text{S.E.(random)}$$

where S.E.(optimal)= Standard error of the stratified mean obtained with *optimal* allocation of sample sizes
S.E.(proportional)= Standard error of the stratified mean obtained with *proportional* allocation
S.E.(random)= Standard error of the mean obtained for the whole population using *simple random sampling*

Thus comes the simple recommendation: *Always stratify your samples!* Unless you are perverse or very unlucky and choose strata that are very heterogeneous, you will always gain by using stratified sampling.

8.2.3 Construction of Strata

How many strata should you use if you are going to use stratified random sampling? The answer to this simple question is not easy. It is clear in the real world that a point of diminishing returns is quickly reached, so that the number of strata should normally not exceed 6 (Cochran 1977, 134). Often even fewer strata are desirable (Iachan 1985), but this will depend on the strength of the gradient. Note that in some cases estimates of means are needed for different geographical regions and a larger number of strata can be used. Duck populations in Canada and the United States are estimated using stratified sampling with 49 strata (Johnson and Grier 1988) in order to have regional estimates of production. But in general you should not expect to gain much in precision by increasing the number of strata beyond about 6.

Given that you wish to set up 2–6 strata, how can you best decide on the boundaries of the strata? Stratification may be decided *a priori* from your ecological knowledge of the sampling situation in different microhabitats. If this is the case, you do not need any statistical help. But sometimes you may wish to stratify on the basis of the variable being measured (x) or some auxiliary variable (y) that is correlated with x. For example, you may be measuring population density of clams (x) and you may use water depth (y) as a stratification variable. Several rules are available for deciding boundaries to strata (Iachan 1985); only one is presented here, the *cum* $\sqrt{f}$ *rule*. This is defined as

$$\text{cum}\sqrt{f} = \text{cumulative square root of frequency of quadrats}$$

This rule is applied as follows:

1. **1.** Tabulate the available data in a frequency distribution based on the stratification variable. Table 8.4 gives some data for illustration.
2. **2.** Calculate the square root of the observed frequency and accumulate these square roots down the table.
3. **3.** Obtain the upper stratum boundaries for L strata from the equally spaced points:

$$\text{Boundary of stratum } i = \left(\frac{\text{Maximum cumulative } \sqrt{f}}{L}\right) i \qquad (8.31)$$

TABLE 8.4 DATA ON THE ABUNDANCE OF SURF CLAMS OFF THE COAST OF NEW JERSEY IN 1981 ARRANGED IN ORDER BY DEPTH OF SAMPLES[a]

Class	Depth, y (m)	No. of samples, f	$\sqrt{f}$	cum $\sqrt{f}$	Observed no. of clams, x	
1	14	4	2.00000	2.000	34, 128, 13, 0	
2	15	1	1.00000	3.000	27	
3	18	2	1.41421	4.414	361, 4	Stratum 1
4	19	3	1.73205	6.146	0, 5, 363	
5	20	4	2.00000	8.146	176, 32, 122, 41	
6	21	1	1.00000	9.146	21	
7	22	2	1.41421	10.560	0, 0	
8	23	5	2.23607	12.796	9, 112, 255, 3, 65	Stratum 2
9	24	4	2.00000	14.796	122, 102, 0, 7	
10	25	2	1.41421	16.210	18, 1	
11	26	2	1.41421	17.625	14, 9	
12	27	1	1.00000	18.625	3	
13	28	2	1.41421	20.039	8, 30	Stratum 3
14	29	3	1.73205	21.771	35, 25, 46	
15	30	1	1.00000	22.771	15	
16	32	1	1.00000	23.771	11	
17	33	4	2.00000	25.771	9, 0, 4, 19	
18	34	2	1.41421	27.185	11, 7	
19	35	3	1.73205	28.917	2, 10, 97	Stratum 4
20	36	2	1.41421	30.332	0, 10	
21	37	3	1.73205	32.064	2, 1, 10	
22	38	2	1.41421	33.478	4, 13	
23	40	3	1.73205	35.210	0, 1, 2	
24	41	4	2.00000	37.210	0, 2, 2, 15	
25	42	1	1.00000	38.210	13	Stratum 5
26	45	2	1.41421	39.624	0, 0	
27	49	1	1.00000	40.624	0	
28	52	1	1.00000	41.624	0	

Source: Iachan 1985.

[a] Stratification is carried out on the basis of the auxiliary variable depth in order to increase the precision of the estimate of clam abundance for this region.

For example, in Table 8.4 if you wished to use five strata, the upper boundaries of strata 1 and 2 would be:

$$\text{Boundary of stratum 1} = \left(\frac{41.624}{5}\right)(1) = 8.32$$

$$\text{Boundary of stratum 2} = \left(\frac{41.624}{5}\right)(2) = 16.65$$

These boundaries are in units of cum $\sqrt{f}$. In this example, 8.32 is between depths 20 and 21, and the boundary 20.5 meters can be used to separate samples belonging to stratum 1 from those in stratum 2. Similarly, the lower boundary of the second stratum is 16.65 cum $\sqrt{f}$ units, which falls between depths 25 and 26 meters in Table 8.4. Using the cum $\sqrt{f}$ rule, you can stratify your samples *after* they are collected, an important practical advantage in sampling. You need to have measurements on a stratification variable (like depth in this example) in order to use the cum $\sqrt{f}$ rule.

8.2.4 Proportions and Percentages

Stratified random sampling can also be applied to the estimation of a proportion like the sex ratio of a population. Again the rule-of-thumb is to construct strata that are relatively homogeneous, if you are to achieve the maximum benefit from stratification. Since the general procedures for proportions are similar to those outlined above for continuous and discrete variables, I will just present the formulas here that are specific for proportions. Cochran (1977, 106) summarizes these and gives more details.

We estimate the proportion of x-types in each of the strata from equation (8.12). Then we have for the stratified mean proportion,

$$\hat{p}_{ST} = \frac{\Sigma N_h \hat{p}_h}{N} \tag{8.32}$$

where $\hat{p}_{ST}$ = Stratified mean proportion
N_h = Size of stratum h
$\hat{p}_h$ = Estimated proportion for stratum h (from equation [8.12])
N = Total population size (total number of sample units)

The standard error of this stratified mean proportion is

$$\text{S.E.}(\hat{p}_{ST}) = \frac{1}{N}\sqrt{\sum\left[\frac{N_h^2(N_h - n_h)}{N_h - 1}\,\frac{\hat{p}_h\hat{q}_h}{n_h - 1}\right]} \tag{8.33}$$

where $\text{S.E.}(\hat{p}_{ST})$ = Standard error of the stratified mean proportion
$\hat{q}_h = 1 - \hat{p}_h$
n_h = Sample size in stratum h

and all other terms are as defined above.

Confidence limits for the stratified mean proportion are obtained using the t-distribution as outlined above for equation (8.20).

Optimal allocation can be achieved when designing a stratified sampling plan for proportions using all of the equations given above (8.26–8.30) and replacing the estimated standard deviation by

$$s_h = \sqrt{\frac{\hat{p}_h\hat{q}_h}{n_h - 1}} \tag{8.34}$$

where s_h = Standard deviation of the proportion p in stratum h
$\hat{p}_h$ = Fraction of x types in stratum h
$\hat{q}_h = 1 - \hat{p}_h$
n_h = Sample size in stratum h

Program-group SAMPLING in Appendix 2 does all these calculations for stratified random sampling and computes proportional and optimal allocations from specified input to assist in planning a stratified sampling program.

8.3 ADAPTIVE SAMPLING

Most of the methods discussed in sampling theory are limited to sampling designs in which the selection of the samples can be done before the survey, so that none of the decisions about sampling depend in any way on what is observed as one gathers the data. A new method of sampling that makes use of the data gathered is called *adaptive sampling*. For example, in doing a survey of a rare plant, a botanist may feel inclined to sample more intensively in an area where one individual is located to see if others occur in a clump. The primary purpose of adaptive sampling designs is to take advantage of spatial pattern in the population to obtain more precise measures of population abundance. Adaptive sampling is often much more efficient for a given amount of effort than the conventional random sampling designs discussed above. Thompson (1992) presents a summary of these methods.

8.3.1 Adaptive Cluster Sampling

When organisms are rare and highly clustered in their geographical distribution, many randomly selected quadrats will contain no animals or plants. In these cases it may be useful to consider sampling clusters in a nonrandom way. Adaptive cluster sampling begins in the usual way with an initial sample of quadrats selected by simple random sampling with replacement, or simple random sampling without replacement. When one of the selected

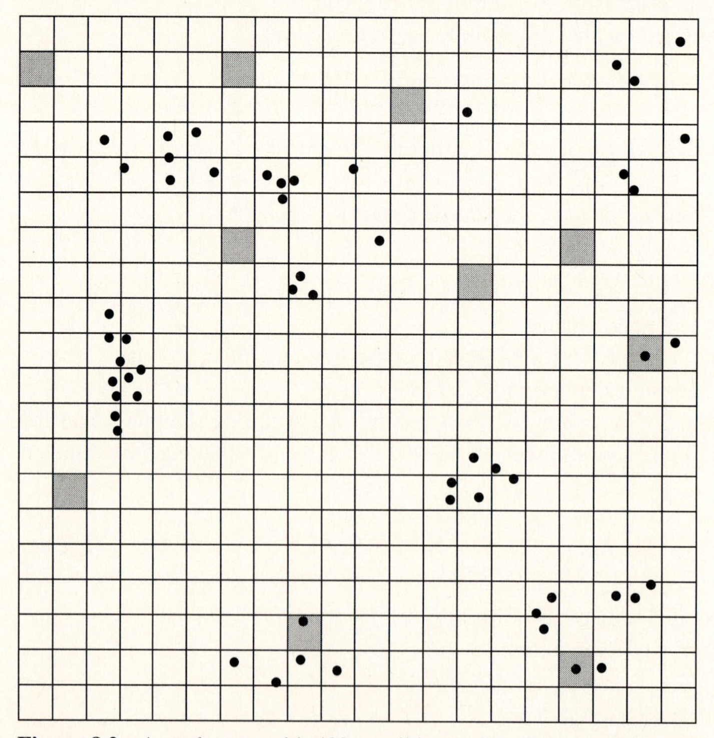

Figure 8.2 A study area with 400 possible quadrats from which a random sample of $n = 10$ quadrats (shaded) has been selected using simple random sampling without replacement. Of the 10 quadrats, 7 contain no organisms, and 3 are occupied by one or more individuals. This hypothetical population of 60 plants is highly clumped.

quadrats contains the organism of interest, additional quadrats in the vicinity of the original quadrat are added to the sample. Adaptive cluster sampling is ideally suited to populations that are highly clumped. Figure 8.2 illustrates a hypothetical example.

To use adaptive cluster sampling we must first make some definitions of the sampling universe:

- *Condition of selection of a quadrat*: A quadrat is selected if it contains at least y organisms (often $y = 1$).
- *Neighborhood of quadrat x*: All quadrats having one side in common with quadrat x.
- *Edge quadrats*: Quadrats that do not satisfy the condition of selection but are next to quadrats that do satisfy the condition (i.e., empty quadrats).
- *Network*: A group of quadrats such that the random selection of any one of the quadrats would lead to all of them being included in the sample.

These definitions are shown more clearly in Figure 8.3, which is identical to Figure 8.2 except that the networks and their edge quadrats are all shown as shaded.

It is clear that we cannot simply calculate the mean of the 37 quadrats counted in this example to get an unbiased estimate of mean abundance. To estimate the mean abundance from adaptive cluster sampling without bias we proceed as follows (Thompson 1992):

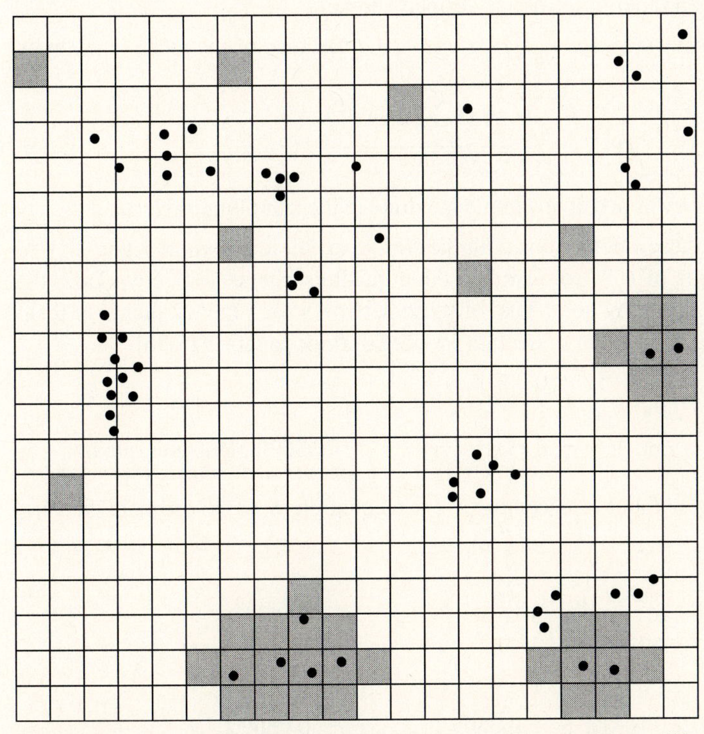

Figure 8.3 The same study area shown in Figure 8.2 with 400 possible quadrats from which a random sample of $n = 10$ quadrats has been selected. All the clusters and edge quadrats are shaded. The observer would count plants in all of the 37 shaded quadrats.

Step 1. Calculate the average abundance of each of the networks:

$$w_i = \frac{\sum_{j=1}^{k} yi}{m_i} \tag{8.35}$$

where w_i = Average abundance of the i-th network
y_j = Abundance of the organism in each of the k quadrats in the i-th network
m_i = Number of quadrats in the i-th network

Step 2. From these values we obtain an estimator of the mean abundance as follows:

$$\bar{x} = \frac{\Sigma_i w_i}{n} \tag{8.36}$$

where $\bar{x}$ = Unbiased estimate of mean abundance from adaptive cluster sampling
n = Number of initial sampling units selected via random sampling

If the initial sample is selected *with replacement*, the variance of this mean is given by:

$$\hat{\text{var}}(\bar{x}) = \frac{\sum_{i=1}^{n} (w_i - \bar{x})^2}{n(n - 1)} \tag{8.37}$$

where $\hat{\text{var}}(\bar{x})$ = Estimated variance of mean abundance for sampling with replacement, and all other terms are as defined above.

If the initial sample is selected *without replacement*, the variance of the mean is given by

$$\hat{\text{var}}(\bar{x}) = \frac{(N - n) \sum_{i=1}^{n} (w_i - \bar{x})^2}{Nn(n - 1)} \tag{8.38}$$

where N = Total number of possible sample quadrats in the sampling universe

We can illustrate these calculations with the simple example shown in Figure 8.3. In the initial random sample of $n = 10$ quadrats, three quadrats intersected networks in the lower and right side of the study area. Two of these networks each have 2 plants in them and one network has 5 plants. From these data we obtain, from equation (8.36),

$$\bar{x} = \frac{\Sigma_i w_i}{n} = \frac{\left(\frac{2}{7} + \frac{2}{8} + \frac{5}{15} + \frac{0}{1} + \frac{0}{1} + \cdots\right)}{10} = 0.08690 \text{ plants per quadrat}$$

Since we were sampling without replacement, we use equation (8.38) to estimate the variance of this mean:

$$\hat{\text{var}}(\bar{x}) = \frac{(N - n) \sum_{i=1}^{n} (w_i - \bar{x})^2}{Nn(n - 1)}$$

$$= \frac{(400 - 10)\left[\left(\frac{2}{7} - 0.0869\right)^2 + \left(\frac{2}{8} - 0.0869\right)^2 + \cdots\right]}{(400)(10)(10 - 1)}$$

$$= 0.00137429$$

We can obtain confidence limits from these estimates in the usual way:

$$\bar{x} \pm t_\alpha \sqrt{\hat{\text{var}}(\bar{x})}$$

For this example with $n = 10$, for 95% confidence limits $t_\alpha = 2.262$, and the confidence limits become:

$$0.0869 \pm (2.262)(\sqrt{0.00137429}) = 0.0869 \pm 0.08385$$

or from 0.0 to 0.171 plants per quadrat. The confidence limits extend below 0.0, but since this is biologically impossible, the lower limit is set to 0. The wide confidence limits reflect the small sample size in this hypothetical example.

When should one consider using adaptive sampling? Much depends on the abundance and the spatial pattern of the animals or the plants being studied. In general the more clustered the population and the rarer the organism, the more efficient it will be to use adaptive cluster sampling. Thompson (1992) shows, for example, from the data in Figure 8.2, that adaptive sampling is about 12% more efficient than simple random sampling for $n = 10$ quadrats and nearly 50% more efficient when $n = 30$ quadrats. In any particular situation it may well pay to conduct a pilot experiment with simple random sampling and adaptive cluster sampling to determine the size of the resulting variances.

8.3.2 Stratified Adaptive Cluster Sampling

The general principle of adaptive sampling can also be applied to situations that are well enough studied to utilize stratified sampling. In stratified adaptive sampling, random samples are taken from each stratum in the usual way, with the added condition that whenever a sample quadrat satisfies some initial conditions (e.g., an animal is present), additional quadrats from the neighborhood of that quadrat are added to the sample. This type of sampling design would allow one to take advantage of the fact that a population may be well stratified but clustered in each stratum in an unknown pattern. Large gains in efficiency are possible if the organisms are clustered within each stratum. Thompson (1992, Chap. 26) discusses the details of the estimation problem for stratified adaptive cluster sampling. The conventional stratified sampling estimators cannot be used for this adaptive design, since the neighborhood samples are not selected randomly.

8.4 SYSTEMATIC SAMPLING

Ecologists often use systematic sampling in the field. For example, mouse traps may be placed on a line or a square grid at 50 m intervals, or the point-quarter distance method might be applied along a compass line with 100 m between points. There are many reasons why systematic sampling is used in practice, but the usual reasons are *simplicity* of application in the field, and the desire to *sample evenly* across a whole habitat.

The most common type of systematic sampling used in ecology is the *centric systematic area-sample* illustrated in Figure 8.4. The study area is subdivided into equal squares, and a sampling unit is taken from the center of each square. The samples along the outer edge are thus half as far from the boundary as they are from the nearest sample. Note that once the number of samples has been specified, there is only one centric sample for any area—all others would be eccentric samples (Milne 1959).

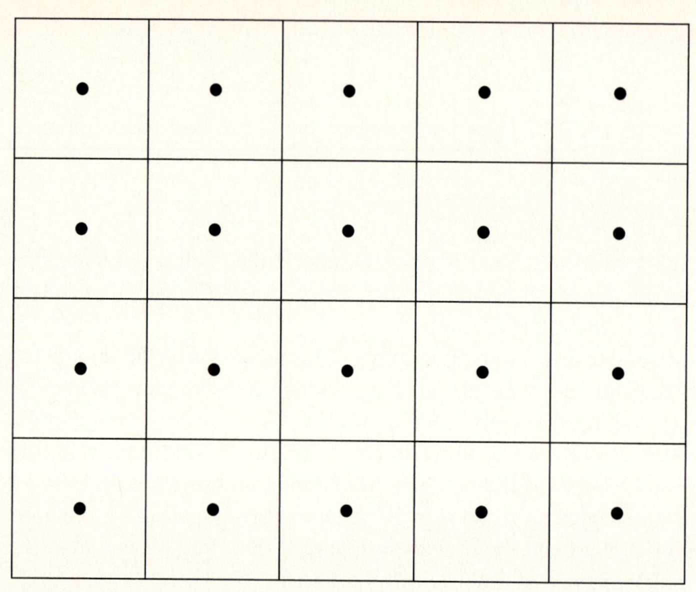

Figure 8.4 Example of a study area subdivided into 20 equal-size squares with one sample taken at the center of each square. This is a *centric systematic area-sample.*

Statisticians have usually condemned systematic sampling in favor of random sampling and have cataloged all the pitfalls that may accompany systematic sampling (Cochran 1977). The most relevant problem for an ecologist is the possible existence of periodic variation in the system under analysis. Figure 8.5 illustrates a hypothetical example in which an environmental variable (soil water content, for example) varies in a sine-wave over the study plot. If you are unlucky and happen to sample at the same periodicity as the sine wave, you can obtain a biased estimate of the mean and the variance (A in Figure 8.5).

But what is the likelihood that problems like periodic variation will occur in actual field data? Milne (1959) attempted to answer this question by looking at systematic samples taken on biological populations that had been completely enumerated (so that the true mean and variance were known). He analyzed data from 50 populations and found that, in practice, there was no error introduced by assuming that a centric systematic sample is a simple random sample, and using all the appropriate formulas from random sampling theory.

Periodic variation like that in Figure 8.5 does not seem to occur in ecological systems. Rather, most ecological patterns are highly clumped and irregular, so that in practice the statistician's worry about periodic influences seems to be a misplaced concern (Milne 1959). Thus the practical recommendation is, *You can use systematic sampling, but watch for possible periodic trends.*

Caughley (1977b) discusses the problems of using systematic sampling in aerial surveys. He simulated a computer population of kangaroos, using some observed aerial counts, and then sampled this computer population with several sampling designs, as outlined in Chapter 4 (Section 4.4.2). Table 8.5 summarizes the results based on 20,000 replicate estimates done by a computer on the hypothetical kangaroo population. All sampling designs

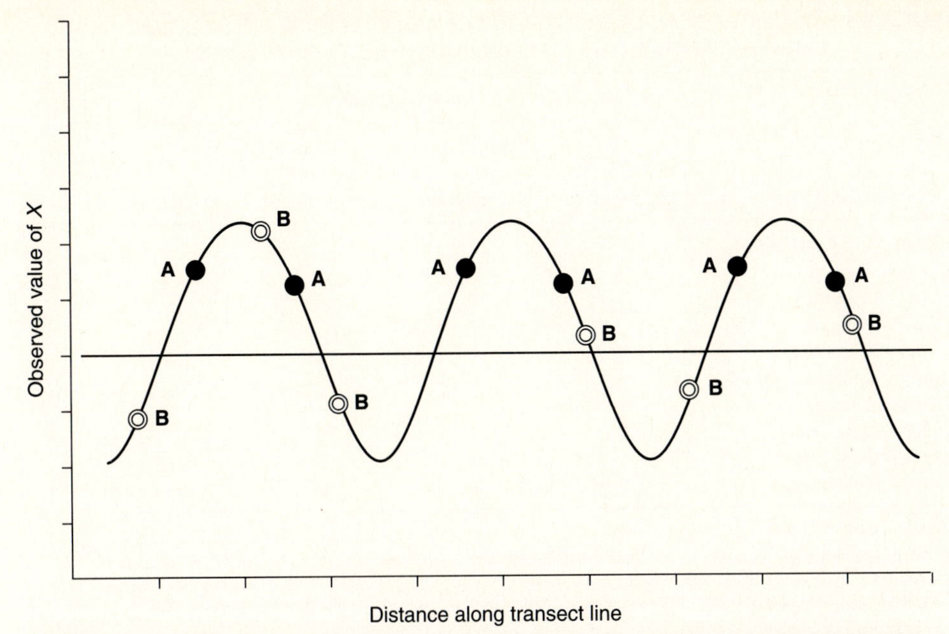

Figure 8.5 Hypothetical illustration of periodic variation in an ecological variable and the effects of using systematic sampling on estimating the mean of this variable. If you are unlucky and sample at *A,* you always get the same measurement and obtain a highly biased estimate of the mean. If you are lucky and sample at *B,* you get exactly the same mean and variance as if you had used random sampling. The important question is whether such periodic variation exists in the ecological world.

provided equally good estimates of the mean kangaroo density, and all means were unbiased, but the standard error estimated from systematic sampling was underestimated, compared with the true value. This bias would reduce the size of the confidence belt in systematic samples, so that confidence limits based on systematic sampling would not be valid because they would be too narrow. The results of Caughley (1977b) should not be generalized to all aerial surveys, but they inject a note of warning into the planning of aerial counts if systematic sampling is used.

There is probably no issue on which field ecologists and statisticians differ more than on the use of random vs. systematic sampling in field studies. If gradients across the study area are important to recognize, systematic sampling like that shown in Figure 8.4 will be more useful than random sampling to an ecologist. This decision will be strongly affected by the exact ecological questions being studied. Some combination of systematic and random sampling may be useful in practice and the most important message for a field ecologist is to avoid haphazard or judgmental sampling.

The general conclusion with regard to ecological variables is that systematic sampling can often be applied, and the resulting data treated as random sampling data, without bias. But there will always be a worry that periodic effects may influence the estimates, so that *if you have a choice of taking a random sample or a systematic one, always choose random sampling*. But if the cost and inconvenience of randomization is too great, you may lose little by sampling in a systematic way.

TABLE 8.5 SIMULATED COMPUTER SAMPLING OF AERIAL TRANSECTS OF EQUAL LENGTH FOR A KANGAROO POPULATION IN NEW SOUTH WALES[a]

	Sampling system			
Method of analysis	PPS[b] with replacement	Random with replacement	Random without replacement	Systematic
Coefficient of variation				
2% sampling rate				
(n = 10 transects)	9	9	9	9
20% sampling rate				
(n = 100 transects)	3	3	2	3
Bias in standard error (%)				
2% sampling rate				
(n = 10 transects)	+1	+1	0	−3
20% sampling rate				
(n = 100 transects)	0	0	0	−23

Source: Caughley 1977b.

[a]Data from actual transects were used to set up the computer population, which was then sampled 20,000 times at two different levels of sampling. The percentage coefficient of variation of the population estimates and the relative bias of the calculated standard errors of the population estimates were compared for random and systematic sampling.

[b]PPS, Probability-proportional-to-size sampling, discussed previously in Chapter 4.

8.5 MULTISTAGE SAMPLING

Ecologists often subsample. For example, a plankton sample may be collected by passing 100 liters of water through a net. This sample may contain thousands of individual copepods or cladocerans, and to avoid counting the whole sample, a limnologist will count 1/100 or 1/1000 of the sample.

Statisticians describe subsampling in two ways. We can view the *sampling unit* in this case to be the 100-liter sample of plankton and recognize that this sample unit can be divided into many smaller samples, called *subsamples* or *elements*. Figure 8.6 shows schematically how subsampling can be viewed. The technique of subsampling has also been called *two-stage sampling* because the sample is taken in two steps:

1. Select a sample of *units* (called the *primary units*)
2. Select a sample of *elements* within each unit.

Many examples can be cited:

Type of data	Primary sample unit	Subsamples, or elements
Aphid infestation	Sycamore tree	Leaves within a tree
DDT contamination	Clutch of eggs	Individual eggs
Plankton	100-liter sample	1 ml subsample
Pollen profiles in peat	1 cm^3 peat at given depth	Microscope slides of pollen grains
Fish population in streams	Entire stream	Habitat section of stream

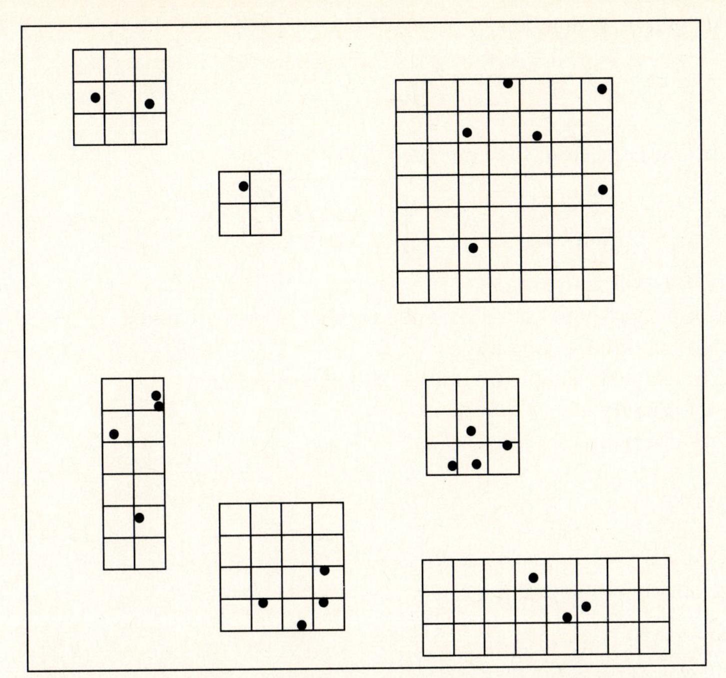

Figure 8.6 Schematic illustration of two-stage sampling. In this example, 7 primary sampling units occur in the study area, and they contain different numbers of elements (from 4 to 49). For example, the 7 primary units could be 7 lakes of varying size, and the elements could be benthic areas of 10 m^2. Alternatively they could be 7 woodlands varying in size from 4 to 49 ha. Dots (•) represent sample elements selected for counting.

If every primary sampling unit contains the same number of elements, subsampling is relatively easy and straightforward (Cochran 1977, Chapter 10). But in most ecological situations, the primary sample units are of unequal size, and sampling is more complex. For example, different sycamore trees will have different numbers of leaves.

Clearly subsampling could be done at several levels; thus, two-stage sampling can be generalized to three-stage sampling, and the general term *multistage sampling* is used to describe any design in which there are two or more levels of sample selection (Hankin 1984).

8.5.1 Sampling Units of Equal Size

Consider first the simplest case of multistage sampling in which n primary sample units are picked, and m subsamples are taken in each unit. For example, you might take 20 plankton samples (each of 100 liters of lake water) and from each of these 20 samples count four subsamples (elements) of 1 ml each. We adopt this notation:

$$x_{ij} = \text{Measured value for the } j \text{ element in primary unit } i$$

$$x_i = \text{Mean value per element in primary unit } i = \sum_{j=1}^{m} \left(\frac{x_{ij}}{m} \right)$$

The mean of the total sample is given by

$$\bar{\bar{x}} = \sum_{i=1}^{n} \left(\frac{\bar{x}_i}{n} \right) \tag{8.39}$$

The standard error of this mean is (from Cochran 1977, 277):

$$\text{S.E.}(\bar{\bar{x}}) = \sqrt{\left(\frac{1 - f_1}{n} \right) s_1^2 + \left[\frac{f_1(1 - f_2)}{mn} \right] s_2^2} \tag{8.40}$$

where f_1 = Sampling fraction in first stage
= Number of primary units sampled/Total number of primary units
f_2 = Sampling fraction in second stage
= Number of elements sampled/Number of elements per unit
n = Number of primary units sampled
m = Number of elements sampled per unit

$$s_1^2 = \frac{\sum^{n} (\bar{x}_i - \bar{\bar{x}})^2}{(n - 1)} \tag{8.41}$$

= Variance among primary unit means

$$s_2^2 = \sum^{n} \sum^{m} \frac{(x_{ij} - \bar{x}_i)^2}{n(m - 1)} \tag{8.42}$$

= Variance among elements within primary units

If the sampling fractions are small, the finite population corrections (f_1, f_2) can be omitted. Note that the standard error can be easily decomposed into two pieces, the first piece due to variation among primary sampling units, and the second piece due to variation within the units (among the subsamples).

Box 8.4 illustrates the use of these formulas in subsampling. Program-group SAMPLING (Appendix 6) will do these calculations.

Box 8.4 Multistage Sampling: Subsampling with Primary Units of Equal Size

A limnologist estimated the abundance of the cladoceran *Daphnia magna* by filtering 1000 liters of lake water (= sampling unit) and subdividing it into 100 subsamples (= elements), of which 3 were randomly chosen for counting. One day when he sampled, he got these results (number of *Daphnia* counted):

Subsample	Sample 9.1	Sample 9.2	Sample 9.3	Sample 9.4
1	46	33	27	39
2	30	21	14	31
3	42	56	65	45
Mean	39.33	36.67	35.33	38.33

In this example, $n = 4$ primary units sampled, and N is very large ($> 10^5$), so the sampling fraction in the first stage (f_1) is effectively 0.0. In the second stage $m = 3$ of a total of $M = 100$ possible elements, so the sampling fraction f_2 is 0.03.

The mean number of *Daphnia* per 10 liters is given by equation (8.39):

$$\bar{\bar{x}} = \sum_{i=1}^{n} \left(\frac{\bar{x}_i}{n} \right)$$

$$= \frac{39.33}{4} + \frac{36.67}{4} + \frac{35.33}{4} + \frac{38.33}{4} = 37.42 \text{ } Daphnia$$

Note that this is the same as the mean that would be estimated if the entire data set were treated as a simple random sample. This would *not* be the case if the number of subsamples varied from primary sample unit to unit.

The standard error of the estimated mean is from equation (8.40):

$$\text{S.E.}(\bar{\bar{x}}) = \sqrt{\frac{1 - f_1}{n} s_1^2 + \frac{f_1(1 - f_2)}{mn} s_2^2}$$

First, calculate s_1^2 and s_2^2:

$$s_1^2 = \frac{\Sigma (\bar{x}_i - \bar{\bar{x}})^2}{n - 1} = \frac{(39.33 - 37.42)^2}{3} + \frac{(36.67 - 37.42)^2}{3} + \cdots$$

$$= 3.1368 \text{ (variance among primary unit means)}$$

$$s_2^2 = \frac{\Sigma \Sigma (x_{ij} - \bar{x}_i)^2}{n(m - 1)} = \left[\frac{(46 - 39.33)^2}{4(2)} + \frac{(30 - 39.33)^2}{4(2)} + \cdots \right]$$

$$= 284.333 \text{ (variance among subsamples)}$$

It is clear from the original data that there is much more variation among the subsamples than there is among the primary samples. Since f_1 is nearly zero, the second term disappears, and

$$\text{S.E.}(\bar{\bar{x}}) = \sqrt{\frac{1}{4}(3.1368)} = 0.8856$$

The 95% confidence interval would be ($t_\alpha = 3.182$ for 3 d.f.):

$$\bar{\bar{x}} \pm t_\alpha \text{S.E.}(\bar{\bar{x}})$$

$$37.42 \pm 3.182(0.8856)$$

or from 34.6 to 40.2 *Daphnia* per 10 liters of lake water. If you wish to express these values in terms of the original sampling unit of 1000 liters, multiply them by 100.

Program-group SAMPLING (Appendix 2) will do these calculations.

8.5.2 Sampling Units of Unequal Size

If the sampling units are of varying size, calculations of the estimated means and variances are more complex. The specific details are too complex to be condensed simply in this book (see Cochran 1977, Chapter 11).

There are two basic choices that you must make in selecting a multistage sampling design model: whether to choose the primary sampling units with *equal probability* or with *probability proportional to size* (PPS). For example, in Figure 8.6 we could choose two of the 7 primary sampling units at random, assigning probability of 1 in 7 to each. Alternatively, we could note that there are 123 elements in the study area in Figure 8.6, and that the largest unit has 49 elements, so its probability of selection should be 49/123 or 0.40, while the smallest unit has only 4 elements, so its probability of selection should be 0.03.

It is usually more efficient to sample a population with some type of PPS sampling design (Cochran 1977). But the problem is that before you can use PPS sampling, you must know (or have a good estimate of) all the numbers of elements in each of the primary units in the population (the equivalent to the information in Figure 8.6). If you do not know this (as is often the case in ecology), you must revert to simple random sampling or stratified sampling, or do a pilot study to get the required information. Cochran (1977) shows that often there is little loss in precision by making a rough estimate of the size of each primary unit, and using probability-proportional-to-estimated-size (PPES) sampling.

Cochran (1977, Chapter 11) has a clear discussion of the various methods of estimation that are applied to multistage sampling designs. Hankin (1984) discusses the application of multistage sampling designs to fisheries research, and notes the need for a computer to calculate estimates from the more complex models (Chao 1982). We have already applied a relatively simple form of PPS sampling to aerial census (Section 4.4.2). Ecologists with more complex multistage sampling designs should consult a statistician. Program-group SAMPLING (Appendix 2) will do the calculations in Cochran (1977, Chapter 11) for unequal size sampling units for equal probability, PPS, or PPES sampling.

With multistage sampling, you must choose the sample sizes to be taken at *each* stage of sampling. How many sycamore trees should you choose as primary sampling units? How many leaves should you sample from each tree? The usual recommendation is to sample the same fraction of elements in each sampling unit, since this will normally achieve a near-optimal estimate of the mean (Cochran 1977, 323). To choose the number of primary units to sample in comparison to the number of elements to sample within units, you need to know the relative variances of the two levels. For example, the total aphid population per sycamore tree may not vary much from tree to tree, but there may be great variability from leaf to leaf in aphid numbers. If this is the case, you should sample relatively few trees and sample more leaves per tree. Cochran (1977) should be consulted for the detailed formulas, which depend somewhat on the relative costs of sampling more units compared with sampling more elements. Schweigert and Sibert (1983) discuss the sample size problem in multistage sampling of commercial fisheries. One useful rule-of-thumb is to sample an average of m elements per primary sampling unit, where

$$m \approx \sqrt{\frac{s_2^2}{s_U^2}} \tag{8.43}$$

and where m = Optimal number of elements to sample per primary unit

s_2^2 = Variance among elements within primary units

$$= \sum^{n} \sum^{m} \frac{(x_{ij} - \bar{x}_i)^2}{n(m-1)} \text{ as defined in equation (8.42)}$$

$s_U^2 = s_1^2 - (s_2^2/M)$ = component of variance among unit means

s_1^2 = Variance among primary units

$$= \frac{\sum_{i=1}^{n} (\bar{x}_i - \bar{\bar{x}})^2}{(n-1)} \text{ as defined in equation (8.41)}$$

For example, from the data in Box 8.4, $s_1^2 = 3.14$, $s_2^2 = 284.3$, and $M = 100$ possible subsamples per primary unit; thus

$$s_U^2 = 3.14 - \left(\frac{284.3}{100}\right) = 0.29$$

$$m \approx \sqrt{\frac{284.3}{0.29}} = 31.3 \text{ elements}$$

There are 100 possible subsamples to be counted, and this result suggests you should count 31 of the 100. This result reflects the large variance between subsample counts in the data of Box 8.4.

Once you have an estimate of the optimal number of subsamples (m in equation [8.43]) you can determine the sample size of the primary units (n) from knowing what standard error you desire in the total mean $\bar{\bar{x}}$ (equation [8.39]) from the approximate formula

$$\text{S.E.}(\bar{\bar{x}}) = \sqrt{\frac{1}{n}\left(s_1^2 - \frac{s_2^2}{M}\right) + \left(\frac{1}{mn}\right) s_2^2 - \frac{1}{N} s_1^2} \tag{8.44}$$

where S.E.($\bar{\bar{x}}$) = Desired standard error of mean

n = Sample size of primary units needed

s_1^2 = Variance among primary units (equation [8.41])

s_2^2 = Variance among elements (equation [8.42])

M = Total number of elements per primary unit

m = Optimal number of elements to subsample (equation [8.43])

This equation can be solved for n if all the other parameters have been estimated in a pilot survey or guessed from prior knowledge.

8.6 SUMMARY

If you cannot count or measure the entire population, you must sample. Several types of sampling designs can be used in ecology. The more complex the design, the more efficient it is, but to use complex designs correctly, you must already know a great deal about your population.

Simple random sampling is the easiest and most common sampling design. Each possible sample unit must have an equal chance of being selected to obtain a random sample. All the formulas of statistics are based on random sampling, and probability theory is the foundation of statistics. Thus *you should always sample randomly* when you have a choice.

In some cases the statistical population is finite in size, and the idea of a *finite population correction* must be added into formulas for variances and standard errors. These formulas are reviewed for measurements, ratios, and proportions.

Often a statistical population can be subdivided into homogeneous subpopulations, and random sampling can be applied to each subpopulation separately. This is *stratified random sampling*, and represents the single most powerful sampling design that ecologists can adopt in the field with relative ease. Stratified sampling is almost always more precise than simple random sampling, and every ecologist should use it whenever possible.

Sample size allocation in stratified sampling can be determined using *proportional* or *optimal* allocation. To use optimal allocation, you need rough estimates of the variances in each of the strata and the cost of sampling each strata. Optimal allocation is more precise than proportional allocation, and is to be preferred. Some simple rules are presented to allow you to estimate the optimal number of strata you should define in setting up a program of stratified random sampling.

If organisms are rare and patchily distributed, you should consider using *adaptive cluster sampling* to estimate abundance. When a randomly placed quadrat contains a rare species, adaptive sampling adds quadrats in the vicinity of the original quadrat to sample the potential cluster. This additional nonrandom sampling requires special formulas to estimate abundance without bias.

Systematic sampling is easier to apply in the field than random sampling, but may produce biased estimates of means and confidence limits if there are periodicities in the data. In field ecology this is usually not the case, and systematic samples seem to be the equivalent of random samples in many field situations. If a gradient exists in the ecological community, systematic sampling will be better than random sampling for describing it.

More complex *multistage sampling* designs involve sampling in two or more stages, often called *subsampling*. If all the sample units are equal in size, calculations are simple. But in many ecological situations the sampling units are not of equal size, and the sampling design can become very complex, so you should consult a statistician. Multistage sampling requires considerable background information, and unless this is available, ecologists are usually better off using stratified random sampling.

SELECTED READING

Anganuzzi, A. A., and Buckland, S. T. 1993. Post-stratification as a bias reduction technique. *Journal of Wildlife Management* 57: 827–834.

Caughley, G. 1977. Sampling in aerial survey. *Journal of Wildlife Management* 41: 605–615.

Cochran, W. G. 1977. *Sampling Techniques*. 3d ed. John Wiley and Sons, New York.

Downing, J. A., and Anderson, M. R. 1985. Estimating the standing biomass of aquatic macrophytes. *Canadian Journal of Fisheries and Aquatic Sciences* 42: 1860–1869.

Iachan, R. 1985. Optimum stratum boundaries for shellfish surveys. *Biometrics* 41: 1053–1062.

Milne, A. 1959. The centric systematic area-sample treated as a random sample. *Biometrics* 15: 270–297.

Morin, A. 1985. Variability of density estimates and the optimization of sampling programs for stream benthos. *Canadian Journal of Fisheries and Aquatic Sciences* 42: 1530–1534.

Schweigert, J. F., Haegele, C. W., and Stocker, M. 1985. Optimizing sampling design for herring spawn surveys in the Strait of Georgia, B.C. *Canadian Journal of Fisheries and Aquatic Sciences* 42: 1806–1814.

Thompson, S. K. 1992. *Sampling*. John Wiley and Sons, New York.

Van Valen, L. M. 1982. A pitfall in random sampling. *Nature* 295: 171.

QUESTIONS AND PROBLEMS

8.1. Reverse the ratio calculations for the wolf-moose data in Box 8.2 and calculate the estimated ratio of *moose to wolves* for these data, along with the 95% confidence interval. Are these limits the reciprocal of those calculated in Box 8.2? Why or why not?

(a) How would these estimates change if the population was considered infinite instead of finite?

8.2. Assume that the total volume of the lake in the example in Box 8.4 is 1 million liters (N). Calculate the confidence limits that occur under this assumption and compare them with those in Box 8.4, which assumes N is infinite.

8.3. In the wood lemming (*Myopus schisticolor*) in Scandinavia there are two kinds of females: normal females that produce equal numbers of males and females, and special females that produce only female offspring. In a spruce forest with an estimated total population of 72 females, a geneticist found in a sample of 41 females, 15 individuals were female-only types. What is the estimate of the fraction of normal females in this population? What are the 95% confidence limits?

8.4. Hoisaeter and Matthiesen (1979) report the following data for the estimation of seaweed (*Ulva*) biomass for a reef flat in the Philippines (quadrat size 0.25 m^2)

Stratum	Area (m^2)	Sample size	Mean (g)	Variance
I (near shore)	2175	9	0.5889	0.1661
II	3996	14	19.3857	179.1121
III	1590	7	2.1429	3.7962
IV	1039	6	0.2000	0.1120

Estimate the total *Ulva* biomass for the study zone, along with its 95% confidence limits. Calculate proportional and optimal allocations of samples for these data, assuming the cost of sampling is equal in all strata, and you require a confidence belt of ±25% of the mean.

8.5. Tabulate the observed number of clams (x) in column 6 of Table 8.4 in a cumulative frequency distribution. Estimate the optimal strata boundaries for this variable, based on 3 strata, using the cum $\sqrt{f}$ procedure. How do the results of this stratification differ from those stratified on the depth variable (as in Table 8.4)?

8.6. A plant ecologist subsampled 0.25 m^2 areas within 9 different deer-exclosures of 16 m^2. She subsampled 2 areas within each exclosure, and randomly selected 9 of 18 exclosures that had been established in the study area. She got these results for herb biomass (g dry weight per 0.25 m^2):

	Exclosure no.								
Subsample	3	5	8	9	12	13	15	16	18
A	2	5	32	23	19	16	23	25	13
B	26	3	6	9	8	7	9	3	9

Estimate the mean biomass per 0.25 m^2 for these exclosures, along with 95% confidence limits. How would the confidence limits for the mean be affected if you assumed all these estimates were replicates from simple random sampling with $n = 18$? What recommendation would you give regarding the optimal number of subsamples for these data?

8.7. Describe an ecological sampling situation in which you would *not* recommend using stratified random sampling. In what situation would you *not* recommend using adaptive cluster sampling?

8.8. How does multistage sampling differ from stratified random sampling?

8.9. Use the marked 25 × 25 grid on Figure 6.1 of the redwood seedlings to set out an adaptive sampling program to estimate density of these seedlings. From a random number table, select 15 of the possible 625 plots and apply adaptive cluster sampling to estimate density. Compare your results with simple random sampling of $n = 15$ quadrats.

CHAPTER 9

Sequential Sampling

Ecologists usually sample populations and communities with the classical statistical approach outlined in the last chapter. Sample sizes are thus fixed in advance by the dictates of logistics and money, or by some forward planning, as outlined in Chapter 7. Classical statistical analysis is then performed on these data, and the decision is made whether to accept or reject the null hypothesis. All of this is very familiar to ecologists as the classical problem of statistical inference.

But there is another way. *Sequential sampling* is a statistical procedure whose characteristic feature is that *sample size is not fixed in advance*. Instead, you make observations or measurements one at a time, and after each observation, you ask the accumulated data whether or not a conclusion can be reached. Sample size is thus minimized; in some cases only half the number of observations required with classical sampling are needed for sequential sampling (Mace 1964). The focus of sequential sampling is thus decision making, so it is most useful in ecological situations that demand a decision—should I spray this crop or not, should I classify this stream as polluted or not, should I sample this population for another night or not? If you must make these kinds of decisions, you should know about sequential sampling. The focus of sequential sampling is to minimize work, and for ecologists it can be a useful tool for some field studies.

The critical differences between classical and sequential sampling are shown in the following table:

	Sample size	Statistical inference
Classical statistical analysis	Fixed in advance	Two possibilities Accept null hypothesis Reject null hypothesis
Sequential analysis	Not fixed	Three possibilities Accept null hypothesis Reject null hypothesis Uncertainty (take another sample)

The chief advantage of sequential sampling is that it minimizes sample size and thus saves time and money. It has been applied in ecological situations in which sampling is done serially, so that the results of one sample are completed before the next one is analyzed. For example, the number of pink bollworms per cotton plant may be counted, and after each plant is counted, a decision can be made whether the population is dense (= spray pesticide) or sparse (do not spray). If you do not sample serially, or do not have an ecological situation in which sampling can be terminated on the basis of prior samples, then you must use the fixed sample approach outlined in Chapter 7. For example, plankton samples obtained on oceanographic cruises are not usually analyzed immediately, and the cost of sampling is so much larger than the costs of counting samples on shore later, that a prior decision of a fixed sample size is the best strategy.

There is a good discussion of sequential sampling in Dixon and Massey (1983) and Mace (1964), and a more theoretical discussion in Wetherill and Glazebrook (1986). Morris (1954) describes one of the first applications of sequential sampling to insect surveys, and Nyrop and Binns (1991) provide a recent overview for insect pest management.

9.1 TWO ALTERNATIVE HYPOTHESES

We shall consider first the simplest type of sequential sampling, in which there are two alternative hypotheses, so that the statistical world is black or white. For example, we may need to know if insect density is *above or below* 10 individuals/leaf, or whether the sex ratio is *more* than 40% males or *less* than 40% males. These are called one-sided alternative hypotheses in statistical jargon because the truth is either A or B, and if it is not A, it must be B.

9.1.1 Means from a Normal Distribution

To illustrate the general principles of sequential sampling, I will describe first the application of sequential methods to the case of variables that have a normal, bell-shaped distribution. As an example, suppose that you have measured the survival time of rainbow trout fry exposed to the effluent from a coal-processing plant. If the plant is operating correctly, you know from previous laboratory toxicity studies that mean survival time should not be less than 36 hours. To design a sequential sampling plan for this situation, proceed as follows.

Step 1. *Set up the alternatives you need to distinguish.* These must always be phrased as *either-or* and are stated statistically as two hypotheses; for example,

$$H_1\text{: mean survival time} \leq 36 \text{ hours}$$

$$H_2\text{: mean survival time} \geq 40 \text{ hours}$$

Note that these two alternatives must not be the same, although they could be very close (e.g., 36 and 36.1 hours instead of 36 and 40 hours).

The alternatives to be tested must be based on prior biological information. For example, toxicity tests could have established that trout fry do not survive on average longer than 36 hours when pollution levels exceed the legal maximum. The alternatives selected must be carefully chosen in keeping with the ecological decisions that will flow from accepting H_1 or H_2. If the two alternatives are very close, larger sample sizes will be needed on average to discriminate between them.

Step 2. *Decide on the acceptable risks of error α and β.* These probabilities are defined in the usual way: α is the chance of rejecting H_1 (and accepting H_2) when in fact H_1 is correct, and β is the chance of rejecting H_2 (and accepting H_1) when H_2 is correct. Often $\alpha = \beta = 0.05$ but this should not be decided automatically, since it depends on the risks you wish to take. In the rainbow trout example, when legal action might occur, you might assign $\alpha = 0.01$ to reduce Type I errors and be less concerned about Type II errors, and assign $\beta = 0.10$.*

Step 3. *Estimate the statistical parameters needed.* In the case of means, you must know the standard deviation to be expected from the particular measurements you are taking. You may know, for example, that for rainbow trout, survival time, $s = 16.4$ hours, from previous experiments. If you do not have an estimate of the standard deviation, you can conduct a pilot study to estimate it (see page 232).

All sequential sampling plans are characterized graphically by one or more sets of parallel lines, as illustrated in Figure 9.1. The equations for these two lines are

$$\text{Lower line: } Y = bn + h_1 \quad (9.1)$$

$$\text{Upper line: } Y = bn + h_2 \quad (9.2)$$

*The usual situation in statistics is that α and β are related through properties of the test statistic and thus cannot be set independently. The reason that this is not the case here is that the two alternative hypotheses have been set independently.

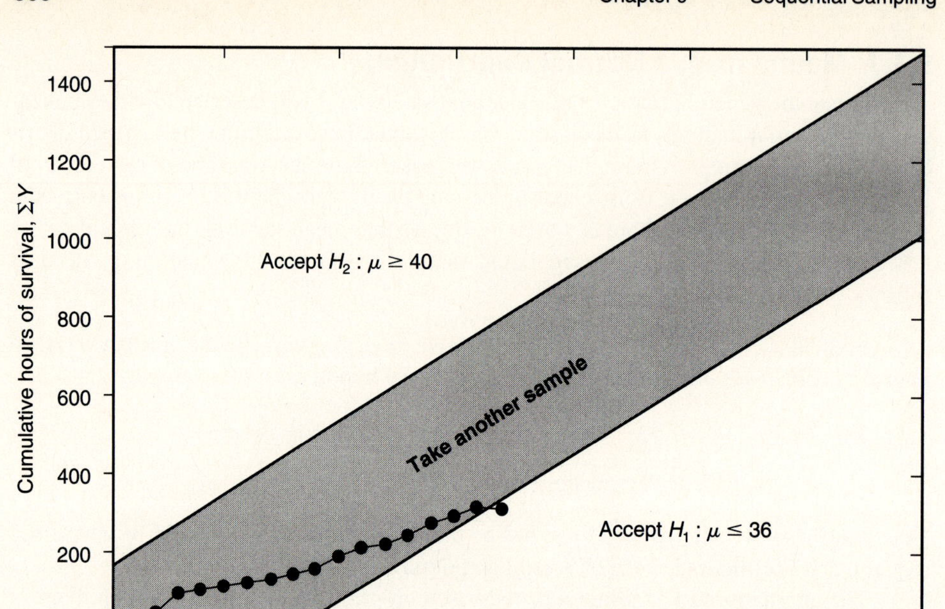

Figure 9.1 Sequential sampling plan for rainbow trout survival experiment discussed in the text. The hypothesis H_1 that mean survival time is less than or equal to 36 hours is tested against the alternative hypothesis H_2 that survival time is 40 hours or more. In the shaded area of statistical uncertainty, continue taking samples. An example is graphed in which the decision to accept H_1 is taken after $n = 17$. Survival time is assumed to follow a normal distribution.

where b = Slope of lines
h_1 = y-intercept of lower line
h_2 = y-intercept of upper line
n = Sample size
Y = Measured variable

The slope of these lines for means from a normal distribution is estimated from

$$b = \frac{\mu_1 + \mu_2}{2} \tag{9.3}$$

where b = Slope of the sequential sampling lines
μ_1 = Mean value postulated in H_1
μ_2 = Mean value postulated in H_2

For our rainbow trout example above, H_1 is that $\mu_1 = 36$, and H_2 is that $\mu_2 = 40$, so

$$b = \frac{36 + 40}{2} = 38$$

The y-intercepts are estimated from the following equations:

$$h_1 = \frac{Bs^2}{\mu_1 - \mu_2} \tag{9.4}$$

$$h_2 = \frac{As^2}{\mu_2 - \mu_1} \tag{9.5}$$

where h_1 = y-intercept of lower line

h_2 = y-intercept of upper line

$$A = \log_e\left(\frac{1-\alpha}{\beta}\right)$$

$$B = \log_e\left(\frac{1-\beta}{\alpha}\right)$$

μ_1 = Mean value postulated in H_1

μ_2 = Mean value postulated in H_2

Note that when $\alpha = \beta$, these two equations are identical, so that $h_1 = -h_2$. For the rainbow trout example above, $\alpha = 0.01$ and $\beta = 0.10$, so

$$A = \log_e\left(\frac{1 - 0.01}{0.10}\right) = 2.29253$$

$$B = \log_e\left(\frac{1 - 0.10}{0.01}\right) = 4.49981$$

Thus, from equations (9.4) and (9.5) with the standard deviation estimated to be 16.4 from previous work,

$$h_1 = \frac{4.49981(16.4)^2}{36 - 40} = -302.6$$

$$h_2 = \frac{2.29253(16.4)^2}{40 - 36} = +154.1$$

The two sequential lines thus become

$$Y = 38n - 302.6$$

$$Y = 38n + 154.1$$

These lines are graphed in Figure 9.1. Note that the y-axis of this graph is the accumulated sum of the observations (e.g., sum of survival times) for the n samples plotted on the x-axis. You can plot a graph like Figure 9.1 by calculating three points on the lines (one as a check); for this example,

Lower line		Upper line	
If n =	ΣY =	If n =	ΣY
0	−302.6	0	154.1
10	77.4	10	534.1
20	457.4	20	914.1

This graph can be used directly in the field to plot sequential samples with the simple decision rule to *stop sampling as soon as you leave the zone of uncertainty*. If you wish to use computed decision rules for this example, they are

1. Accept H_1 if ΣY is less than $(38n - 302.6)$
2. Accept H_2 if ΣY is more than $(38n + 154.1)$
3. Otherwise, take another sample and go back to (1).

You would expect to take relatively few samples, if the true mean is much lower than that postulated in H_1, and also to sample relatively little if the true mean is much larger than that postulated in H_2. It is possible to calculate the sample size you may *expect* to need before a decision is reached in sequential sampling. For means from a normal distribution, the expected sample sizes for the three points μ_1, μ_2, and $(\mu_1 + \mu_2/2)$ are

For μ_1,

$$n_1 = 2\left[\frac{h_2 + (1 - \alpha)(h_1 - h_2)}{\mu_1 - \mu_2}\right] \tag{9.6}$$

For μ_2,

$$n_2 = 2\left[\frac{h_2 + \beta(h_1 - h_2)}{\mu_2 - \mu_1}\right] \tag{9.7}$$

For $(\mu_1 + \mu_2)/2$,

$$n_M = \frac{-h_1 h_2}{s^2} \tag{9.8}$$

where n_1 = Expected sample size required when mean = μ_1
n_2 = Expected sample size required when mean = μ_2
n_M = Expected sample size required when mean = $(\mu_1 + \mu_2)/2$
h_2 = y-intercept of upper line (equation [9.5])
h_1 = y-intercept of lower line (equation [9.4])
s^2 = Variance of measured variable

For example, in the rainbow trout example above,

For $\mu_1 = 36$,

$$n_1 = 2\left[\frac{154.1 + (1 - 0.01)(-302.6 - 154.1)}{36 - 40}\right] = 149.0$$

For $\mu_2 = 40$,

$$n_2 = 2\left[\frac{154.1 + 0.10(-302.6 - 154.1)}{40 - 36}\right] = 54.2$$

For $\mu = 38$,

$$n_M = \frac{-(-302.6)(154.1)}{(16.4)^2} = 173.4$$

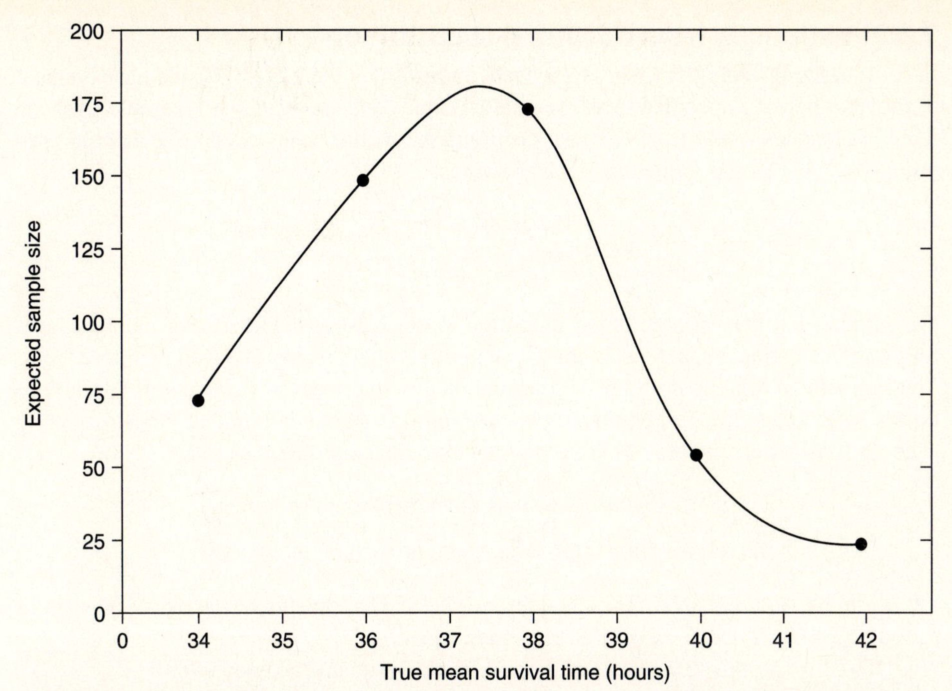

Figure 9.2 The theoretical curve for expected sample sizes in relation to the true mean of the population for the rainbow trout example in the text. Given $\alpha = .01$ and $\beta = .10$, with $s = 16.4$, this is the sample size you would expect to have to take before you could make a sequential sampling decision about the mean. In any particular study you would not know the true mean survival time, and you would get an estimate of this mean only after doing the sampling. These theoretical expected sample size curves are useful however in planning sampling work, since they show you how much work you might have to do under different circumstances.

These values are plotted in Figure 9.2. Note that the expected sample size curve will be symmetric if $\alpha = \beta$, which is not the case in the example shown in Figure 9.2.

You can compare these sample sizes with that expected in a single sampling design of fixed n determined as in Chapter 7. Mace (1964, 134) gives the expected sample size in fixed sampling for the same level of precision as

$$n_s = \left[\frac{(z_\alpha + z_\beta)s}{\mu_1 - \mu_2} \right]^2 \tag{9.9}$$

where z_α and z_β are estimated from the standard normal distribution (i.e., $z_{0.05} = 1.96$). For this trout example, with $\alpha = 0.01$, $\beta = 0.10$ and $s = 16.4$:

$$n_s = \left[\frac{(2.576 + 1.645)(16.4)}{40 - 36} \right]^2 = 299.5 \text{ samples}$$

which is more than twice the number of observations expected to be taken under sequential sampling.

This sequential sampling test assumes that you have an accurate estimate of the standard deviation of the measurement for your population. Since this is never the case in practice, you should view this as an approximate procedure.

9.1.2 Variances from a Normal Distribution

In some cases an ecologist wishes to test a hypothesis that the variability of a measurement is above or below a specified level. Several possible tests are available if sample sizes are fixed in advance (Van Valen 1978), but the problem can also be attacked by sequential sampling. We consider the following two hypotheses:

$$H_1\colon s^2 \le \sigma_1^2$$

$$H_2\colon s^2 \ge \sigma_2^2$$

We assume that the variable being measured is normally distributed. For example, you might be evaluating a new type of chemical method for analyzing nitrogen in moose feces. You know the old method produces replicate analyses that have a variance of 0.009. You do not want to buy the new equipment needed unless you are certain that the new method is about 10% better than the old. You thus can express the problem as follows:

$$H_1\colon s^2 \le 0.008 \quad \text{(and thus the new method is better)}$$

$$H_2\colon s^2 \ge 0.009 \quad \text{(and thus the new method is not better)}$$

Let $\alpha = 0.01$, and $\beta = 0.05$. These values are specified according to the risks you wish to take of rejecting H_0 or H_1 falsely, as explained above.

To carry out this test, calculate the two lines (Dixon and Massey 1983):

$$Y = bn + h_1$$

$$Y = bn + h_2$$

The slope of these lines is

$$b = \frac{\log_e(\sigma_2^2/\sigma_1^2)}{1/\sigma_1^2 - 1/\sigma_2^2} \tag{9.10}$$

The y-intercepts are given by

$$h_1 = \frac{-2B}{1/\sigma_1^2 - 1/\sigma_2^2} \tag{9.11}$$

$$h_2 = \frac{2A}{1/\sigma_1^2 - 1/\sigma_2^2} \tag{9.12}$$

where h_1 = y-intercept of lower line
h_2 = y-intercept of upper line
$A = \log_e[(1 - \alpha)/\beta]$
$B = \log_e[(1 - \beta)/\alpha]$
σ_1^2 = Postulated variance for H_1
σ_2^2 = Postulated variance for H_2

When $\alpha = \beta$, $h_1 = -h_2$, and the lines are symmetric about the origin. For the nitrogen analysis problem above,

$$b = \frac{\log_e(0.009/0.008)}{1/0.008 - 1/0.009} = \frac{0.11778}{13.8889} = 0.0084802$$

$$h_1 = \frac{-2\log_e[(1 - 0.05)/0.01]}{1/0.008 - 1/0.009} = \frac{-9.10775}{13.8889} = -0.65576$$

$$h_2 = \frac{2\log_e[(1 - 0.05)/0.01]}{1/0.008 - 1/0.009} = \frac{5.97136}{13.8889} = 0.429938$$

These lines are plotted in Figure 9.3. Note that the y-axis in this graph is the *sums of squares* of X measured by

$$\sum (x - \mu)^2 = \text{Sum of squares of the measurements } x$$

If you know the true mean of the population, this sum is easy to compute as an accumulated sum. If you do not know the true mean, you must use the observed mean $\bar{x}$ as an estimate of μ, recompute the sums of squares at each sample, and plot these values against $(n - 1)$ rather than n.

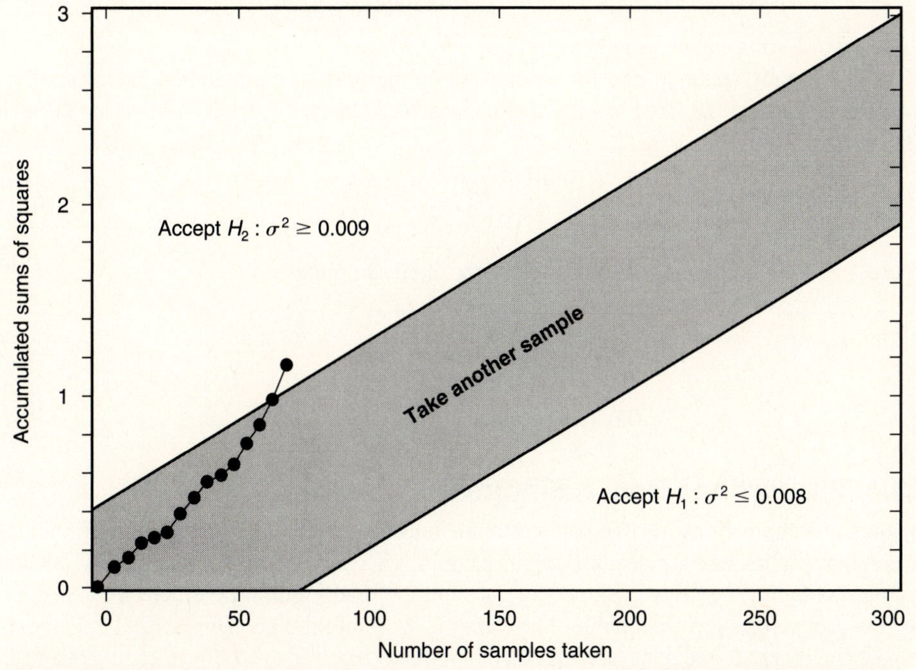

Figure 9.3 Sequential sampling plan for variances from a normal distribution, illustrated for the nitrogen analysis experiment discussed in the text. The hypothesis H_1 that the variance of replicate samples is less than or equal to 0.008 is tested against the alternative hypothesis H_2 that the variance is 0.009 or more. A sample data run in which a decision is made after $n = 70$ is shown for illustration.

The expected sample sizes can be calculated from these equations (Dixon and Massey 1983):

For σ_1^2,

$$n_1 = \frac{(1 - \alpha)h_1 + \alpha h_2}{\sigma_1^2 - b} \tag{9.13}$$

For σ_2^2,

$$n_2 = \frac{\beta h_1 + (1 - \beta)h_2}{b - \sigma_1^2} \tag{9.14}$$

For $\sigma^2 = b$,

$$n_M = \frac{-h_1 h_2}{2b^2} \tag{9.15}$$

where n_1 = Expected sample size required when $s^2 = \sigma_0^2$
n_2 = Expected sample size required when $s^2 = \sigma_1^2$
n_M = Expected sample size required when $s^2 = b$
h_2 = y-intercept of upper line (equation [9.12])
h_1 = y-intercept of lower line (equation [9.11])
b = Slope of the sequential lines (equation [9.10])

These sample sizes can be plotted as in Figure 9.2.

The sample size required for a single sampling plan of comparable discriminating power, according to the fixed sample size approach of Chapter 7, is given by Mace (1964, 138) as

$$n_s \cong \frac{3}{2} + \frac{1}{2}\left(\frac{z_\alpha + Rz_\beta}{R - 1}\right)^2 \tag{9.16}$$

where n_s = Expected sample size under fixed sampling
z_α, z_β = Standard normal deviates (e.g., $z_{0.05} = 1.96$)
R = Ratio of the two standard deviations

$$= \frac{\sigma_1}{\sigma_2} = \sqrt{\frac{\text{Larger variance}}{\text{Smaller variance}}}$$

9.1.3 Proportions from a Binomial Distribution

Sequential sampling can also be applied to attributes like sex ratios (proportion of males) or incidence of disease studies, as long as sampling is conducted serially. It is possible to examine samples in groups in all sequential sampling plans, and little efficiency is lost as long as group sizes are reasonable. For example, you might take samples of 10 fish and inspect each for external parasites, and then plot the results, rather than proceeding one fish at a time.

For proportions, we consider the following two hypotheses:

$$H_1\colon p \le \pi_1$$

$$H_2\colon p \ge \pi_2$$

where p = Observed proportions
π_1 = Expected lower estimate of the population proportion
π_2 = Expected upper estimate of the population proportion

For example, you might hypothesize that if more than 10% of fish are parasitized, treatment is required; but if less than 5%, no treatment is needed. Thus,

$$H_1: p \leq 0.05 \quad \text{(no treatment of fish needed)}$$

$$H_2: p \geq 0.10 \quad \text{(treatment is required)}$$

You must decide α and β, and for this example, assume $\alpha = \beta = 0.05$.

The two lines to be calculated are

$$Y = bn + h_1$$

$$Y = bn + h_2$$

where Y = Number of individual x-types in sample of size n

The slope of the sequential lines for proportions is given by

$$b = \frac{\log_e(q_1/q_2)^2}{\log_e(p_2q_1/p_1q_2)} \tag{9.17}$$

where b = Slope of the sequential sampling lines (see Figure 9.4)

and the other terms are as defined below:

$$h_1 = \frac{-A}{\log_e(p_2q_1/p_1q_2)} \tag{9.18}$$

$$h_2 = \frac{B}{\log_e(p_2q_1/p_1q_2)} \tag{9.19}$$

where h_1 = y-intercept of lower line
h_2 = y-intercept of upper line
$A = \log_e[(1 - \alpha)/\beta]$
$B = \log_e[(1 - \beta)/\alpha]$
$p_1 = \pi_1$ = Expected proportion under H_1
$p_2 = \pi_2$ = Expected proportion under H_2
$q_1 = 1 - p_1$
$q_2 = 1 - p_2$

Note that the denominator is the same in these three formulas (Mace 1964, 141).

For the fish disease example, if $\alpha = \beta = 0.05$, $p_1 = 0.05$, and $p_2 = 0.10$, then

$$b = \frac{\log_e(0.95/0.90)}{\log_e[0.10(0.95)/0.05(0.90)]} = \frac{0.054067}{0.747214} = 0.07236$$

$$h_1 = \frac{-\log_e[(1 - 0.05)/0.05]}{0.747214} = -3.94$$

$$h_2 = \frac{\log_e[(1 - 0.05)/0.05]}{0.747214} = 3.94$$

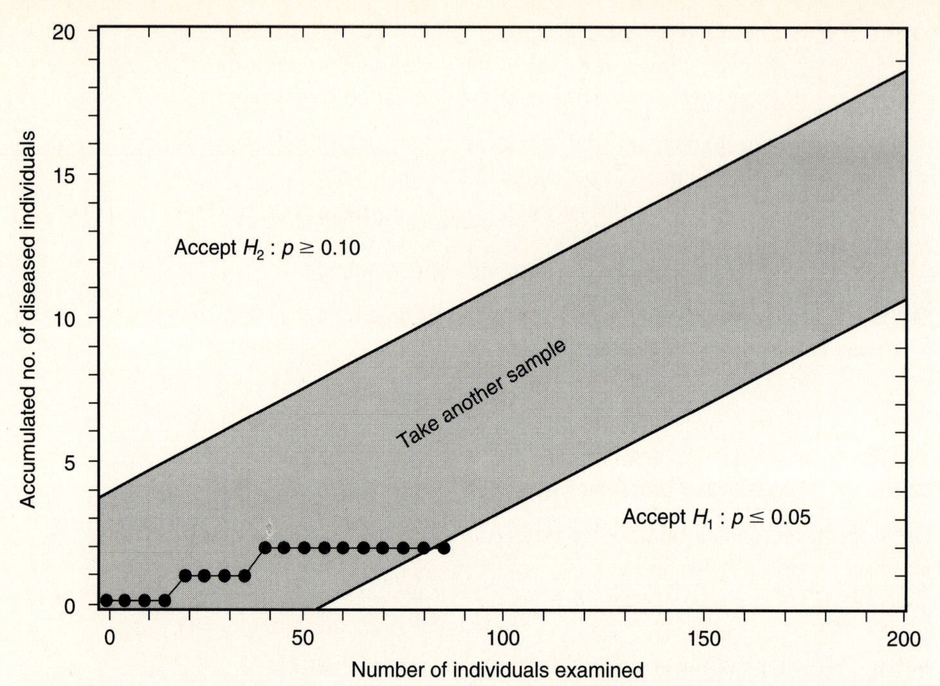

Figure 9.4 Sequential sampling plan for proportions from a binomial distribution, illustrated for the fish disease example discussed in the text. The hypothesis H_1 that the fraction of diseased fish is less than or equal to 0.05 is tested against the alternative hypothesis H_2 that the fraction is 0.10 or more. A sample data run in which a decision is reached after $n = 85$ is shown for illustration.

These lines are illustrated in Figure 9.4 for this disease example. Note that the y-axis of this graph is the accumulated number of diseased individuals in the total sample of n individuals.

You can calculate the expected sample size curve (as in Figure 9.2) for the proportion in the population from the following points:

For p_1,

$$n_1 = \frac{-A(1 - \alpha) + \alpha\beta}{p_1[\log_e(p_2/p_1)] + q_1[\log_e(q_2/q_1)]} \tag{9.20}$$

For p_2,

$$n_2 = \frac{-A\beta + (1 - \beta)B}{p_2[\log_e(p_2/p_1)] + q_2[\log_e(q_2/q_1)]} \tag{9.21}$$

For $p = 1.0$,

$$n_U = \frac{B}{\log_e(p_2/p_1)} \tag{9.22}$$

For $p = 0.0$,

$$n_l = \frac{-A}{\log_e(q_2/q_1)} \tag{9.23}$$

For $p = \dfrac{p_1 + p_2}{2}$,

$$n_M = \frac{-A(B)}{\log_e(p_2/p_1)\log_e(q_2/q_1)} \tag{9.24}$$

where all terms are as defined above. The maximum expected sample size will occur around the midpoint between p_1 and p_2, as defined in equation (9.24).

The sample size required for a fixed sample of comparative statistical precision as determined by the methods outlined in Chapter 7 is given by Mace (1964, 142) as

$$n_s = \left(\frac{z_\alpha + z_\beta}{2 \arcsin \sqrt{p_2} - 2 \arcsin \sqrt{p_1}}\right)^2 \tag{9.25}$$

where z_α, z_β = Standard normal deviates (e.g., $z_{0.05} = 1.96$)
p_1 = Expected proportion under H_1
p_2 = Expected proportion under H_2

and the arcsines are expressed in radians (*not* in degrees).

It is sometimes convenient to specify a sequential sampling plan in tabular form instead of in a graph like that of Figure 9.4. Table 9.1 illustrates one type of table that could be used by field workers to classify a population based on proportion of plants infested.

9.1.4 Counts from a Negative Binomial Distribution

In many sampling programs, organisms are aggregated, so that quadrat counts will often be best described by a negative binomial distribution (Chapter 4, Section 4.2.2). A sequential sampling scheme for negative binomial counts can be designed if you know the exponent k of the variable being measured.

For example, counts of the green peach aphid (*Myzus persicae*) on sugar beet plants follow the negative binomial distribution with an average k-value of 0.8 (Silvester and Cox 1961). You wish to test two hypotheses:

H_1: Mean aphid abundance $\leq$ 10 per leaf

H_2: Mean aphid abundance $\geq$ 20 per leaf

Assume that $\alpha = \beta = 0.05$ for this example. Note that for a negative binomial distribution,

$$\mu = kc$$

and so we define

$$c_1 = \frac{\text{Mean value postulated under } H_1}{k}$$

$$c_2 = \frac{\text{Mean value postulated under } H_2}{k}$$

For this particular example, $c_1 = 10/0.8 = 12.5$, and $c_2 = 20/0.8 = 25$.

TABLE 9.1 EXAMPLE OF SEQUENTIAL SAMPLING PLAN IN TABULAR FORMAT[a]

Plant number	Lower limit	Running total	Upper limit	Plant number	Lower limit	Running total	Upper limit
1	N.D.[b]	—	4	51	18	—	25
2	N.D.	—	4	52	18	—	26
3	N.D.	—	5	53	19	—	26
4	N.D.	—	5	54	19	—	26
5	N.D.	—	6	55	20	—	27
6	N.D.	—	6	56	20	—	27
7	N.D.	—	7	57	21	—	28
8	N.D.	—	7	58	21	—	28
9	N.D.	—	7	59	21	—	29
10	1	—	8	60	22	—	29
11	1	—	8	61	22	—	29
12	2	—	9	62	23	—	30
13	2	—	9	63	23	—	30
14	2	—	9	64	24	—	31
15	3	—	10	65	24	—	31
16	3	—	10	66	24	—	32
17	4	—	11	67	25	—	32
18	4	—	11	68	25	—	32
19	5	—	12	69	26	—	33
20	5	—	12	70	26	—	33
21	5	—	12	71	27	—	34
22	6	—	13	72	27	—	34
23	6	—	13	73	28	—	34
24	7	—	14	74	28	—	35
25	7	—	14	75	28	—	35
26	7	—	15	76	29	—	36
27	8	—	15	77	29	—	36
28	8	—	15	78	30	—	37
29	9	—	16	79	30	—	37
30	9	—	16	80	30	—	37
31	10	—	17	81	31	—	38
32	10	—	17	82	31	—	38
33	10	—	18	83	32	—	39
34	11	—	18	84	32	—	39
35	11	—	18	85	32	—	40
36	12	—	19	86	33	—	40
37	12	—	19	87	33	—	40
38	13	—	20	88	34	—	41
39	13	—	20	89	34	—	41
40	13	—	21	90	35	—	42
41	14	—	21	91	35	—	42
42	14	—	21	92	35	—	43
43	15	—	22	93	36	—	43
44	15	—	22	94	36	—	43
45	16	—	23	95	37	—	44
46	16	—	23	96	37	—	44
47	16	—	23	97	38	—	45
48	17	—	24	98	38	—	45
49	17	—	24	99	38	—	46
50	18	—	25	100	39	—	46

Source: Pieters and Sterling 1974.

[a]Field workers often find it easier to use a table like this instead of a graph like Figure 9.4. This plan is based on the proportion of plants infested with the cotton leafhopper.

[b]N.D. = no decision until running total is obtained from a minimum of ten plants.

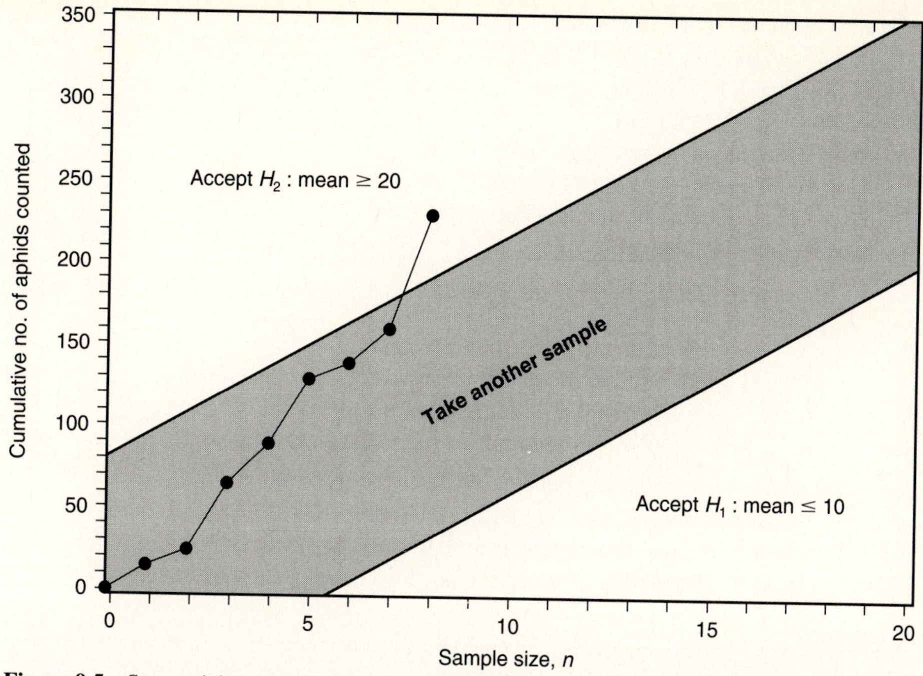

Figure 9.5 Sequential sampling plan for counts from a negative binomial distribution, illustrated for the green peach aphid data in Box 9.1. The hypothesis H_1 that mean aphid abundance is less than or equal to 10 aphids per leaf is tested against the alternative hypothesis H_2 that the mean is more than 20 per leaf.

To carry out this statistical test, calculate the slope and y-intercepts of the two sequential sampling lines:

$$Y = bn + h_1$$

$$Y = bn + h_2$$

The y-axis of this graph is the cumulative number of individuals counted (see Figure 9.5).

The slope of these lines is, for the negative binomial case,

$$b = \frac{(k)\log_e[(c_2 + 1)/(c_1 + 1)]}{\log_e[(c_2 + c_1c_2)/(c_1 + c_1c_2)]} \tag{9.26}$$

where b = Slope of sequential sampling lines
k = Negative binomial exponent
c_1 = (Mean value postulated under H_1)/k
c_2 = (Mean value postulated under H_2)/k

The y-intercepts are calculated as follows:

$$h_1 = \frac{-A}{\log_e[(c_2 + c_1c_2)/(c_1 + c_1c_2)]} \tag{9.27}$$

$$h_2 = \frac{B}{\log_e[(c_2 + c_1c_2)/(c_1 + c_1c_2)]} \tag{9.28}$$

where h_1 = y-intercept of lower line
h_2 = y-intercept of upper line
$A = \log_e[(1 - \alpha)/\beta]$
$B = \log_e[(1 - \beta)/\alpha]$
c_1 = (Mean value postulated under H_0)/k
c_2 = (Mean value postulated under H_1)/k
k = Negative binomial exponent

Box 9.1 illustrates the use of these formulas for negative binomial data.

The expected sample size curve (see Figure 9.2) for negative binomial counts can be estimated from the following general equation:

$$\left\{\begin{array}{c}\text{Expected sample size}\\ \text{to reach a decision if}\\ \text{the true mean is } \mu\end{array}\right\} = \frac{h_2 + (h_1 - h_2)L(c)}{\mu - b} \qquad (9.29)$$

where h_1 and h_2 = y-intercepts (equations [9.27] and [9.28])
μ = Population mean (defined in equation [9.31])
b = Slope of sequential sampling lines (equation [9.26])
$L(c)$ = Probability of accepting H_0 if true mean is μ (equation [9.30])

Box 9.1 Sequential Sampling for Counts from a Negative Binomial Distribution

An insect ecologist wishes to test the abundance of green peach aphids on sugar beets to distinguish two hypotheses:

H_1: Mean aphid abundance $\leq$ 10 per leaf

H_2: Mean aphid abundance $\geq$ 20 per leaf

She decides to set $\alpha = \beta = 0.05$, and she knows from previous work that the counts fit a negative binomial distribution with exponent $k = 0.8$ and that k is reasonably constant over all aphid densities.

1. Calculate c_1 and c_2:

$$c_1 = \frac{\text{Mean value under } H_1}{k} = \frac{10}{0.8} = 12.5$$

$$c_2 = \frac{\text{Mean value under } H_2}{k} = \frac{20}{0.8} = 25.0$$

2. Calculate the slope of the sequential lines using equation (9.26):

$$b = \frac{(k)\log_e[(c_2 + 1)/(c_1 + 1)]}{\log_e[(c_2 + c_1c_2)/(c_1 + c_1c_2)]}$$

$$b = \frac{0.8 \log_e(26.0/13.5)}{\log_e(337.5/325)} = \frac{0.52432}{0.03774} = 13.893$$

3. Calculate the y intercepts using equations (9.27) and (9.28); note that the denominator of these equations has already been calculated in step 2 above.

$$h_1 = \frac{-\log_e[(1-\alpha)/\beta]}{\log_e[(c_2 + c_1c_2)/(c_1 + c_1c_2)]}$$

$$= \frac{-\log_e(0.95/0.05)}{0.03774} = -78.02$$

$$h_2 = \frac{\log_e[(1-\beta)/\alpha]}{\log_e[(c_2 + c_1c_2)/(c_1 + c_1c_2)]}$$

$$= \frac{\log_e(0.95/0.05)}{0.03774} = 78.02$$

Note that if $\alpha = \beta$, these two intercepts are equal in absolute value.

4. Calculate three points on the lines:

$$y_0 = 13.9n - 78.0$$

$$y_1 = 13.9n + 78.0$$

Lower line		Upper line	
n	ΣY	n	ΣY
0	−78	0	78
10	61	10	217
20	200	20	256

These are plotted in Figure 9.5 for this example.

5. Suppose you were field sampling and obtained these results:

	Sample							
	1	2	3	4	5	6	7	8
No. of aphids	20	19	39	10	15	48	45	41
Accumulated no. of aphids	20	39	78	88	103	151	196	237

You would be able to stop sampling after sample 7 and conclude that the study zone had an aphid density above 20/leaf. These points are plotted in Figure 9.5 for illustration.

You may then use these counts in the usual way to get a mean density and 95% confidence interval for counts from the negative binomial distribution, as described in Chapter 4, Section 4.2.2. For these 8 samples, the mean is 29.6, and the 95% confidence limits are 16.3 to 41.3.

Program-group SEQUENTIAL SAMPLING (Appendix 2) can do these calculations.

This equation is solved by obtaining pairs of values of μ and $L(c)$ from the following paired equations:

$$L(c) = \frac{A^h - 1}{A^h - B^h} \qquad \text{(for } h \neq 0\text{)} \tag{9.30}$$

$$\mu = k\left[\frac{1 - (q_1/q_2)^h}{(c_2 q_1/c_1 q_2)^h - 1}\right] \qquad \text{(for } h \neq 0\text{)} \tag{9.31}$$

where $A = (1 - \beta)/\alpha$
$B = \beta/(1 - \alpha)$
$q_1 = c_1 + 1$
$q_2 = c_2 + 1$

and other variables are defined as above. The variable h is a dummy variable and is varied over a range of (say) -6 to $+6$ to generate pairs of μ and $L(c)$ values (Allen et al. 1972), which are then used in equation (9.29).

9.2 THREE ALTERNATIVE HYPOTHESES

Not all decisions can be cast in the form of *either-or*, and in many ecological situations we must choose among three or more courses of action. For example, spruce budworm larvae in conifer forests damage trees by their feeding activities, and budworm infestations need to be classified as light, moderate, or severe (Waters 1955). The simplest way to approach three alternatives is to construct two separate sequential sampling plans, one between each pair of neighboring hypotheses. For example, suppose in the budworm case that

H_1: Mean density $\leq$ 1 larvae/branch (light infestation)

H_2: Mean density $\geq$ 5 larvae/branch but $\leq$ 10 larvae/branch (moderate infestation)

H_3: Mean density $\geq$ 20 larvae/branch (severe infestation)

Using the procedures just discussed, you can calculate a sequential sampling plan for the alternatives H_1 and H_2, and a second sampling plan for the alternatives H_2 and H_3. For example,

Plan A	Plan B
H_1: $x \leq 1$	H_2: $x \leq 10$
H_2: $x \geq 5$	H_3: $x \geq 20$

The result is a graph like that in Figure 9.6. In some cases the results of combining two separate sequential sampling plans may lead to anomalies that must be decided somewhat arbitrarily by continued sampling (Wetherill and Glazebrook 1986).

Note that in constructing Figure 9.6 we have used only two of the three possible sequential sampling plans. For example, for the spruce budworm,

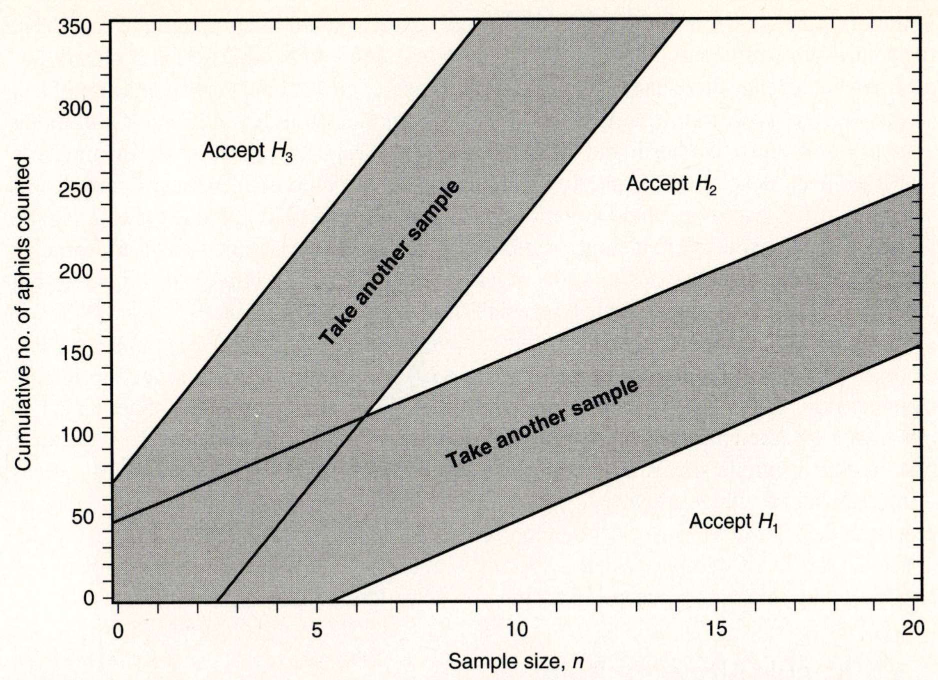

Figure 9.6 Hypothetical example of a sequential sampling plan to distinguish three hypotheses. Two separate plans like that in Figure 9.1 are superimposed defining two regions of uncertainty between the three hypotheses H_1, H_2, and H_3. A plan like this, for example, could be set up to distinguish low, moderate, and severe insect infestations.

Plan A	Plan B	Plan C
H_1: $x \leq 1$	H_2: $x \leq 10$	H_1: $x \leq 1$
H_2: $x \geq 5$	H_3: $x \geq 20$	H_3: $x \geq 20$

We did not include Plan C because the alternatives are covered in Plans A and B. Armitage (1950), however, has suggested that all three sequential plans be computed. One advantage of using all three plans is that some possible sampling anomalies are avoided (Wetherill and Glazebrook 1986), but in practice these may rarely occur anyway.

There is at present no way to calculate the expected sample sizes needed for three alternative hypotheses (Wetherill and Glazebrook 1986). The only approach available is to calculate them separately for each pair of sequential lines and use these as an approximate guideline for field work.

9.3 STOPPING RULES

Sequential sampling plans provide for a quick resolution of two or more competing hypotheses in all cases where the population mean differs from the mean value postulated in the alternative hypotheses, but problems will arise whenever the population mean is near the critical limits postulated in H_1 or H_2. For example, suppose that the true mean is 38.0

in the example shown in Figure 9.1. If this is the case, you are likely to remain forever in the central zone of uncertainty and take an infinite number of samples. This is clearly undesirable, and hence there has been an attempt to specify closed boundaries for sample size, so that there is a maximum sample size at which sampling stops, even if one is still in the zone of uncertainty. Wetherill and Glazebrook (1986, Chapter 6) discusses some methods that have been developed to specify closed boundaries. None of them seem particularly easy to apply in ecological field situations. The simplest practical approach is to calculate the sample size expected in a single sampling scheme of fixed sample size from formulas like equation (9.9), and to use this value of n as an upper limit for sample size. If no decision has been made by the time you have sampled this many quadrats, quit.

An alternative approach to specifying an upper limit to permissible sample size was devised by Kuno (1969, 1972), and Green (1970) suggested a modified version of this stopping rule. Both approaches are similar to those discussed in Chapter 7, Section 7.1.1, in which a specified precision of the mean is selected, but they are applied to the sequential case in which sample size is not fixed in advance. These stopping rules are useful to field ecologists because they allow us to minimize sampling effort and yet achieve a level of precision decided in advance. A common problem in estimating abundance, for example, is to take too many samples when a species is abundant and too few samples when it is rare. Stopping rules can be useful in telling you when to quit sampling.

9.3.1 Kuno's Stopping Rule

Kuno (1969) suggested a fixed-precision stopping rule based on obtaining an estimate of the mean with a fixed confidence belt. To determine the stop line from this method, proceed as follows.

Step 1. Fit the quadratic equation to data previously obtained for this population:

$$s^2 = a_1\bar{x} + a_2\bar{x}^2 + b \tag{9.32}$$

where
s^2 = Observed variance of a series of measurements or counts
$\bar{x}$ = Observed mean of a series of measurements or counts
a_1, a_2 = Regression coefficients for the quadratic equation
b = y-intercept

Standard statistical textbooks provide the techniques for fitting a quadratic equation (e.g., Steel and Torrie 1980, 338; Sokal and Rohlf 1995, 665). If there is not a quadratic relationship between the mean and the variance in your samples, you cannot use this technique. Note that many different samples are needed to calculate this relationship, each sample being one point in the regression.

Step 2. Specify the size of the standard error of the mean density that you wish to obtain. Note that this will be about one-half of the width of the 95% confidence interval you will obtain.

Step 3. From previous knowledge, estimate the mean density of your population and decide on the level of precision you wish to achieve:

$$D = \frac{s_{\bar{x}}}{\bar{x}} = \frac{\text{Desired standard error}}{\text{Estimated mean}} \tag{9.33}$$

D is sometimes called the coefficient of variation of the mean and will be approximately one-half the width of the 95% confidence belt.

Step 4. For a range of sample sizes (n) from 1 to (say) 200, solve the following equation:

$$\sum_{i=1}^{n} Y_i = \frac{a_1}{D^2 - a^2/n} \tag{9.34}$$

where $\sum Y_i$ = Total accumulated count in n quadrats
a_1, a_2 = Slope parameters of quadratic equation (9.32)
D = Desired level of precision as defined in equation (9.33)

Equation (9.34) has been called the "stop line" by Kuno (1969), since it gives the accumulated count needed in n samples to give the desired precision of the mean. The sampler can stop once the $\sum Y_i$ is exceeded, and this will in effect set an upper boundary on sample size. Allen et al. (1972) illustrate the use of this stop rule on field populations of the cotton bollworm in California. Program-group SEQUENTIAL-SAMPLING (Appendix 2) can calculate the stop line from these equations.

9.3.2 Green's Stopping Rule

If you have counts from quadrats and your sampling universe can be described by Taylor's Power Law, you can use Green's (1970) method as an alternative to Kuno's approach.

To calculate the stop line for this sequential sampling plan, you must have estimates of the two parameters of Taylor's Power Law (a, b, see below, Section 9.4.2), which specifies that the log of the means and the log of the variances are related linearly (instead of the quadratic assumption made by Kuno above). You must decide on the level of precision you would like. Green (1970) defines precision as a fixed ratio in the same manner as Kuno (1969):

$$D = \frac{s_{\bar{x}}}{\bar{x}} = \frac{\text{Standard error}}{\text{Mean}} \tag{9.35}$$

Note that D is expressed as one standard error; approximate 95% confidence levels would be $2D$ in width.

The stop line is defined by the log-log regression:

$$\log\left(\sum_{i=1}^{n} Y_i\right) = \frac{\log(D^2/a)}{b - 2} + \left[\frac{b - 1}{b - 2}\right]\log(n) \tag{9.36}$$

where $\sum_{i=1}^{n} Y_i$ = Cumulative number of organisms counted in n samples
D = Fixed level of precision (defined above in equation [9.35])
a = y-intercept of Taylor's Power Law regression
b = Slope of Taylor's Power Law regression
n = Number of samples counted

Figure 9.7 illustrates the stop line for a sampling program to estimate density of an intertidal snail to a fixed precision of 15% ($D = 0.15$).

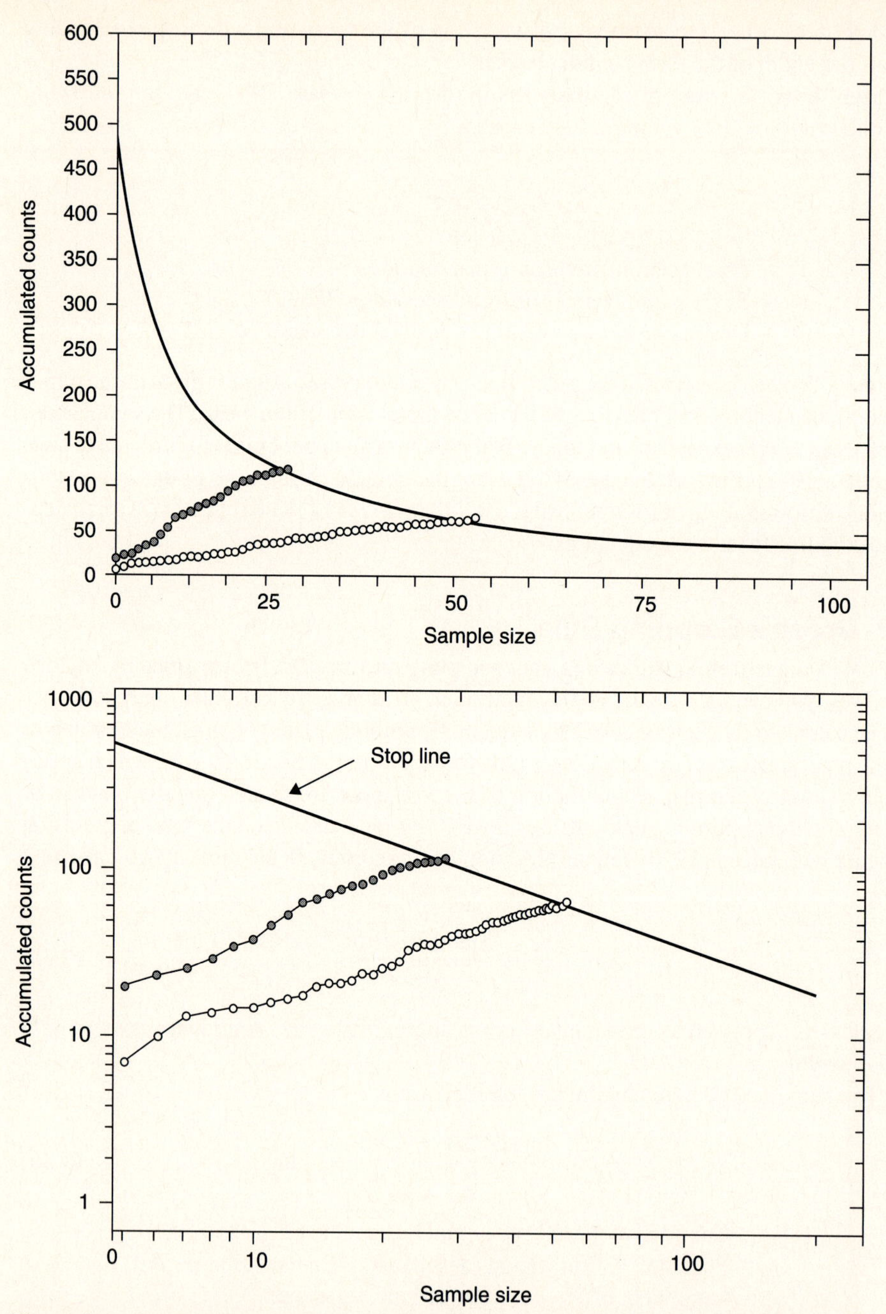

Figure 9.7 Green's fixed precision sampling plan for counts of the intertidal snail *Littorina sitkana.* For this population, Taylor's Power Law has a slope of 1.47 and a y-intercept of 1.31. The desired precision is 0.15 (S.E./mean). The linear and the log-log plot are shown to illustrate that the stop line is linear in log-log space. Two samples are shown for one population with a mean of 4.83 (stopping after 28 samples) and a second with a mean of 1.21 (stopping after 53 samples).

Green's stopping rule can be used effectively to reduce the amount of sampling effort needed to estimate abundance in organisms sampled by quadrats. It is limited to sampling situations in which enough data has accumulated to have prior estimates of Taylor's Power Law, which specifies the relationship between means and variances for the organism being studied.

9.4 ECOLOGICAL MEASUREMENTS

In all the cases discussed so far, we have used sequential sampling as a method of testing statistical hypotheses. An alternative use of sequential sampling is to estimate population parameters. For example, you may wish to estimate the abundance of European rabbits in an area (as opposed to trying to classify the density as low, moderate, or high). There is a great deal of statistical controversy over the use of standard formulas to calculate means and confidence intervals for data obtained by sequential sampling (Wetherill and Glazebrook 1986, Chapter 8). The simplest advice is to use the conventional formulas and to assume that the stopping rule as defined sequentially does not bias estimates of means and confidence intervals obtained in the usual ways.

There are special sequential sampling designs for abundance estimates based on mark-recapture and on quadrat counts.

9.4.1 Sequential Schnabel Estimation of Population Size

In a Schnabel census, we sample the marked population on several successive days (or weeks) and record the fraction of marked animals at each recapture (see Section 2.2). An obvious extension of this method is to sample sequentially until we reach a predetermined number of recaptures. Sample size thus becomes a variable. The procedure is to stop the Schnabel sampling at the completion of sample s, with s being defined such that

$$\sum_{t=1}^{s-1} R_t < L \quad \text{but} \quad \sum_{t=1}^{s} R_t \geq L$$

where ΣR_t = Cumulative number of recaptures to time t
L = Predetermined number of recaptures

L can be set either from prior information on the number of recaptures desired, or from equation (7.19), in which L is fixed once you specify the coefficient of variation you desire in your population estimate (see Box 9.2).

There are two models available for sequential Schnabel estimation (Seber 1982, 188).

Goodman's Model This model is appropriate only for large sample sizes from a large population. The population estimate derived by Goodman (1953) is

$$\hat{N}_G = \frac{\left(\sum_{i=1}^{t} Y_i\right)^2}{2L} \tag{9.37}$$

where $\hat{N}_G$ = Goodman's estimate of population size for a sequential Schnabel experiment
Y_i = Total number of individuals caught at time i
L = Predetermined stopping-rule value for ΣR_t as defined above

Box 9.2 Sequential Estimation of Population Size by the Schnabel Method

A wildlife officer carries out a population census of European rabbits. He wishes to continue sampling until the coefficient of variation of population density is less than 8%. From equation (7.19),

$$\text{CV}(\hat{N}) \cong \frac{1}{\sqrt{\Sigma R_t}} = 0.08$$

Thus,

$$\sqrt{\sum R_t} = \frac{1}{0.08} \quad \text{or} \quad \sum R_t = 156.25$$

Solving for ΣR_t, he decides that he needs to sample until the total number of recaptures is 157 (L). He obtains these data:

Day	Total caught, Y_t	No. marked, R_t	Cumulative no. marked caught, ΣR_t	No. of accidental deaths	No. marked at large, M_t
1	35	0	0	0	0
2	48	3	3	2	35
3	27	8	11	0	78
4	39	14	25	1	97
5	28	9	34	2	121
6	41	12	46	1	138
7	32	15	61	0	166
8	19	9	69	0	183
9	56	26	95	1	194
10	42	18	113	0	223
11	39	22	135	1	247
12	23	12	147	1	263
13	29	16	163	0	273

At sample 13 he exceeds the predetermined total of recaptures, so he stops sampling. At the end of day 13 the number of actual recaptures L is 163, in excess of the predetermined number of 156.3.

Goodman's Method

Since more than 10% of the rabbits are marked, we use Goodman's method. From equation (9.37),

$$\hat{N}_G = \frac{(\Sigma\ Y_t)^2}{2L} = \frac{(458)^2}{2(163)} = 643.4 \text{ rabbits}$$

The 95% confidence limits from equation (9.39) are

$$\frac{2(\Sigma\ Y_t)^2}{(1.96 + \sqrt{4L - 1})^2} < \hat{N}_G < \frac{2(\Sigma\ Y_t)^2}{(-1.96 + \sqrt{4L - 1})^2}$$

$$\frac{2(458)^2}{(1.96 + 25.51)^2} < \hat{N}_G < \frac{2(458)^2}{(-1.96 + 25.51)^2}$$

$$555.8 < \hat{N}_G < 756.1$$

For the same rabbit data, the Schnabel estimate of population size (equation [2.9]) is 419 rabbits with 95% confidence limits of 358 to 505 rabbits. The Goodman sequential estimate is nearly 50% higher than the Schnabel one, and the confidence limits do not overlap. This is likely caused by unequal catchability among the individual rabbits with marked animals avoiding the traps. This type of unequal catchability reduces the number of marked individuals recaptured, increases the Goodman estimate, and reduces the Schnabel estimate. You would be advised in this situation to use a more robust closed estimator that can compensate for heterogeneity of trap responses.

Program-group SEQUENTIAL SAMPLING (Appendix 2) can do these calculations.

The large-sample variance of Goodman's estimate of population size is

$$\text{Variance of } \tilde{N}_G = \frac{N_G^2}{L} \tag{9.38}$$

If the total number of recaptures exceeds 30, an approximate 95% confidence interval for $\hat{N}_G$ is given by this equation (Seber 1982, 188):

$$\frac{2(\Sigma\ Y_i)^2}{(1.96 + \sqrt{4L - 1})^2} < \hat{N}_G < \frac{2(\Sigma\ Y_i)^2}{(-1.96 + \sqrt{4L - 1})^2} \tag{9.39}$$

where all terms are as defined above. For the simple case when approximately the same number of individuals are caught in each sample (Y_i = a constant), you can calculate the expected number of samples you will need to take from

$$\left\{\begin{matrix}\text{Expected number}\\ \text{of samples to reach}\\ L \text{ recaptures}\end{matrix}\right\} \cong \frac{\sqrt{2\hat{N}L}}{Y_i} \tag{9.40}$$

where $\hat{N}$ = Approximate guess of population size
L = Predetermined stopping rule for $\Sigma\ R_t$
Y_i = Number of individuals caught in each sample (a constant)

Thus if you can guess, for example, that you have a population of 2000 fish and L should be 180, and you catch 100 fish each day, you can calculate

$$\left\{\begin{matrix}\text{Expected number}\\ \text{of samples to reach}\\ \text{180 recaptures}\end{matrix}\right\} \cong \frac{\sqrt{2(2000)(180)}}{100} = 8.5 \text{ samples}$$

This is only an approximate guideline to help in designing your sampling study.

Chapman's Model This model is more useful in smaller populations but should not be used if a high fraction (>10%) of the population is marked. The population estimate derived by Chapman (1954) is

$$\hat{N}_C = \frac{\sum_{i=1}^{t} Y_i M_i}{L} \tag{9.41}$$

where $\hat{N}_C$ = Chapman's estimate of population size for a sequential Schnabel experiment
Y_i = Number of individuals caught in sample i ($i = 1, 2, \ldots, t$)
M_i = Number of marked individuals in the population at the instant before sample t is taken
L = Predetermined stopping rule for ΣR_t as defined above

The variance of this population estimate is the same as that for Goodman's estimate (equation [9.38]). For a 95% confidence interval when the total number of recaptures exceeds 30, Chapman (1954) recommends

$$\frac{4B}{(1.96 + \sqrt{4L - 1})^2} < \hat{N}_C < \frac{4B}{(-1.96 + \sqrt{4L - 1})^2} \tag{9.42}$$

where $B = \Sigma Y_i M_i$ (for all samples)

Box 9.2 works out an example of sequential Schnabel estimation for a rabbit population.

In all the practical applications of sequential Schnabel sampling, the only stopping rule used has been that based on a predetermined number of recaptures. Samuel (1969) discusses several alternative stopping rules. It is possible, for example, to stop sampling once the ratio of marked to unmarked animals exceeds some constant like 1.0. Samuel (1969) discusses formulas appropriate to Schnabel estimation by this stopping rule (Rule C in her terminology).

9.4.2 Sampling Plans for Count Data

Some count data are adequately described by the negative binomial distribution (Kuno 1972; Taylor et al. 1979; Binns and Nyrop 1992). If we wish to develop sequential sampling plans for count data, we must have available the negative binomial model as well as other more general models to describe count data. Three models can be applied to count data, and one of these will fit most ecological populations.

Negative Binomial Model If the parameter k (exponent of the negative binomial) is constant and independent of population density, we can use the sampling plan developed above (Section 9.1.4) for these populations. For negative binomial populations, the variance is related to the mean as follows:

$$s^2 = \mu + \frac{\mu^2}{k} \tag{9.43}$$

Unfortunately populations are not always so simple (see Figure 9.8). If negative binomial k is not constant, you may have a sampling universe in which k is linearly related to mean density, and you will have to use the next approach to develop a sampling plan.

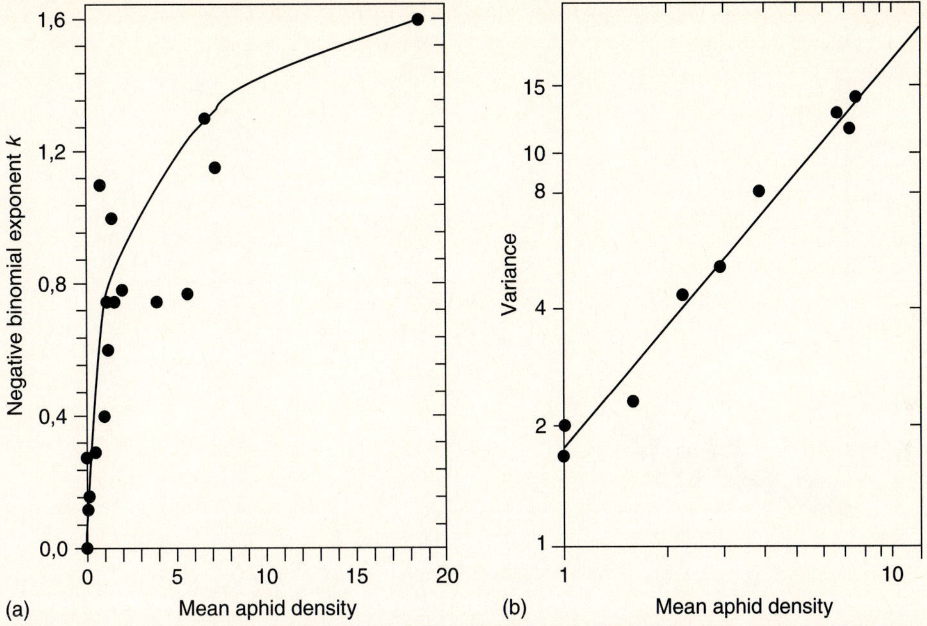

Figure 9.8 Counts of the numbers of green peach aphids (*Myzus persicae*) on sugar beet plants in California. (a) The negative binomial exponent k varies with density through the season, so a constant k does not exist for this system. (b) Taylor's Power Law regression for this species is linear for the whole range of densities encountered over the season. The regression is [variance = 1.46 (mean $^{1.60}$)]. (Data from Iwao 1975.)

Taylor's Power Law Taylor (1961) observed that many quadrat counts for insects could be summarized relatively simply because the variance of the counts was related to the mean. High-density populations had high variances, and low-density populations had low variances. Taylor (1961) pointed out that the most common relationship was a power curve:

$$s^2 = a\bar{x}^b \tag{9.44}$$

where s^2 = Observed variance of a series of population counts
$\bar{x}$ = Observed mean of a series of population counts
a = Constant
b = Power exponent

By taking logarithms, this power relationship can be converted to a straight line:

$$\log(s^2) = \log(a) + b \log(\bar{x}) \tag{9.45}$$

where the power exponent b is the slope of the straight line.

To fit Taylor's Power Law to a particular population, we need a series of samples (counts) of organisms in quadrats. Each series of counts will have a mean and a variance and will be represented by one point on the graph (see Figure 9.9). A set of samples must be taken over a whole range of densities from low to high to span the range, as illustrated in Figure 9.9. If the organisms being counted have a random spatial pattern, then for each series of samples the variance will equal the mean, and the slope of Taylor's Power Law

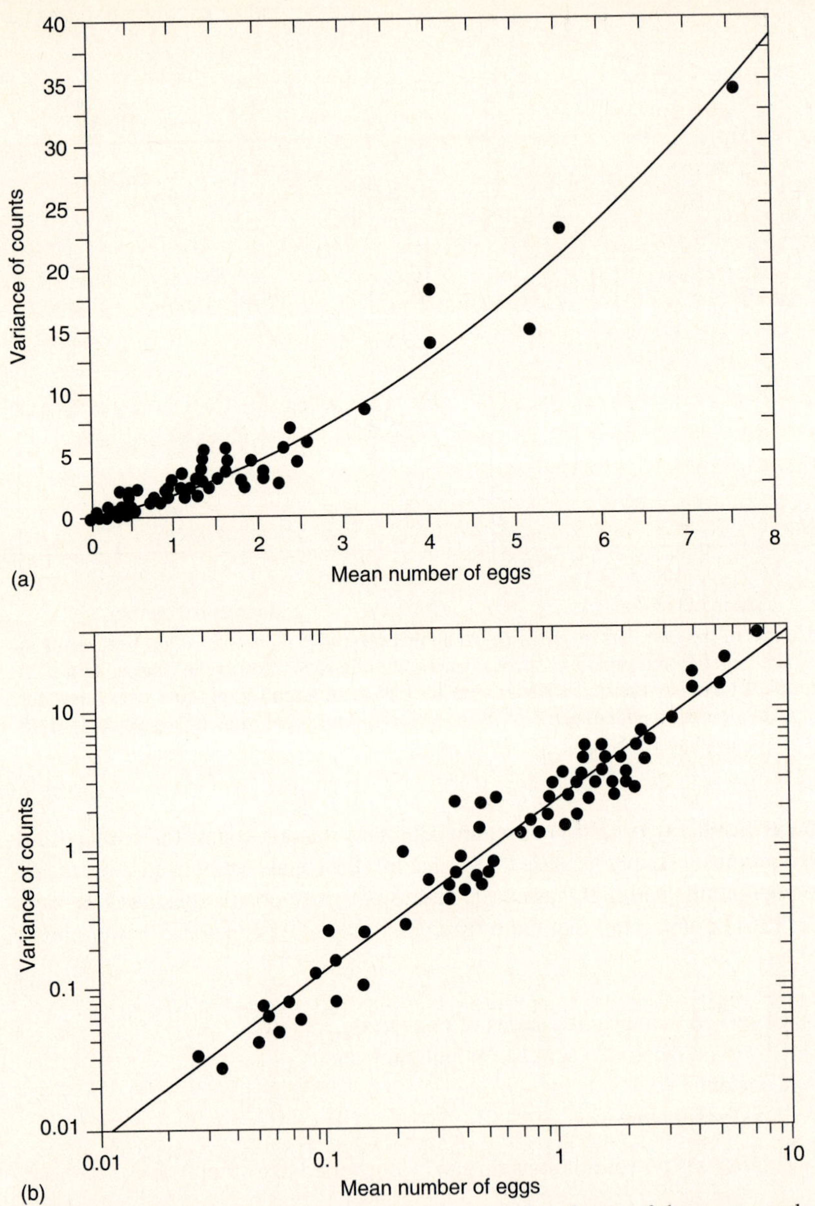

Figure 9.9 The relationship between the mean density of eggs of the gypsy moth *Lymantria dispar* and the variance of these counts. The top graph is an arithmetic plot and illustrates that the mean and variance are not linearly related. The lower graph is Taylor's Power Law for these same data, showing the good fit of these data to the equation $s^2 = 1.08\ m^{1.27}$, which is a straight line in a log-log plot. (Data from Taylor 1984, 329.)

will be 1.0. Because most populations are aggregated, the slope (b) will usually be greater than 1.0.

Taylor's Power Law is a useful way of summarizing the structure of a sampling universe (Taylor 1984; Nyrop and Binns 1991; Binns and Nyrop 1992). Once it is validated for a particular ecosystem, you can predict the variance of a set of counts once you know the mean density and design a sampling plan to obtain estimates of specified precision, as will be shown in the next section. There is a large literature on the parameters of Taylor's Power Law for insect populations (Taylor 1984), and it is clear that for many populations of the same insect in different geographical regions, the estimated slopes and intercepts are similar but not identical (Trumble et al. 1987). It is safest to calibrate Taylor's Power Law for your local population rather than relying on some universal constants for your particular species.

Taylor's Power Law is probably the most common model that describes count data from natural populations, and it is the basis of the general sequential sampling procedure for counts given below (in Section 9.4.3).

Empirical Models If neither of the previous models is an adequate description of your sampling universe, you can develop a completely empirical model. Nyrop and Binns (1992) suggest, for example, a simple model:

$$\ln[-\ln(p_0)] = a + b \ln(\bar{x}) \tag{9.46}$$

where p_0 = Proportion of samples with no organisms
$\bar{x}$ = Mean population density
a = y-intercept of regression
b = Slope of linear regression

They provide a computer program to implement a sequential sampling plan based on this empirical relationship. If you need to develop an empirical model for your population, it may be best to consult a professional statistician to determine the best model and the best methods for estimating the parameters of the model.

9.4.3 General Models for Two Alternative Hypotheses from Quadrat Counts

Sequential decision making from count data may often not fit the simple models described above for deciding between two alternative hypotheses. Iwao (1975) suggested an alternative approach that is more general because it will adequately describe counts from the binomial, Poisson, or negative binomial distributions, as well as a variety of clumped patterns that do not fit the usual distributions.

Iwao's original method assumes that the spacing pattern of the individuals being counted can be adequately described by the linear regression of mean crowding on mean density (Krebs 1989, 260). Taylor (1984) and Nyrop and Binns (1991) pointed out that Iwao's original method makes assumptions that could lead to errors, and they recommended a more general method based on Taylor's Power Law.

To determine a sequential sampling plan from Taylor's Power Law using Iwao's method, proceed as follows:

Step 1. Calculate the *slope* (b) and *y-intercept* (a) of the log-log regression for Taylor's Power Law (equation [9.45]) in the usual way (e.g., Sokal and Rohlf 1995, 466).

Step 2. Determine the critical density level μ_0 to set up the two-sided alternative:

$$H_0\text{: Mean density} = \mu_0$$

$$H_1\text{: Mean density} < \mu_0 \quad \text{(lower density)}$$

$$H_2\text{: Mean density} > \mu_0 \quad \text{(higher density)}$$

Step 3. Calculate the upper and lower limits of the cumulative number of individuals counted on n quadrats from the following equations:

$$\begin{aligned} \text{Upper limit: } U_n &= n\mu_0 + 1.96A \\ \text{Lower limit: } L_n &= n\mu_0 - 1.96A \end{aligned} \tag{9.47}$$

where
U_n = Upper limit of cumulative counts for n quadrats at $\alpha = 0.05$
L_n = Lower limit of cumulative counts for n quadrats at $\alpha = 0.05$
μ_0 = Postulated critical density (mean per quadrat)
$A = \sqrt{n[\text{var}(\mu_0)]}$
$\text{var}(\mu_0)$ = Variance of the critical density level μ_0

Calculate these two limits for a range of sample sizes from $n = 1$ up to (say) 100. If you wish to work at a 99% confidence level, use 2.576 instead of 1.96 in the equations (9.47), and if you wish to work at a 90% confidence level, use 1.645 instead of 1.96.

Step 4. Plot the limits U_n and L_n (equation [9.47]) against sample size (n) to produce a graph like that in Figure 9.10. Note that the lines in this sequential sampling graph will *not* be straight lines, as has always been the case so far, but will in fact diverge.

Step 5. If the true density μ is close to the critical density μ_0, sampling may continue indefinitely (e.g., see Figure 9.5). This is a weak point in all sequential sampling plans, and we need a stopping rule. If we decide in advance that we will quit when the confidence interval is $\pm d$(at $\mu = \mu_0$), we can determine the maximum sample size from the following equation (Iwao 1975):

$$n_M \cong \frac{4}{d^2}[\text{var}(\mu_0)] \tag{9.48}$$

where
n_M = Maximum sample size to be taken
d = Absolute half-width of the confidence interval for 95% level
μ_0 = Critical density
$\text{var}(\mu_0)$ = Variance expected from Taylor's Power Law for the critical density

If you wish to use 99% confidence belts, replace 4 with 6.64 in this equation, and if you wish 90% confidence belts, replace 4 with 2.71.

The method of Iwao (1975) can be easily extended to cover three or more alternative hypotheses (e.g., see Figure 9.6) just by computing two or more sets of limits defined in equation (9.47) above. This method can also be readily extended to two-stage sampling, and Iwao (1975) gives the necessary formulas.

The major assumption is that, for the population being sampled, there is a linear regression between the log of the variance and the log of the mean density (Taylor's Power

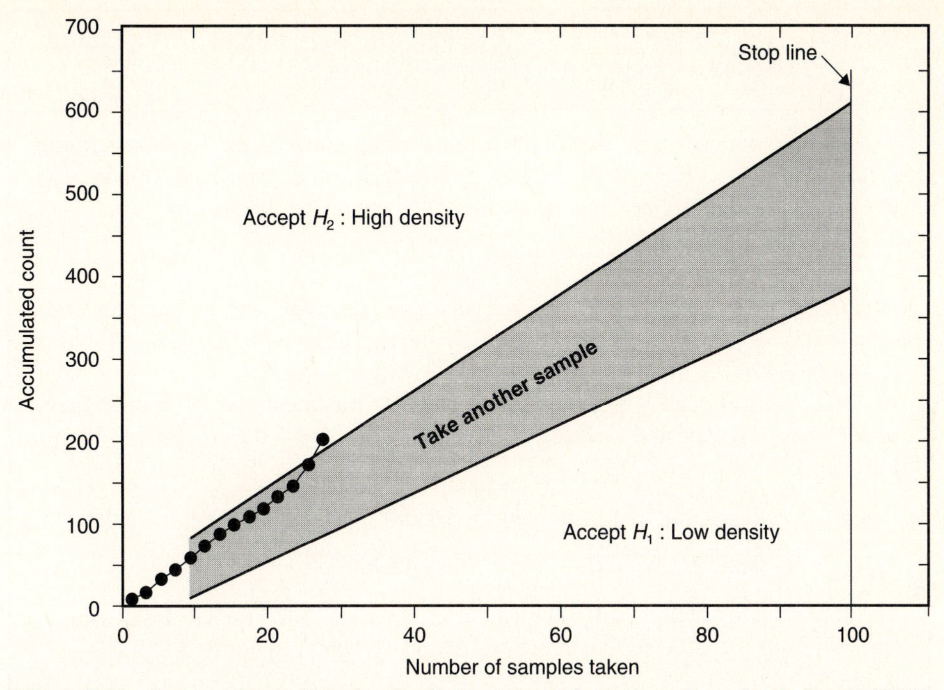

Figure 9.10 Sequential sampling plan for the European red mite in apple orchards. Iwao's (1975) method was used to develop this sampling plan, based on Taylor's Power Law specifying the relationship between the mean and the variance for a series of sample counts. For this population, the slope of the Power Law was 1.42, and the intercept was 4.32. At any given density, mite counts fit a negative binomial distribution, but k is not a constant, so we cannot use the methods in Section 9.1.4. The critical density for biological control is 5.0 mites per quadrat, and 10% levels were used for α and β. A maximum of 100 samples specifies the stop line shown. One sampling session is illustrated, in which sampling stopped at $n = 28$ samples for a high-density population with a mean infestation of 7.6 mites per sample. (Data from Nyrop and Binns 1992.)

Law). A considerable amount of background data is required before this can be assumed, and consequently you could not use Iwao's method to sample a new species or situation in which the relationship between density and dispersion was unknown. Box 9.3 illustrates the application of the Iwao method for green peach aphids.

Program-group SEQUENTIAL-SAMPLING (Appendix 2) can calculate the sequential sampling lines for all of the types of statistical distributions discussed in this chapter.

9.5 VALIDATING SEQUENTIAL SAMPLING PLANS

Sequential sampling plans are now in wide use for monitoring insect pest populations, and this has led to an important series of developments in sequential sampling. Nyrop and Simmons (1984) were the first to recognize that although ecologists set α-levels for sequential sampling at typical levels (often 5%), the actual Type I error rates can be much larger than this. This is because in sequential sampling you never know ahead of time what the mean will be, and since variances are not constant (and are often related to the mean in some

Box 9.3 Iwao's Method for Sequential Sampling to Classify Population Density Estimated by Quadrat Counts

The first requirement is to describe the relationship between the variance and the mean with Taylor's Power Law. For the green peach aphid, prior work (Figure 9.8) has shown that the variance and the mean density are related linearly by

$$\log(s^2) = \log(4.32) + 1.42 \log(\bar{x})$$

in which density is expressed as aphids per sugar beet plant, and the sampling unit (quadrat) is a single plant (Silvester and Cox 1961). Thus $a = 4.32$, and $b = 1.42$ for equation (9.45).

The critical density for beginning insecticide treatment was 5.0 aphids/plant, so $\mu = 5.0$. The expected variance of density at a mean of 5.0 from Taylor's Power Law above is

$$\log(s^2) = \log(4.32) + 1.42 \log(\bar{x}) = 0.63548 + 1.42[\log(5.0)] = 1.62802$$

$$s^2 = \text{antilog}(1.62802) = 10^{1.62802} = 42.464$$

The upper and lower limits are calculated from equation (9.47) for a series of sample sizes (n) from 1 to 200. To illustrate this, for $n = 10$ plants, and $\alpha = \beta = 0.10$,

$$U_n = n\mu_0 + 1.64A$$

$$U_{10} = 10(5.0) + 1.64A_{10}$$

where

$$A = \sqrt{n[\text{var}(\mu_0)]}$$

where $\text{var}(\mu_0)$ = variance of the critical density level μ_0

$$A_{10} = \sqrt{10(42.464)} = 20.607$$

and thus

$$U_{10} = 10(5.0) + 1.64(20.607) = 83.8 \text{ aphids}$$

$$L_n = n\mu_0 - 1.64A$$

$$L_{10} = 10(5.0) - 1.64(20.607) = 16.2 \text{ aphids}$$

Similarly, calculate

n	Lower limit	Upper limit
20	51.9	148.1
30	91.1	208.9
40	132.0	268.0
50	174.0	326.0
60	216.7	383.3

Note that these limits are in units of *cumulative* total number of aphids counted on all n plants. These limits can now be used to plot the sequential lines, as in Figure 9.1 or set out in a table that lists the decision points for sequential sampling (e.g., Table 9.1).

If the true density is nearly equal to 5.0, you could continue sampling for a very long time. It is best to fix an upper limit from equation (9.48). Assume you wish to achieve a confidence interval of $\pm 20\%$ of the mean with 90% confidence (if the true mean is 5.0 aphids/plant). This means for $\mu_0 = 5.0$, $d = \pm 1.0$ aphids per plant, and thus from equation (9.48):

$$n_M \cong \frac{2.71}{d^2}[\mathrm{var}(\mu_0)]$$

$$= \frac{2.71}{1.0^2}[42.464] = 115$$

and about 115 plants should be counted as a maximum.

Program-group SEQUENTIAL SAMPLING (Appendix 2) can do these calculations.

way) and count data are not normally distributed, the resulting error rates are not easy to determine.

To determine actual error rates for field sampling situations, we must rely on computer simulation. Two approaches have been adopted. You can assume a statistical model (like the negative binomial) and use a computer to draw random samples from this statistical distribution at a variety of mean densities. This approach is discussed by Nyrop and Binns (1992), who have developed an elegant set of computer programs to do these calculations. The results of computer simulations can be summarized in two useful graphs shown in Figure 9.11. The operating characteristic curve (OC) estimates the probability of accepting the null hypothesis when it is actually true. Statistical power, as discussed in Chapter 7, is simply (1.0 − OC). For sequential sampling, we typically have two hypotheses (low density, high density), and the OC curve will typically show high values when mean density is low and low values when mean density is high. The critical point for the field ecologist is where the steep part of the curve lies. The second curve is for the average sample number (ASN), which we have already seen in Figure 9.2. By doing 500 or more simulations in a computer, we can determine on average how many samples we will have to take to make a decision.

A second approach to determining actual error rates for field sampling programs is to ignore all theoretical models like the negative binomial and to use resampling methods to estimate the operating characteristic curve and the expected sample size curve. Naranjo and Hutchison (1994) have developed the computer program RVSP (*Resampling Validation of Sampling Plans*) to do these calculations. Resampling methods (see Chapter 15, Section 15.5) use a set of observations as the basic data and take random samples from this set to generate data for analysis. Resampling methods are completely empirical, and make no assumptions about statistical distributions, although they require prior information on

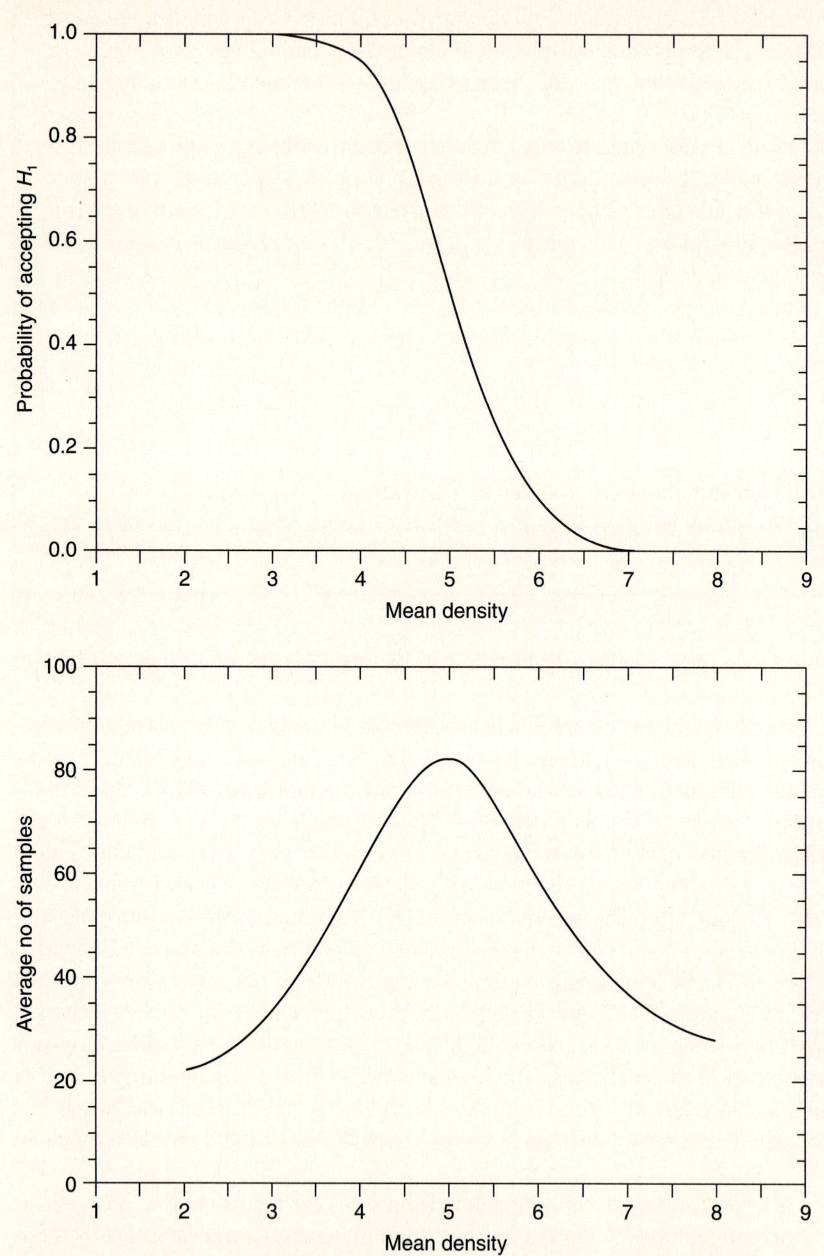

Figure 9.11 Sequential sampling scheme for the European red mite on apple trees. The critical density is 5.0 mites per leaf. Simulations at each density were run 500 times, given Taylor's Power Law for this population ($a = 4.32$, $b = 1.42$) and assuming a negative binomial distribution of counts and ($\alpha = \beta = 0.10$). The operating characteristic curve gives the probability of accepting the hypothesis of low mite density (H_1) when the true density is given on the x-axis, and is essentially a statistical power curve. The curve for the average number of samples gives the expected intensity of effort needed to reach a decision about high or low mite density for each level of the true mean density. (After Nyrop and Binns 1992.)

the mean-variance relationship in the population being studied (e.g., Taylor's Power Law). There is at present no clear indication of whether the more formal statistical models or the more empirical resampling models give better insight for field sampling. In practice the two methods often give similar pictures of the possible errors and the effort required for sequential sampling schemes. Field ecologists will be more at ease with the empirical resampling schemes for designing field programs.

9.6 SUMMARY

Sequential sampling differs from classical statistical methods because *sample size is not fixed in advance*. Instead, you continue taking samples, one at a time, and after each sample, ask if you have enough information to make a decision. Sample size is minimized in all sequential sampling schemes. Sequential methods are useful only for measurements that can be made in sequential order, so that one can stop sampling at any time. They have been used in ecology principally in resource surveys and in insect pest-control problems.

Many sequential sampling schemes are designed to test between two alternative hypotheses, such as whether a pest infestation is light or severe. You must know in advance how the variable being measured is distributed (e.g., normal, binomial, or negative binomial). Formulas are given for specifying the slope and y-intercepts of sequential sampling lines for variables from these different statistical distributions. The expected sample size can be calculated for any specific hypothesis in order to judge in advance how much work may be required.

If there are three or more alternative hypotheses that you are trying to distinguish by sequential sampling, you simply repeat the sequence for two alternatives a number of times.

One weakness of sequential sampling is that if the true mean is close to the critical threshold mean, you may continue sampling indefinitely. To prevent this, various arbitrary stopping rules have been suggested, so that sample size never exceeds some upper limit.

Sequential sampling schemes have been designed for two common ecological situations. The Schnabel method of population estimation lends itself readily to sequential methods. Quadrat counts for estimating density to a fixed level of precision can be specified to minimize sampling effort.

Sequential decision making has become important in practical problems of insect pest control. To test between two alternative hypotheses, Iwao developed a general method for quadrat counts that are not adequately described by the negative binomial distribution. If the relationship between the mean and the variance can be described by Taylor's Power Law, Iwao's method can be used to design a sampling plan for sequential decision making from quadrat data.

Field sampling plans for deciding whether populations are at low density or high density should be tested by computer simulation, because the nominal α level in sequential decisions may differ greatly from the actual error rate. Statistical power curves and expected sample size curves can be generated by computer simulation to illustrate how a specified sampling protocol will perform in field usage.

SELECTED READING

Allen, J., Gonzalez, D., and Gokhale, D. V. 1972. Sequential sampling plans for the bollworm, *Heliothis zea*. *Environmental Entomology* 1: 771–780.

Binns, M. R. 1994. Sequential sampling for classifying pest status. Pp. 137–174 in *Handbook of Sampling Methods for Arthropods in Agriculture*, ed. L. P. Pedigo and G. D. Buntin. CRC Press, Boca Raton, FL.

Iwao, S. 1975. A new method of sequential sampling to classify populations relative to a critical density. *Researches in Population Ecology* 16: 281–288.

Kuno, E. 1969. A new method of sequential sampling to obtain the population estimates with a fixed level of precision. *Researches in Population Ecology* 11: 127–136.

Kuno, E. 1991. Sampling and analysis of insect populations. *Annual Review of Entomology* 36: 285–304.

Nyrop, J. P., and M. R. Binns. 1991. Quantitative methods for designing and analyzing sampling programs for use in pest management. Pp. 67–132 in D. Pimentel, ed., *Handbook of Pest Management in Agriculture*. 2d ed., vol. 3, CRC Press, Boca Raton, FL.

Pieters, E. P., and Sterling, W. L. 1974. A sequential sampling plan for the cotton fleahopper, *Pseudatomoscelis seriatus*. *Environmental Entomology* 3: 102–106.

Taylor, L. R. 1984. Assessing and interpreting the spatial distributions of insect populations. *Annual Review of Entomology* 29: 321–357.

Wald, A. 1947. *Sequential Analysis*. John Wiley, New York.

Wetherill, G. B., and Glazebrook, K. D. 1986. *Sequential Methods in Statistics*. 3d ed. Chapman and Hall, London.

QUESTIONS AND PROBLEMS

9.1. Construct a sequential sampling plan for the cotton fleahopper. Pest-control officers need to distinguish fields in which less than 35% of the cotton plants are infested from those in which 50% or more of the plants are infested. They wish to use $\alpha = \beta = 0.10$.

(a) Calculate the sequential sampling lines and plot them.

(b) Calculate the expected sample size curve for various levels of infestation.

(c) Compare your results with those of Pieters and Sterling (1974).

9.2. Calculate a sequential sampling plan for the data in Box 9.1 under the false assumption that these counts are normally distributed with variance = 520. How does this sequential sampling plan differ from that calculated in Box 9.1?

9.3. Construct a stopping rule for a sequential Schnabel population estimate for the data in Box 2.2 for coefficients of variation of population size estimates of 60%, 50%, 25%, and 10%.

9.4. Calculate a sequential sampling plan that will allow you to classify a virus disease attack on small spruce plantations into three classes: $<10\%$ attacked, 15–25% attacked, and $>30\%$ attacked. Assume that it is very expensive to classify stands as $>30\%$ attacked if in fact they are less than this value.

(a) Calculate the expected sample size curves for each of the sequential lines.

(b) How large a sample would give you the same level of precision if you used a fixed-sample-size approach?

9.5. Calculate a sequential sampling plan for the cotton bollworm in which Taylor's Power Law describes the variance-mean regression with a slope (b) of 1.44 and a y-intercept (a) of 0.22, and the critical threshold density is 0.2 worms/plant. Assume $\alpha = \beta = 0.05$.

(a) What stopping rule would you specify for this problem?

(b) What would be the general consequence of defining the sample unit not as 1 cotton plant but as 3 plants or 5 plants? See Allen et al. (1972, 775) for a discussion of this problem.

9.6. Construct a sequential sampling scheme to estimate population size by the Schnabel method, using the data in Table 2.2. Assume that you wish to sample until the coefficient of variation of population size is about 25%. What population estimate does this provide? How does this compare with the Schnabel estimate for the entire data set of Table 2.2?

9.7. Beall (1940) counted beet webworm larvae on 325 plots in each of five areas with these results:

No. of larvae	Plot 1	Plot 2	Plot 3	Plot 4	Plot 5
0	117	205	162	227	55
1	87	84	88	70	72
2	50	30	45	21	61
3	38	4	23	6	54
4	21	2	5	1	12
5	7	2			18
6	2				21
7	2				16
8	0				14
9	1				2

(a) Fit a negative binomial distribution to each of these data sets and determine what spatial pattern these insects show.

(b) Design a sequential sampling scheme for these insects, based on the need to detect low infestations (<0.8 larvae per plot) and high infestations (>1.0 larvae per plot).

(c) Determine a stopping rule for this sequential scheme and plot the resulting sequential sampling scheme.

9.8. Sequential sampling methods are used in a relatively low fraction of ecological studies. Survey the recent literature in your particular field of interest and discuss why sequential methods are not applied more widely.

CHAPTER 10

Experimental Designs

Measurements in ecology must not only be done accurately and precisely, they must also be carried out within the general framework of a good experimental design. As ecologists have attempted to do more field experiments, the difficulties and pitfalls of experimental design have begun to emerge (Hayne 1978; Underwood 1997). There are many excellent textbooks on experimental design (for example, Cox 1958, Winer et al. 1991, Box et al. 1978, Damon and Harvey 1987), and I will not summarize here the detailed discussion you may find in one of these statistical texts. I concentrate on the simple principles of experimental design as they are applied to ecological studies, in order to teach you about the general principles of ecological experimental design so that you can talk intelligently to a statistician or pick up a statistical text on the subject without a grimace.

10.1 GENERAL PRINCIPLES OF EXPERIMENTAL DESIGN

Experimental design is a term describing the logical structure of an experiment. Let us begin by defining some of the terms that are commonly used in discussions of experimental design. An *experiment* is an attempt to test a hypothesis, and for ecologists hypotheses are typically suggested explanations for an ecological pattern or process. There are two broad types of experiments (Hurlbert 1984):

1. *Mensurative experiments* involve making some measurements on ecological units. The ecologist does not apply some treatment to the organisms or the plots but measures what currently exists. For example, you might measure the density of deer in forest and in grassland habitats to test the hypothesis that deer prefer forest habitats.
2. *Manipulative experiments* involve assigning some treatment to the experimental units or plots. At least two treatments or manipulations are required. For example, an ecologist might burn one plot of grassland and leave a second plot unburned to test the hypothesis that burning reduces grasshopper abundance.

Most ecologists apply the word *experiment* only to manipulative experiments, but this is too restrictive in general.

Experiments always involve taking some measurements on the experimental units. The concept of an *experimental unit* is critical for understanding the design of ecological experiments. An experimental unit is defined as *the smallest division of the experimental material such that any two units may receive different treatments*. Note that an experimental unit is *not* always equivalent to the "unit of study." Consider two manipulative experiments and one mensurative experiment.

1. In a fire study, one 10 ha block of grassland may be burned, and one 10 ha block of grassland left unmanipulated. An ecologist might sample 50 1 m^2 plots on each of these two areas. In this case there are two experimental units (the two 10 ha blocks), and the 50 small plots are *subsamples* of each experimental unit.
2. In a plant growth study, four fertilizer treatments (none, N, $N + P$, $N + P + K$) may be applied at random to 50 1 m^2 plots on each of two areas. In this case there are 50 experimental units on each area because any single 1 m^2 plot might be treated with any one of the four fertilizers.
3. In a study of tree growth up an altitudinal gradient, a plant ecologist wishes to determine if growth rates decrease with altitude. The experimental unit is a single tree. A random sample of 100 trees is selected, and the growth rate and altitude of each tree are recorded.

So the first step in specifying your experimental design is to determine the experimental units. A statistician who asks you about the number of *replicates* wants to know how many experimental units you have in each treatment. Most of the difficulty that Hurlbert (1984) has described as *pseudoreplication* arises from a failure to define exactly what the experimental unit is.*

* *Pseudoreplication* occurs when experimental measurements are not independent. If you weigh the same fish twice, you do not have two replicates.

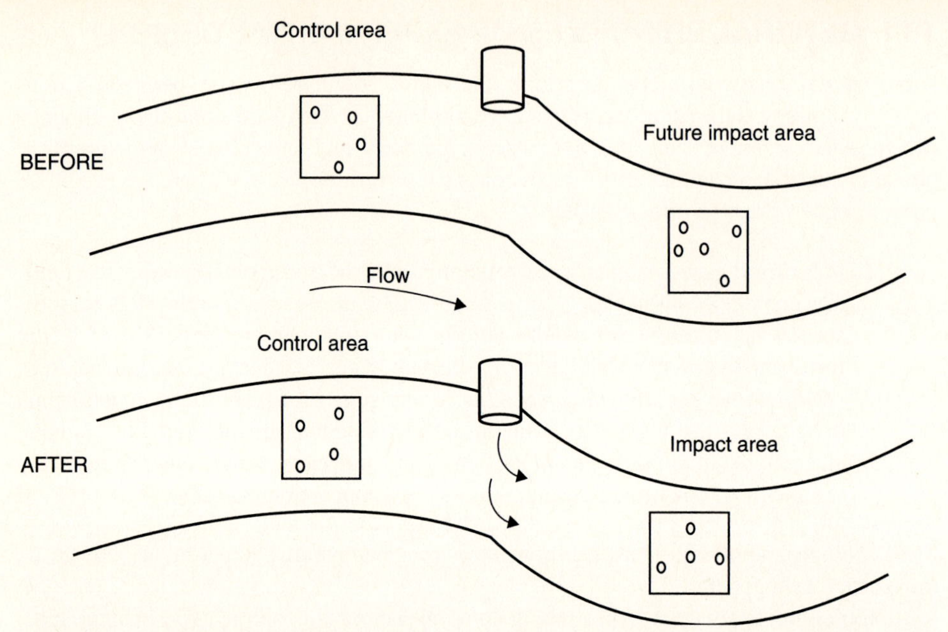

Figure 10.1 Example of the requirements for a control in ecological studies. A stream is to be subjected to nutrient additions from a mining operation. By sampling both the control and the impact sites before and after the nutrient additions, both temporal and spatial controls are utilized. Green (1979) calls this the BACI design (before-after, control-impact) and suggests that it is an optimal impact design. In other situations one cannot sample before the treatment or impact is applied, and only the spatial control of the lower diagram is present. (Modified from Green 1979.)

A general rule of manipulative experimentation is that one should have a *control*, which is usually defined as an experimental unit that has been given no treatment. Thus a control is usually the baseline against which the other treatments are to be compared (see Figure 10.1). In some cases the control unit is subjected to a sham treatment (e.g., spraying with water vs. spraying with a water + fertilizer solution).

Mensurative experiments also require a control in the sense of a set of experimental units that serve as a baseline for comparison. The exact nature of the controls will depend on the hypothesis being tested. For example, if you wish to measure the impact of competition from species A on plant growth, you can measure plants growing in natural stands in a mixture with species A and compare these with plants growing in the absence of species A (the "control" plants).

There is one fundamental requirement of all scientific experimentation:

Every experiment must have a control.

If a control is not present, it is impossible to conclude anything definite about the experiment.* In ecological field experiments there is so much year-to-year variation in communities and ecosystems that an even stronger rule should be adopted:

Every ecological field experiment must have a contemporaneous control.

*In some experiments two or more treatments are applied to determine which one is best. One treatment can act as a "control" for another.

TABLE 10.1 POTENTIAL SOURCES OF CONFUSION IN AN EXPERIMENT, AND MEANS FOR MINIMIZING THEIR EFFECT

Source of confusion	Features of an experimental design that reduce or eliminate confusion
1. Temporal change	Control treatments
2. Procedure effects	Control treatments
3. Experimenter bias	Randomized assignment of experimental units to treatments
	Randomization in conduct of other procedures; "blind" procedures[a]
4. Experimenter-generated variability	Replication of treatments
5. Initial or inherent variability among experimental units	Replication of treatments
	Interspersion of treatments
	Concomitant observations
6. Nondemonic intrusion[b]	Replication of treatments
	Interspersion of treatments

Source: After Hurlbert 1984.

[a]Usually employed only where measurement involves a large subjective element.

[b]Nondemonic intrusion is defined as the impingement of chance events on an experiment in progress.

Because of the need for replication, this rule dictates that field experiments should utilize at least two controls and two experimental areas or units. Hayne (1978) pointed out that this obvious requirement was often missing from ecological field studies. Clearly statistical power will increase if you have three replicates rather than two, so two is a minimum number rather than the most desirable number.

Before-after comparisons are statistically powerful because every experimental unit can serve as its own control. For this reason most experimental design texts recommend this approach. Fisheries and wildlife management groups often rely on time series of data on abundance, in which a management manipulation occurs at a given time. But unless there is a contemporaneous control, all before-after comparisons must assume homogeneity over time, a dubious balance-of-nature model that has been found invalid time and time again in ecological work (Green 1979). Populations and communities change over time in ways we only dimly understand, and to achieve reliable statistical inference, we need spatial controls for all ecological experiments. Figure 10.1 illustrates the need for both temporal and spatial controls in ecological studies.

There are at least six sources of variability that can cloud the interpretation of experiments (Table 10.1). These sources of confusion can be reduced by three statistical procedures: *randomization*, *replication*, and *design control.*

10.1.1 Randomization

Most statistical tests make the assumption that the observations are independent, but as with most statistical assumptions, this is an ideal that can never be achieved. One way to help achieve the goal of independent observations is to randomize by taking a random sample from the population or by assigning treatments at random to the experimental units. If observations are not independent, we cannot achieve the true value of α, the probability of a Type I error.

Randomization is also a device for reducing bias that can invade an experiment inadvertently. Randomization thus increases the accuracy of our estimates of treatment effects.

In many ecological situations complete randomization is not possible. Study sites cannot be selected at random, if only because not all land areas are available for ecological research. Within areas that are available, vehicle access will often dictate the location of study sites. The rule of thumb to use is simple:

Randomize whenever possible.

Systematic sampling is normally the alternative to random sampling (see Chapter 8, Section 8.4). While most statisticians do not approve of systematic sampling, most ecologists working with field plots use some form of systematic layout of plots. Systematic sampling achieves coverage of the entire study area, which ecologists often desire. So far, no good evidence has shown that systematic sampling in complex natural ecosystems leads to biased estimates or unreliable comparisons. But there is always a residue of doubt when systematic sampling is used, and hence the admonition to random sample when possible. A good compromise is to sample semisystematically (e.g., Figure 5.6). Randomization is a kind of statistical insurance.

Randomization should always be used in manipulative experiments when assigning treatments to experimental units. If some subjective procedure is used to assign treatments, the essential touchstone of statistical analysis is lost, and probability statements cannot be assigned to the resulting data because of the possibility of bias.

10.1.2 Replication and Pseudoreplication

Replication means the *repetition* of the basic experiment, and the main idea of replication is enshrined in the first commandment of statistics:

Let $n \geq 2$

Replication is necessary to permit an estimate of "experimental error," which is the basic unit of measurement for assessing statistical significance or for determining confidence limits. Replication is one way of increasing the precision of a statistical estimate, and this simple fact is the basis of the graduate students' credo: "Get as many samples as possible!"

Replication is a type of insurance against the intrusion of chance events on ecological experiments. Chance events are one of the major sources of interference, or "noise," in field ecology, and ecologists often use chance events as a good excuse when observations fail to match theory (Weatherhead 1986). Chance events are most troublesome when they impinge on one experimental unit and not the others. For example, one plot might be accidentally burned during a study, or be subject to an insect outbreak. Replication is the only way to avoid these difficulties.

Where should experimental plots be placed spatially, and in what time sequence should manipulations be done? Ecologists always face these questions about the *interspersion* of treatments in space and in time (Hurlbert 1984). In field studies interspersion is more important than randomization because it deals with the critical question of how the experimental units should be distributed in space. Figure 10.2 illustrates good and poor designs for interspersion of a simple, two-treatment field experiment. Let us look at each design briefly.

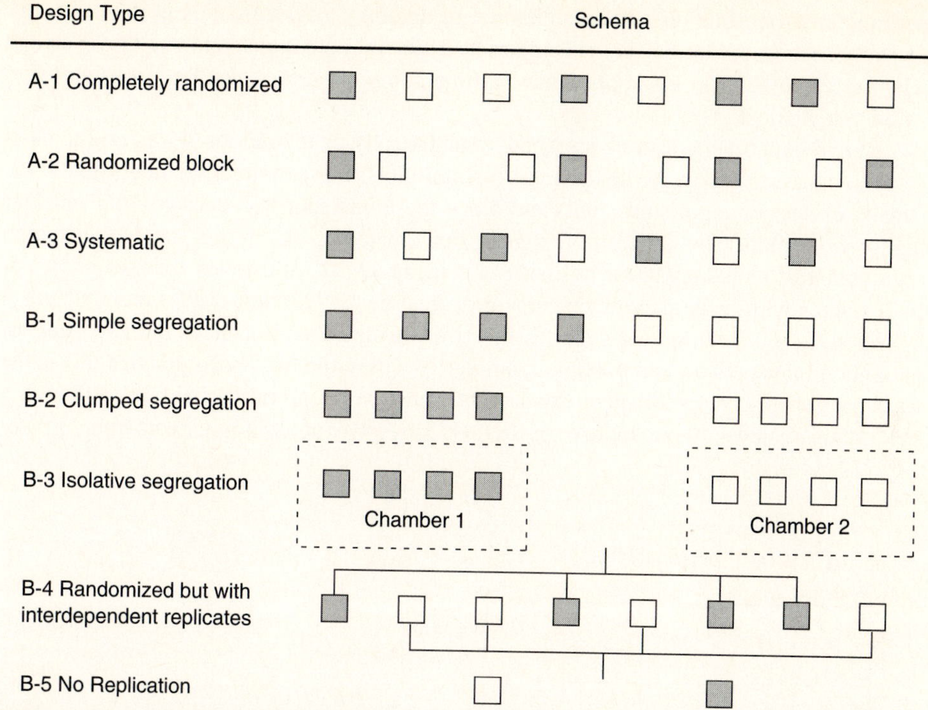

Figure 10.2 Schematic representation of various acceptable modes (A) of interspersing the replicates of two treatments (shaded, unshaded) and various ways (B) in which the principle of interspersion can be violated. (From Hurlbert 1984.)

Completely Randomized Design The simplest design recommended by many statistical tests is completely randomized design. Hurlbert (1984) pointed out that strict randomization can result in treatments being spatially segregated by chance, especially if only a few replicates are possible. Spatial segregation will produce spurious treatment effects when there are preexisting gradients in the study area. For this reason Hurlbert recommends against this statistical design in ecological studies when replicates are few, even though technically this is a perfectly acceptable statistical design.

Randomized Block Design In randomized block design the experimental units are grouped together in *blocks*. In ecological use the blocks may be areas of habitat, time periods, or rooms within a greenhouse. The main point is that the blocks are relatively uniform internally, and the differences between blocks may be large or small. This is an excellent design for most field experiments because it automatically produces an interspersion of treatments (e.g., Figure 10.2) and thus reduces the effect of chance events on the results of the experiment. One additional advantage of the randomized block design is that whole blocks may be lost without compromising the experiment. If a bulldozer destroys one set of plots, all is not lost.

Systematic Design Systematic design achieves maximum interspersion of treatments but incurs the statistical risk of errors arising from a periodic environment. Since spatially periodic environments are almost unknown in natural ecosystems, this problem is nonexistent for most ecological work. Temporal periodicities are quite common, however, and when the treatments being applied have a time component, we must be more careful to avoid systematic designs.

Field experiments can be assigned systematically or in random blocks on the basis of some measured property of the experimental plots. For example, we might know the density of deer on eight study plots and block these into four low-density plots and four high-density plots. The danger of assigning randomized blocks in this way is that spatial segregation of plots may occur as a side effect. Figure 10.3 illustrates some examples of field studies with inadequate interspersion of treatments. Hurlbert (1984) recommends a hybrid approach in these cases that assigns treatments so as to achieve the two goals of maximum interspersion and maximum similarity within the blocks. In practice this is the way many large-scale ecological experiments must be set up in the field, even though a strict statistician would not approve of this subjective way of assigning treatments.

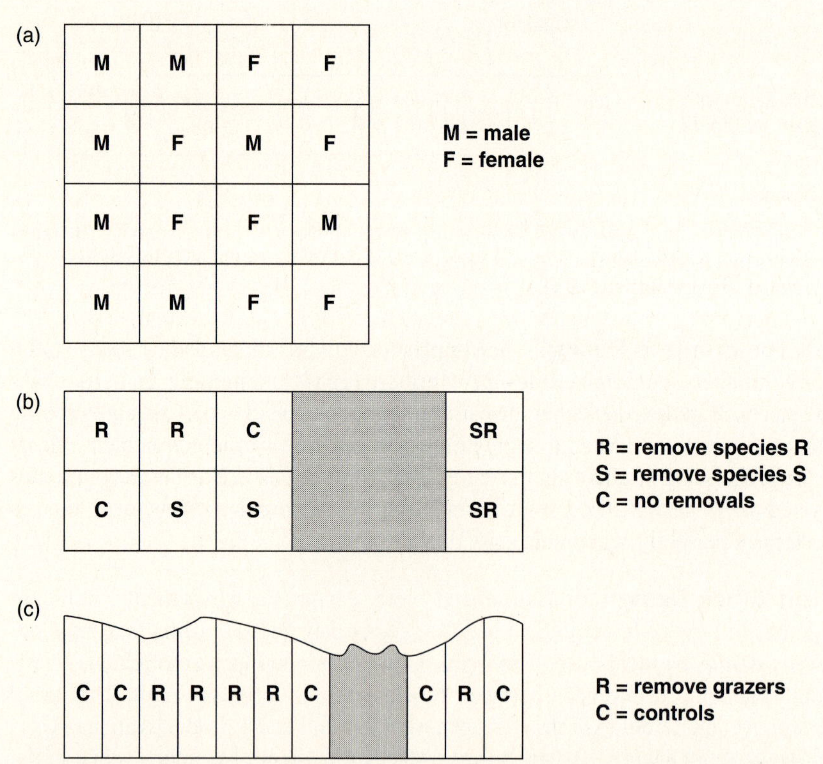

Figure 10.3 Three experimental layouts that show partial but inadequate interspersion of treatments. (a) An experiment to compare predation rates on male vs. female floral parts (Cox 1982). (b) Impact of removals on dispersal rates of two rodent species into field plots (Joule and Cameron 1975). (c) Effects on algal growth of grazer removals in rocky intertidal areas (Slocum 1980). Shaded areas of the diagrams represent unused areas. In all these cases Hurlbert (1984) recommends using subjective assignments of treatments to even out the interspersion. (After Hurlbert 1984.)

Segregated Designs The simple segregated designs shown in Figure 10.2 (B-1 to B-3) are rarely found in ecological field studies but may occur in laboratory experiments. Even if laboratory or greenhouse experiments are set up with the same initial conditions, subsequent chance events may have uneven effects if the treatments are isolated. Hurlbert (1984) gives several examples from laboratory work.

In another type of segregated experimental design that is less easily seen, the replicates are physically interdependent (Figure 10.2, B-4). In this case the replicates may share a common heating duct, or a common filtration or water system, even though they are spatially separated. All replicates of all treatments must be hooked up together to the same heating, water, or other systems, or else each replicate must have its own system.

Randomization and interspersion often conflict in field experiments. Some statistically valid designs will produce on occasion very spatially segregated treatment plots. From an ecological view, the best approach is to reject these segregated layouts and rerandomize until you get a layout with an acceptable amount of interspersion (Cox 1958, 86–87). Segregated layouts are not usually a problem when there are many replications.

Hurlbert (1984) introduced the useful idea of *pseudoreplication* to describe a statistical error of using replicates from experiments that violate the principle of interspersion (Figure 10.2, B-1 to B-5). The basic statistical problem is that in these cases "replicates" are not independent, and the first assumption of statistical inference is violated. Hurlbert reported that in two separate surveys 26% and 48% of the ecological papers surveyed showed the statistical error of pseudoreplication. Underwood (1981) found statistical errors in 78% of the papers he surveyed in marine ecology. Clearly we need to improve the statistical design of ecological studies.

Three types of pseudoreplication can be recognized (see Figure 10.4). The simplest and most common type of pseudoreplication occurs when there is only one replicate per treatment. For example, there may be one large burned area and one unburned area. If several 1 m^2 plots are measured within each area, these 1 m^2 plots are *not* replicates (they are subsamples), and they should not be used in a *t*-test to compare burned vs. unburned areas in general. A *t*-test would only answer the specific question of whether this particular burned area differed from this particular unburned area. Sacrificial pseudoreplication occurs when there is a proper, replicated experimental design, but the data for the replicates are pooled together prior to statistical analysis. This is a simple statistical error and should be a recoverable problem in data analysis. Temporal pseudoreplication is common in ecological experiments in which we accumulate a time series of data. For example, with burned and unburned plots, we might return to sample quadrats each week for two months after the fire. Successive samples over time from a single experimental unit are clearly not independent samples and should be analyzed as a repeated-measure analysis of variance, or ANOVA (see Section 10.2.6).

10.1.3 Balancing and Blocking

One objective of a good experimental design is to reduce the size of experimental error, that is, to make the conclusions more precise. There are four general ways to increase precision in any statistical comparison:

1. Use more homogenous experimental units. This advice is useful for laboratory experimenters but is difficult to use in many field experiments.

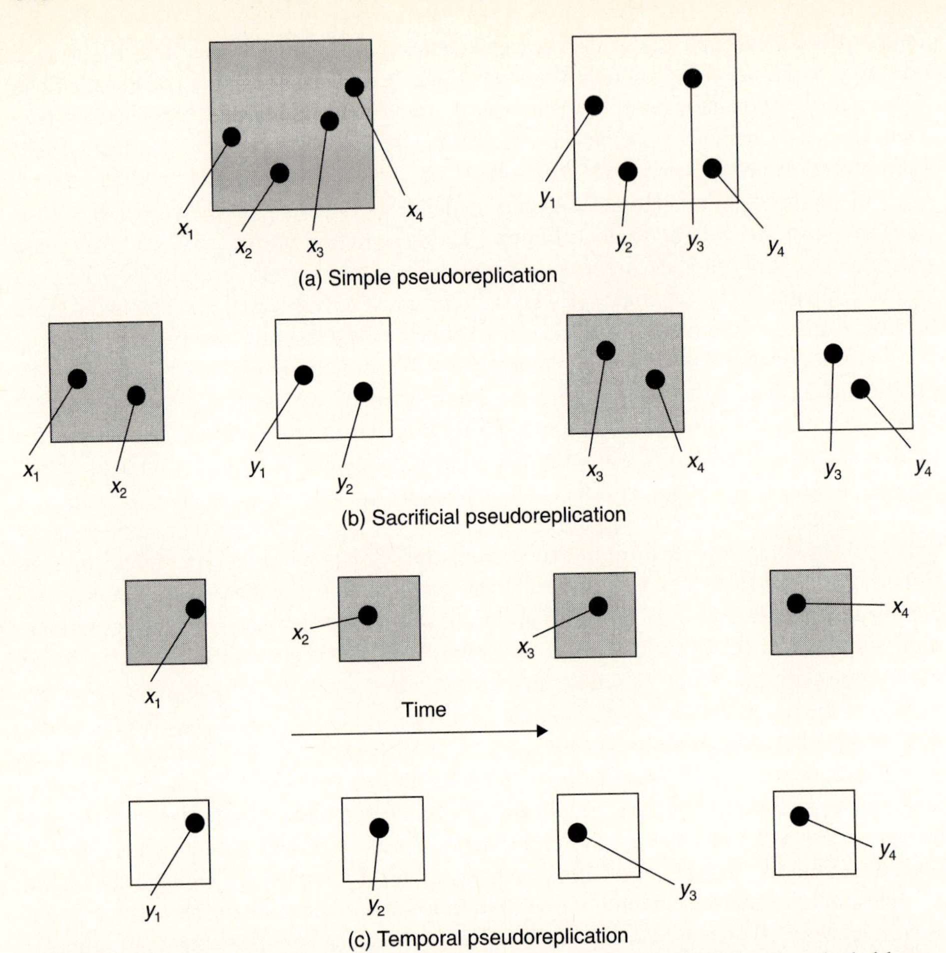

Figure 10.4 The three most common types of pseudoreplication. Shaded and unshaded boxes represent experimental units receiving two different treatments. Each dot represents a sample or measurement. Pseudoreplication is a consequence, in each example, of statistically testing for a treatment effect by means of procedures (e.g., *t*-test, *U*-test) that assume, implicitly, that the four data sets for each treatment have come from four independent experimental units (= treatment replicates). (From Hurlbert 1984.)

2. Use information provided by related variables that can be measured in each experimental unit. The analysis of covariance is the simplest example of this approach (Green 1979).
3. Use more replicates. This is the bulldozer approach to statistical precision. It always works, if you have the time, money, and enough space.
4. Use a more efficient experimental design. By this, statisticians mean the amount of balancing and blocking in the layout of the experiment.

We shall discuss the general idea of grouping the experimental units into homogeneous blocks in Section 10.2.3. Balanced designs, in which an equal number of replicates are used for each treatment, are always more efficient than unbalanced designs of the same type. One of the unfortunate statistical problems of field ecologists is the difficulty of having unbalanced designs routinely. Often we cannot avoid unequal replication, but as a general rule, we should try to achieve balance even in field studies.

10.2 TYPES OF EXPERIMENTAL DESIGNS

There is a wide and ever-increasing range of experimental designs, but fortunately for ecologists they fall into relatively few classes, and many are of limited use for ecological research. I shall describe very briefly and nontechnically five types of designs that are useful in ecology. For technical details please see Damon and Harvey (1987) or Winer et al. (1991).

Before we discuss experimental designs, however, we must define fixed and random classifications. The decision about whether a treatment* in ANOVA is fixed or random is crucial for all hypothesis testing (see Underwood 1997).

> *Fixed Factors:* (1) All levels of the classification are in the experiment, or (2) the only levels of interest to the experimenter are in the experiment, or (3) the levels in the experiment were deliberately and not randomly chosen.
>
> *Random Factors*: All levels in the experiment are a random sample from all possible levels.

Thus sex is a fixed factor because both sexes are studied, and temperature (10°, 16°, 27°) would be a fixed factor if these were the only temperatures the experimenter is interested in or a random factor if these were a random sample of all possible temperature levels. It is important as a first step in experimental design to decide whether the factors you wish to study will be fixed or random factors, since the details of statistical tests differ between fixed and random factor designs.

10.2.1 Linear Additive Models

All of the complex designs used in the ANOVA can be described very simply by the use of linear additive models. The basic assumption underlying all of these models is *additivity*. The measurement obtained when a particular treatment is applied to one of the experimental units is assumed to be

$$\left\{\begin{matrix}\text{A quantity depending}\\ \text{only on the particular}\\ \text{experimental unit}\end{matrix}\right\} + \left\{\begin{matrix}\text{A quantity depending}\\ \text{on the treatment}\\ \text{applied}\end{matrix}\right\}$$

The essential feature is that the treatment effect *adds* on to the unit term, rather than being multiplied by it. Figure 10.5 illustrates this idea graphically. A second critical assumption is that the treatment effects are constant for all experimental units. Finally, you must assume that the experimental units operate independently so that treatment effects do not spill over from one unit to another.

*Treatments are called *factors* in most statistics discussions of ANOVA, and each factor has several *levels*. For example, *sex* can be a factor with two levels, *males* and *females*.

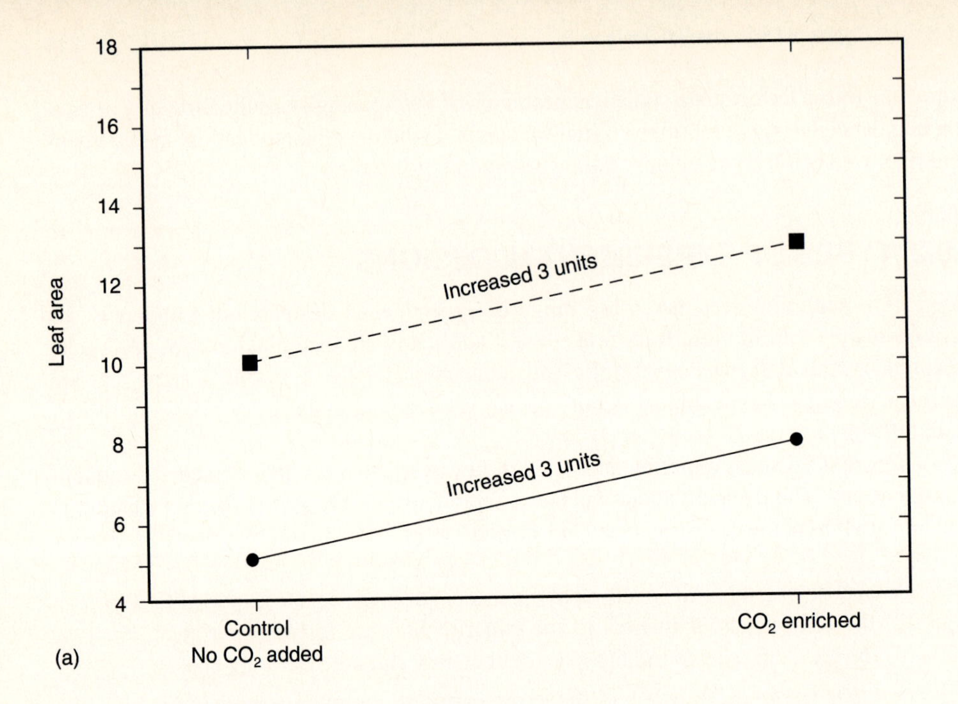

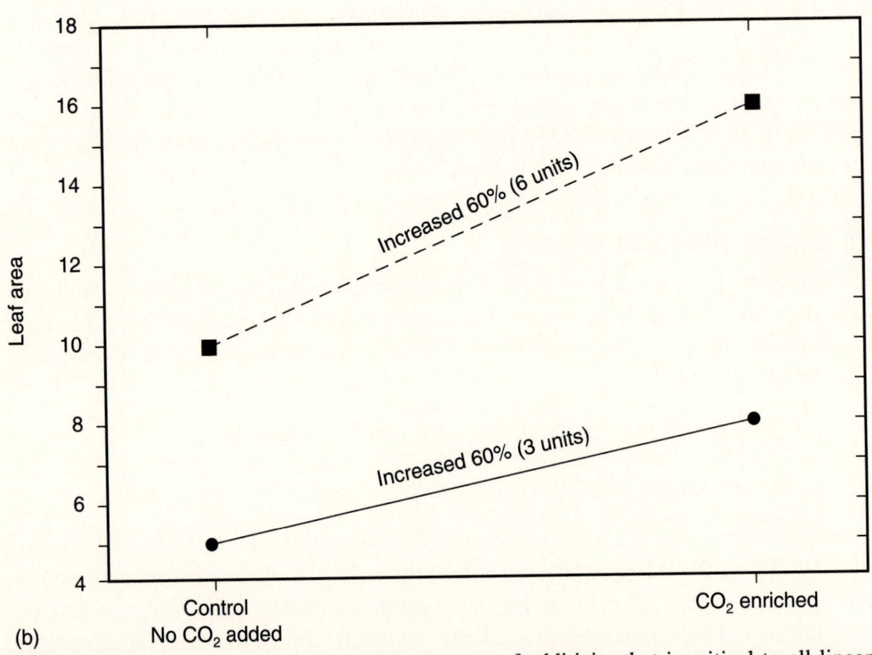

Figure 10.5 Schematic illustration of the concept of additivity that is critical to all linear additive models in the analysis of variance. (a) Addition of CO_2 increases leaf area by the same amount, no matter what the initial size of the plant. This would be highly unlikely in any growth experiment. (b) Addition of CO_2 increases leaf area proportionally by 60%. Thus the treatment effect is not additive but multiplicative.

These are the essential features of *linear additive models*, which form the core of modern parametric statistics. Consider one simple example of a linear additive model. The density of oak seedlings was measured on a series of six burned and six unburned plots. The linear additive model follows:

$$\left\{\begin{matrix}\text{Density of}\\ \text{oak seedlings}\\ \text{on plot}\end{matrix}\right\} = \left\{\begin{matrix}\text{Average density}\\ \text{of oak seedlings}\\ \text{on burned and}\\ \text{unburned plots}\end{matrix}\right\} + \left\{\begin{matrix}\text{Effect of}\\ \text{burning or}\\ \text{not burning}\end{matrix}\right\} + \left\{\begin{matrix}\text{Experimental}\\ \text{error}\end{matrix}\right\}$$

Or more abstractly:

$$Y_{ij} = \mu + T_i + e_{ij}$$

where Y = Variable being measured
μ = Mean value of the Y variable
T = Treatment effect
e = Experimental error
i = Treatment number (1 = burning, 2 = not burning)
j = Replicate number (1, 2, . . . , 6)

Linear additive models are often written as deviations:

$$Y_{ij} - \mu = T_i + e_{ij}$$

Interest usually centers on the treatment effects that can be estimated from the observed means:

$$\left\{\begin{matrix}\text{Effect of burning}\\ \text{on density}\end{matrix}\right\} = \left\{\begin{matrix}\text{Average density}\\ \text{in burned plots}\end{matrix}\right\} - \left\{\begin{matrix}\text{Average density}\\ \text{in all plots}\end{matrix}\right\}$$

Note that the effects of burning in this case are being related to a hypothetical world that is half burned and half unburned. From the data in Table 10.2,

$$\text{Effect of burning on density } (T_1) = 2.0 - 5.0$$
$$= -3.0$$

TABLE 10.2 HYPOTHETICAL DATA FOR OAK SEEDLING DENSITY[a] ON EACH OF SIX BURNED AND SIX UNBURNED PLOTS

Plot no.	Unburned plots	Plot no.	Burned plots
1	6	7	1
2	9	8	2
3	5	9	2
4	8	10	1
5	11	11	4
6	9	12	2
$n = 6$		$n = 6$	
Mean =	8.0		2.0
	Total n = 12		
	Grand mean = 5.0		

[a] Number of trees per square meter.

Thus burning reduces density by 3.0 trees/m^2. The effect of not burning is similarly

$$\begin{Bmatrix}\text{Effect of not} \\ \text{burning } (T_2)\end{Bmatrix} = \begin{Bmatrix}\text{Average density on} \\ \text{unburned plots}\end{Bmatrix} - \begin{Bmatrix}\text{Average density} \\ \text{on all plots}\end{Bmatrix}$$

$$= 8.0 - 5.0$$

$$= +3.0$$

Note that for designs like this with two levels of treatment, the measured effects are identical in absolute value but opposite in sign.

Treatment effects are always relative, and we need to estimate effects of one treatment in comparison with the other so that we can determine the ecological significance of the treatment. For example, in this burning experiment an ecologist wants to know the difference between burned and unburned:

$$\begin{Bmatrix}\text{Difference between} \\ \text{two treatments}\end{Bmatrix} = T_1 - T_2$$

$$\begin{Bmatrix}\text{Difference between} \\ \text{burned and unburned}\end{Bmatrix} = -3.0 - 3.0 = 6.0 \text{ trees/m}^2$$

You can also use these treatment effects to decompose the data from each individual quadrat. For example, for plot 5,

$$Y_{ij} = \mu + t_i + e_{ij}$$

$$11 = 5.0 + 3.0 + 3.0$$

Since you know there were 11 trees/m^2 in this plot, and the overall density μ is 5.0 (the grand mean), and the effect of not burning (T_2) estimated above is +3.0, the experimental-error term must be 3.0 to balance the equation. Note again that the error term measures inherent biological variability among plots, and not error in the sense of "mistake."

Linear additive models may be made as complex as you wish, subject to the constraint of still being linear and additive. Many statistical computer packages will compute an ANOVA for any linear additive model you can specify, assuming you have adequate replication. Linear additive models are a convenient shorthand for describing many experimental designs.

10.2.2 Factorial Designs

When only one factor is of interest, the resulting statistical analysis is simple. But ecologists usually need to worry about several factors at the same time. For example, plankton samples may be collected from several lakes at different months of the year, or rates of egg deposition may be measured at three levels of salinity and two temperatures. Two new concepts arise when one has to deal with several factors: *factorials* and *interaction*.

The concept of factorials is just the commonsense notion that all treatments of one factor should be tried with all treatments of the other factors. Thus if egg-laying rates are measured at three salinities and two temperatures, you should do all three salinities at each of the two temperatures. The test for a factorial arrangement of treatments is simple: make a table.

		Salinity		
		Low	Medium	High
Temperature	Low			
	High			

Statisticians refer to these six boxes as *cells* in an ANOVA. You must have observations in each box, or *cell*, of this table to have a factorial design. In the best of all worlds, you will have equal sample sizes in each cell and thus a *balanced* design. But in many ecological situations sample sizes are unequal, and you will have an *unbalanced* design.

In an ideal world all factors operate independently. Salinity will raise or lower egg production, and temperature will independently change it as well. In the real world, factors enhance or interfere with one another and thus are not independent in their effects. Statisticians say that factors *interact*. The simplest way to look at and understand interactions is graphically. Figure 10.6 illustrates a hypothetical set of data with and without interactions. When there are no interactions in the data, a graph will show only sets of parallel lines, as in Figure 10.6a. In this example level 2 of treatment A (area A_2) always has a higher mean abundance per haul than level 1 (area A_1), regardless of what season it is. When interaction is present, the lines diverge (Figure 10.6b) or cross (Figure 10.6c). In this example area A_2 shows higher abundance in the first two seasons of the year but lower abundance in the second two seasons, relative to area A_1.

Another way to understand what interactions are is to ask a simple question about each of the factors in the experiment. For example, from Figure 10.6, in which the two factors are *area* and *season*, we can ask,

1. What effect does *area* have on population density?
2. What effect does *season* have on population density?

When there is no interaction, the answer to these questions is straightforward: area 2 has twice the density of area 1, or autumn densities are 3 times those of spring, for example. But when there is interaction, these questions have no simple answer, and you must reply, "It all depends . . ." and give a more detailed answer; for example,

> For area 1, season has no effect on population size, but for area 2 there is a strong seasonal peak in autumn.

Interactions produce statistical headaches but interesting ecology, so there is always a conflict in factorial designs between the desire for simplicity with no interactions or for complexity with interactions that require ecological understanding.

Linear additive models for factorial designs must therefore have a term in them for interactions. For example, a two-factor analysis like that in Figure 10.6 can be represented by

$$Y_{ijk} - \mu = A_i + B_j + AB_{ij} + e_{ijk}$$

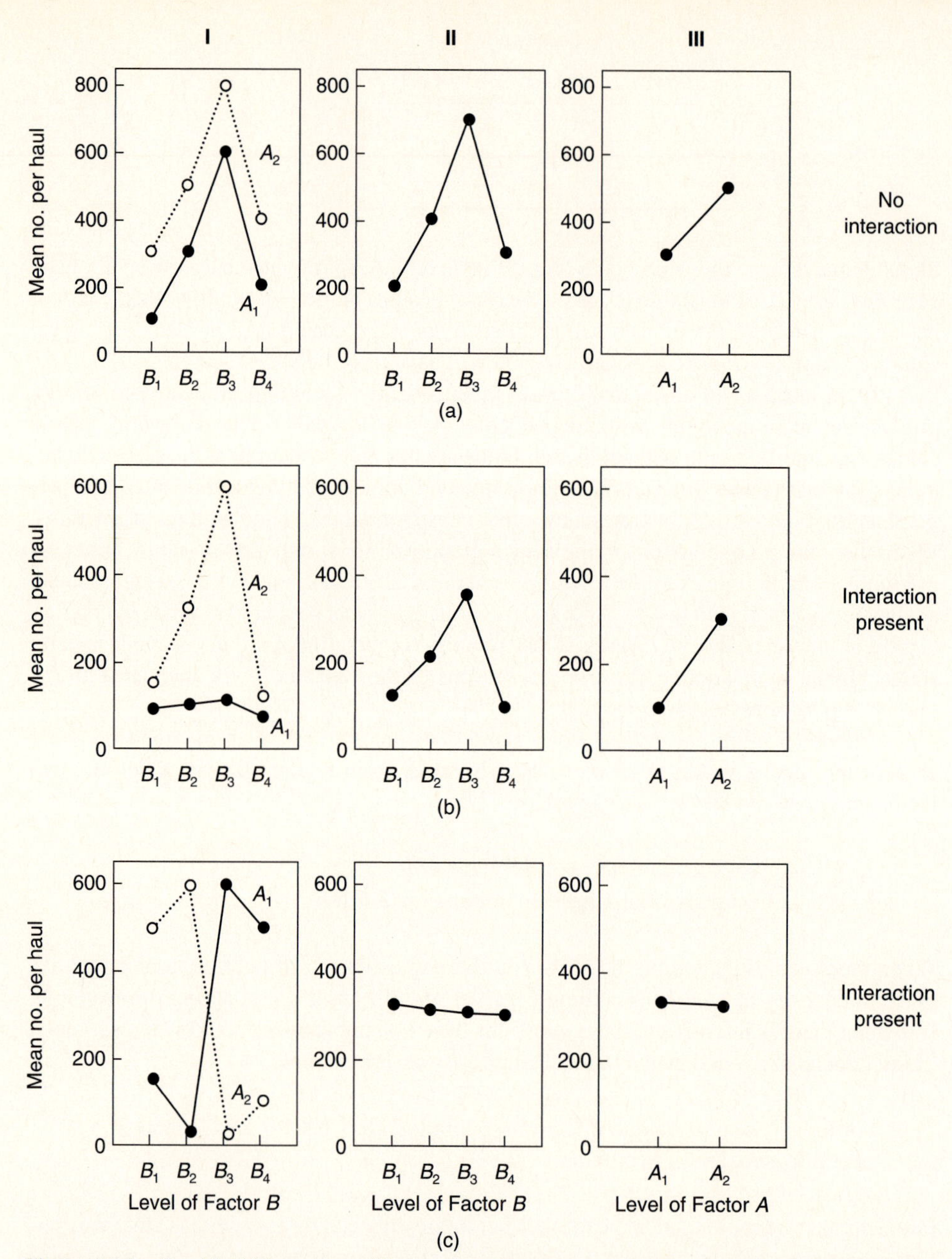

Figure 10.6 Graphical illustration of interactions between two factors. In each set of graphs, the means of hypothetical data are plotted. In column I, the means (number of animals per haul) from two areas (A_1 and A_2) are plotted for four seasons of the year (B_1 to B_4). In column II, the means in each season, averaged from the two areas, are plotted. In column III, the means in each area are averaged from the four seasons. (a) No interaction between the two factors; (b) and (c) different types of interactions present between the two factors. (Modified from Underwood 1981.)

where $Y_{ijk} - \mu$ = Deviation of observed value from the grand mean μ
A_i = Main effect of level i of factor A
B_j = Main effect of level j of factor B
AB_{ij} = Interaction term between A_i and B_j
e_{ijk} = Experimental error

Each of these terms can be estimated, as illustrated in Box 10.1, and the significance of each term determined by the ANOVA. These techniques are thoroughly described in most statistics books and are not repeated here.

It is important to remember that there is a priority of testing in factorial designs. You should first ask whether the interactions are significant statistically. If they are, you should stop there and find out what is happening. It is misleading to present and analyze tests of significance for the main effects in a factorial when the interaction term is significant. The important thing is to explain the interaction. Figure 10.6 illustrates why. Testing for the main effects is the same as asking the simple questions 1 and 2 above.

In order to calculate the complete analysis of variance for a factorial design, each cell (or combination of treatments) must have two or more replicates. Replication allows us to calculate the interaction terms and to judge their statistical significance. But this means that

Box 10.1 Estimation of Main Effects and Interactions in a Two-Factor, Completely Randomized Design

For two variables of classification we can define

$$A_i = \text{Main effect of level } i \text{ of factor } A = \bar{x}_i - \bar{\bar{x}}$$

$$B_j = \text{Main effect of factor } B \text{ level } j = \bar{x}_j - \bar{\bar{x}}$$

$$AB_{ij} = \text{Interaction effect for cell } ij = \bar{x}_{ij} - A_i - B_j + \bar{\bar{x}}$$

Example

Mean growth rates of trees (millimeters per year):

		Habitat (= factor *A*)		
		A	*B*	*C*
Tree species	*X*	50	55	60
	Y	60	67	60
	Z	70	80	90

(Note that these are cell means)

$$\text{Grand mean} = \bar{\bar{x}} = 65.777$$

Habitat means: $\bar{x}_A = 60.00$ Species means: $\bar{x}_X = 55.00$

$\bar{x}_B = 67.33$ $\bar{x}_Y = 62.33$

$\bar{x}_C = 70.00$ $\bar{x}_Z = 80.00$

Main effects of habitats:

1. Habitat $A = 60 - 65.777 = -5.78$
2. Habitat $B = 67.33 - 65.7777 = +1.56$
3. Habitat $C = 70 - 65.777 = +4.22$

Main effects of tree species:

1. Species $X = 55 - 65.777 = -10.78$
2. Species $Y = 62.33 - 65.777 = -3.44$
3. Species $Z = 80 - 65.777 = +14.22$

Interaction effects:

1. Cell A-$X = 50 - 60 - 55 + 65.777 = +0.78$
   (Alternatively, cell A-X deviates from the population mean by -15.78 units, of which -5.78 can be ascribed to a low average growth in habitat A and -10.78 can be ascribed to low average growth of species X; the remainder, $+0.78$, is interaction.)
2. Cell B-$X = 55 - 67.33 - 55 + 65.777 = -1.56$
3. Cell C-$X = 60 - 70 - 55 + 65.777 = +0.78$
4. Cell A-$Y = 60 - 60 - 62.33 + 65.777 = +3.45$
5. Cell B-$Y = 67 - 67.33 - 62.33 + 65.777 = +3.11$
6. Cell C-$Y = 60 - 70 - 62.33 + 65.777 = -6.56$
7. Cell A-$Z = 70 - 60 - 80 + 65.777 = -4.22$
8. Cell B-$Z = 80 - 67.33 - 80 + 65.777 = -1.56$
9. Cell C-$Z = 90 - 70 - 80 + 65.777 = +5.78$

Thus the largest interactions are found in cells C-Y and C-Z. Note that the sum of the main effects and the interaction effects is zero (within rounding errors).

For example,

$$\left\{\begin{array}{l}\text{Main effect of habitat } A \\ + \text{ main effect of habitat } B \\ + \text{ main effect of habitat } C\end{array}\right\} = 0$$

$$-5.78 + 1.56 + 4.22 \approx 0$$

This follows from the definitions of these terms.

The magnitude of the main effects and the interactions are the interesting parts of an ANOVA to an ecologist. The usual ANOVA table of mean squares and F values is singularly uninformative about the ecological effects studied in the experiment and should almost never be presented in a paper.

the total number of replicates grows very large in a factorial design when there are many factors or many levels within a factor. This alone seems to constrain most ecological experiments to two- or three-factor designs with a maximum of 4–5 levels of each factor. While there are no theoretical limits to the complexity of factorial designs, ecology involves very real practical limits that restrain both field and laboratory studies.

10.2.3 Randomized Block Designs

Randomized block designs always begin with the identification of *blocks*, which are relatively homogeneous groups of experimental units. For example, a muskrat litter could be a block of 5 animals. It is well known that littermates are more similar than individual muskrats in a random sample. A grassland on a southeast-facing slope could be a block. In this case the environmental conditions (soil, temperature, rainfall) would be more similar within this block of grassland than between this site and a nearby southwest-facing plot. Blocks can be constructed around any known or suspected source of variation. Plots of habitat are the most obvious type of blocks in field ecology, but a block could also be a room in a greenhouse, a lab of aquariums, a day of the week, a group of animals of similar weight, the measurements taken on instrument X, or the data collected by technician Y. Block-to-block variation is a known or suspected source of variation, and therefore it is not usual to ask in an analysis of variance whether the blocks are significantly different from one another: you already *know* that they are. The important feature of block designs is that the variation between blocks is removed from the experimental-error term in an ANOVA, thereby increasing the precision of the experiment. Box 10.2 illustrates how this is achieved.

There are many types of randomized block designs. The most common one is randomized complete block design, in which each treatment appears once in each block, and each block thus contains t experimental units (where t = Number of treatments). There are many other incomplete block designs, in which each block does not contain the treatments. Steel and Torrie (1980, Chap. 12) and Damon and Harvey (1987, Chap. 7) discuss some incomplete block designs that are commonly used in biological experiments.

10.2.4 Nested Designs

All analyses of variance are concerned with one or more factors or variables of classification. These factors can be of two types:

1. *Main effects:* Every level can be identified independently of any other level (factorial or crossed designs).
2. *Nested effects:* If each of the levels of a main effect can be subdivided into randomly selected subgroups, the classification of the groups is said to be *nested.*

Nested designs always have at least one main effect, which may be a *fixed* or *random* effect. Nested effects are always random effects.

Nested designs can easily be mistaken for factorial designs, and it is important to be sure what type of factor you are using in an ANOVA. Litters in breeding experiments are typically a nested effect, for example, in a nutrition study:

Box 10.2 Comparison of the Relative Efficiency of Randomized Block and Completely Randomized Designs

A plant ecologist measured the growth response of cotton grass (*Eriophorum angustifolium*) to four fertilizer treatments in five tundra locations in northern Alaska. She obtained these data:

	Location					
Fertilizer	B	M	R	S	Q	Means
None	10	6	11	2	5	6.8
N	58	45	55	50	37	49.0
$N + P$	63	43	68	41	39	50.8
$N + P + K$	68	47	63	43	40	52.2
Means	49.75	35.25	49.25	34.0	30.25	39.7

The main factor she wishes to analyze is *fertilizer response*, and she considers two analyses. In the first analysis five tundra locations are considered as *replicates* and the design is a one-way, completely randomized design:

$$Y_{ij} - \mu = T_i + e_{ij}$$

where Y_{ij} = Growth rate of plants receiving fertilizer i in location j
μ = Average growth rate of plants
T_i = Main effect on growth caused by fertilizer i
e_{ij} = Error variability

Using the conventional approach outlined in Sokal and Rohlf (1995, 214) or Zar (1996, 186), she obtains

Source	d.f.	Sum of squares	Mean square	*F*-value
Fertilizer	3	7241.8	2413.9	22.7***
Error	16	1700.4	106.27	
Total	19	8942.2		

Note: *** means $p < .001$.

The second analysis is a randomized complete block design in which the five locations are now considered as *blocks* rather than replicates. The linear additive model is

$$Y_{ij} - \mu = T_i + B_j + e_{ij}$$

where B_j = Effect of blocks on growth rate

and the other terms are as defined above.

Using the randomized block calculations outlined in Zar (1996, 254) or Sokal and Rohlf (1995, 343), she obtains

Source	d.f.	Sum of squares	Mean square	F-value
Fertilizer	3	7241.8	2413.9	79.3***
Locations (blocks)	4	1335.2	333.8	11.0***
Error	12	365.2	30.4333	
Total	19	8942.2		

Note: *** means $p < .001$.

As you can see clearly from the linear additive models, the error term in the completely randomized design contains the sum of squares due to variation among the blocks in the randomized block design.

The gain in efficiency that is obtained from using a randomized block design rather than a completely randomized design can be calculated from equations (6.1) and (6.2) in Damon and Harvey (1987):

$$\left\{\begin{matrix}\text{Relative efficiency of} \\ \text{randomized block}\end{matrix}\right\} = 100\left(\frac{MS_{CR}}{MS_{RB}}\right)\frac{(n_1 + 1)(n_2 + 3)}{(n_1 + 3)(n_2 + 1)}$$

where MS_{CR} = Error mean square obtained in a completely randomized design
MS_{RB} = Error mean square obtained in a randomized complete block design
n_1 = Error degrees of freedom for randomized block design
n_2 = Error degrees of freedom for completely randomized design

For this particular example,

$$\text{Relative efficiency} = 100\left(\frac{106.27}{30.433}\right)\frac{(13)(19)}{(15)(17)}$$

$$= 338\%$$

This means that to achieve the same precision, you would need to take about 3.4 times as many samples in a completely randomized design as you need in a randomized block design. Clearly, it pays to use a block design in this situation.

It is possible for the relative efficiency of a block design to be less than 100%, which means a completely randomized design is better. If blocks have a small effect, you will get a relative efficiency less than 100% because you lose degrees of freedom to blocks. I have never seen an ecological case where this was true.

		Diet 1		Diet 2		Diet 3	
		Litter A	Litter B	Litter C	Litter D	Litter E	Litter F
Littermate	1						
Number	2						
	3						

In this example litters are nested within diets, and we are interested in comparing diets, not litters. Each litter is assigned at random to one of the diets, but once a litter is assigned to a diet, all of the littermates must be in the same diet (i.e., littermates are not independent experimental units).

Other examples of nested designs could be *lakes* within *geological provinces*, or *depths* within *lakes* of different configurations, or *slides* of pollen within each *depth* within each *pollen core*.

The linear additive model for the simplest type of nested design is as follows:

$$Y_{ijk} - \mu = A_i + B_{j(i)} + e_{ijk}$$

where Y_{ijk} = Value for the kth replicate of level j of the nested factor in level i of the main factor
μ = Grand mean of all data
A_i = Main effect of level i of factor A
$B_{j(i)}$ = Effect of level j of the nested factor in level i of factor A
e_{ijk} = Error term

Many nested designs in ecology are the result of subsampling, and one important use of nested designs is to look at sources of variation in measurements so that a cost-benefit analysis can be applied to optimize the allocation of sampling effort. A typical problem is outlined by Underwood (1981, 557). To estimate egg production in marine invertebrates, three levels of sampling are required: replicate animals, replicate slides from the ovaries of each individual, and replicate sections within each slide. How should one allocate sampling effort to sections, slides, and animals? Nested ANOVA provides one way of optimizing allocation of effort when subsampling. Underwood (1997), Sokal and Rohlf (1995, 309), and Cochran (1977) give details of how to proceed with these calculations.

10.2.5 Latin Square Designs

Randomized block designs are useful when there is one source of variation known before the experiments are carried out. In a few cases ecologists can recognize two sources of variation and wish to correct for both sources in doing an experiment. One way to achieve this is to use the latin square design. A latin square design is a simple extension of the randomized complete block design that involves blocking in two directions.

Latin square designs are highly constrained designs because the number of levels of each factor must be equal. For example, assume you wish to compare the population increase rate of three species of aphids at different seasons of the year, and you have available several plots in different field locations. In this case *plots* is one blocking factor (a known source of variation), and *season* is a second blocking factor (a second known source of variation). *Species* is the third factor of classification, and you wish to know whether there is significant variation in species' growth rates. Since there are three species, you are immediately constrained to a 3 × 3 latin square in which

$$\text{Number of plots} = \text{number of seasons} = \text{number of species} = 3$$

If you cannot have this symmetry, you cannot use a latin square, you must use a factorial design.

The linear additive model for a latin square is

$$Y_{ijk} - \mu = A_i + B_j + C_k + e_{ijk}$$

where Y_{ijk} = Observed value of level i of A, level j of B, and level k of C
μ = Grand mean of all observations
A_i = Effect of level i of blocking factor A
B_j = Effect of level j of blocking factor B
C_k = Main effect of treatment factor C
e_{ijk} = Error deviation
$i = j = k$ = Number of levels of each factor

Table 10.3 gives some examples of latin square designs. The smallest possible latin square is 3 × 3 and in practice few are larger than 5 × 5. Note the symmetry of the latin square: Each treatment appears once in each row and once in each column. Thus each row is a complete block, and each column is also a complete block.

The most critical assumption of the latin square design is that there are *no interactions* between any of the three factors A, B, and C. For this reason the linear additive

TABLE 10.3 SOME EXAMPLES OF LATIN SQUARE DESIGNS[a]

3 × 3	4 × 4				5 × 5
A B C	A B C D	A B C D	A B C D	A B C D	A B C D E
B C A	B A D C	B C D A	B D A C	B A D C	B A E C D
C A B	C D B A	C D A B	C A D B	C D A B	C D A E B
	D C A B	D A B C	D C B A	D C B A	D E B A C
					E C D B A

6 × 6	7 × 7	8 × 8
A B C D E F	A B C D E F G	A B C D E F G H
B F D C A E	B C D E F G A	B C D E F G H A
C D E F B A	C D E F G A B	C D E F G H A B
D A F E C B	D E F G A B C	D E F G H A B C
E C A B F D	E F G A B C D	E F G H A B C D
F E B A D C	F G A B C D E	F G H A B C D E
	G A B C D E F	G H A B C D E F
		H A B C D E F G

[a]The first blocking factor (A) is represented by the columns, and the second blocking factor (B) by the rows of these tables. The treatment levels of factor C are represented by capital letters in the tables. Note the symmetry of these designs: each row is a complete block, and each column is a complete block. These examples of latin squares are abstracted from a much larger set published by Cochran and Cox (1957).

model is very simple, and the analysis itself is simple. If you do not know enough to make this assumption of no interactions, you should use a full factorial design with replication so that interactions can be estimated.

The latin square design is not commonly found in field ecological research, but when its restrictive assumptions can be met, it is a very efficient experimental design to apply, particularly in laboratory or common garden experiments.

10.2.6 Repeated-Measure Designs

Ecological experiments often involve repeated measurements over time on individual organisms or individual plots. For example, growth studies may involve measuring the height of individual plants at several times over months or years. *Replicates* in repeated-measure designs are the multiple measurements made on each individual subject. Repeated measurement violates the basic assumption of ANOVA that each observation is independent. Clearly two measurements of the height of the same plant cannot be independent, especially if time intervals are close.

The analysis of data from repeated-measure designs is difficult because a series of statistical decisions must be made before tests of significance can be calculated (Barcikowski and Robey 1984; Zar 1996, 259). Rowell and Walters (1976) discuss the validity of the assumptions made in most repeated-measure experiments and provide an alternative approach when the assumptions are not satisfied. Gurevitch and Chester (1986) discuss another approach to repeated-measure data that attempts to analyze the time trends in ecological variables. Most of the common statistical packages (SYSTAT, SAS, SPSS, JMP, NCSS) provide computations for repeated-measure designs.

10.3 ENVIRONMENTAL IMPACT STUDIES

Environmental impact studies form an important component of applied ecological research, and it is important that these studies be done properly. The basic types of environmental impact studies have been summarized by Green (1979) and are illustrated in Figure 10.7. Populations typically vary in abundance from time to time, and this variability, or "noise," means that detecting impacts can be difficult. By replicating measurements of abundance several times before the anticipated impact and several times after, you can determine statistically if an impact has occurred (Figure 10.7a). If there is only one control area and one impact area, all differences will be attributed to the impact, even if other ecological factors may, for example, affect the control area but not the impact area. For this reason Underwood (1994) pointed out that a spatial replication of control sites is also highly desirable (Figure 10.7b). Note that the different control sites may differ among themselves in average population abundance because of habitat variation. This variation among areas can be removed in an ANOVA, and the relevant ecological measure is whether the impact area changes after the impact, relative to the controls. This will show up in the ANOVA as a significant interaction between locations and time, and should be obvious visually, as shown in Figure 10.7.

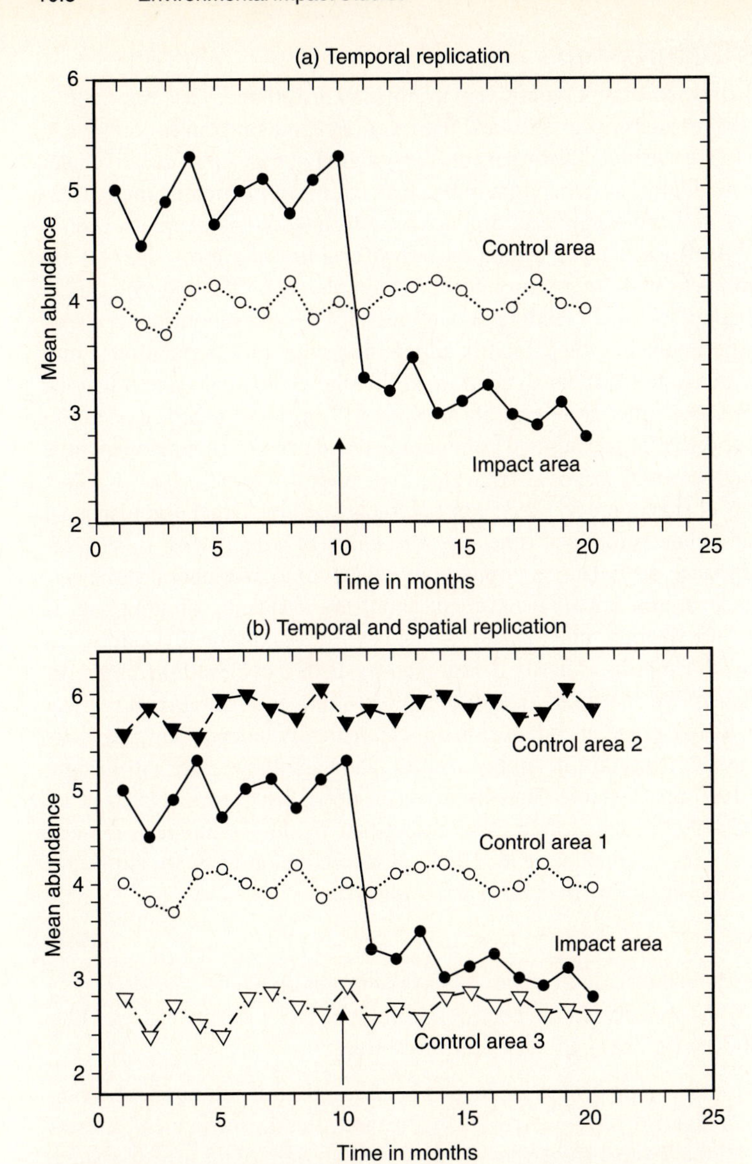

Figure 10.7 Environmental impact assessment using the BACI design (before-after, control-impact). The abundance of some indicator species is measured several times before and after impact (indicated by arrow). The statistical problem is to determine if the changes in the impact zone represent significant changes in abundance or merely random variations in abundance. (a) The simplest design with one control site and one impact site. (b) By adding spatial replication to the design, unique changes on one control site can be recognized and the power of the ecological inference increased.

10.3.1 Types of Disturbances

It is useful to distinguish two different types of environmental disturbances or experimental perturbations. *Pulse* disturbances occur once and then stop. *Press* disturbances continue to occur, and because they are sustained disturbances, they ought to be easier to detect. Under press disturbances, an ecological system ought to reach an equilibrium after the initial transition changes. Pulse disturbances may also produce long-term ecological effects, even if the actual disturbance itself was short. A flood on a river or a fire in a forest are two examples of pulse disturbances with long-term impacts. A supplemental feeding experiment could be a press experiment in which feeding is continued for several months or years, or it could be a pulse experiment in which food is added only once to a population. Pulse disturbances typically generate transient dynamics, so that the ecological system may be changing continuously rather than at an equilibrium point. Transient dynamics occur on different time scales for different species and communities, and can be recognized by time trends in the data being measured.

Disturbances may change not only the average abundance of the target organisms but also the variability of the population over time. The variability of a population is often directly related to its abundance, so that larger populations will have larger temporal variances. Decisions must be made in any impact assessment about the frequency of sampling. If samples are taken too infrequently, pulse disturbances could go undetected (Underwood 1994). The time course of sampling is partly determined by the life cycles of the target organisms; thus, for each ecological system, there is an optimal frequency of sampling. If you sample too frequently, the observed abundances are not statistically independent. The conclusion is that impact assessment must be highly species- and habitat-specific in its design.

Statistical tests for impact studies are discussed in detail by Stewart-Oaten et al. (1992) and Underwood (1994). Many impact assessments use randomization tests to measure changes at impact sites (Carpenter et al. 1989). Randomization tests are similar to t-tests in measuring differences between control and impact sites:

$$d_t = C_t - I_t \tag{10.1}$$

where d_t = Difference in the ecological variable of interest at time t
C_t = Value of the variable on the control site at time t
I_t = Value of the variable on the impact site at time t

A two-sample t test would compare these d_t values before and after the impact. A randomization test would compare the average d value before and after the impact with a series of random samples of all the d-values regardless of their timing. Both of these tests assume that the successive d-values are independent samples, that variances are the same in the control and impact populations, and that the effects of the impact are additive. Stewart-Oaten et al. (1992) discuss the implications of violating these assumptions, which can render the formal statistical tests invalid. A simple illustration of these methods is shown in Box 10.3.

10.3.2 Transient Response Studies

Ecological systems are often responding to transient effects such as the impact of a cold winter or a dry growing season. These transient effects add noise to environmental assessments, and the results of an experiment or a management manipulation may become

Box 10.3 Use of the Welch *t*-Test and the Randomization Test to Assess Environmental Impacts in a BACI Design (before-after, control-impact)

Estimates of chironomid abundance in sediments were taken at one station above and one below a pulpmill outflow pipe for 3 years before plant operation and for 6 years after with the following results:

	Year	Control site	Impact site	Difference (control-impact)
	1988	14	17	−3
	1989	12	14	−2
	1990	15	17	−3
Impact begins	1991	16	21	−5
	1992	13	24	−11
	1993	12	20	−8
	1994	15	21	−6
	1995	17	26	−9
	1996	14	23	−9

We analyze the differences between the control and impact sites. Using B for the before data, and A for the after-impact data, and the usual statistical formulas, we obtain

$$n_B = 3, \ \bar{d}_B = -2.67, \ s_B^2 = 0.33$$

$$n_A = 6, \ \bar{d}_A = -8.00, \ s_A^2 = 4.80$$

$$\text{Estimated difference due to impact} = \bar{d}_B - \bar{d}_A = -2.67 - (-8.00) = 5.33$$

Compute Welch's *t*-test as described in Zar (1996, 129) or Sokal and Rohlf (1995, 404) as

$$t = \frac{\bar{d}_B - \bar{d}_A}{\sqrt{\dfrac{s_B^2}{n_B} + \dfrac{s_A^2}{n_A}}} = \frac{-2.67 - (-8.00)}{\sqrt{\dfrac{0.33}{3} + \dfrac{4.80}{6}}} = \frac{5.33}{\sqrt{0.9111}} = 5.59$$

Because the Welch test allows for unequal variances before and after impact, the degrees of freedom are adjusted by the following formula from Zar (1996, 129):

$$d.f. = \frac{\left(\dfrac{s_B^2}{n_B} + \dfrac{s_A^2}{n_A}\right)^2}{\dfrac{\left(\dfrac{s_B^2}{n_B}\right)^2}{n_B - 1} + \dfrac{\left(\dfrac{s_A^2}{n_A}\right)^2}{n_A - 1}} = \frac{(0.9111)^2}{\dfrac{(0.111)^2}{2} + \dfrac{(0.80)^2}{5}} = 6.19 \text{ (rounded to 6)}$$

The critical value from Student's t-distribution for 6 degrees of freedom at $\alpha = 0.05$ is 2.447 and consequently we reject the null hypothesis that chironomid abundance is not impacted by the pulpmill effluent.

These data can also be analyzed by a randomization test (Sokal and Rohlf 1995, 803). Randomization tests are computer-intensive, but are simple in principle and proceed as follows:

1. All the 9 difference values are put into a hat, and random samples of $n = 3$ are drawn to represent the *before* data and $n = 6$ to represent the *after* data. This sampling is done without replacement.
2. This procedure is repeated 500–1000 times or more in a computer to generate a distribution of difference values that one would expect under the null hypothesis of no difference in the before-after abundance values. Manly (1991) provides a computer program to do these computations.
3. For these particular data, a randomization test of 5000 samples gave only 1.32% of the differences equal to or larger than the observed -5.33 difference in mean abundance. We would therefore reject the null hypothesis of no impact.

This example is simplified by having only one control and one impact site. If several control sites are used, as recommended for increased power, ANOVA methods should be used, as described by Underwood (1994, 6).

confounded with this noise. An example of this confounding is discussed by Walters et al. (1988). The survival of hatchery-raised Pacific salmon can be affected by ocean temperature as well as by hatchery-related factors like the accumulation of diseases or parasites in hatcheries. Coho salmon survival, measured by the number of adults returning for each fingerling going to sea, declined dramatically for hatchery fish from British Columbia from high levels when the hatcheries first opened in the 1970s to very low levels ten years later, and they have remained low. Two possible explanations are confounded: (1) a hypothesis about an ocean change that coincided by chance with the opening of the hatcheries; and (2) a hatchery-environment hypothesis that looks to changes in the parasites, diseases, and genetics of the hatchery stocks over time. Management agencies need to decide which of these hypotheses is correct in order to take remedial action. How might they do this?

Transient responses can be diagnosed with a modified experimental design called the staircase design (Figure 10.8). The critical feature of staircase designs is that the treatments are begun at different time periods so that environmental changes can be separated from treatment effects and the time-times-treatment interaction can be estimated statistically. The staircase design is not simple statistically, and yet it is important for ecologists to recognize that environmental impacts or experimental treatments are always carried out against a variable environment in which effects may be confounded.

10.3.3 Variability of Measurements

The power of any test for environmental impacts is constrained by the variability of the data being gathered, the magnitude of the environmental impact, and the number of replicates obtained in time and space. The choice of the type of data to be gathered is the first

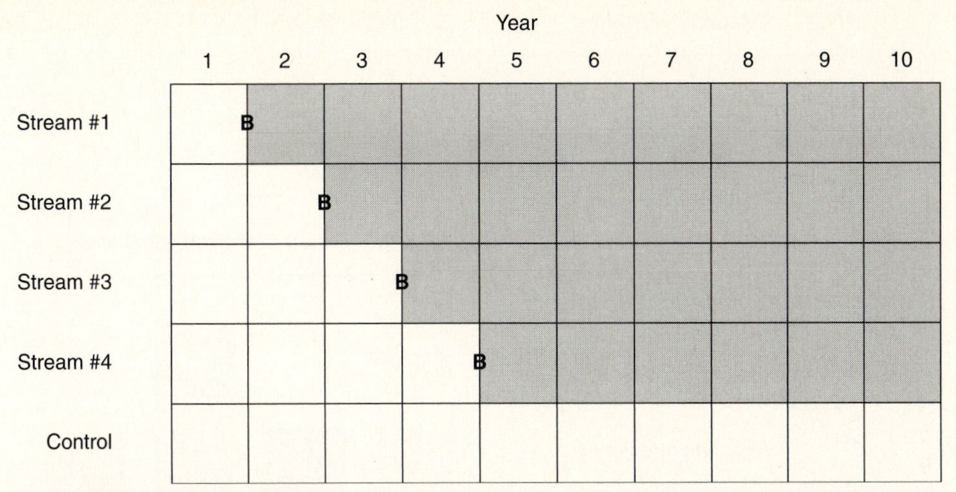

Figure 10.8 Staircase design for environmental impact studies in which environmental time trends may confound treatment effects. In this example, all salmon streams are monitored for at least one year before a hatchery is operated at point *B,* and the four hatcheries begin operation in four successive years. (Modified from Walters et. al. 1988.)

choice in any ecological study, and we need to know the variability to design powerful impact studies. Osenberg et al. (1994) considered three types of data that might be gathered: physical-chemical data (e.g., temperature, nutrient concentrations), individual-based data (e.g., body size, gonad weight, condition), and population-based measures (e.g., density). Individual and population data can be gathered on many different species in the community. Osenberg et al. provide an interesting data set for a marine study, and provide a model for the kind of approach that would be useful for other impact studies. First, they measured impacts as the difference in the parameter between the control site and the impact site at a specific time (equation [10.1] above). Population-based parameters were most highly variable (see Figure 10.9a), and physical-chemical parameters were least variable. All else being equal, physical-chemical parameters should provide more reliable measures of environmental impact. But a second critical variable intervenes here: the effect size, or how large a change occurs in response to the impact. For this particular system, Figure 10.9b shows that the effect size is much larger for population parameters than for chemical parameters. By combining these two measures into a measure of standardized effect size, we can determine the best variables statistically:

$$\text{Standardized effect size} = \frac{\text{Absolute effect size}}{\text{Standard deviation of differences}} = \frac{\bar{d}_B - \bar{d}_A}{s_d} \qquad (10.2)$$

where $\bar{d}_B$ = Mean difference between sites before impact begins
$\bar{d}_A$ = Mean difference between sites after impact begins
s_d = Standard deviation of observed differences between sites

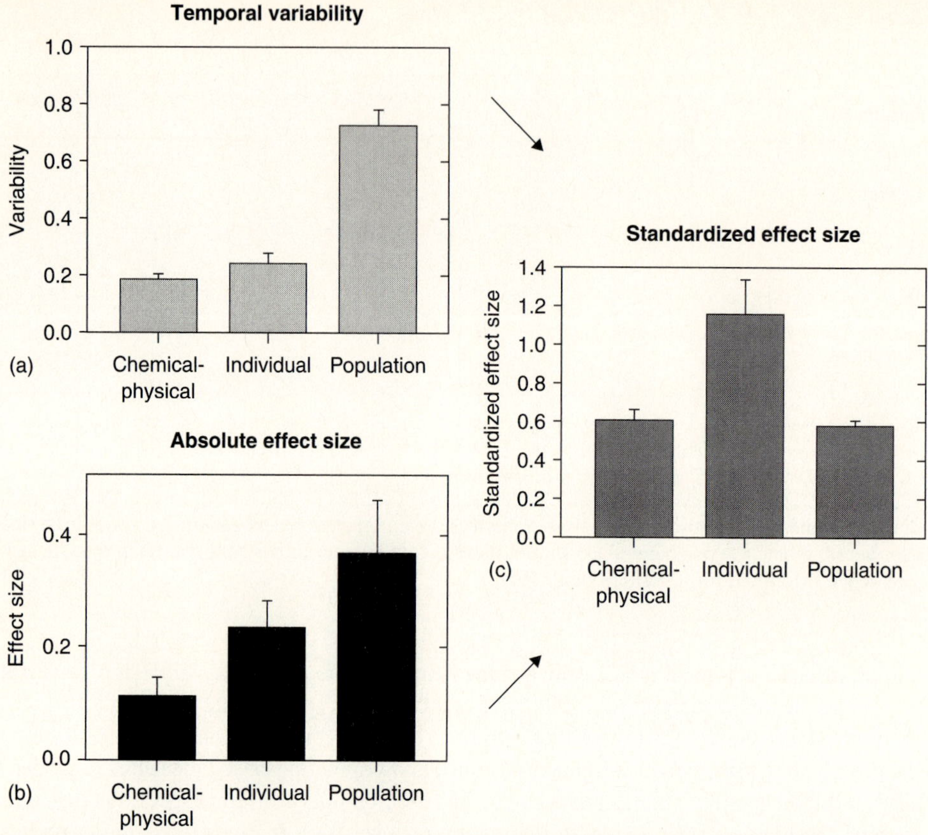

Figure 10.9 Estimates of parameter variability for a California marine coastal environmental study. All measures of variables refer to the differences between control and impact sites (Equation 10.1). (a) Temporal variability for 30 chemical-physical variables, 7 individual-based variables, and 25 population-based variables. (b) Absolute effect sizes (mean difference before and after) to measure the impact of the polluted water on chemical-physical, individual-, and population-based variables. (c) Standardized effect sizes that combine (a) and (b) for these variables. Variables with larger standardized effect sizes are more powerful for monitoring impacts. (Data from Osenberg et. al. 1994.)

Figure 10.9c shows the standardized effect sizes for this particular study. Individual-based parameters are better than either physical-chemical or population-based measures for detecting impacts. Approximately four times as many samples would be needed to detect an impact with population-based measures, compared with individual-based measures.

This example must not be generalized to all environmental impact assessments, but it does provide a template of how to proceed to evaluate what variables might be the best to measure in order to detect impacts. It emphasizes the critical point in experimental design that measurements with high variability can be useful indicators if they also show large effect sizes. What is critical in all ecological research is not statistical significance but ecological significance, which is measured by effect sizes. How large an impact has occurred is the central issue, and *large* can be defined only with respect to a specific ecosystem.

10.4 WHERE SHOULD I GO NEXT?

The analysis of variance and the design of experiments are complex and rapidly growing subdisciplines of statistics, and you could spend a lifetime becoming an expert in one of these areas. This chapter has covered 0.1% of the material available, and at this stage you need to do two things:

1. Take a course in experimental design or read a good elementary text like Cox (1958), Box et al. (1978), or Winer et al. (1991).
2. Think out your experimental design and take it to a professional statistician *before* you begin work. If you have a simple design, you may find the texts listed above adequate.

You can also get some guidance on design by looking at previous studies in your area of ecology, but read Underwood (1994), Hurlbert (1984), and Peters (1991) before you accept what is already published too easily. Alternatively, you may get some advice from a statistical computer package such as SigmaStat by Jandel Software. Unfortunately most statistical packages assume that you already have a fully implemented experimental design and thus give little guidance, and I do not recommend using computer packages until you have thought out and planned your design.

Finally, remember that good and interesting ecology comes from exciting ideas *and* good experimental design. Being a first-class statistician will not make you a good ecologist, but many exciting ideas have floundered on poor experimental design.

10.5 SUMMARY

The general principles of experimental design are often overlooked in the rush to set up ecological experiments or ignored in the rush to utilize computer packages of statistical programs. The first step in designing a good experiment is to define carefully the *experimental units*. These units must be sampled *randomly* to satisfy the assumption that all observations are independent and to reduce bias. *Replication* is needed to estimate experimental error, the yardstick of statistical significance. Treatments should be *interspersed* in space and in time to minimize the possibility that chance events will affect the results of the experiment. If interspersion is not used, replicates may not be independent, and the error of *pseudoreplication* may occur.

Many experimental designs are available to ecologists to assist in increasing the precision of field experiments. All of them can be represented by a linear additive model. The elementary strategy for increasing precision is to group together experimental units into *blocks* that are homogeneous. Any known or suspected source of variation can be blocked out of the experiment by this simple grouping.

Other experimental designs are less commonly used. *Nested* designs can be used when subsampling occurs to find the optimal way to measure a system. *Repeated-measure* designs should be used when a time series of data is obtained on a set of treatments.

Environmental impact studies present design problems similar to those of any ecological manipulative experiment. The BACI (before-after control-impact) design provides a mechanism for analyzing impacts. If environmental time trends occur, they may be confounded with experimental impacts unless care is taken to provide temporal replication. Different variables can be measured to assess impacts, and some provide more statistical power than others.

There is an enormous, complex literature on experimental design. Refer to a comprehensive text like Winer et al. (1991) or consult your local statistician for details before you launch into any project with a complex design.

SELECTED READING

Cox, D. R. 1958. *Planning of Experiments*. John Wiley and Sons, New York.

Dutilleul, P. 1993. Spatial heterogeneity and the design of ecological field experiments. *Ecology* 74: 1646–1658.

Heffner, R. A., Butler, M. J., and Reilly, C. K. 1996. Pseudoreplication revisited. *Ecology* 77: 2558–2562.

Hurlbert, S. H. 1984. Pseudoreplication and the design of ecological field experiments. *Ecological Monographs* 54: 187–211.

Rice, W. R., and Gaines, S. D. 1994. "Heads I win, tails you lose": Testing directional alternative hypotheses in ecological and evolutionary research. *Trends in Ecology and Evolution* 9: 235–237.

Shaw, R. G., and Mitchell-Olds, T. 1993. ANOVA for unbalanced data: An overview. *Ecology* 74: 1638–1645.

Stewart-Oaten, A., Bence, J. R., and Osenberg, C. W. 1992. Assessing effects of unreplicated perturbations: No simple solutions. *Ecology* 73: 1396–1404.

Underwood, A. J. 1994. On beyond BACI: Sampling designs that might reliably detect environmental disturbances. *Ecological Applications* 4: 3–15.

Walters, C. J. 1993. Dynamic models and large scale field experiments in environmental impact assessment and management. *Australian Journal of Ecology* 18: 53–62.

Winer, B. J., Brown, D.R., and Michels, K. M. 1991. *Statistical Principles in Experimental Design*. 3d ed. McGraw-Hill, New York.

Yates, F. 1964. Sir Ronald Fisher and the Design of Experiments. *Biometrics* 20: 307–321.

QUESTIONS AND PROBLEMS

10.1. Examine a sample of 40 papers in an ecological journal of your choice and classify the frequency of errors in the use of the analysis of variance, following the breakdown in Table II of Underwood (1981). Compare your results with those of Underwood (1981) for marine ecology journals.

10.2. Hurlbert (1984, 200) gives the following problem: A beetle population in a large field has a mean density of 51 beetles/m^2. We wish to test if a herbicide has an effect on beetle density. Two possible designs are

(a) The field is divided into two parts, and a preapplication sampling of 20 quadrats in each part shows no significant difference between them. The herbicide is applied to one part of the field, and the other part is kept as a control. After 48 hours, each part is sampled with 20 quadrats.

(b) The field is partitioned into a grid of 1000 4×4 m plots. Twenty of these are selected at random and used as controls, and another 20 plots are selected randomly as herbicide plots.

A preapplication sampling is done on each plot, and no significant difference is found. The herbicide is applied, and 48 hours later each plot is sampled again.

Are designs (a) and (b) equivalent? Why or why not?

10.3. Pseudoreplication was found in 50% of the field experiments done on small rodents and only 10% of the field experiments done on freshwater plankton (Hurlbert 1984, Table 3). Discuss why this might occur.

10.4. Hayne (1978) states in discussing experimental design,

> If you insist on a replicate area being an exact duplication of your first (study area), then you've restricted your scientific interest to a very narrow subset of the whole universe. It is a misconception that replicates have to be identical. There's no such thing. . . . Any decent naturalist-type investigator will find differences between any two areas you choose.

Do you agree? What are the implications for field experiments in ecology?

10.5. Is the experimental design shown in Figure 10.8 pseudoreplicated? If it is, how could you change the design to avoid this problem?

10.6. List the sources of variability in measurements of abundance for the design shown in Figure 10.1. Which sources of variability can you reduce or eliminate? Describe exactly how these sources of variability change if you increase the number of control areas from one to four.

10.7. Discuss the implications of doing a manipulative experiment in which there are no *before* data, and all the measurements of control and experimental areas are carried out *after* the manipulation has already begun. Can such experiments be analyzed statistically?

10.8. How would you design a field experiment of the impact of burning on tree regeneration for a heterogeneous environment? Discuss the situation in which you have a spatial map of the heterogeneity and the alternative situation in which you do not know where the heterogeneous patches are located. Does it matter if the heterogeneity is a gradient or a complex set of patches? What is the effect on your experimental design if your map of the spatial heterogeneity is not correct?

PART FOUR

Estimating Community Parameters

Community ecologists face a special set of statistical problems in attempting to characterize and measure the properties of communities of plants and animals. Some community studies, such as energetic analyses, need only apply the general principles discussed in Part 1 to estimate the abundance of the many species that comprise a community, but others need to utilize new parameters applicable only at the community level. One community parameter is similarity, and Chapter 11 discusses how to measure the similarity among communities. Similarity is the basis of classification, and this chapter discusses cluster analysis as one method of objectively defining the relationships among many community samples.

Plant ecologists in particular have developed a wide array of multivariate statistical techniques to assist in the analysis of community patterns and to help in defining the environmental controls of community patterns. Gradient analysis and ordination techniques are among the statistical tools of all community ecologists. Although these methods are briefly mentioned in Chapter 11, I do not treat them in this book because there are several good texts devoted specifically to multivariate statistical methods in ecology (Pielou 1984; Digby and Kempton 1987).

Species diversity is one of the most obvious and characteristic features of a community. From the earliest observations about the rich diversity of tropical communities in comparison with impoverished polar communities, ecologists have tried to quantify the diversity concept. Chapter 12 summarizes the accumulated wisdom of ways to measure biological diversity in plant and animal communities.

Niche theory has been one of the most powerful methods for analyzing community structure after the pioneering work of MacArthur (1968). Analyses of the structure of a community and the dynamic interactions of competing species all depend on the measurement of the niche parameters of species. Chapter 13 presents the methods developed for the measurement of niche breadth and niche overlap in natural communities. The measurement of dietary preference is similar to the problem of measuring niche breadth, and Chapter 13 also discusses the measures that have been suggested for quantifying the simple idea of preference in animals. Other community concepts such as trophic structure and succession are analyzed by various combinations of the methods outlined in the earlier parts of this book.

Community dynamics is an important area of analysis in modern ecology and a challenging focus of experimental work. To study communities rigorously, ecologists must use a wide array of population and community methods, all arranged in an experimental design that will satisfy a pure statistician. To achieve this goal is perhaps the most challenging methodological problem in modern ecology.

CHAPTER 11

Similarity Coefficients and Cluster Analysis

In many community studies ecologists obtain a list of the species that occur in each of several communities and, if quantitative sampling has been done, some measure of the relative abundance of each species. Often the purpose of this sampling is to determine if the communities can be classified together or need to be separated. For the designation of conservation areas, we often wish to ask how much separate areas differ in their flora and fauna. To begin answering these complex questions of community classification, we now ask how we can measure the *similarity* between two such community samples.

11.1 MEASUREMENTS OF SIMILARITY

There are more than two dozen measures of similarity available (Legendre and Legendre 1983; Wolda 1981; Sneath and Sokal 1973), and much confusion exists about which measure to use. Similarity measures are peculiar kinds of coefficients because they are mainly

TABLE 11.1 NUMBER OF SEABIRDS NESTING ON TWO OF THE PRIBILOF ISLANDS OF THE BERING SEA

	St. Paul Island		St. George Island	
Seabird	No. of individuals	Proportion	No. of individuals	Proportion
Northern fulmar	700	0.0028	70,000	0.0278
Red-faced cormorant	2,500	0.0099	5,000	0.0020
Black-legged kittiwake	31,000	0.1221	72,000	0.0286
Red-legged kittiwake	2,200	0.0087	220,000	0.0873
Common murre	39,000	0.1537	190,000	0.0754
Thick-billed murre	110,000	0.4334	1,500,000	0.5955
Parakeet auklet	34,000	0.1340	150,000	0.0595
Crested auklet	6,000	0.0236	28,000	0.0111
Least auklet	23,000	0.0906	250,000	0.0992
Horned puffin	4,400	0.0173	28,000	0.0111
Tufted puffin	1,000	0.0039	6,000	0.0024
Total	253,800	1.0000	2,519,000	0.9999

Source: Data from Hunt et al. 1986.

descriptive coefficients, not estimators of some statistical parameter. It is difficult to give reliable confidence intervals for most measures of similarity and probable errors can be estimated only by some type of randomization procedure (Ricklefs and Lau 1980, Chapter 15, Section 15.5).

There are two broad classes of similarity measures. *Binary* similarity coefficients are used when only presence/absence data are available for the species in a community; thus they are appropriate for the nominal scale of measurement. *Quantitative* similarity coefficients require that some measure of relative abundance also be available for each species. Relative abundance may be measured by *number* of individuals, *biomass*, *cover*, *productivity*, or any measure that quantifies the importance of the species in the community. Table 11.1 illustrates data on the relative abundance of 11 species of seabirds on two islands.

There are two desirable attributes of all similarity measures. First, the measure should be independent of sample size and of the number of species in the community (Wolda 1981). Second, the measure should increase smoothly from some fixed minimum to a fixed maximum, as the two community samples become more similar. Wolda (1981) has done an extensive analysis of the properties of 21 different similarity coefficients to see if they all behave desirably, and his work forms the basis of many of the recommendations I give here.

11.1.1 Binary Coefficients

The simplest similarity measures deal only with presence/absence data. The basic data for calculating binary (or association) coefficients is a 2×2 table:

		Sample A	
		No. of species present	No. of species absent
Sample B	No. of species present	a	b
	No. of species absent	c	d

where a = Number of species in sample A and sample B (joint occurrences)
b = Number of species in sample B but not in sample A
c = Number of species in sample A but not in sample B
d = Number of species absent in both samples (zero-zero matches)

There is considerable disagreement in the literature about whether d is a biologically meaningful number. It could be meaningful in an area where the flora or fauna is well known and the absence of certain species is relevant. But at the other extreme, elephants are always absent from plankton samples and clearly they should not be included in d when plankton are being studied. For this reason most users of binary similarity measures ignore species that are absent in both samples.

There are more than 20 binary similarity measures available in the literature (Cheetham and Hazel 1969), and they have been reviewed by Clifford and Stephenson (1975, Chapter 6) and by Romesburg (1984, Chapter 10). I will describe here only four of the most often used similarity coefficients for binary data.

Coefficient of Jaccard The coefficient of Jaccard is expressed as follows:

$$S_j = \frac{a}{a + b + c} \tag{11.1}$$

where S_j = Jaccard's similarity coefficient
a, b, c = As defined above in presence/absence matrix

Coefficient of Sorensen Sorensen's coefficient is very similar to the Jaccard measure; it was first used by Czekanowski in 1913 and discovered anew by Sorensen (1948):

$$S_S = \frac{2a}{2a + b + c} \tag{11.2}$$

where S_S = Sorensen's similarity coefficient

This coefficient weights matches in species composition between the two samples more heavily than mismatches. Whether or not one thinks this weighting is desirable will depend on the quality of the data. If many species are present in a community but not present in a sample from that community, it may be useful to use Sorensen's coefficient rather than Jaccard's.

Simple Matching Coefficient This is the simplest coefficient for binary data that makes use of *negative* matches as well as positive matches:

$$S_{SM} = \frac{a + d}{a + b + c + d} \tag{11.3}$$

where S_{SM} = Simple matching similarity coefficient

Baroni-Urbani and Buser Coefficient A more complex similarity coefficient that also makes use of *negative* matches is the Baroni-Urbani and Buser coefficient:

$$S_B = \frac{\sqrt{ad} + a}{a + b + c + \sqrt{ad}} \tag{11.4}$$

where S_B = Baroni-Urbani and Buser similarity coefficient

This was proposed by Baroni-Urbani and Buser (1976). Faith (1983) proposed a very similar binary similarity index.

The range of all similarity coefficients for binary data is supposed to be from 0 (no similarity) to 1.0 (complete similarity), but this is not true for all coefficients. Wolda (1981) investigated how sample size and species richness affected the maximum value that could be obtained with similarity coefficients. He used a simulated community of known species richness with 100,000 individuals whose species abundances were distributed according to the log series (see Krebs 1994, Chap. 23). Figure 11.1 shows how the coefficient of Sorensen and the Baroni-Urbani and Buser coefficient are affected by sample size and by species richness. Sample size effects are very large indeed. For example, the maximum value of the Sorensen coefficient when 750 species are present in the community and each community sample contains 200 individuals is 0.55, not 1.0 as one might expect.

Given this dependence of binary similarity coefficients on sample size, you have two possible choices: (1) When doing a similarity analysis, use samples of nearly equal size in all communities; (2) calculate the expected maximum value of the binary coefficient from

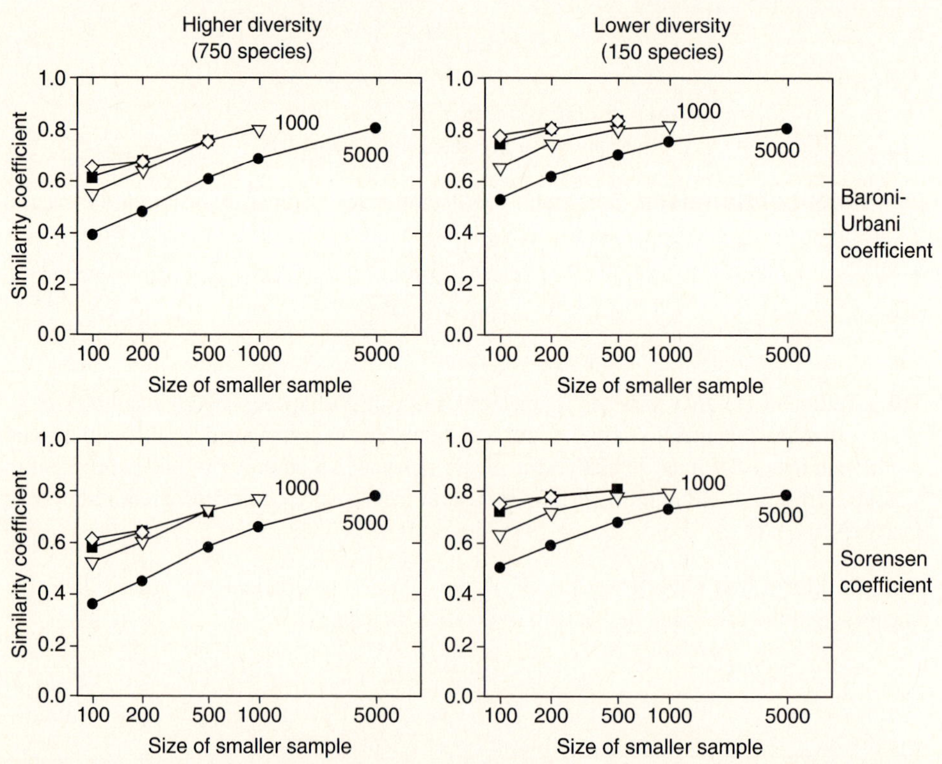

Figure 11.1 Expected maximum values of two commonly used binary coefficients of similarity as a function of sample size. The number of individuals in the smaller of the two community samples is given on the x-axis, and the lines connect samples of equal size for the larger community sample (n = 5000, 1000, 500, and 200; a different symbol is used for each of these sample sizes). A highly diverse community is shown on the left and a lower-diversity community on the right. Although the theoretical maximum value of each of these coefficients is 1.0, the expected maximum is much less than 1.0 when samples are small. (From Wolda 1981.)

TABLE 11.2 EQUATIONS APPROXIMATING THE RELATION BETWEEN THE MAXIMUM VALUES OF SOME SIMILARITY INDICES AND SAMPLE SIZE[a]

Index	Species diversity[b]	Regression equation
Baroni-Urbani and Buser (S_B)	150	$S_B = 1.190 - 1.563S^{-0.265} - 389 \times 10^{-7}L$
	380	$S_B = 1.190 - 2.108S^{-0.310} - 389 \times 10^{-7}L$
	580	$S_B = 1.208 - 2.204S^{-0.288} - 432 \times 10^{-7}L$
	750	$S_B = 1.213 - 2.651S^{-0.312} - 438 \times 10^{-7}L$
Sorensen (S_S)	150	$S_S = 1.148 - 2.146S^{-0.322} - 301 \times 10^{-7}L$
	380	$S_S = 1.130 - 3.292S^{-0.364} - 264 \times 10^{-7}L$
	580	$S_S = 1.137 - 3.375S^{-0.347} - 281 \times 10^{-7}L$
	750	$S_S = 1.125 - 4.170S^{-0.375} - 251 \times 10^{-7}L$
Percentage similarity; Renkonen (P)	150	$P = 1 - 1.642S^{0.405} - 4.282L^{-0.866}$
	380	$P = 1 - 2.410S^{0.384} - 2.754L^{-0.719}$
	580	$P = 1 - 2.810S^{0.375} - 0.645L^{-0.438}$
	750	$P = 1 - 3.111S^{0.375} - 0.640L^{-0.470}$
Horn (R_0)	150	$R_0 = 1 - 1.247S^{-0.631} - 6.486L^{-0.835}$
	380	$R_0 = 1 - 1.799S^{-0.539} - 9.393L^{-0.772}$
	580	$R_0 = 1 - 1.802S^{-0.485} - 5.825L^{-0.639}$
	750	$R_0 = 1 - 2.556S^{-0.517} - 7.040L^{-0.646}$

Source: Wolda (1981).

[a] The number of individuals in the smaller (S) and the larger (L) of the two samples compared.

[b] Number of species in community.

the formulas given in Table 11.2 from Wolda (1981) and rescale all the coefficients to the theoretical 0–1 scale of similarity. No reputable statistician would recommend the second choice, but for a practical ecologist, it is better to be approximate than to ignore a serious problem. Box 11.1 illustrates the calculation of these coefficients for binary data and the estimation of their true maximum value. Wolda (1981) did not investigate the sampling properties of the Jaccard coefficient, but they would probably be similar to those for the Sorensen coefficient (see Figure 11.1).

Binary similarity coefficients are the most crude measures available for judging similarity between communities because they do not take commonness and scarcity into consideration. Binary coefficients thus weight rare species the same as common species, and should be used whenever one wishes to weight all species on an equal footing. More commonly, binary similarity measures are used because only lists of species names are available for particular communities, and comparisons are possible only at this lower level of resolution.

11.1.2 Distance Coefficients

Distance coefficients are intuitively appealing to an ecologist because we can visualize them. Note that distance coefficients are measures of *dissimilarity*, rather than similarity. When a distance coefficient is zero, communities are identical. We can visualize distance measures of similarity by considering the simplest case of two species in two community samples. Distance coefficients typically require some measure of *abundance* for each

Box 11.1 Calculation of Similarity Measures for Binary Data

The crustacean zooplankton of the Great Lakes were sampled by Watson (1974), who obtained these data:

		Lake Erie: No. of species present	Lake Erie: No. of species absent
Lake Ontario	No. of species present	18	1
	No. of species absent	1	5

A total of 25 species of crustacean zooplankton occur in all the Great Lakes, and of these, five species do not occur in either Lake Erie or Lake Ontario.

Coefficient of Jaccard:

$$S_J = \frac{a}{a + b + c}$$

$$= \frac{18}{18 + 1 + 1} = 0.90$$

Coefficient of Sorensen:

$$S_S = \frac{2a}{2a + b + c}$$

$$= \frac{2(18)}{2(18) + 1 + 1} = 0.95$$

Simple matching coefficient:

$$S_{SM} = \frac{a + d}{a + b + c + d}$$

$$= \frac{18 + 5}{18 + 1 + 1 + 5} = 0.92$$

Baroni-Urbani and Buser coefficient:

$$S_B = \frac{\sqrt{ad} + a}{a + b + c + \sqrt{ad}}$$

$$= \frac{\sqrt{(18)(5)} + 18}{18 + 1 + 1 + \sqrt{(18)(5)}} = \frac{27.487}{29.487} = 0.93$$

From Table 11.2 we can calculate the expected maximum value of the Sorensen and Baroni-Urbani and Buser coefficients for these data. Assuming a sample size of 1000 individuals for Lake Erie and 1500 individuals for Lake Ontario and using the closest species richness value ($S = 150$ spp.), we obtain

For Sorensen coefficient:

$$\text{Maximum expected value} = 1.148 - 2.146S^{-0.322} - 0.301 \times 10^{-4}L$$

where S = Number of individuals sampled in smaller sample (= 1000)

L = Number of individuals sampled in larger sample (= 1500)

$$\text{Maximum value} = 1.148 - 2.146(1000^{-0.322}) - 0.301 \times 10^{-4}(1500) = 0.87$$

For Baroni-Urbani and Buser coefficient:

$$\text{Maximum expected value} = 1.190 - 1.563S^{-0.265}) - 0.389 \times 10^{-4}L$$

where S and L are defined above;

$$\text{Maximum value} = 1.190 - 1.563(1000^{-0.265}) - 0.389 \times 10^{-4}(1500) = 0.88$$

Because we have only 25 species instead of 150 species, these maximum expected values are underestimates. This is shown most clearly for the calculated Baroni-Urbani and Buser coefficient of 0.93, which is above the expected theoretical maximum of 0.88. With only 25 species and large sample sizes, the theoretical expected values are very near the theoretical limit of 1.0, and this will be the case in most situations in which relatively few species (<50) are sampled with more than 500 individuals in each sample.

species in the community. Figure 11.2 illustrates the simplest case, in which the number of individuals in the samples is used to measure abundance. The original data follow:

		Sample A	Sample B
Number of	Species 1	35	18
individuals of	Species 2	12	29

Euclidean Distance The distance between these two samples is clearly seen from Figure 11.2 as the hypotenuse of a triangle and is calculated from simple geometry as

$$\text{Distance} = \sqrt{x^2 + y^2}$$

$$= \sqrt{(35 - 18)^2 + (29 - 12)^2} = 24.04 \text{ (indiv.)}$$

This distance is formally called *Euclidian distance* and could be measured from Figure 11.2 with a ruler. More formally,

$$\Delta_{jk} = \sqrt{\sum_{i=1}^{n} (X_{ij} - X_{ik})^2} \tag{11.5}$$

where Δ_{jk} = Euclidean distance between samples j and k
X_{ij} = Number of individuals (or biomass) of species i in sample j
X_{ik} = Number of individuals (or biomass) of species i in sample k
n = Total number of species

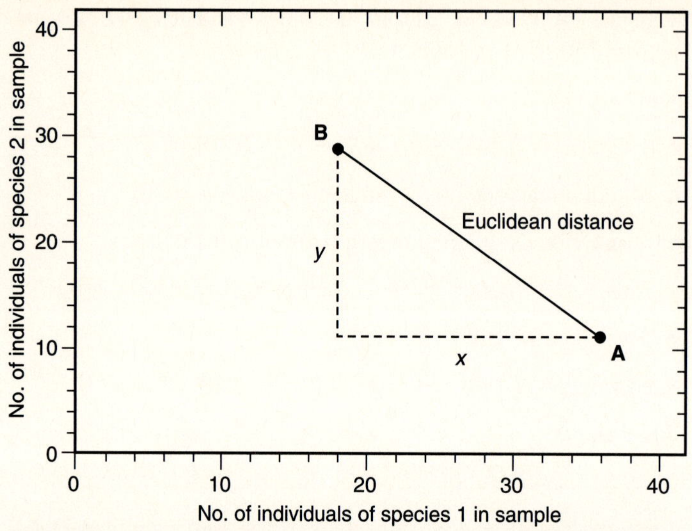

Figure 11.2 Hypothetical illustration of the Euclidean distance measure of similarity. Two communities, A and B, each with two species, are shown to illustrate the concept. As more species are included in the community, the dimensionality increases, but the basic principle does not change. Note that the *smaller* the distance, the *more similar* the two communities, so that Euclidean distance is a measure of *dissimilarity*.

Euclidean distance increases with the number of species in the samples, and to compensate for this, the average distance is usually calculated:

$$d_{jk} = \sqrt{\frac{\Delta_{jk}^2}{n}} \tag{11.6}$$

where d_{jk} = Average Euclidean distance between samples j and k
Δ_{jk} = Euclidean distance (calculated in equation 11.5)
n = Number of species in samples

Both Euclidean distance and average Euclidean distance vary from 0 to infinity; the larger the distance, the *less* similar the two communities.

Euclidean distance is a special case of a whole class of metric functions, and just as there are many ways to measure distance on a map, there are many other distance measures. One of the simplest metric functions is called the Manhattan, or city-block, metric:

$$d_M(j, k) = \sum_{i=1}^{n} |X_{ij} - X_{ik}| \tag{11.7}$$

where $d_M(j, k)$ = Manhattan distance between samples j and k
X_{ij}, X_{ik} = Number of individuals in species i in each sample j and k
n = Number of species in samples

This function measures distances as the length of the path you have to walk in a city—hence the name. Two measures based on the Manhattan metric have been used widely in plant ecology to measure similarity.

Bray-Curtis Measure Bray and Curtis (1957) standardized the Manhattan metric so that it has a range from 0 (similar) to 1 (dissimilar).

$$B = \frac{\sum_{i=1}^{n} |X_{ij} - X_{ik}|}{\sum_{i=1}^{n} (X_{ij} + X_{ik})} \tag{11.8}$$

where B = Bray-Curtis measure of dissimilarity
X_{ij}, X_{ik} = Number of individuals in species i in each sample (j, k)
n = Number of species in samples

Some authors (e.g., Wolda 1981) prefer to define this as a measure of similarity by using the complement of the Bray-Curtis measure $(1.0 - B)$.

The Bray-Curtis measure ignores cases in which the species is absent in both community samples, and it is dominated by the abundant species, so that rare species add very little to the value of the coefficient.

Canberra Metric Lance and Williams (1967) standardized the Manhattan metric over species instead of individuals and invented the Canberra metric:

$$C = \frac{1}{n}\left[\sum_{i=1}^{n}\left(\frac{|X_{ij} - X_{ik}|}{X_{ij} + X_{ik}}\right)\right] \tag{11.9}$$

where C = Canberra metric coefficient of dissimilarity between samples j and k
n = Number of species in samples
X_{ij}, X_{ik} = Number of individuals in species i in each sample (j, k)

The Canberra metric is not affected as much by the more abundant species in the community, and thus differs from the Bray-Curtis measure. The Canberra metric has two problems. It is undefined when there are species that are absent from both community samples, and consequently missing species can contribute no information and must be ignored. When no individuals of a species are present in one sample, but are present in the second sample, the index is at maximum value (Clifford and Stephenson 1975). To avoid this second problem, many ecologists replace all zero values by a small number (like 0.1) when doing the summations. The Canberra metric ranges from 0 to 1.0 and, like the Bray-Curtis measure, can be converted into a similarity measure by using the complement $(1.0 - C)$. Box 11.2 illustrates the calculation of these three distance measures for two small mammal communities in Colorado.

Both the Bray-Curtis measure and the Canberra metric measure are strongly affected by sample size (Wolda 1981). Figure 11.3 shows that in diverse communities with large sample sizes these two distance coefficients are particularly poor because their expected maximum value is low. They would appear to be best used in situations with low species diversity and small sample size.

11.1.3 Correlation Coefficients

One frequently used approach to the measurement of similarity is to use correlation coefficients of the standard kind described in every statistics book (e.g., Sokal and Rohlf 1995,

Box 11.2 Calculation of Distance Coefficients

Armstrong (1977) trapped nine species of small mammals in the Rocky Mountains of Colorado and obtained relative abundance (percentage of total catch) estimates for two habitat types ("communities") as follows:

	Small mammal species								
Habitat type	Sc	Sv	Em	Pm	Cg	Pi	Ml	Mm	Zp
Willow overstory	70	58	5	0	4	0	31	5	35
No overstory	10	11	20	20	9	8	11	46	44

Euclidean Distance

From equation (11.5),

$$\Delta_{jk} = \sqrt{\sum (X_{ij} - X_{ik})^2}$$

$$= \sqrt{(70 - 10)^2 + (58 - 11)^2 + (5 - 20)^2 + \cdots}$$

$$= \sqrt{8685} = 93.19$$

Average Euclidean distance (from equation 11.6):

$$d_{jk} = \sqrt{\frac{\Delta_{jk}^2}{n}}$$

$$= \sqrt{\frac{8685}{9}} = 31.06$$

Bray-Curtis Measure

From equation (11.8),

$$B = \frac{\Sigma\,|X_{ij} - X_{ik}|}{\Sigma\,(X_{ij} + X_{ik})}$$

$$= \frac{(70 - 10) + (58 - 11) + (20 - 5) + (20 - 0) + (9 - 4) + \cdots}{70 + 10 + 58 + 11 + 20 + 5 \cdots}$$

$$= \frac{225}{387} = 0.58$$

To use as a measure of similarity calculate the complement of B:

$$1 - B = 1 - 0.58 = 0.42$$

Canberra Metric

From equation (11.9),

$$C = \frac{1}{n}\left[\sum_{i=1}^{n}\left(\frac{|X_{ij} - X_{ik}|}{X_{ij} + X_{ik}}\right)\right]$$

$$= \frac{1}{9}\left(\frac{70 - 10}{70 + 10} + \frac{58 - 11}{58 + 11} + \frac{5 - 20}{5 + 20} + \frac{0.1^* - 20}{0.1 + 20}\right) + \cdots$$

$$= \frac{1}{9}(5.775) = 0.64$$

To use the Canberra metric as a measure of similarity calculate its complement:

$$1 - C = 1 - 0.64 = 0.36$$

Program-group SIMILARITY COEFFICIENTS (Appendix 2) can do these calculations.

*Note that zero values are replaced by 0.1.

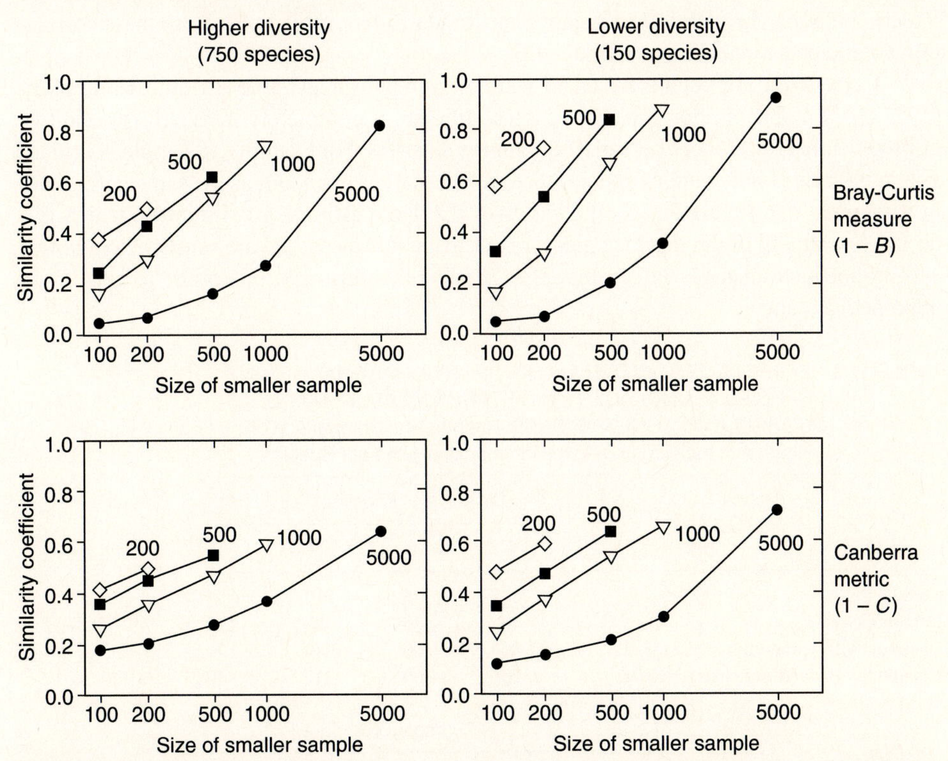

Figure 11.3 Expected maximum values of two commonly used distance measures as a function of sample size. The number of individuals in the smaller of two community samples is given on the x-axis, and the lines connect samples of equal size for the larger community sample (n = 5000, 1000, 500, and 200; a different symbol is used for each of these sample sizes). A highly diverse community is shown on the left and a less diverse community on the right. Both the Bray-Curtis measure and the Canberra metric are strongly affected by sample size, and their expected maximum is often far from the theoretical value. For this reason Wolda (1981) recommends against using these distance measures. (From Wolda 1981.)

Chap. 15; Zar 1996, Chap. 18). In the terminology used in this chapter, the Pearson correlation coefficient is given by

$$r = \frac{\Sigma\, xy}{\sqrt{\Sigma\, x^2\, \Sigma\, y^2}} \tag{11.10}$$

where

$$\Sigma\, xy = \text{Sum of cross products} = \Sigma_i\, X_{ij} X_{ik} - \frac{\Sigma_i\, X_{ij}\, \Sigma_i\, X_{ik}}{n}$$

$$\Sigma\, x^2 = \text{Sum of squares of } x = \Sigma_i\, X_{ij}^2 - \frac{(\Sigma_i\, X_{ij})^2}{n}$$

$$\Sigma\, y^2 = \text{Sum of squares of } y = \Sigma_i\, X_{ik}^2 - \frac{(\Sigma_i\, X_{ik})^2}{n}$$

X_{ij}, X_{ik} = Number of individuals of species i in each sample (j, k)

In order to use the Pearson product-moment correlation coefficient r as a similarity measure, one must make the usual assumption of a linear relationship between species abundances in the two communities. If you do not wish to make this assumption, you can use Spearman's rank correlation coefficient r_s or Kendall's tau instead of r. Both these correlation coefficients range from -1.0 to $+1.0$.

Correlation coefficients have one desirable and one undesirable attribute. Romesburg (1984, 107) points out that the correlation coefficient is completely insensitive to additive or proportional differences between community samples. For example, if sample A is identical to sample B but contains species that are one-half as abundant as the same species are in sample B, the correlation coefficient gives the same estimate of similarity, which is a desirable trait. All of the distance measures we have just discussed are sensitive to additive and proportional changes in communities. Table 11.3 illustrates this problem with some hypothetical data.

TABLE 11.3 EFFECTS OF ADDITIVE AND PROPORTIONAL CHANGES IN SPECIES ABUNDANCES ON DISTANCE MEASURES AND CORRELATION COEFFICIENTS. Hypothetical Comparison of Number of Individuals in Two Communities with Four Species

	Species			
	1	2	3	4
Community A	50	25	10	5
Community B	40	30	20	10
Community B_1 (proportional change, 2×)	80	60	40	20
Community B_2 (additive change, +30)	70	60	50	40

	Samples compared		
	A − B	A − B_1	A − B_2
Average Euclidean distance	7.90	28.50	33.35
Bray-Curtis measure	0.16	0.38	0.42
Canberra metric	0.22	0.46	0.51
Pearson correlation coefficient	0.96	0.96	0.96
Spearman rank correlation coefficient	1.00	1.00	1.00

Conclusion: If you wish your measure of similarity to be independent of proportional or additive changes in species abundances, you should not use a distance coefficient to measure similarity.

Correlation coefficients may be undesirable measures of similarity because they are all strongly affected by sample size, especially in high-diversity communities (see Figure 11.4). Field (1970) recognized this problem and recommended that when more than half of the abundances are zero in a community sample, the correlation coefficient should not be used as a measure of similarity. Figure 11.4 suggests that correlation coefficients may be most useful in low-diversity communities when samples are reasonably large.

11.1.4 Other Similarity Measures

Many other measures of similarity have been proposed. I select here only three to illustrate the most useful measures of similarity for quantitative data on communities.

Percentage Similarity The measure of percentage similarity was proposed by Renkonen (1938) and is sometimes called the Renkonen index. In order to calculate this measure of similarity, each community sample must be standardized in terms of *percentages*, so that

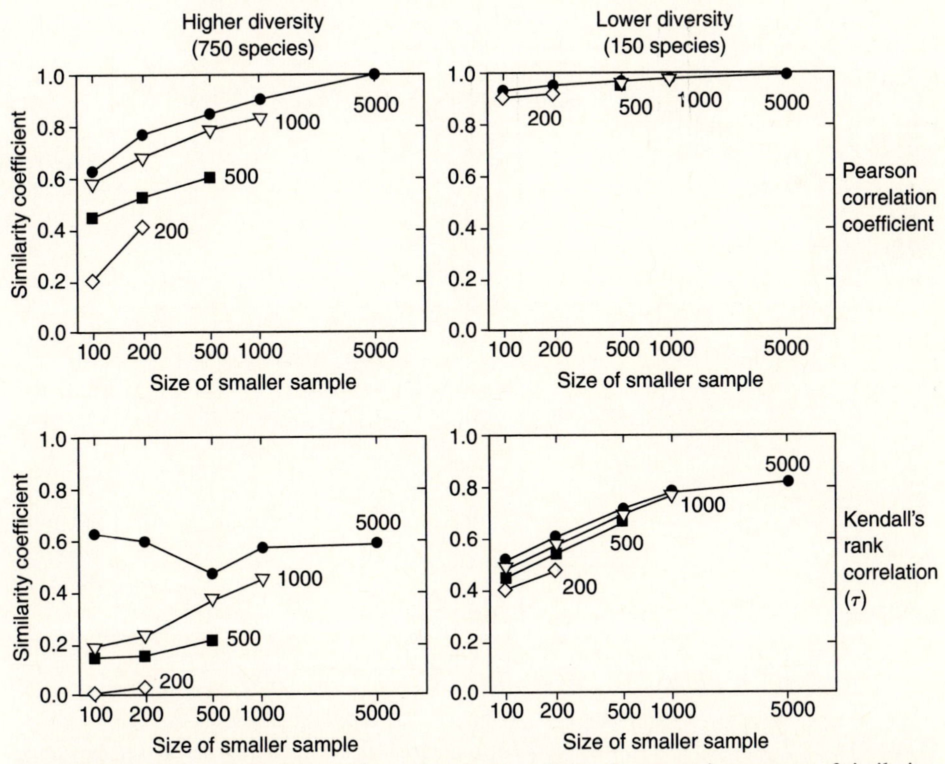

Figure 11.4 Expected maximum values of two commonly used parametric measures of similarity, the parametric correlation coefficient r and the nonparametric correlation coefficient Kendall's *tau*. The number of individuals in the smaller of the two community samples is given on the x-axis, and the lines connect samples of equal size for the larger community sample (n = 5000, 1000, 500, and 200; a different symbol is used for each of these sample sizes). A highly diverse community is shown on the left and a lower-diversity community on the right. In highly diverse communities the maximum expected value may diverge greatly from the theoretical value of 1.0 (From Wolda 1981.)

the relative abundances all sum to 100% in each sample. The index is then calculated as

$$P = \sum_i \text{minimum } (p_{1i}, p_{2i}) \tag{11.11}$$

where P = Percentage similarity between sample 1 and 2
p_{1i} = Percentage of species i in community sample 1
p_{2i} = Percentage of species i in community sample 2

In spite of its simplicity, the percentage similarity measure is one of the best quantitative similarity coefficients available (Wolda 1981). Figure 11.5 shows that it is relatively unaffected by sample size and by species diversity (compare with Figure 11.3). Percentage similarity is not affected by proportional differences in abundance between the samples but is sensitive to additive changes (c.f. page 386). Box 11.3 illustrates the calculation of percentage similarity. This index ranges from 0 (no similarity) to 100 (complete similarity). Table 11.2 gives Wolda's (1981) equations for estimating the maximum expected value for percentage similarity at different sample sizes.

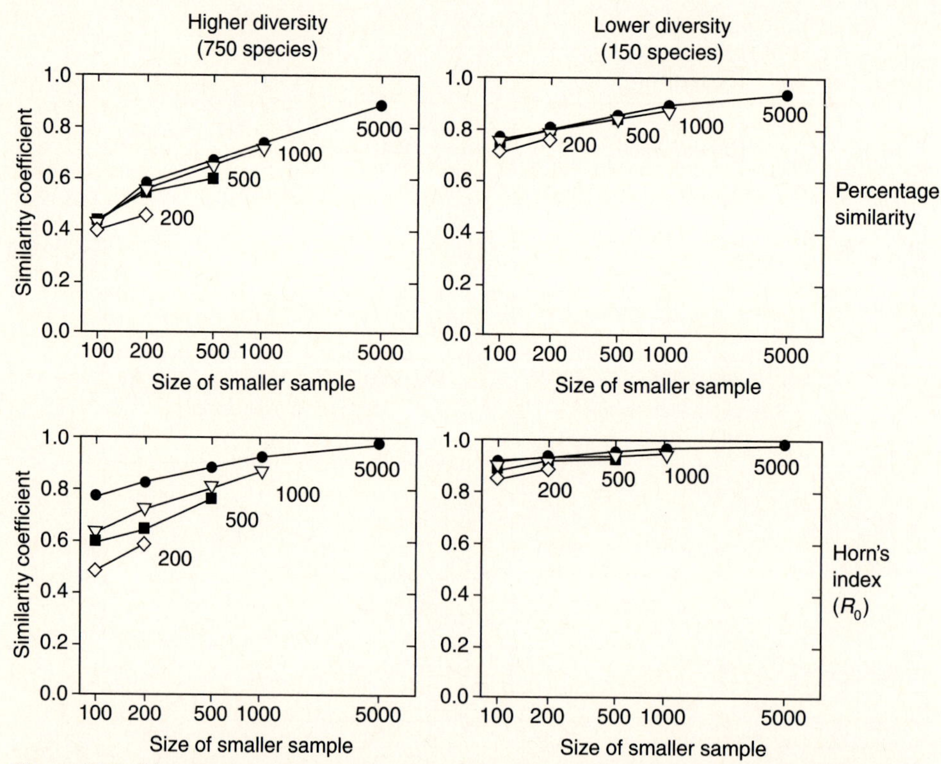

Figure 11.5 Expected maximum values of two other measures of similarity, the percentage similarity and Horn's index. The number of individuals in the smaller of the two community samples is given on the x-axis, and the lines connect samples of equal size for the larger community sample (n = 5000, 1000, 500, and 200; a different symbol is used for each of these sample sizes). A highly diverse community is shown on the left and a lower-diversity community on the right. Both of these measures are relatively unaffected by sample size, and they are recommended as possible alternatives to Morisita's index. (From Wolda 1981.)

Box 11.3 Calculation of Percentage Similarity, Morisita, and Horn Indices of Similarity

Nelson (1955) gave the basal areas of the trees in a watershed of western North Carolina for 2 years before and 17 years after chestnut blight had removed most of the chestnuts:

	Basal area (ft^2)		Percentage composition	
Tree species	1934	1953	1934	1953
Chestnut	53.3	0.9	49.2	1.1
Hickory	18.8	20.7	17.3	25.1
Chestnut oak	10.5	14.2	9.7	17.2
Northern red oak	9.8	5.2	9.0	6.3
Black oak	9.6	17.9	8.9	21.7
Yellow poplar	2.9	13.0	2.7	15.8
Red maple	2.0	3.7	1.8	4.5
Scarlet oak	1.5	6.9	1.4	8.4
Total	108.4	82.5	100.0	100.1

The first step is to express the abundances of the different species as relative abundances (or percentages), which must sum to 100%.

Percentage Similarity

From equation (11.11),

$$P = \sum \text{minimum}(p_{1i}, p_{2i})$$
$$= 1.1 + 17.3 + 9.7 + 6.3 + 8.9 + 2.7 + 1.8 + 1.4$$
$$= 49.2\%$$

Morisita's Index of Similarity

From equation (11.12),

$$C_\lambda = \frac{2 \Sigma X_{ij} X_{ik}}{(\lambda_1 + \lambda_2) N_j N_k}$$

$$\lambda_1 = \frac{(53.3)(52.3) + (18.8)(17.8) + (10.5)(9.5) + (9.8)(8.8) + \cdots}{108.4(107.4)}$$

$$= 0.292$$

$$\lambda_2 = \frac{(0.9)(0) + (20.7)(19.7) + (14.2)(13.2) + \cdots}{82.5(81.5)} = 0.167$$

$$C_\lambda = \frac{2[(53.3)(0.9) + (18.8)(20.7) + (10.5)(14.2) + \cdots]}{(0.292 + 0.167)(108.4)(82.5)} = \frac{1728.96}{4104.837} = 0.42$$

From equation (11.15) we can calculate the simplified Morisita index as

$$C_H = \frac{2 \sum X_{ij}X_{ik}}{[(\sum X_{ij}^2/N_j^2) + (\sum X_{ik}^2/N_k^2)]N_jN_k}$$

$$C_H = \frac{1728.96}{\left[\left(\frac{53.3^2 + 18.8^2 + \cdots}{108.4^2}\right) + \left(\frac{0.9^2 + 20.7^2 + \cdots}{82.5^2}\right)\right](108.4)(82.5)}$$

$$= \frac{1728.96}{4257.55} = 0.41$$

Horn's Index of Similarity

From equation (11.16), using the raw data on basal areas and using logs to base 10,

$$R_0 = \frac{\sum [(X_{ij} + X_{ik})\log(X_{ij}X_{ik})] - \sum (X_{ij} \log X_{ij}) - \sum (X_{ik} \log X_{ik})}{[(N_J + N_K)\log(N_J + N_K)] - (N_J \log N_J) - (N_K \log N_K)}$$

Breaking down the terms of summation in the numerator:

$$\sum [(X_{ij} + X_{ik})\log(X_{ij}X_{ik})]$$

$$= (53.3 + 0.9)(\log 54.2) + (18.8 + 20.7)(\log 39.5)$$

$$+ (10.5 + 14.2)(\log 24.7) + \cdots$$

$$= 279.846$$

$$\sum (X_{ij} \log X_{ij}) = 53.3(\log 53.3) + 18.8(\log 18.8)$$

$$+ 10.5(\log 10.5) + \cdots = 148.062$$

$$\sum (X_{kj} \log X_{kj}) = 0.9(\log 0.9) + 20.7(\log 20.7)$$

$$+ 14.2(\log 14.2) + \cdots = 92.083$$

$$R_0 = \frac{279.846 - 148.062 - 92.083}{(108.4 + 82.5)(\log 190.9) - 108.4(\log 108.4) - 82.5(\log 82.5)} = 0.70$$

Program-group SIMILARITY COEFFICIENTS (Appendix 2) can do these calculations.

Morisita's Index of Similarity This measure was first proposed by Morisita (1959) to measure similarity between two communities. It should not be confused with Morisita's index of dispersion (Section 6.4.4). It is calculated as

$$C_\lambda = \frac{2 \sum X_{ij}X_{ik}}{(\lambda_1 + \lambda_2)N_jN_k} \tag{11.12}$$

where C_λ = Morisita's index of similarity between sample j and k
X_{ij}, X_{ik} = Number of individuals of species i in sample j and sample k
$N_j = \Sigma X_{ij}$ = Total number of individuals in sample j
$N_k = \Sigma X_{ik}$ = Total number of individuals in sample k

$$\lambda_1 = \frac{\Sigma [X_{ij}(X_{ij} - 1)]}{N_j(N_j - 1)} \tag{11.13}$$

$$\lambda_2 = \frac{\Sigma [X_{ik}(X_{ik} - 1)]}{N_k(N_k - 1)} \tag{11.14}$$

The Morisita index is most easily interpreted as a probability:

$$C_\lambda = \frac{\left\{\begin{array}{c}\text{Probability that an individual drawn from sample } j \text{ and one}\\ \text{drawn from sample } k \text{ will belong to the same species}\end{array}\right\}}{\left\{\begin{array}{c}\text{Probability that two individuals drawn from either}\\ j \text{ or } k \text{ will belong to the same species}\end{array}\right\}}$$

The Morisita index varies from 0 (no similarity) to about 1.0 (complete similarity). Box 11.3 illustrates the calculation of the Morisita index. The Morisita index was formulated for counts of individuals and not for other abundance estimates based on biomass, productivity, or cover. Horn (1966) proposed a simplified Morisita index in which all the (−1) terms in equations (11.13) and (11.14) are ignored:

$$C_H = \frac{2 \Sigma X_{ij} X_{ik}}{[\Sigma X_{ij}^2 / N_j^2) + (\Sigma X_{ik}^2 / N_k^2)] N_j N_k} \tag{11.15}$$

where C_H = Simplified Morisita index of similarity (Horn 1966)

and all other terms are as defined above. This formula is appropriate when the original data are expressed as *proportions* rather than numbers of individuals and should be used when the original data are not numbers but biomass, cover, or productivity.

The Morisita index of similarity is nearly independent of sample size, except for samples of very small size. Morisita (1959) did extensive simulation experiments to show this, and these results were confirmed by Wolda (1981), who recommended Morisita's index as the best overall measure of similarity for ecological use.

Horn's Index of Similarity Horn (1966) developed another index of similarity based on information theory. It can be calculated directly from raw data (numbers) or from relative abundances (proportions or percentages).

$$R_0 = \frac{\Sigma [(X_{ij} + X_{ik}) \log (X_{ij} X_{ik})] - \Sigma (X_{ij} \log X_{ij}) - \Sigma (X_{ik} \log X_{ik})}{[(N_J + N_K) \log (N_J + N_K)] - (N_J \log N_J) - (N_K \log N_K)} \tag{11.16}$$

where R_0 = Horn's index of similarity for samples j and k
X_{ij}, X_{ik} = Number of individuals of species i in sample j and sample k
$N_J = \Sigma X_{ij}$ = Total number of individuals in sample j
$N_K = \Sigma X_{ik}$ = Total number of individuals in sample k

and all summations are over all the n species. Horn's index can be calculated from this equation using numbers or using proportions to estimate relative abundances. Note that the value obtained for Horn's index is the same whether numbers or proportions are used in equation (11.16) and is *not* affected by the base of logarithms used.

Horn's index is relatively little affected by sample size (see Figure 11.5), although it is not as robust as Morisita's index. Table 11.2 gives equations for estimating the maximum value for Horn's index for samples of different size and species diversity. Box 11.3 illustrates the calculation of Horn's index.

11.2 DATA STANDARDIZATION

Data to be used for community comparisons may be provided in several forms, and we have already seen examples of data as numbers and proportions (or percentages; see Table 11.1). Here we discuss briefly some rules of thumb that are useful in deciding how data should be standardized and when. Romesburg (1984, Chap. 8) and Clifford and Stephenson (1975, Chap. 7) have discussed this problem in more detail.

A considerable amount of judgment is involved in deciding how data should be summarized before similarity values are calculated. Three broad strategies exist: apply *transformations*, use *standardization*, and do nothing. No one strategy can be universally recommended; much depends upon your research objectives.

Transformations may be applied to the numbers of individuals counted in each species.* Transformations are typically used to replace each of the original counts (X) with $\sqrt{X}$ or $\sqrt{X + 1}$, or in extreme cases by log (X + 1.0). These transformations will reduce the importance of extreme values, for example, if one species is extremely abundant in one sample. In general, transformations are also used to reduce the contributions of the common species and to enhance the contributions of the rare species. Transformations also affect how much weight is given to habitat patchiness. If a single patch contains one highly abundant species, fox example, this one patch may produce a strong effect on the calculated similarity values. A transformation can help to smooth out these variations among patches, if you wish to do this smoothing on ecological grounds. If a transformation is used, it is applied before the similarity index is calculated.

Standardization of data values can be done in several ways. The most familiar standardization is to convert absolute numbers to proportions (see Table 11.1). Note that in doing this, all differences in population sizes between sites are lost from the data. Whether or not you wish to omit such differences from your analysis will determine your use of standardization. Romesburg (1984, Chap. 7) discusses other types of standardization.

The two most critical questions you must answer before you can decide on the form your data should take are the following:

1. Are a few species excessively common or rare in your samples, so that these extreme values distort the overall picture? If yes, use a transformation. You will have to use ecological intuition to decide what *excessively* means. A tenfold difference in abundance between the most common and the next most common species might be used as a rule of thumb for defining *excessively common*.

*See Chapter 15, Section 15.1, for more discussion of transformations.

2. Do you wish to include the absolute level of abundance as a part of the measurement of similarity between communities? If not, use *standardization* to proportions to express relative abundances. If you do not use either of these strategies, you should remember that if you do nothing to your raw data, you are still making a decision about what components of similarity to emphasize in your analysis.

One additional question about data standardization concerns when data may be deleted from the analysis. Many ecologists routinely eliminate rare species from their data before they apply a similarity measure. This practice is rooted in the general ecological feeling that species represented by only one or two individuals in a large community sample cannot be an important and significant component of the community (Clifford and Stephenson 1975, 86). The original rationalization for eliminating species was to reduce the computer time needed to calculate similarity and subsequent classifications for large sets of data (e.g., >700 samples, >300 species). This reason is no longer so important, since computers have become faster. It is important that we try to use only ecological arguments for eliminating species from data sets and that we try to eliminate as few species as possible (Stephenson et al. 1972).

The most important point to remember is that *data transformation changes the values of almost all of the coefficients of similarity*. You should decide *before* you begin your analysis on the appropriate type of data standardization for the questions you are trying to answer. You must do this standardization for *a priori* ecological reasons before you start your analyses, so that you do not bias the results. Table 11.4 illustrates how data standardization can affect the value of various similarity coefficients: all measures of similarity except the Spearman rank correlation coefficient are affected by the decisions made about data standardization. Table 11.4 shows clearly why you must decide on the type of data standardization on ecological grounds before you begin your analysis rather than fishing around for the kinds of values that will verify your preconceptions.

11.3 CLUSTER ANALYSIS

The measurement of similarity between samples from communities may be useful as an end in itself, especially when there are very few samples or only a few communities. In other cases we have many samples to analyze, and we now discuss techniques for grouping similar samples.

Clustering methods are ways to achieve a classification of a series of samples. Classification may not be a desirable goal for all ecological problems, and we may wish to treat variation as continuous instead of trying to classify samples into a series of groups. We will continue on the assumption that this methodological decision to classify has been made. There are four major decisions we must make before we can decide on a method of classification (Pielou 1969). The method of classification can be of the following kinds.

1. *Hierarchical* or *reticulate*: Hierarchical classifications are like trees; reticulate classifications overlap like a net. Ordinary taxonomic classifications are hierarchical; everyone uses hierarchical classifications because they are easy to understand.
2. *Divisive* or *agglomerative*: In a divisive classification we begin with the whole set of samples and divide it up into classes; in agglomerative classification we start

TABLE 11.4 EFFECTS OF DATA STANDARDIZATION ON THE VALUE OF SIMILARITY MEASURES FOR A HYPOTHETICAL DATA SET

	Sample A				Sample B			
	No. of individuals		Transformation		No. of individuals		Transformation	
Species	n_i (1)	Proportion (2)	$\sqrt{n_i}$ (3)	$\log(n_i + 1)$ (4)	n_i (5)	Proportion (6)	$\sqrt{n_i}$ (7)	$\log(n_i + 1)$ (8)
1	100	.388	10.0	2.004	10	.017	3.2	1.041
2	60	.233	7.7	1.785	30	.051	5.5	1.491
3	50	.194	7.1	1.708	60	.103	7.7	1.785
4	20	.078	4.5	1.322	400	.684	20.0	2.603
5	10	.039	3.2	1.041	50	.085	7.1	1.708
6	10	.039	3.2	0.778	20	.034	4.5	1.322
7	5	.019	2.2	0.041	0[a]	0	0.3	0.041
8	0[a]	0	0.3	0.301	15	.026	3.9	1.204
9	1	.004	1.0	0.301	0[a]	.000	0.3	0.041
10	1	.004	1.0	0.301	0[a]	.000	0.3	0.041
11	1	.004	1.0	0.301	0[a]	.000	0.3	0.041
	258				585			

If we assume two species are available but not present in either sample A or B, we obtain two tables:

1. All of sample included:

		A +	A −
B	+	6	1
B	−	4	2

2. Excluding "singletons":

		A +	A −
B	+	6	1
B	−	1	2

Binary similarity coefficients	All of sample included	Excluding "singletons" (species 9, 10, 11)
Jaccard	0.54	0.75
Sorensen	0.71	0.86
Simple matching	0.61	0.80
Baroni-Urbani and Buser	0.65	0.83

For the quantitative measures of similarity and dissimilarity,

	Type of data standardization			
Coefficient	Raw data (1), (5)	Proportions (2), (6)	$\sqrt{n_i}$ (3), (7)	$\log(n_i + 1)$ (4), (8)
Euclidean distance	118.88	0.22	5.48	0.70
1 − Bray-Curtis	0.31	0.32	0.59	0.71
1 − Canberra metric	0.26	0.26	0.50	0.51
Pearson correlation	0.02	0.02	0.32	0.60
Spearman correlation	0.60	0.60	0.60	0.60
Percentage similarity	—	32	58[b]	72[b]
Morisita's index	0.26	0.26	0.65	0.85
Horn index	0.56	0.56	0.81	0.85

[a] Zero values replaced by 0.1 for calculations of transformations.

[b] Percentages calculated on the transformed values instead of the raw data.

at the bottom and work upward, beginning with the individual samples. Divisive techniques ought to be more accurate because chance anomalies with individual samples may start agglomerative techniques off with some bad combinations that snowball as more agglomeration proceeds.

3. *Monothetic* or *polythetic*: In a monothetic classification two sister groups are distinguished by a single attribute, such as the presence of one species. In a polythetic classification overall similarity is used, based on all the attributes (species). Monothetic classifications are simple to understand and easy to determine but ignore information and may be poor if we choose the wrong attribute.
4. *Qualitative* or *quantitative* data: The main argument for using quantitative data is to avoid weighting the rare species as much as the common ones. This is a question of ecological judgment for each particular situation. In some cases only qualitative (binary) data are available.

The most important point to note at this stage is that *there is no single kind of classification that is the "best" system of grouping samples.* We must rely on our ecological knowledge to evaluate the end results of any classification.

Cluster analysis is the general term applied to the many techniques used to build classifications. Many of these are reviewed by Romesburg (1984), Clifford and Stephenson (1975), Sneath and Sokal (1973), and Pielou (1969). I will discuss here only a couple of simpler techniques, all of which are *hierarchical, agglomerative, polythetic* techniques. Virtually all of the techniques of cluster analysis demand a great deal of calculation and hence have become useful only with the advent of computers. Paradoxically, some of the best techniques available are still impractical because they require too much computation even today!

11.3.1 Single Linkage Clustering

Single linkage clustering technique is the simplest form of hierarchical, agglomerative cluster analysis. It has been called the *nearest neighbor* method. We will use the data in Table 11.5 to illustrate the calculations involved in cluster analysis.

Begin (as in all cluster analysis of an agglomerative type) with a matrix of similarity (or dissimilarity) coefficients. Table 11.6 gives the similarity matrix for the seabird data in Table 11.5 and uses the complement of the Canberra metric as the similarity measure. Given the matrix in Table 11.6, the rules for single linkage clustering are as follows:

1. To start, find the most similar pair(s) of samples—this is defined to be the first cluster.
2. Next, find the second most similar pair(s) of samples *or* the highest similarity between a sample and the first cluster, whichever is greater.

   Definitions: For single linkage clustering,

$$\left\{\begin{matrix}\text{Similarity between a sample} \\ \text{and an existing cluster}\end{matrix}\right\} = \left\{\begin{matrix}\text{Similarity between the sample and the} \\ \textit{nearest} \text{ member of that cluster}\end{matrix}\right\}$$

$$\left\{\begin{matrix}\text{Similarity between two} \\ \text{existing clusters}\end{matrix}\right\} = \left\{\begin{matrix}\text{Similarity between the two } \textit{nearest} \\ \text{members of the clusters}\end{matrix}\right\}$$

TABLE 11.5 RELATIVE ABUNDANCES (PROPORTIONS) OF 23 SPECIES OF SEABIRDS ON 9 COLONIES IN NORTHERN POLAR AND SUBPOLAR AREAS

	Cape Hay, Bylot Island	Prince Leopold Island, eastern Canada	Coburg Island, eastern Canada	Norton Sound, Bering Sea	Cape Lisburne, Chukchi Sea	Cape Thompson, Chukchi Sea	Skomer Island, Irish Sea	St. Paul Island, Bering Sea	St. George Island, Bering Sea
Northern fulmar	0	.3422	0	0	0	0	.0007	.0028	.0278
Glaucous-winged gull	.0005	.0011	.0004	.0051	.0004	.0007	0	0	0
Black-legged kittiwake	.1249	.1600	.1577	.1402	.1972	.0634	.0151	.1221	.0286
Red-legged kittiwake	0	0	0	0	0	0	0	.0087	.0873
Thick-billed murre	.8740	.4746	.8413	.0074	.2367	.5592	0	.4334	.5955
Common murre	0	0	0	.7765	.5522	.3728	.0160	.1537	.0754
Black guillemot	.0006	.0220	.0005	0	.0013	.00001	0	0	0
Pigeon guillemot	0	0	0	0	0	.00003	0	0	0
Horned puffin	0	0	0	.0592	.0114	.0036	0	.0173	.0111
Tufted puffin	0	0	0	.0008	.0002	0	0	.0039	.0024
Atlantic puffin	0	0	0	0	0	0	.0482	0	0
Pelagic cormorant	0	0	0	.0096	.0006	.0001	.0001	0	0
Red-faced cormorant	0	0	0	0	0	0	0	.0099	.0020
Shag	0	0	0	0	0	0	.0001	0	0
Parakeet auklet	0	0	0	.0012	0	0	0	.1340	.0595
Crested auklet	0	0	0	0	0	0	0	.0236	.0111
Least auklet	0	0	0	0	0	0	0	.0906	.0992
Razorbill	0	0	0	0	0	0	.0130	0	0
Manx shearwater	0	0	0	0	0	0	.7838	0	0
Storm petrel	0	0	0	0	0	0	.0389	0	0
Herring gull	0	0	0	0	0	0	.0229	0	0
Great black-backed gull	0	0	0	0	0	0	.0001	0	0
Lesser black-backed gull	0	0	0	0	0	0	.0603	0	0

Source: Data from Hunt et al. 1986.

TABLE 11.6 MATRIX OF SIMILARITY COEFFICIENTS FOR THE SEABIRD DATA IN TABLE 11.5. ISLANDS ARE PRESENTED IN SAME ORDER AS IN TABLE 11.5[a]

	CH	PLI	CI	NS	CL	CT	SI	SPI	SGI
CH	1.0	0.88	0.99	0.66	0.77	0.75	0.36	0.51	0.49
PLI		1.0	0.88	0.62	0.70	0.71	0.36	0.51	0.49
CI			1.0	0.66	0.78	0.75	0.36	0.50	0.48
NS				1.0	0.73	0.64	0.28	0.53	0.50
CL					1.0	0.76	0.29	0.51	0.49
CT						1.0	0.34	0.46	0.45
SI							1.0	0.19	0.20
SPI								1.0	0.80
SGI									1.0

[a]The complement of the Canberra metric $(1.0 - C)$ is used as the index of similarity. Note that the matrix is symmetrical about the diagonal.

3. Repeat the cycle specified in (2) until all the samples are in one big cluster. Box 11.4 and Figure 11.6 illustrate the application of these rules to the data in Table 11.6. The advantage of single linkage clustering is that it is simple to calculate. Its major disadvantage is that one inaccurate sample may compromise the entire clustering process.

11.3.2 Complete Linkage Clustering

This technique has been called *farthest neighbor* clustering. It is conceptually the exact opposite of single linkage clustering, although it proceeds in the same general way, with the exception of the definitions of similarity.

Definitions: For complete linkage clustering,

$$\left\{\begin{array}{c}\text{Similarity between a sample}\\\text{and an existing cluster}\end{array}\right\} = \left\{\begin{array}{c}\text{Similarity between the sample and the}\\\textit{farthest}\text{ member of that cluster}\end{array}\right\}$$

$$\left\{\begin{array}{c}\text{Similarity between two}\\\text{existing clusters}\end{array}\right\} = \left\{\begin{array}{c}\text{Similarity between the two }\textit{farthest}\\\text{members of the clusters}\end{array}\right\}$$

One of the possible difficulties of single linkage clustering is that it tends to produce long, strung-out clusters. Complete linkage clustering often tends to the opposite extreme, producing very tight, compact clusters. Like single linkage clustering, complete linkage clustering is very easy to compute. Because neither of these two extremes is usually desirable, most researchers using cluster analysis have suggested modifications of single and complete linkage clustering to produce intermediate results.

11.3.3 Average Linkage Clustering

The techniques of average linkage clustering were developed to avoid the extremes introduced by single linkage and complete linkage clustering. All types of average linkage clustering require additional computation at each step in the clustering process, so they are normally done with a computer (Romesburg 1984). In order to compute the average similarity between a sample and an existing cluster, we must define more precisely the types of averages to be used.

Box 11.4 Single Linkage Clustering of the Data in Tables 11.5 and 11.6 on Seabird Communities

1. From tables 11.5 and 11.6, we can see that the most similar pair of communities is *Cape Hay* and *Coburg Island*, and they join to form cluster 1 at similarity 0.99.
2. The next most similar community is *Prince Leopold Island*, which is similar to *Cape Hay* and *Coburg Island*. From the definition,

$$\left\{\begin{array}{c}\text{Similarity between a sample}\\ \text{and an existing cluster}\end{array}\right\} = \left\{\begin{array}{c}\text{Similarity between the sample}\\ \text{and the } \textit{nearest} \text{ member of}\\ \text{the cluster}\end{array}\right\}$$

this occurs at similarity 0.88, and we have now a single cluster containing three communities: *Prince Leopold Island*, *Cape Hay*, and *Coburg Island*.
3. The next most similar pair is *St. Paul* and *St. George Islands*, and they join to form a second cluster at similarity 0.80.
4. The next most similar community is *Cape Lisburne*, which joins the first cluster at similarity 0.78 (the similarity between Cape Lisburne and Coburg Island). The first cluster now has four islands in it.
5. The next most similar community is *Cape Thompson*, which joins this large cluster at similarity 0.76 because this is the Cape Thompson-Cape Lisburne similarity. This cluster now has five communities in it.
6. *Norton Sound* joins this large cluster next because it has similarity 0.66 with both Coburg Island and Cape Hay.
7. The two clusters (CH, PLI, CI, NS, CL, CT) and (SPI, SGI) now join together at similarity 0.53, the similarity between the closest two members of these clusters (St. Paul Island and Norton Sound). This large cluster now has eight communities in it.
8. Finally, the last joining occurs between Skomer Island and this large cluster at similarity 0.36, and all nine bird communities are now combined.

Figure 11.6 shows the tree summarizing this cluster analysis.

The most frequently used clustering strategy is called by the impressive name *unweighted pair-group method using arithmetic averages*, abbreviated UPGMA (Sneath and Sokal 1973; Romesburg 1984). This clustering strategy proceeds exactly as before, with the single exception of the definition:

Definitions: For arithmetic average clustering by the unweighted pair-group method,

$$\left\{\begin{array}{c}\text{Similarity between a sample}\\ \text{and an existing cluster}\end{array}\right\} = \left\{\begin{array}{c}\text{Arithmetic mean of similiarities}\\ \text{between the sample and all}\\ \text{the members of the cluster}\end{array}\right\}$$

$$S_{J(K)} = \frac{1}{t_J t_K}\left(\sum S_{JK}\right) \tag{11.17}$$

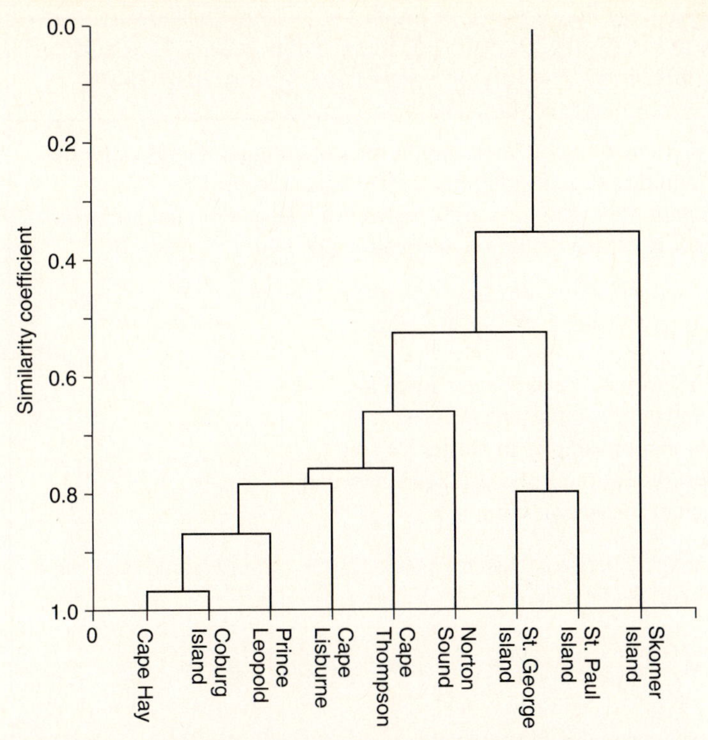

Figure 11.6 Tree diagram resulting from a single linkage cluster analysis of the seabird community data given in Table 11.5 and analyzed in Box 11.4.

where $S_{J(K)}$ = Similarity between clusters J and K
t_J = Number of samples in cluster J (≥ 1)
t_K = Number of samples in cluster K (≥ 2)

The same formula applies to dissimilarity coefficients, such as Euclidian distances.

Box 11.5 and Figure 11.7 illustrate the calculations for average linkage clustering by the UPGMA method. Normally, all these tedious calculations are done by computer. There are many different clustering programs available. Davies (1971) gives a simple program, and Romesburg (1984) describes a more complex series of programs. Because of the use of cluster analysis in both taxonomy and plant community analysis, specialized programs of cluster analysis are readily available in these areas (Orloci 1978; Feoli and Orloci 1991; Rohlf 1995). One of the most useful of these programs is the Numerical Taxonomy System of computer programs (NTSYS), available for IBM PCs from F. J. Rohlf (Exeter Software, Setauket, New York).

Several additional methods of cluster analysis are available, and I have only scratched the surface of a complex technique in this chapter. It is encouraging that Romesburg (1984), after a detailed analysis of various methods of clustering, comes out recommending the UPGMA method for most types of cluster applications. Cluster analysis should be used to increase our ecological insight and not to baffle the reader, and for this reason simpler methods are often preferable to very complex ones.

Box 11.5 Average Linkage Clustering of the Data in Tables 11.5 and 11.6 Using the Unweighted Pair-Group Method (UPGMA)

1. From the data in Table 11.6, the most similar pair of communities is *Cape Hay* and *Coburg Island*, and they join at similarity 0.99 to make cluster 1.

2. We now recompute the entire similarity matrix for the seven remaining communities and cluster 1, using the definition in equation (11.17):

$$S_{J(K)} = \frac{1}{t_J t_K}\left(\sum S_{JK}\right)$$

where $S_{J(K)}$ = Similarity between clusters J and K

t_J = Number of samples in cluster J

t_K = Number of samples in cluster K

S_{JK} = Observed similarity coefficients between each of the samples in J and K

For example, the similarity between cluster J (Cape Hay + Coburg Island) and cluster K (St. George Island) is given by

$$S_{J(K)} = \frac{1}{t_J t_K}\left(\sum S_{JK}\right) = \frac{1}{2(1)}(0.49 + 0.48) = 0.485$$

The largest similarity value in this recomputed matrix is that between Prince Leopold Island and cluster 1:

$$S_{J(K)} = \frac{1}{(1)(2)}(0.88 + 0.88)$$

$$= 0.88$$

3. We recompute the entire similarity matrix for the seven groups. The next largest similarity coefficient is that between *St. Paul Island* and *St. George Island* at similarity 0.80, forming cluster 2. We now have two clusters and four remaining individual community samples.

4. We recompute the similarity matrix for the six groups, and the next largest similarity coefficient is for *Cape Lisburne* and *Cape Thompson*, which join at similarity 0.76, forming cluster 3. We now have three clusters and two remaining individual community samples.

5. We again recompute the similarity matrix for the five groups, and the next largest similarity coefficient is for cluster 3 (CL, CT) and cluster 1 (CH, CI, and PLI):

$$S_{J(K)} = \frac{1}{(2)(3)}(0.77 + 0.78 + 0.70 + 0.75 + 0.75 + 0.71)$$

$$= 0.74$$

so cluster 1 now has five members formed at similarity 0.74.

6. We again recompute the similarity matrix for the four groups, and the largest similarity coefficient is for *Norton Sound* and cluster 1 (CH, CI, PLI, CL, CT):

$$S_{J(K)} = \frac{1}{(1)(5)}(0.66 + 0.62 + 0.66 + 0.73 + 0.64)$$
$$= 0.66$$

so cluster 1 now has six members.

7. We recompute the similarity matrix for the three groups from equation (11.17) and obtain

	Cluster 1	Skomer Island	Cluster 2
Cluster 1	1.0	0.33	0.49
Skomer Island	—	1.0	0.19
Cluster 2	—	—	1.0

For example, the similarity between cluster 1 and cluster 2 is

$$S_{J(K)} = \frac{1}{(6)(2)}(0.51 + 0.51 + 0.50 + 0.53 + 0.51 + 0.46 + 0.49 + 0.49 + 0.48 + 0.50 + 0.49 + 0.45)$$
$$= 0.493$$

Thus, clusters 1 and 2 are joined at similarity 0.49. We now have two groups: Skomer Island and all the rest in one big cluster.

8. The last step is to compute the average similarity between the remaining two groups:

$$S_{J(K)} = \frac{1}{(1)(8)}(0.36 + 0.36 + 0.36 + 0.28 + 0.29 + 0.34 + 0.19 + 0.20)$$
$$= 0.297$$

so the final clustering is at similarity 0.3.

The clustering tree for this cluster analysis is shown in Figure 11.7. It is very similar to that shown in Figure 11.6 for single linkage clustering.

11.4 RECOMMENDATIONS FOR CLASSIFICATIONS

You should begin your search for a classification with a clear statement of your research goals. If a classification is very important as a step to achieving these goals, you should certainly read a more comprehensive book on cluster analysis, such as Romesburg (1984), Legendre and Legendre (1983), or Clifford and Stephenson (1975).

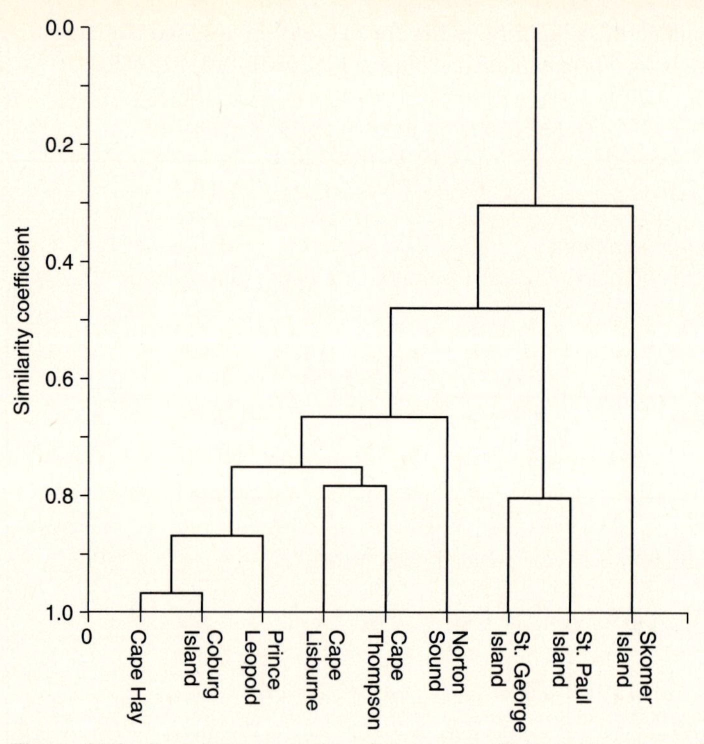

Figure 11.7 Tree diagram resulting from average linkage clustering using the unweighted pair-group method (UPGMA) on the seabird community data given in Table 11.5 and analyzed in Box 11.5.

You must first decide on the type of similarity measure you wish to use. Measures of similarity based on binary data are adequate for some classification purposes but are much weaker than quantitative measures. Of all the similarity measures available, Morisita's index of similarity is clearly to be preferred because it is not dependent on sample size (Wolda 1981). Most ecologists seem to believe that the choice of a similarity measure is not a critical decision, but it certainly is, particularly when we realize that a joint decision about data standardization or transformation and the index to be used can greatly affect the resulting cluster analysis. If data are transformed with a log transformation, Wolda (1981) suggests using the simplified Morisita index (equation 11.15) or the percentage similarity index (equation 11.11).

On top of all these decisions, the choice of a clustering algorithm can also affect the resulting tree. Romesburg (1984, 110–114) discusses an interesting taxonomic cluster analysis using bone measurements from several species of hominoids. Each similarity coefficient produced a different taxonomic tree, and the problem of which one was closer to the true phylogenetic tree was not immediately clear without independent data. *The critical point is that, given a set of data, there is no single, objective, "correct" cluster analysis.* If you are to evaluate these different cluster analyses, it must be done with additional data, or ecological insight. This is and must remain the central paradox of clustering methods—that each method is exact and objective once the subjective decisions have been made about the similarity index and data standardization.

Finally, virtually none of the similarity measures has a statistical probability distribution; hence, you cannot readily set confidence intervals on these estimates of similarity. It is therefore not possible to assess probable error without taking replicate community samples. There is no general theory to guide you in the sample size you require from each community. Wolda (1981) suggests that more than 100 individuals are always needed before it is useful to calculate a similarity index (unless the species diversity of the community is very low). A reasonable community sample would probably be 200–500 individuals for low-diversity communities, and 10 times the number of species for high-diversity communities. These are only rule-of-thumb guesses, and a rigorous statistical analysis of sampling for similarity is waiting to be done.

11.5 OTHER MULTIVARIATE TECHNIQUES

Classification by means of cluster analysis is by no means the only way to analyze community data. Plant ecologists have developed a series of multivariate techniques that are useful for searching for patterns in community data. Three basic strategies have been developed to analyze community data.

11.5.1 Direct Gradient Analysis

Direct gradient analysis is used to study the distribution of species along environmental gradients like depth in a lake or soil moisture in a forest. One simple type of direct gradient analysis is to measure the altitudinal distribution of trees up a mountainside (see Figure 11.8). Gradient analysis is the simplest type of multivariate community analysis. Whittaker (1967) and Gauch (1982) provide a detailed summary of this approach.

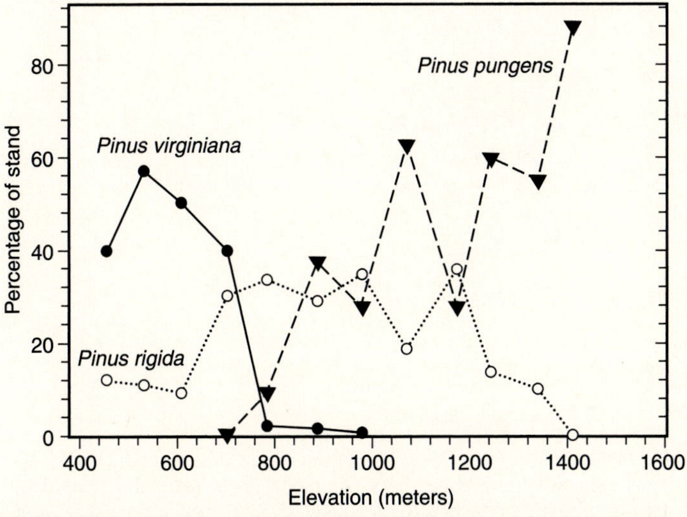

Figure 11.8 Example of direct gradient analysis. A transect of the elevation gradient along dry, south-facing slopes in the Great Smoky Mountains of Tennessee. *Pinus virginiana* dominates the forest at low elevations, *Pinus rigida* at middle elevations, and *Pinus pungens* at high elevations. (Data from Whittaker 1956, Table 7.)

11.5.2 Ordination

Ordination is a method for arranging species and samples along 1–3 dimensions such that similar species or samples are close together, and dissimilar species or samples are far apart. Ordination summarizes community data of many species and many samples by collapsing it on to a single graph that summarizes the patterns in the data, as shown in Figure 11.9. Since ordination scores all community data on a continuous scale, it does not produce a classification of community types. Ordination is useful for recognizing the pattern present in community data. It must then be combined with environmental information and classification techniques to gain a more complete description and understanding of the community.

Many ordination methods are available, and considerable controversy exists over which technique is best (Gauch 1982; Digby and Kempton 1987). All ordination techniques are computer-intensive and many ordination programs are available (Orloci 1978, Gauch 1982). Good discussions of ordination techniques can be found in several books (Orloci 1978, Gauch 1982, Pielou 1984, Digby and Kempton 1987).

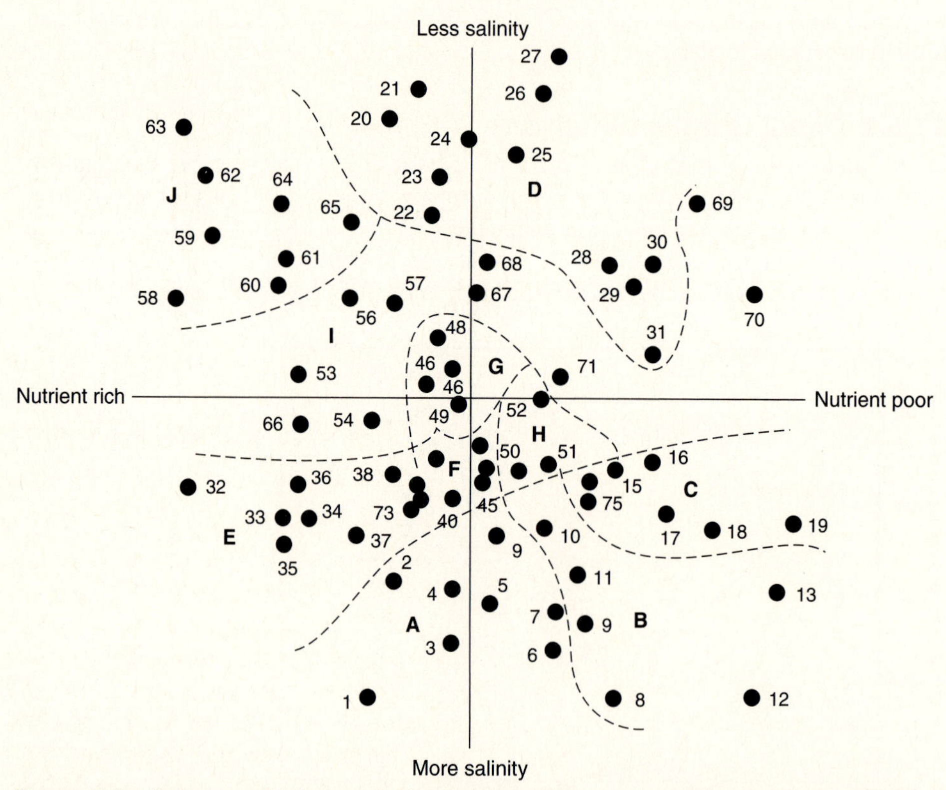

Figure 11.9 Ordination of 76 plant species from sea cliffs on the island of Anglesey, Wales, by principal components analysis. The species are arranged in an ecological space defined by two axes. The vertical axis is associated with salinity and the horizontal axis with soil fertility. Species cluster into groups separated by the dashed lines, defined by natural history observations. For example, species in group A (bottom center) are characteristic of bird colonies, those in group D (top center) are characteristic of wet, acid heath. (Modified from Goldsmith 1973.)

11.5.3 Classification

Classification is often the final goal of community analyses, so that ecologists can assign names to classes or groups. Classification is especially important in applied ecology and conservation. Ecologists have classified plant communities on the basis of many different characteristics, and since the advent of computers, there has been a growing literature on objective, quantitative methods of classification. Cluster analysis has been one of the most common methods of classification used on community data. Gauch (1982) provides a good discussion of alternative methods of classification developed by plant ecologists.

I have chosen not to discuss the detailed methods of gradient analysis and ordination, not because they are not important but because excellent books already exist that describe these complex multivariate methods. Gradient analysis and ordination have nearly become separate disciplines within ecology, and you should consult Digby and Kempton (1987), Pielou (1984), or Gauch (1982) for detailed discussion of these important methods.

11.6 SUMMARY

Communities may be more or less similar, and ecologists often wish to express this similarity quantitatively and to classify communities on the basis of this similarity. Similarity measures may be *binary*, based on presence/absence data, or *quantitative*, based on some measure of importance like population size, biomass, cover, or productivity. There are more than two dozen similarity measures, and I describe four binary coefficients and eight quantitative measures that are commonly used. Some measures emphasize the common species in the community, others emphasize the rare species. Many of the commonly used measures of similarity are strongly dependent on sample size, and should be avoided if possible. Morisita's index of similarity is recommended for quantitative data because it is not affected by sample size.

Cluster analysis is a method for generating classifications from a series of community samples. Many different types of cluster analysis have been developed, and there is no single, "correct," or ideal system. Most ecological data can be classified simply by average linkage clustering (UPGMA), and this technique is recommended for general usage.

Data for cluster analysis may be input as raw numbers, transformed by square root or logarithmic transformations, or expressed as proportions (relative abundance). Decisions about the type of data to be used, the similarity index, and the clustering algorithm should be made before any analysis is done on the basis of the research objectives you wish to achieve. Cluster analysis and the measurement of ecological similarity are two parts art and one part science, and ecological intuition is essential to success.

SELECTED READING

Baroni-Urbani, C., and Buser, M. W. 1976. Similarity of binary data. *Systematic Zoology* 25: 251–259.

Field, J. G. 1970. The use of numerical methods to determine benthic distribution patterns from dredgings in False Bay. *Transactions of the Royal Society of South Africa* 39: 183–200.

Gower, J. C., and P. Legendre. 1986. Metric and Euclidean properties of dissimilarity coefficients. *Journal of Classification* 3: 5–48.

Huhta, V. 1979. Evaluation of different similarity indices as measures of succession in arthropod communities of the forest floor after clear-cutting. *Oecologia* 41: 11–23.

Jackson, D. A., Somers, K. M., and Harvey, H. H. 1989. Similarity coefficients: measures of co-occurrence and association or simply measures of occurrence? *American Naturalist* 133: 436–453.

Legendre, L., and Legendre, P. 1983. *Numerical Ecology*. Chapter 6. Measures of ecological resemblance, 171–218. Elsevier, New York.

Morisita, M. 1959. Measuring of interspecific association and similarity between communities. *Memoirs of the Faculty of Science, Kyushu University, Series E (Biology)* 3: 65–80.

Pielou, E. C. 1984. *The Interpretation of Ecological Data: A Primer on Classification and Ordination*. John Wiley and Sons, New York.

Romesburg, H. C. 1984. *Cluster Analysis for Researchers*. Lifetime Learning Publications, Belmont, California.

Stephenson, W., Williams, W. T., and Cook, S. 1972. Computer analyses of Petersen's original data on bottom communities. *Ecological Monographs* 42: 387–415.

Wolda, H. 1981. Similarity indices, sample size and diversity. *Oecologia* 50: 296–302.

QUESTIONS AND PROBLEMS

11.1. A Christmas bird count in three areas of Manitoba in 1985 produced the following data:

	No. of individuals counted		
	Brandon	Delta Marsh	Winnipeg
Canada goose	0	0	2
Mallard	0	0	5
Northern goshawk	0	1	2
Golden eagle	0	1	0
American kestrel	0	0	3
Merlin	2	0	1
Gray partridge	29	45	112
Ruffed grouse	4	0	0
Sharp-tailed grouse	48	11	1
Rock dove	907	8	6,179
Mourning dove	0	0	2
Great horned owl	6	1	5
Snowy owl	3	1	2
Barred owl	0	0	1
Black-backed woodpecker	0	[a]	0
Downy woodpecker	12	4	79
Hairy woodpecker	13	8	51
Horned lark	0	1	0
Bluejay	29	18	99
Black-billed magpie	89	28	31
American crow	0	3	2
Common raven	2	1	0
Blackcapped chickadee	134	26	595
Red-breasted nuthatch	3	0	7

	No. of individuals counted		
	Brandon	Delta Marsh	Winnipeg
White-breasted nuthatch	36	11	169
Brown creeper	0	0	1
Golden crowned kinglet	2	0	0
American robin	4	0	2
Varied thrush	1	0	0
Bohemian waxwing	30	45	192
Cedar waxwing	35	0	0
Northern shrike	2	4	3
European starling	55	52	982
White-throated sparrow	0	0	1
Dark-eyed junco	4	5	11
Lapland longspur	0	3	63
Snow bunting	2	4,019	68
Red-winged blackbird	0	1	2
Rusty blackbird	0	0	5
Brewer's blackbird	0	7	0
Common grackle	[a]	0	1
Pine grosbeak	150	48	701
Purple finch	[a]	0	2
Red crossbill	0	[a]	0
White-winged crossbill	1	0	0
Common redpoll	499	1,191	859
Hoary redpoll	5	35	16
Pine siskin	0	2	14
American goldfinch	0	2	3
Evening grosbeak	136	46	30
House sparrow	3,024	855	11,243
Total individuals	5,267	6,483	21,547
Total species	30	31	40
Total species in Manitoba = 61			

[a] Species known to be in the area but not seen on the day of the count.

The amount of effort expended in these counts cannot be assumed to be equal in the three areas.

(a) Choose a binary coefficient that you think should be used for data of this type and calculate the similarity between these three winter bird communities.

(b) Discuss what type of data standardization should be done before a quantitative similarity index is calculated.

(c) Calculate the value of the most appropriate quantitative index of similarity for these three bird communities.

11.2. Calculate the Sorensen coefficient for the data in Table 11.1 and also the simplified Morisita index. Why do these two measures differ in value?

11.3. Compare and contrast the evaluations of similarity indices by Huhta (1979) and Wolda (1981).

11.4. Species absent from community samples have usually been ignored in calculating binary similarity coefficients for ecological data. Baroni-Urbani and Buser (1976) argue forcefully that this is a mistake. Evaluate their arguments in favor of including the absent species in the calculations. Compare their resolution of this problem with that of Faith (1983).

11.5. The species composition of the groundfish community off Nova Scotia was measured by Mahon et al. (1984), who obtained these data for two parts of the Bay of Fundy:

Species	Average no. of individuals per tow	Percent
	West Fundy area	
Redfish	12.7	14.1
White hake	11.3	12.6
Haddock	10.1	11.3
Cod	9.8	10.9
Spiny dogfish	9.3	10.4
Thorny skate	5.8	6.5
Plaice	4.9	5.5
Witch	4.5	5.0
Longhorn sculpin	4.2	4.7
Pollock	2.6	2.9
Sea raven	2.0	2.9
Silver hake	2.6	2.9
Winter flounder	2.2	2.5
Ocean pout	2.0	2.2
Smooth skate	1.8	2.0
Little skate	1.1	1.3
Winter skate	0.6	0.7
Cusk	0.5	0.6
Angler	0.3	0.3
Wolffish	0.2	0.3
	South Fundy area	
Haddock	125.1	58.4
Cod	18.6	8.7
Longhorn sculpin	17.9	8.3
Winter flounder	15.0	7.0
Spiny dogfish	7.6	3.6
Sea raven	5.9	2.8
Pollock	2.7	1.2
Thorny skate	3.3	1.5
White skate	2.4	1.1
Redfish	2.4	1.1
Ocean pout	1.9	0.9
Plaice	2.0	0.9
Witch	1.6	0.8
Winter skate	1.3	0.6
Yellowtail	1.3	0.6
Wolffish	1.0	0.5

Discuss how best to measure similarity among the groundfish communities of these two regions.

11.6. The following data were obtained on the proportions of different shrubs available for winter feeding by snowshoe hares on nine areas of the southwestern Yukon:

Plot name	*Salix glauca*	*Picea glauca*	*Betula glandulosa*	*Shepherdia canadensis*	Other
Silver Creek	0.34	0.55	0.00	0.00	0.10
Beaver Pond	0.49	0.09	0.42	0.00	0.10
Kloo Lake	0.63	0.31	0.02	0.03	0.00
1050	0.57	0.08	0.35	0.00	0.00
Microwave Food	0.21	0.00	0.79	0.00	0.00
Jacquot Control	0.14	0.73	0.00	0.10	0.03
Jacquot Food	0.50	0.33	0.00	0.02	0.14
Gribble's control	0.26	0.38	0.29	0.01	0.06
Dezadeash Food	0.66	0.22	0.00	0.08	0.05

Calculate a cluster analysis of these nine areas and plot the tree showing which areas are most similar.

CHAPTER 12

Species Diversity Measures

A biological community has an attribute that we can call *species diversity*, and many different ways have been suggested for measuring this concept. Recent interest in conservation biology has generated a strong focus on how to measure biodiversity in plant and animal communities. Different authors have used different indices to measure species diversity, and the whole subject area has become confused with poor terminology and an array of possible measures. Peet (1974) and Washington (1984) have reviewed the problem, and Magurran (1988) has provided a unified framework for the study of species diversity. Note that *biodiversity* has a broader meaning than species diversity: it includes both *genetic diversity* and *ecosystem diversity*. Nevertheless species diversity is a large part of the focus of biodiversity at the local and regional scale, and we will concentrate here on how to measure species diversity.

12.1 BACKGROUND PROBLEMS

One must make a whole series of background assumptions in order to measure species diversity for a community. Ecologists tend to ignore most of these difficulties, but this is untenable if we are to achieve a coherent theory of diversity.

The first assumption is that the subject matter is well defined. Measurement of diversity first of all requires a clear taxonomic classification of the subject matter. In most cases ecologists worry about *species diversity*, but there is no reason why *generic* diversity or *subspecific* diversity could not be analyzed as well. Within the classification system, all the individuals assigned to a particular class are assumed to be identical. This can cause problems. For example, if males are smaller than females, should they be grouped together or kept as two groups? Should larval stages count the same as an adult stage? This sort of variation is usually ignored in species diversity studies.

Most measures of diversity assume that the classes (species) are all equally different. There seems to be no easy way around this limitation. In an ecological sense sibling species may be very similar functionally while more distantly related species may play other functional roles. Measures of diversity cannot address these kinds of functional differences among species.

Diversity measures require an estimate of species importance in the community. The simple choices are numbers, biomass, cover, or productivity. The decision depends in part on the question being asked, and—as in all questions about methods in ecology—you should begin by asking yourself what the problem is and what hypotheses you are trying to test. Numbers are used in most cases as a measure of species importance, although Dickman (1968) found that with lake samples of plankton the best measure was productivity.

A related question is, *How much of the community should we include in our sampling?* We must define precisely the collection of species we are trying to describe. Most authors pick one segment—bird species diversity or tree species diversity—and in doing so ignore soil nematode diversity and bacterial diversity. Rarely do diversity measures cross trophic levels, and only rarely are they applied to whole communities. Colwell (1979) argues convincingly that ecologists should concentrate their analyses on parts of the community that are functionally interacting, the "guilds" described by Root (1973). These guilds often cross trophic levels and include taxonomically unrelated species. The choice of what to include in a community is critical to achieving ecological understanding, yet no rules are available to help you make this decision. The functionally interacting networks can be determined only by detailed natural history studies of the species in a community.

12.2 CONCEPTS OF SPECIES DIVERSITY

Early naturalists very quickly observed that tropical areas contained more species of plants and animals than did temperate areas. But as ecological ideas matured and ideas of quantitative measurement were introduced, it became clear that the idea of species diversity involved two quite distinct concepts.

12.2.1 Species Richness

The oldest and the simplest concept of species diversity is *species richness:* the number of species in the community. McIntosh (1967) coined the name *species richness* to describe this concept. The basic measurement problem is that it is often impossible to enumerate all of the species in a natural community.

12.2.2 Heterogeneity

If a community has 10 equally abundant species, should it have the same diversity as another community with 10 species, one of which comprises 99% of the total individuals? No, answered Simpson (1949), who proposed a second concept of diversity that combines two separate ideas: species richness and evenness. In a forest with 10 equally abundant tree species, two trees picked at random are likely to be different species. But in a forest with 10 species, one of which is dominant and contains 99% of all the individuals, two trees picked at random are unlikely to be different species. Figure 12.1 illustrates this concept.

The term *heterogeneity* was first applied to this concept by Good (1953), and for many ecologists this concept is synonymous with *diversity* (Hurlbert 1971). Heterogeneity is a popular concept in ecology partly because it is relatively easily measured.

12.2.3 Evenness

Since heterogeneity contains two separate ideas—species richness and evenness—it was only natural to try to measure the evenness component separately. Lloyd and Ghelardi (1964) were the first to suggest this concept. For many decades field ecologists had known that most communities of plants and animals contain a few dominant species and many species that are relatively uncommon. Evenness measures attempt to quantify this unequal representation against a hypothetical community in which all species are equally common. Figure 12.1 illustrates this idea.

12.3 SPECIES RICHNESS MEASURES

Some communities are simple enough to permit a complete count of the number of species present, and this is the oldest and simplest measure of species richness. Complete counts can often be done for bird communities in small habitat blocks, mammal communities, and often for temperate and polar communities of higher plants, reptiles, amphibians, and fish, but it is often impossible to enumerate every species in communities of insects, intertidal invertebrates, soil invertebrates, or tropical plants, fish, or amphibians. How can we measure species richness when we only have a sample of the community's total richness? Three approaches have been used in an attempt to solve this sampling problem.

12.3.1 Rarefaction Method

One problem that frequently arises in comparing community samples is that they are based on different sample sizes. The larger the sample, the greater the expected number of species. If we observe one community with 125 species in a collection of 2200 individuals and a second community with 75 species in a collection of 750 individuals, we do not know immediately which community has higher species richness. One way to overcome this

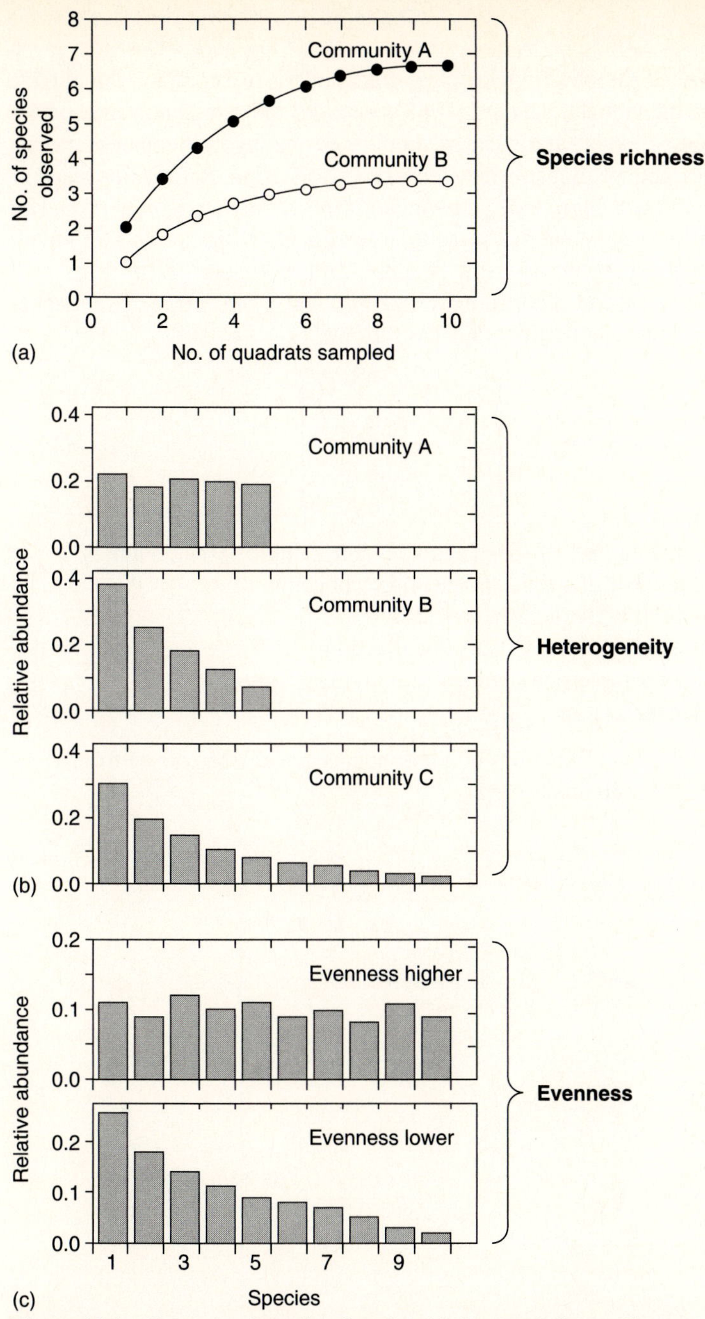

Figure 12.1 Concepts of species diversity. (a) *Species richness*: Community A has more species than community B and thus higher species richness. (b) *Heterogeneity*: Community A has the same number of species as community B but the relative abundances are more even, so by a heterogeneity measure A is more diverse than B. Community C has the same abundance pattern as B but has more species, so it is more diverse than B. (c) *Evenness*: When all species have equal abundances in the community, evenness is maximal.

problem is to standardize all samples from different communities to a common sample size of the same number of individuals. Sanders (1968) proposed the rarefaction method for achieving this goal. Rarefaction is a statistical method for estimating the number of species expected in a random sample of individuals taken from a collection. Rarefaction answers this question: *If the sample had consisted of* n *individuals* ($n < N$), *what number of species* (s) *would likely have been seen*? Note that if the total sample has S species and N individuals, the rarefied sample must always have $n < N$ and $s < S$ (see Figure 12.2).

Sanders's (1968) rarefaction algorithm was wrong, and it was corrected independently by Hurlbert (1971) and Simberloff (1972) as follows:

$$E(\hat{S}_n) = \sum_{i=1}^{s} \left[1 - \frac{\binom{N - N_i}{n}}{\binom{N}{n}} \right] \tag{12.1}$$

where $E(\hat{S}_n)$ = Expected number of species in a random sample of n individuals
S = Total number of species in the entire collection
N_i = Number of individuals in species i
N = Total number of individuals in collection = ΣN_i
n = Value of sample size (number of individuals) chosen for standardization ($n \leq N$)
$\binom{N}{n}$ = Number of combinations of n individuals that can be chosen from a set of N individuals
= $N!/n!(N - n)!$

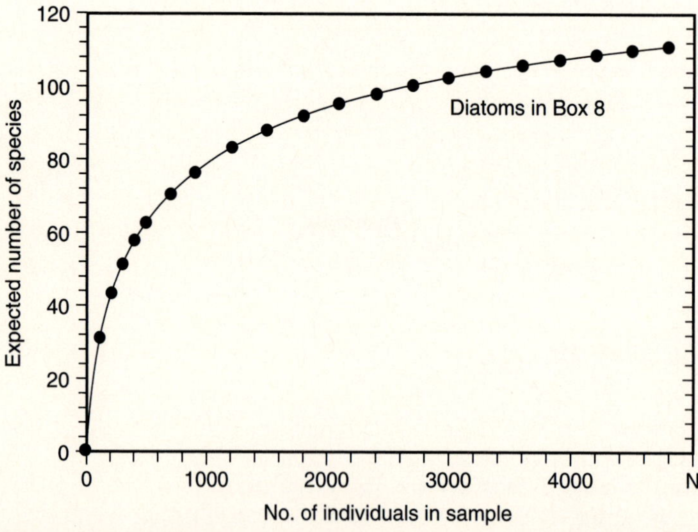

Figure 12.2 Rarefaction curve for the diatom community data from Patrick (1968). There were 4874 individuals in 112 species in this sample (Box 8). Original data appear in Table 12.1. If a sample of 2000 individuals were taken, we would expect to find only 94 species, for example. This illustrates the general principle that the larger the sample size of individuals, the more species we expect to enumerate.

The large-sample variance of this estimate was given by Heck et al. (1975) as

$$\operatorname{var}(\hat{S}_n) = \binom{N}{n}^{-1} \left[\begin{array}{l} \displaystyle\sum_{i=1}^{s} \binom{N-N_i}{n} \left[1 - \frac{\binom{N-N_i}{n}}{\binom{N}{n}} \right] + \\ \displaystyle 2 \sum_{i=1}^{s-1} \sum_{j=i+1}^{s} \left[\binom{N-N_i-N_i}{n} - \frac{\binom{N-N_i}{n}\binom{N-N_j}{n}}{\binom{N}{n}} \right] \end{array} \right] \tag{12.2}$$

where $\operatorname{var}(\hat{S}_n)$ = Variance of the expected number of species in a random sample of n individuals

and all other terms are as defined above.

Box 12.1 illustrates the calculation of the rarefaction method for some rodent data. Because these calculations are so tedious, a computer program should normally be used for the rarefaction method. Program-group SPECIES DIVERSITY (Appendix 2) can do these calculations. It contains a modified version of the program given by Simberloff (1978).

There are important ecological restrictions on the use of the rarefaction method. Since rarefaction is not concerned with species names, the communities to be compared by rarefaction should be taxonomically similar. As Simberloff (1979) points out, if community A has the larger sample primarily of butterflies and community B has the smaller sample mostly of moths, no calculations are necessary to tell you that the smaller sample is not a random sample of the larger set.

Sampling methods must also be similar for two samples to be compared by rarefaction (Sanders 1968). For example, you should not compare insect light trap samples with insect sweep net samples, since whole groups of species may be captured by one technique but not by the other. Most sampling techniques are species-selective, and it is important to standardize collection methods. Sanders (1968) argued that rarefaction should be used only on samples from the same or similar habitats, since different habitats—coniferous forest vs. deciduous forest, for example—will have different species diversities.

Rarefaction curves cannot be extrapolated beyond the number of individuals (N) in the larger sample. The only way one can extrapolate beyond the limits of the samples is by assuming an underlying statistical distribution, like the log-series or the lognormal (see Sections 12.4.1 and 12.4.2).

One assumption that rarefaction does make is that all individuals in the community are randomly dispersed with respect to other individuals of their own or of different species. In practice most distributions are clumped (see Chapter 4, Section 4.2.2) within a species and there may be positive or negative association between species. Fager (1972) used computer simulations to investigate what effect clumping would have on the rarefaction estimates, and observed that the more clumped the populations were, the greater the overestimation of the number of species by the rarefaction method. The only way to reduce this bias in practice is to use large samples spread widely throughout the community being analyzed.

Box 12.1 Expected Number of Species by the Rarefaction Method

A sample of Yukon rodents produced four species in a collection of 42 individuals. The species abundances were 21, 16, 3, and 2 individuals. We wish to calculate the expected species richness for samples of 30 individuals.

Expected Number of Species

From equation (12.1)

$$E(\hat{S}_n) = \sum_{i=1}^{s} \left[1 - \frac{\binom{N - N_i}{n}}{\binom{N}{n}} \right]$$

$$E(\hat{S}_{30}) = \left[1 - \frac{\binom{42 - 21}{30}}{\binom{42}{30}} \right] + \left[1 - \frac{\binom{42 - 16}{30}}{\binom{42}{30}} \right] + \left[1 - \frac{\binom{42 - 3}{30}}{\binom{42}{30}} \right] + \left[1 - \frac{\binom{42 - 2}{30}}{\binom{42}{30}} \right]$$

$$\binom{42 - 21}{30} = \binom{21}{30} = 0 \quad \text{(by definition)}$$

$$\binom{42}{30} = \frac{42!}{30!(42 - 30)!} = 1.1058 \times 10^{10}$$

$$\binom{42 - 16}{30} = \binom{26}{30} = 0 \quad \text{(by definition)}$$

$$\binom{42 - 3}{30} = \frac{39!}{30!(39 - 30)!} = 2.1192 \times 10^{8}$$

$$\binom{42 - 2}{30} = \frac{40!}{30!(40 - 30)!} = 8.4766 \times 10^{8}$$

$$\begin{aligned} E(\hat{S}_{30}) &= 1 + 1 + \left(1 - \frac{2.1192 \times 10^{8}}{1.1058 \times 10^{10}}\right) + \left(1 - \frac{8.4766 \times 10^{8}}{1.1058 \times 10^{10}}\right) \\ &= 1 + 1 + 0.981 + 0.923 \\ &= 3.90 \text{ species} \end{aligned}$$

These calculations are for illustration only, since you would never use this method on such a small number of species.

Large-Sample Variance of the Expected Number of Species

From equation (12.2)

$$\text{var}(\hat{S}_n) = \binom{N}{n}^{-1} \left[\begin{aligned} &\sum_{i=1}^{s} \binom{N - N_i}{n} \left[1 - \frac{\binom{N - N_i}{n}}{\binom{N}{n}} \right] + \\ &2 \sum_{i=1}^{s-1} \sum_{j=i+1}^{s} \left[\binom{N - N_i - N_i}{n} - \frac{\binom{N - N_i}{n}\binom{N - N_j}{n}}{\binom{N}{n}} \right] \end{aligned} \right]$$

$$\operatorname{var}(\hat{S}_{30}) = \binom{42}{30}^{-1} \left\{ \begin{aligned} &\binom{21}{30}\left(1 - \frac{\binom{21}{30}}{\binom{42}{30}}\right) + \binom{26}{30}\left(1 - \frac{\binom{26}{30}}{\binom{42}{30}}\right) \\ &+ \binom{39}{30}\left(1 - \frac{\binom{39}{30}}{\binom{42}{30}}\right) + \binom{40}{30}\left(1 - \frac{\binom{40}{30}}{\binom{42}{30}}\right) \\ &+ 2\left[\left(\binom{42-21-16}{30} - \frac{\binom{42-21}{30}\binom{42-16}{30}}{\binom{42}{30}}\right)\right. \\ &+ \left(\binom{42-21-3}{30} - \frac{\binom{42-21}{30}\binom{42-3}{30}}{\binom{42}{30}}\right) \\ &+ \left(\binom{42-21-3}{30} - \frac{\binom{42-21}{30}\binom{42-2}{30}}{\binom{42}{30}}\right) \\ &+ \left(\binom{42-16-3}{30} - \frac{\binom{42-16}{30}\binom{42-3}{30}}{\binom{42}{30}}\right) \\ &+ \left(\binom{42-16-2}{30} - \frac{\binom{42-16}{30}\binom{42-2}{30}}{\binom{42}{30}}\right) \\ &+ \left.\left(\binom{42-3-2}{30} - \frac{\binom{42-3}{30}\binom{42-2}{30}}{\binom{42}{30}}\right)\right] \end{aligned} \right\}$$

Note that for this particular example almost all of the terms are zero.

$$\begin{aligned} \operatorname{var}(\hat{S}_{30}) &= (1.1058 \times 10^{-10})\left[\begin{aligned} 2.0785 \times 10^8 + 7.8268 \times 10^8 \\ + (2)(-5.9499 \times 10^6) \end{aligned}\right] \\ &= 0.0885 \end{aligned}$$

$$\begin{aligned} \text{Standard deviation of } (\hat{S}_{30}) &= \sqrt{\operatorname{var}(\hat{S}_{30})} \\ &= \sqrt{0.0885} = 0.297 \end{aligned}$$

These tedious calculations can be done by Program-group SPECIES DIVERSITY (see Appendix 2).

TABLE 12.1 TWO SAMPLES OF A DIATOM COMMUNITY OF A SMALL CREEK IN PENNSYLVANIA IN 1965[a]

Species	Number of individuals	
	Box 8	Box 7
Nitzxchia frustulum v. *perminuta*	1446	1570
Synedra parasitica v. *subconstricta*	456	455
Navicula cryptocephala	450	455
Cyclotella stelligera	330	295
Navicula minima	318	305
N. secreta v. *apiculata*	306	206
Nitzschia palea	270	225
N. frustulum	162	325
Navicula luzonensis	132	78
Nitzschia frustulum v. *indica*	126	180
Melosira varians	118	140
Nitzschia amphibia	93	95
Achnanthes lanceolata	75	275
Stephanodiscus hantzschii	74	59
Navicula minima v. *atomoides*	69	245
N. viridula	68	72
Rhoicosphenia curvata v. *minor*	61	121
Navicula minima v. *atomoides*	59	47
N. pelliculsa	54	19
Melosira granulata v. *angustissima*	54	73
Navicula seminulum	52	36
N. gregaria	40	34
Nitzschia capitellata	40	16
Achnanthes subhudsonis v. *kraeuselii*	39	51
A minutissima	35	61
Nitzschia diserta	35	53
Amphora ovalis v. *pediculus*	33	53
Cymbella tumida	29	95
Synedra parasitica	24	42
Cymbella ventricosa	21	27
Navicula paucivisitat	20	12
Nitzschia kutzingiana	19	70
Gomphonema parvulum	18	66
Rhoicosphenia curvata	18	22
Synedra ulna	18	36
Surirella angustata	17	11
Synedra ulna v. *danica*	17	37
Navicula pupula	17	27
Achnanthes biporoma	16	32
Stephanodiscus astraea v. *minutula*	16	21
Navicula germainii	13	19
Denticula elegans	12	4
Gomphonema sphaerophorum	11	40
Synedra rumpens	11	13
S. vaucheriae	11	14
Cocconeis placentula v. *euglypta*	10	5
Navicula menisculus	10	5
Nitzschia linearis	10	18
Melosira italica v. *valida*	6	15
Navicula cryptoocephala v. *veneta*	6	6
Cymbella turgida	5	8
Fragilaria intermedia	5	5
Gomphonema augustatum v. *obesa*	5	16
G. angustatum v. *producta*	5	4
G. ongiceps v. *subclavata*	5	9
Meridion circulare	5	4
Melosira ambigua	5	—
Nitzschia acicularis	5	—
Synedra rumpens v. *familiaris*	5	37
Cyclotella meneghiniana	4	8
Gyrosigma spencerii	4	2
Fragilaria construens v. *venter*	3	—
Gomphonema gracile	3	10
Navicula cincta	3	2
N. gracilis fo. *Minor*	3	—
Navicula decussis	3	2
N. pupula v. *capitat*	3	10
N. symmetrica	3	—
Nitzxchia dissipata v. *media*	3	4
N. tryblionella v. *debilis*	3	1
N. sigmoidea	3	—
Anomoeoneis exilis	2	—
Caloneis hyalina	2	2
Diatoma vulgare	2	—
Eunotia pectinalis v. *minor*	2	1
Fragilaria leptostauron	2	3
Gomphonema constrictum	2	—
G. intricatum v. *pumila*	2	10
Navicula hungarica v. *capitat*	2	5
N. protraccta	2	3
Synedra acus v. *angustissima*	2	—
Bacillaria paradoxa	1	—
Cyclotella kutzingiana	1	—
Cymbella triangulum	1	—
Cocconeis sp.	1	—
Caloneis bacillum	1	3
Fragilaria bicapitat	1	—
Frustulia vularis	1	—
Gomphonema carolinese	1	1
G. sp. [MH IV Ridley]	1	—
Navicula capitat v. *hungarica*	1	1
N. contenta f. *biceps*	1	1
N. cincta v. *rostrata*	1	—
N. americana	1	—
Nitzschia hungarica	1	—
N. sinuata v. *tabularia*	1	—

TABLE 12.1 (Cont.)

Species	Number of individuals		Species	Number of individuals	
	Box 8	Box 7		Box 8	Box 7
Stephanoddiscus invisitatus	10	22	*N. confinis*	1	5
Amphora ovalis	9	16	*Synedra pulchella* v. *lacerata*	1	1
Cymbella sinuata	9	5	*Surirella ovata*	1	3
Gyrosigma wormleyii	9	5	*Achnanthes cleveii*	—	2
Nitzschia fonticola	9	6	*Amphora submontana*	—	1
N. bacata	9	7	*Caloneis silicula* v. *ventricosa*	—	3
Synedra rumpens v. *meneghiniana*	9	17	*Eunotia lunaris*	—	2
Cyclotella meneghiniana small	8	4	*E. tenella*	—	1
Nitzschia gracilis v. *minor*	8	10	*Fragilaria pinnata*	—	3
N. frustulum v. *subsalina*	7	10	*Gyrosigma scalproides*	—	1
N. subtilis	7	16	*Gomphonema sparsistriata*	—	—
Cymbella affinis	6	3	*Meridion circulara* v. *constricta*	—	3
Cocconeis placetula v. *lineata*	6	13	*Navicula tenera*	—	3
			N. omissa	—	1
			N. ventralis	—	1
			N. mutica	—	1
			N. sp. [LL 30]	—	1
			N. mutica v. *cohnii*	—	1
			Nitzschia brevissima	—	1
			N. frequens	—	1

[a] The numbers of individuals settling on glass slides were counted. Data from Patrick (1968).

The variance of the expected number of species (equation [12.2]) is appropriate only with reference to the sample under consideration. If you wish to ask a related question—for example, *Given a sample of* N *individuals from a community, how many species would you expect to find in a second, independent sample of n ($n < N$) individuals?*—Smith and Grassle (1977) give the variance estimate appropriate for this more general question, and have a computer program for generating these variances. Simberloff (1979) showed that the variance given in equation (12.2) provides estimates only slightly smaller than the Smith and Grassle (1977) estimator.

Figure 12.2 illustrates a rarefaction curve for the diatom community data in Table 12.1. James and Rathbun (1981) provide additional examples from bird communities.

12.3.2 Jackknife Estimate

When quadrat sampling is used to sample the community, it is possible to use another nonparametric approach, the jackknife,* to estimate species richness. This estimate is based on the observed frequency of rare species in the community and is obtained as follows (Heltshe and Forrester 1983a). Data from a series of random quadrats are tabulated in the form shown in Table 12.2, recording only the presence (1) or absence (0) of the species in each quadrat. Tally the number of *unique species* in the quadrats sampled. A *unique species* is defined as a species that occurs in one and only one quadrat. Unique species are

*For a general discussion of jackknife estimates, see Chapter 15, Section 15.5.

TABLE 12.2 QUADRAT SAMPLING DATA SUMMARIZED IN A FORM NEEDED FOR THE JACKKNIFE ESTIMATE OF SPECIES RICHNESS[a]

	Quadrat						
Species	A	B	C	D	E	F	Row sum
1	1	0	0	1	1	0	3
2	0	1	0	0	0	0	1
3	1	1	1	1	1	0	5
4	0	1	0	0	1	0	2
5	1	1	1	1	1	1	6
6	0	0	0	0	1	0	1
7	0	0	1	1	1	1	4
8	1	1	0	0	1	1	4

[a]Only presence/absence data are required. *Unique species* are those whose row sums are 1 (species 2 and 6 in this example). 0 = absent; 1 = present.

spatially rare species and are not necessarily numerically rare, since they could be highly clumped. From Heltshe and Forrester (1983a), the jackknife estimate of the number of species is

$$\hat{S} = s + \left(\frac{n-1}{n}\right)k \tag{12.3}$$

where $\hat{S}$ = Jackknife estimate of species richness
s = Observed total number of species present in n quadrats
n = Total number of quadrats samples
k = Number of unique species

The variance of this jackknife estimate of species richness is given by

$$\text{var}(\hat{S}) = \left(\frac{n-1}{n}\right)\left[\sum_{j=1}^{s}(j^2 f_j) - \frac{k^2}{n}\right] \tag{12.4}$$

where $\text{var}(\hat{S})$ = Variance of jackknife estimate of species richness
f_j = Number of quadrats containing j unique species ($j = 1, 2, 3, \cdots, s$)
k = Number of unique species
n = Total number of quadrats samples

This variance can be used to obtain confidence limits for the jackknife estimator,

$$\hat{S} \pm t_\alpha \sqrt{\text{var}(\hat{S})} \tag{12.5}$$

where $\hat{S}$ = Jackknife estimator of species richness (equation [12.3])
t_α = Student's t value for $n - 1$ degrees of freedom for the appropriate value of α
$\text{var}(\hat{S})$ = Variance of $\hat{S}$ from equation (12.4)

Box 12.2 gives an example of these calculations.

There is some disagreement about whether the jackknife estimator of species richness tends to be biased. Heltshe and Forrester (1983a) state that it has a positive bias; that is, it tends to *overestimate* the number of species in a community. Palmer (1990) found that the jackknife estimator had a slight negative bias in his data. This bias was much less than the

Box 12.2 Jackknife Estimate of Species Richness from Quadrat Samples

Ten quadrats from the benthos of a coastal creek were analyzed for the abundance of 14 species (Heltshe and Forrester 1983a). For a sample taken from a subtidal marsh creek, Pettaquamscutt River, Rhode Island, April 1978, we have the following data:

	Quadrat									
Species	1	2	3	4	5	6	7	8	9	10
Streblospio benedicti		13	21	14	5	22	13	4	4	27
Nereis succines	2	2	4	4	1	1	1		1	6
Polydora ligni	—	1	—	—	—	—	—	1	—	—
Scoloplos robustus	1	—	1	2	—	6	—	—	1	2
Eteone heteropoda	—	—	1	2	—	—	1	—	—	1
Heteromastus filiformis	1	1	2	1	—	1	—	—	1	5
*Capitella capitata**	1	—	—	—	—	—	—	—	—	—
*Scolecolepides viridis**	2	—	—	—	—	—	—	—	—	—
*Hypaniola grayi**	—	1	—	—	—	—	—	—	—	—
*Branis clavata**	—	—	1	—	—	—	—	—	—	—
Macoma balthica	—	—	3	—	—	—	—	—	—	2
Ampelisca abdita	—	—	5	1	—	2	—	—	—	3
*Neopanope texana**	—	—	—	—	—	—	—	1	—	—
Tubifocodies sp.	8	36	14	19	3	22	6	8	5	41

Note: Blank entries (—) in table are absent from quadrat.

Five species (marked with *) occur in only one quadrat and are thus defined as *unique species*. Thus, from equation (12.3),

$$\hat{S} = s + \left(\frac{n-1}{n}\right)k$$

$$\hat{S} = 14 + \left(\frac{9}{10}\right)(5)$$

$$= 18.5 \text{ species}$$

The variance, from equation (12.4), is

$$\text{var}(\hat{S}) = \left(\frac{n-1}{n}\right)\left[\sum_{j=1}^{s}(j^2 f_j) - \frac{k^2}{n}\right]$$

From the table we tally

No. of unique spp., j	No. of quadrats containing j unique species, f_j
1	3 (i.e., quadrats 2, 3, and 8)
2	1 (i.e., quadrat 1)
3	0
4	0
5	0

Thus,

$$\text{var}(\hat{S}) = \left(\frac{9}{10}\right)\left[(1)^2(3) + 2^2(1) - \frac{5^2}{10}\right]$$

$$= 4.05$$

For this small sample, for 95% confidence $t_\alpha = 2.26$, and thus the 95% confidence interval would be approximately

$$18.5 \pm (2.26)(\sqrt{4.05})$$

or 14 to 23 species.

Program-group SPECIES DIVERSITY (Appendix 2) can do these calculations for quadrat data.

negative bias of the observed number of species (s). Palmer found that the jackknife estimator was the most accurate of the eight estimators he used for his data.

Note from equation (12.3) that the maximum value of the jackknife estimate of species richness is twice the observed number of species. Thus this approach cannot be used on communities with exceptionally large numbers of rare species or on communities that have been sampled too little (so s is less than half the species present).

12.3.3 Bootstrap Procedure

One alternative method of estimating species richness from quadrat samples is to use the bootstrap procedure (Smith and Van Belle 1984).* The bootstrap method is related to the jackknife, but it requires simulation on a computer to obtain estimates. The essence of the bootstrap procedure is as follows: given a set of data of species presence/absence in a series of q quadrats (as in Table 12.2),

1. Draw a random sample of size n from the q quadrats within the computer, using sampling *with* replacement; this is the "bootstrap sample."
2. Calculate the estimate of species richness from the following equation (Smith and Van Belle 1984):

$$B(\hat{S}) = S + \sum (1 - p_i)^n \tag{12.6}$$

where $B(\hat{S})$ = Bootstrap estimate of species richness
S = Observed number of species in original data
p_i = Proportion of the n bootstrap quadrats that have species i present

3. Repeat steps 1 and 2 N times in the computer, where N is between 100 and 500.

The variance of this bootstrap estimate is given by

$$\text{var}[B(\hat{S})] = \sum_i (1 - p_i)^n[1 - (1 - p_i)^n] + \sum_j \sum_{i \neq j} \{q_{ij}^n - [(1 - p_i)^n - (1 - p_j)^n]\} \tag{12.7}$$

* See Chapter 15, Section 15.5 for more discussion of the bootstrap method.

where $\text{var}[B(\hat{S})]$ = Variance of the bootstrap estimate of species richness
n, p_i, p_j = As defined above
q_{ij} = Proportion of the n bootstrap quadrats that have both species i and species j absent

Smith and Van Belle (1984) recommend the jackknife estimator when the number of quadrats is small and the bootstrap estimator when the number of quadrats is large. The empirical meaning of *small* and *large* for natural communities remains unclear; perhaps $n = 100$ quadrats is an approximate division for many community samples, but at present this is little more than a guess. For Palmer's data (1990) with $n = 40$ quadrats, the bootstrap estimator had twice the amount of negative bias as did the jackknife estimator. Both the bootstrap and the jackknife estimators are limited to maximum values that are twice the number of observed species, so they cannot be used on sparsely sampled communities.

12.3.4 Species-Area Curve Estimates

One additional way of estimating species richness is to extrapolate the species-area curve for the community. Since the number of species tends to rise with the area sampled, one can fit a regression line and use it to predict the number of species on a plot of any particular size. This method is useful only for communities that have enough data to compute a species-area curve, and so it could not be used on sparsely sampled sites. Figure 12.3 illustrates a species-area curve for birds from the West Indies.

There is much disagreement about the exact shape of the species-area curve. Two models are most common. Gleason (1922) suggested that species richness is proportional to the logarithm of the area sampled, so that one should compute a semilog regression of the form

$$S = a + \log(A) \tag{12.8}$$

where S = Number of species (= species richness)
A = Area sampled
a = y-intercept of the regression

Preston (1962) argued that the species-area curve is a log-log relationship of the form

$$\log(S) = a + \log(A) \tag{12.9}$$

where all terms are as defined above. Palmer (1990) found that both these regressions overestimated species richness in his samples, and that the log-log regression of equation (12.9) was a particularly poor estimator. The semilog form of the species-area regression was highly correlated with species richness, in spite of its bias, and thus could serve as an index of species richness (Colwell and Coddington 1994).*

12.4 HETEROGENEITY MEASURES

The measurement of diversity by means of heterogeneity indices has proceeded along two relatively distinct paths. The first approach is to use statistical sampling theory to investigate how communities are structured. The logarithmic series was first applied by Fisher, Corbet, and Williams (1943) to a variety of community samples. Preston (1948, 1962) applied the lognormal distribution to community samples. Because of the empirical nature of these

*Robert K. Colwell has provided a computer program for this approach. See http://viceroy.eeb.uconn.edu

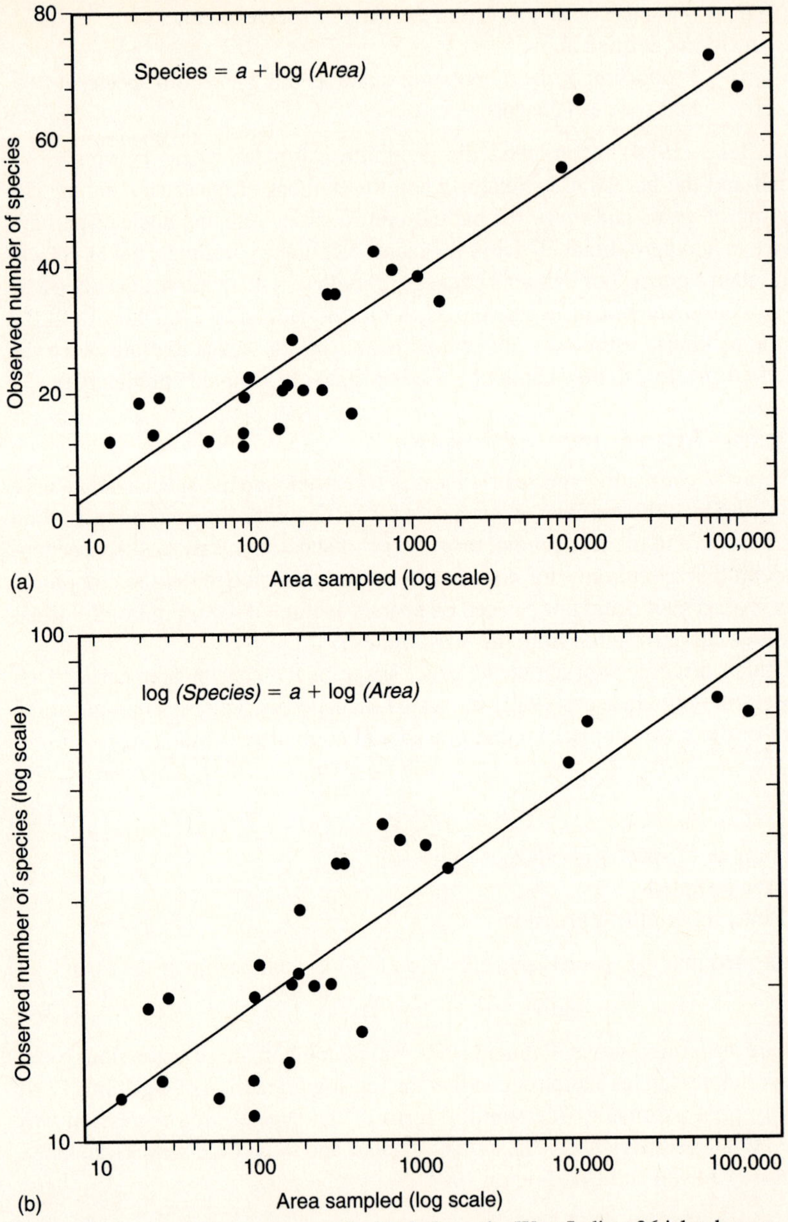

Figure 12.3 A species-area curve for birds from the West Indies: 26 islands ranging in size from Saba (13 km^2) to Cuba (112,000 km^2). Data are from Wright 1981, 746. (a) The exponential function $S = \log a + z \log A$ (equation [12.8]). (b) The more typical power function $\log S = \log a + z \log A$ (equation [12.9]). For these data, both curves fit about equally well.

statistical distributions, other workers looked to information theory for appropriate measures of diversity. Arguments continue about the utility of both of these approaches since they are not theoretically justified (Washington 1984; Hughes 1986). Both approaches are widely used in diversity studies, and it would be premature to dismiss any measure because it lacks comprehensive theoretical justification, since a diversity measure could be used as an index to diversity for practical studies.

It is important to keep in mind the ecological problem for which we wish to use these measures of heterogeneity. The key is to obtain some measure of community organization related to how the relative abundances vary among the different species in the community. Once we can measure community organization, we can begin to ask questions about patterns shown by different communities and processes that can generate differences among communities.

12.4.1 Logarithmic Series

One characteristic feature of communities is that they contain comparatively few species that are common and comparatively large numbers of species that are rare. Since it is relatively easy to determine for any given area the *number of species* in the area and the *number of individuals* in each of these species, a great deal of information of this type has accumulated (Williams 1964). The first attempt to analyze these data was made by Fisher, Corbet, and Williams (1943).

In many faunal samples the number of species represented by a single specimen is very large; species represented by two specimens are less numerous, and so on, until only a few species are represented by many specimens. Fisher, Corbet, and Williams (1943) plotted the data and found that they fit a "hollow curve" (Figure 12.4). Fisher concluded

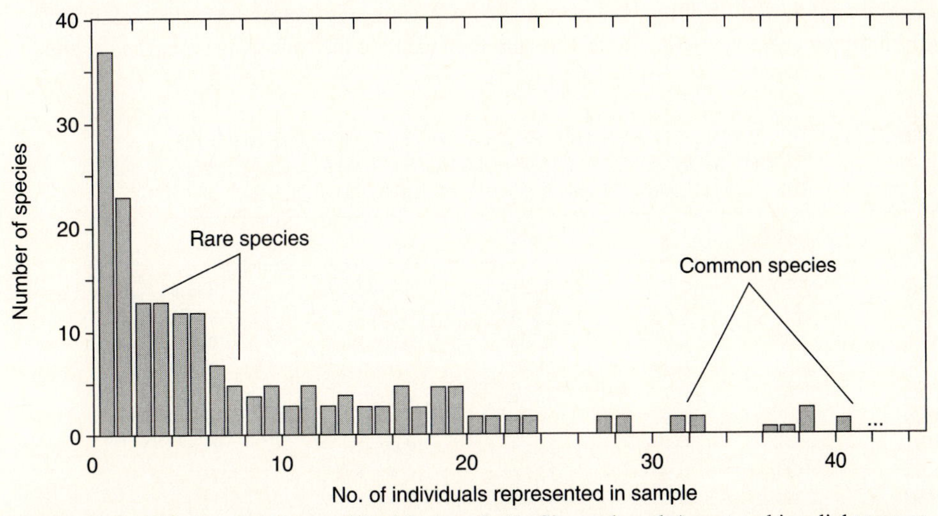

Figure 12.4 Relative abundance of Lepidoptera (butterflies and moths) captured in a light trap at Rothamsted, England, in 1935. Not all of the abundant species are shown. There were 37 species represented in the catch by only a single specimen (rare species); one very common species was represented by 1799 individuals in the catch (off the graph to the right!). A total of 6814 individuals were caught, representing 197 species. Six common species made up 50 percent of the total catch. (From Williams 1964).

that the data available were best fitted by the logarithmic series, which is a series with a finite sum whose terms can be written as a function of two parameters:

$$\alpha x, \frac{\alpha x^2}{2}, \frac{\alpha x^3}{3}, \frac{\alpha x^4}{4}, \ldots \tag{12.10}$$

where αx = Number of species in the total catch represented by *one* individual

$\frac{\alpha x^2}{2}$ = Number of species represented by two individuals, and so on

The sum of the terms in the series is equal to $\alpha \log_e(1 - x)$, which is the total number of species in the catch. The logarithmic series for a set of data is fixed by two variables, the *number of species* in the sample and the *number of individuals* in the sample. The relationship between these is

$$S = \alpha \log_e\left(1 + \frac{N}{\alpha}\right) \tag{12.11}$$

where S = Total number of species in the sample
N = Total number of individuals in the sample
α = Index of diversity

The constant α is an expression of species diversity in the community. It is low when the number of species is low and high when the number of species is high.

There are several methods of fitting a logarithmic series to a set of species-abundance data (Williams 1964, Appendix A). Only two variables are needed to fit a logarithmic series: the total number of species in the sample (S) and the total number of individuals (N). Williams (1964, 311) and Southwood (1978, 431) provide nomograms from which α may be read directly from values of N and S. A more accurate procedure is to estimate an approximate value of x from Table 12.3 and then to solve the following equation iteratively

TABLE 12.3 RELATION BETWEEN VALUES OF *x* AND THE AVERAGE NUMBER OF UNITS PER GROUP (*N/S*) IN SAMPLES FROM POPULATIONS DISTRIBUTED ACCORDING TO THE LOGARITHMIC SERIES

x	*N/S*	*x*	*N/S*	*x*	*N/S*
0.50	1.000	0.97	9.214	0.9990	144.6
0.60	1.637	0.980	12.53	0.9992	175.1
0.70	1.938	0.985	15.63	0.9994	224.5
0.80	2.483	0.990	21.47	0.9996	319.4
0.85	2.987	0.991	23.38	0.9998	586.9
0.90	3.909	0.992	25.68	0.99990	1086
0.91	4.198	0.993	28.58	0.99995	2020
0.92	4.551	0.994	32.38	0.999990	8696
0.93	4.995	0.995	37.48	0.999995	16,390
0.94	5.567	0.996	45.11	0.9999990	71,430
0.95	6.340	0.997	57.21	—	—
0.96	7.458	0.998	80.33	—	—

Source: Williams 1964, 308.

for a more accurate value of x:

$$\frac{S}{N} = \frac{1 - x}{x}[-\log_e(1 - x)] \tag{12.12}$$

where S = Total number of species in the sample
N = Total number of individuals in the sample
x = Parameter of logarithmic series (equation [12.10])

Trial values of x are used until this equation balances. Given this estimate of x, we obtain $\hat{\alpha}$ from

$$\hat{\alpha} = \frac{N(1 - x)}{x} \tag{12.13}$$

where $\hat{\alpha}$ = Index of diversity from logarithmic series
N = Total number of individuals in sample

Program-group SPECIES DIVERSITY (Appendix 2) can do these calculations. Given α and x, the theoretical values of the entire logarithmic series can be calculated from equation (12.10).

The large sample variance of the diversity index α was given by Anscombe (1950) as

$$\text{var}(\hat{\alpha}) = \frac{0.693147\alpha}{\left[\log_e\left(\frac{x}{1 - x}\right) - 1\right]^2} \tag{12.14}$$

where all terms are defined as above. Taylor et al. (1976) pointed out that many authors, including Williams (1964), have used the wrong formula to calculate the variance of α.

To analyze any set of empirical community data, the first thing you should do is to plot a *species-abundance curve*. Species-abundance curves can be plotted in three different ways (May 1975). On arithmetic or log scales,

- y-axis: relative abundance, density, cover, or some measure of the importance of a species
- x-axis: rank of the n species from 1 (most abundant species) to n (most rare species)

Species-abundance plots may thus be arithmetic (y)-arithmetic (x), log-log, or log(y)-arithmetic (x). By taking logs of the y- or the x-axis you can vary the shape of the resulting curves. Figure 12.5 illustrates a standard plot of species abundances, after Whittaker (1965). I call these *Whittaker plots* and recommend that the standard species-abundance plot utilize log-relative abundance (y)-arithmetic species ranks (x). The expected form of this curve for the logarithmic series is nearly a straight line and is shown in Figure 12.5.

The theoretical Whittaker plot for a logarithmic series (e.g., Figure 12.5a) can be calculated as indicated in May (1975) by solving the following equation for n:

$$R = \alpha E_1\left[n \log_e\left(1 + \frac{\alpha}{N}\right)\right] \tag{12.15}$$

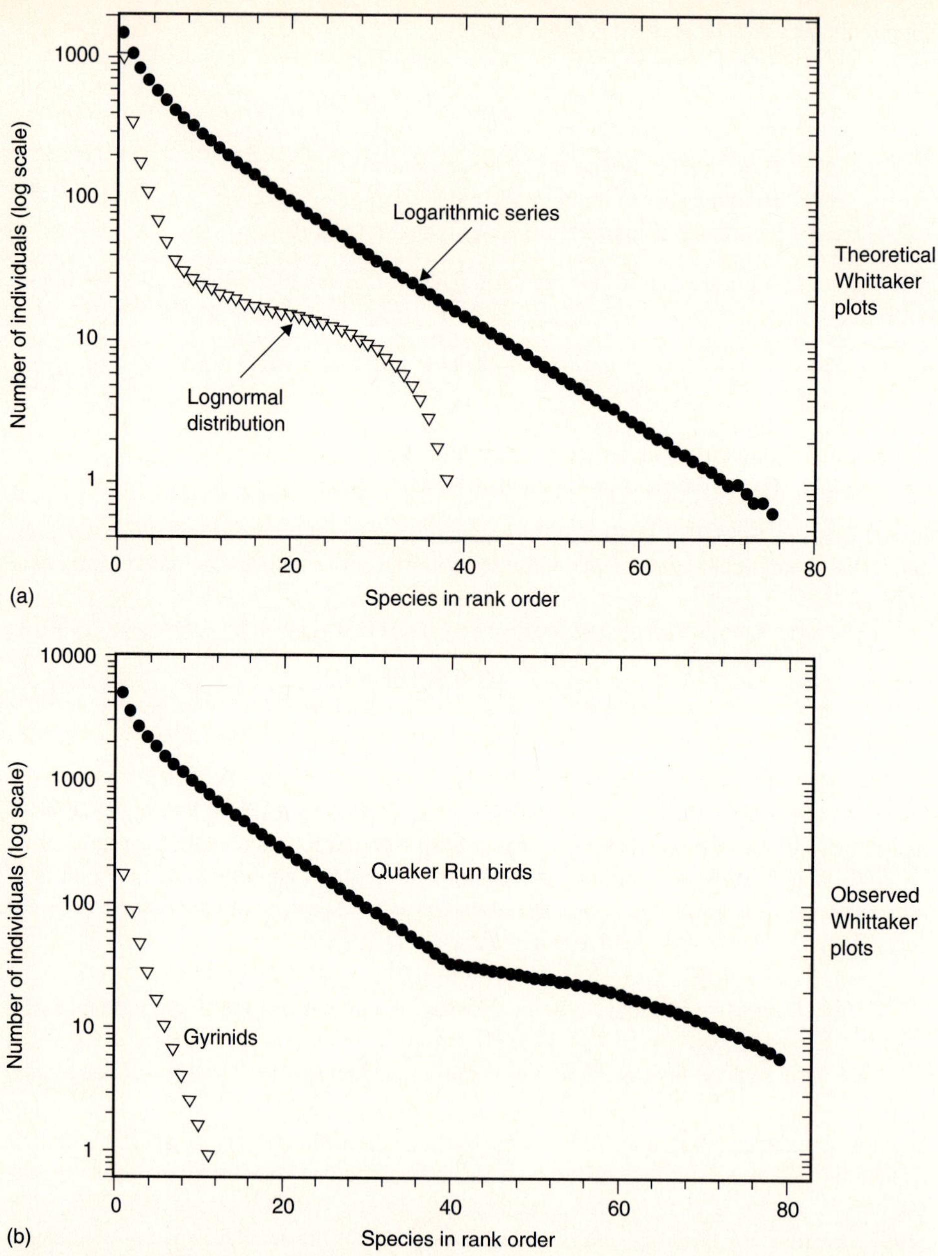

Figure 12.5 Whittaker plots of species-abundance data. (a) Theoretical plots. The logarithmic series produces a nearly straight line, while the lognormal distribution predicts a reverse S-shaped curve. (b) Observed data. The relative abundances of 11 species of Gyrinids from Mount Tremblant Park, Quebec (Lake des Fammes), is quite well described by the logarithmic series (data from Williams 1964, 271). The relative abundances of 79 species of birds from Quaker Run Valley, New York, is better described by the lognormal distribution (data from Williams 1964, 49).

where R = Species rank (x-axis, Figure 12.5; i.e., 1, 2, 3, 4, $\cdots$, s)
α = Index of diversity calculated in equation (12.13)
n = Number of individuals expected for specified value of R (y-axis of Figure 12.5)
N = Total number of individuals in sample
E_1 = Standard exponential integral (Abramowitz and Stegun 1964, Chapter 5)

By solving this equation for n using integer values of R, you can reconstruct the expected Whittaker plot and compare it to the original data. Program-group SPECIES DIVERSITY (Appendix 2) has an option to calculate these theoretical values for a Whittaker plot.

Considerable disagreement exists in the ecological literature about the usefulness of the logarithmic series as a good measure of heterogeneity. Taylor et al. (1976) analyzed light-trap catches of Macrolepidoptera from 13 sites in Britain, each site with 6–10 years of replicates. They showed that the logarithmic series parameter α was the best measure of species diversity for these collections. Hughes (1986), in contrast, examined 222 samples from many taxonomic groups and argued that the logarithmic series was a good fit for only 4% of these samples, primarily because the abundant species in the samples were more abundant than predicted by a logarithmic series. May (1975) attempted to provide some theoretical justification for the logarithmic series as a description of species-abundance patterns, but in most cases the logarithmic series is treated only as an empirical description of a sample from a community. Wolda (1983) concluded that α of the logarithmic series was the best measure of species diversity available. Box 12.3 illustrates the calculation of the logarithmic series for a community of rodents.

The goodness-of-fit of the logarithmic series to a set of community data can be tested by the usual chi-squared goodness-of-fit test (Taylor et al. 1976). But this chi-squared test is of low power and thus many samples are accepted as fitting the logarithmic series when the fit is actually not good (Routledge 1980b). Thus in most cases the decision as to whether or not to use the logarithmic series to describe the diversity of a data set must be made on ecological grounds (Taylor et al. 1976; Hughes 1986), rather than statistical goodness-of-fit criteria.

Koch (1987) used the logarithmic series to answer a critical methodological question in paleoecology: If two samples are taken from exactly the same community, how many species will be found in both data sets, and how many species will appear to be unique to one data set? Sample-size effects may be critical in paleoecological studies, since absent species are typically classed as extinct. Koch used the logarithmic series and simple probability theory to predict the expected number of unique species in large samples from paleocommunities. These predictions can serve as a null model to compare with observed differences between samples. Figure 12.6 illustrates that the percentage of unique species can be very large when samples differ in size, even when the samples are taken from the same community. Rare species are inherently difficult to study in ecological communities, and sample-size effects should always be evaluated before differences are assumed between two collections.

12.4.2 Lognormal Distribution

The logarithmic series implies that the greatest number of species has minimal abundance and that the number of species represented by a single specimen is always maximal. This is not the case in all communities. Figure 12.7 shows the relative abundance of breeding

Box 12.3 Fitting a Logarithmic Series to Species-Abundance Data

Krebs and Wingate (1976) sampled the small-mammal community in the Kluane region of the southern Yukon and obtained these results:

	No. of individuals
Deer mouse	498
Northern red-backed vole	495
Meadow vole	111
Tundra vole	61
Long-tailed vole	45
Singing vole	40
Heather vole	23
Northern bog lemming	5
Meadow jumping mouse	5
Brown lemming	4
	$N = 1287$

$$S = 10$$
$$N/S = 128.7$$

From Table 12.3, an approximate estimate of x is 0.999. From equation (12.12), using this provisional estimate of x, we have

$$\frac{S}{N} = \left(\frac{1 - x}{x}\right)[-\log_e(1 - x)]$$

$$\frac{10}{1287} = \left(\frac{1 - 0.999}{0.999}\right)[-\log_e(1 - 0.999)]$$

$$0.007770 \neq 0.006915$$

Since the term on the right side is too small, we reduce the estimate of x. Try 0.99898:

$$0.007770 = \left(\frac{1 - 0.99898}{0.99898}\right)[-\log_e(1 - 0.99898)]$$

$$0.007770 \neq 0.007033$$

The right side of the equation is still too small, so we reduce x to 0.99888:

$$0.007770 = \left(\frac{1 - 0.99888}{0.99888}\right)[-\log_e(1 - 0.99888)]$$

$$0.007770 \neq 0.0076183$$

The right side of the equation is still slightly too small, so we repeat this calculation with $x = 0.998854$ to obtain, using equation (12.12),

$$0.007770 \cong 0.007769$$

We accept 0.998854 as an estimate of the parameter x of the logarithmic series.

From equation (12.13),

$$\hat{\alpha} = \frac{1287(1 - 0.998854)}{0.998854}$$

$$= 1.4766$$

The variance of this estimate of α is, from equation (12.14),

$$\text{var}(\hat{\alpha}) = \frac{\hat{\alpha}}{-\log_e(1 - x)}$$

$$= \frac{1.4766}{-\log_e(1 - 0.998854)}$$

$$= 0.2181$$

The individual terms of the logarithmic series are given by equation (12.10):

$$\alpha x, \frac{\alpha x^2}{2}, \frac{\alpha x^3}{3}, \ldots$$

i	No. of species represented by i individuals
1	1.475
2	0.737
3	0.491
4	0.367
5	0.294
6	0.244
7	0.209
.	.
.	.
.	.

The sum of the terms of this series (which is infinite) is the number of species in the sample ($S = 10$).

These data are used for illustration only. One would not normally fit a logarithmic series to a sample with such a small number of species.

Program-group SPECIES DIVERSITY (Appendix 2) can do these calculations.

birds in Quaker Run Valley, New York. The greatest number of bird species are represented by ten breeding pairs, and the relative abundance pattern does not fit the hollow-curve pattern of Figure 12.4. Preston (1948) suggested expressing the x-axis (number of individuals represented in sample) on a geometric (logarithmic) scale rather than an arithmetic scale. One of several geometric scales can be used, since they differ only by a constant multiplier; a few scales are indicated in Table 12.4.

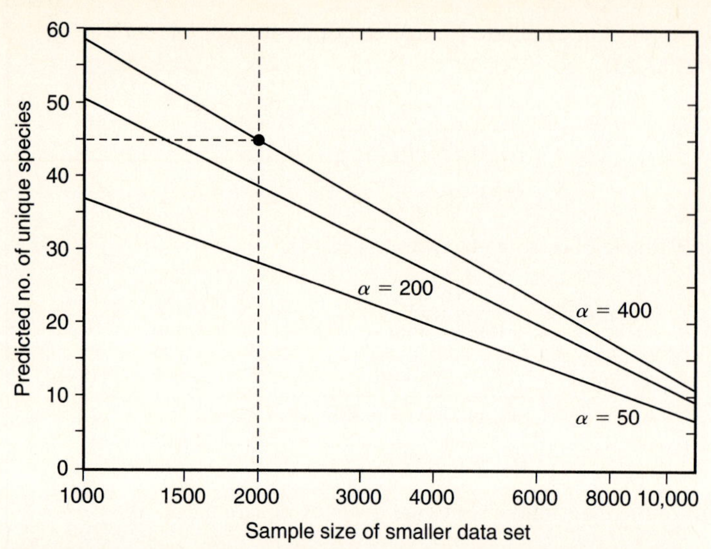

Figure 12.6 Use of the logarithmic series to predict the percentage of unique species in the larger of two data sets from exactly the same hypothetical community. Three values of α, the diversity parameter of the logarithmic series, are plotted to indicate low-, moderate-, and high-diversity communities. The larger sample is 10,000 individuals. The point illustrates an independent sample of $n = 2000$, which is predicted to have about 45 unique species in spite of being a sample from the identical community. These curves illustrate how difficult it is to sample the rare species in a diverse biological community. (Modified from Koch 1987.)

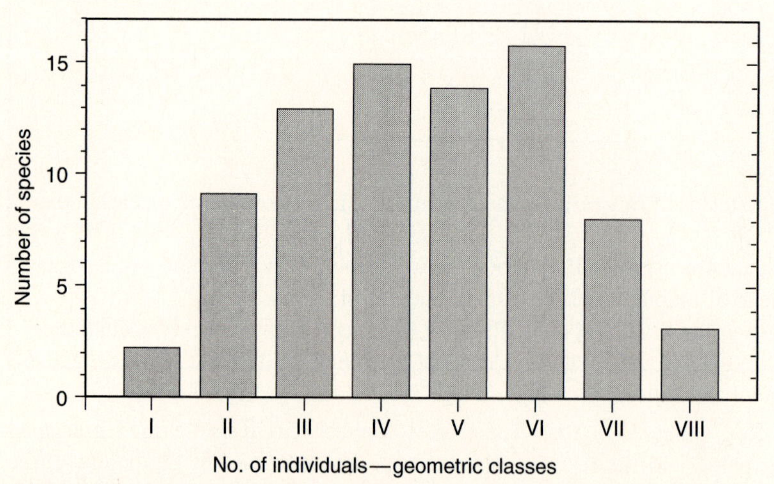

Figure 12.7 Relative abundance of nesting bird species in Quaker Run Valley, New York on a geometric scale with ×3 size groupings (1–2, 3–8, 9–26, 27–80, 81–242, etc.). These data do not fit a hollow curve like that described by the logarithmic series. (From Williams 1964.)

TABLE 12.4 GROUPINGS OF ARITHMETIC SCALE UNITS OF ABUNDANCE INTO GEOMETRIC SCALE UNITS FOR THREE TYPES OF GEOMETRIC SCALES[a]

Geometric scale no.	Arithmetic numbers grouped according to ×2 Scale[b]	×3 Scale[c]	×10 Scale[d]
1	1	1–2	1–9
2	2–3	3–8	10–99
3	4–7	9–26	100–999
4	8–15	27–80	1,000–9,999
5	16–31	81–242	10,000–99,999
6	32–63	243–728	100,000–999,999
7	64–127	729–2,186	—
8	128–255	2,187–6,560	—
9	256–511	6,561–19,682	—

[a] This type of grouping is used in Figure 12.7.

[b] Octave scale of Preston (1948), equivalent to $\log_2$ scale.

[c] Equivalent to $\log_3$ scale.

[d] Equivalent to $\log_{10}$ scale.

When this conversion of scale is done, relative abundance data take the form of a bell-shaped, normal distribution, and because the *x*-axis is expressed on a geometric or logarithmic scale, this distribution is called *lognormal*. The lognormal distribution has been analyzed comprehensively by May (1975). The lognormal distribution is completely specified by two parameters, although as May (1975) shows, there are several ways of expressing the equation:

$$\hat{S}_T = \frac{1.772454}{a} S_0 \tag{12.16}$$

where $\hat{S}_T$ = Total number of species in the community
a = Parameter measuring the spread of the lognormal distribution
S_0 = Number of species in the largest class

The lognormal distribution fits a variety of data from surprisingly diverse communities (Preston 1948, 1962).

The shape of the lognormal curve is supposed to be characteristic for any particular community. Additional sampling of a community should move the lognormal curve to the right along the abscissa but not change its shape. Few communities have been sampled enough to test this idea, and Figure 12.8 shows some data from moths caught in light traps, which suggests that additional sampling moves the curve out toward the right. Since we cannot collect one-half or one-quarter of an animal, there will always be some rare species that are not represented in the catch. These rare species appear only when very large samples are taken.

Preston (1962) showed that data from lognormal distributions from biological communities commonly took on a particular configuration that he called the *canonical*

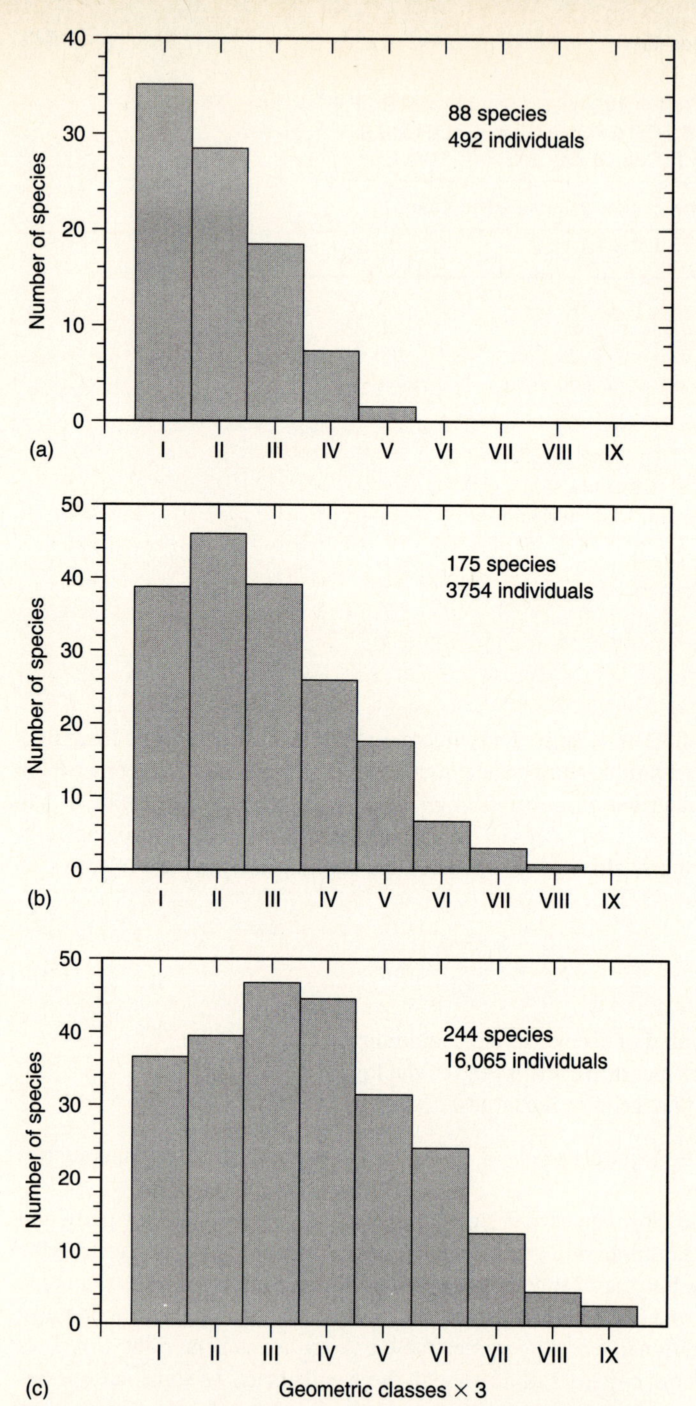

Figure 12.8 Lognormal distributions of the relative abundances of Lepidopteran insects captured in light traps at Rothamsted Experimental Station, England, in periods ranging from (a) 1/8 years to (b) 1 year to (c) 4 years. Note that the lognormal distribution slides to the right as the sample size is increased. (From Williams 1964.)

distribution. Preston showed that for many cases $a = 0.2$, so that the entire lognormal distribution could be specified by *one* parameter:

$$\hat{S}_T = 5.11422\, S_0 \tag{12.17}$$

where $\hat{S}_T$ = Total number of species in the community
S_0 = Parameter measuring number of species in the modal (largest) class of the lognormal as defined above

Note that when the species-abundance distribution is lognormal, it is possible to estimate the total number of species in the community, including rare species not yet collected. This is done by extrapolating the bell-shaped curve below the class of minimal abundance and measuring the area. Figure 12.9 illustrates how this can be done. This can be a useful property for communities where all the species cannot readily be seen and tabulated.

Although the lognormal distribution is an attractive model for species-abundance relationships, in practice it is a very difficult distribution to fit to ecological data (Hughes 1986). In practice a sample should be described by a truncated lognormal only if there is some evidence of a mode or maximum in the species-abundance curve (e.g., Figures 12.7, 12.9). Many authors have calculated a lognormal distribution from data like those in Figure 12.8a, which have no mode, but this should not be done. Hughes showed that parameter estimates from artificial lognormal distributions that did not include the mode were wildly inaccurate. The shape of the "true" lognormal distribution cannot be calculated from small samples, unless you have independent evidence that the first octave of your sample is close to the true mode for that community.

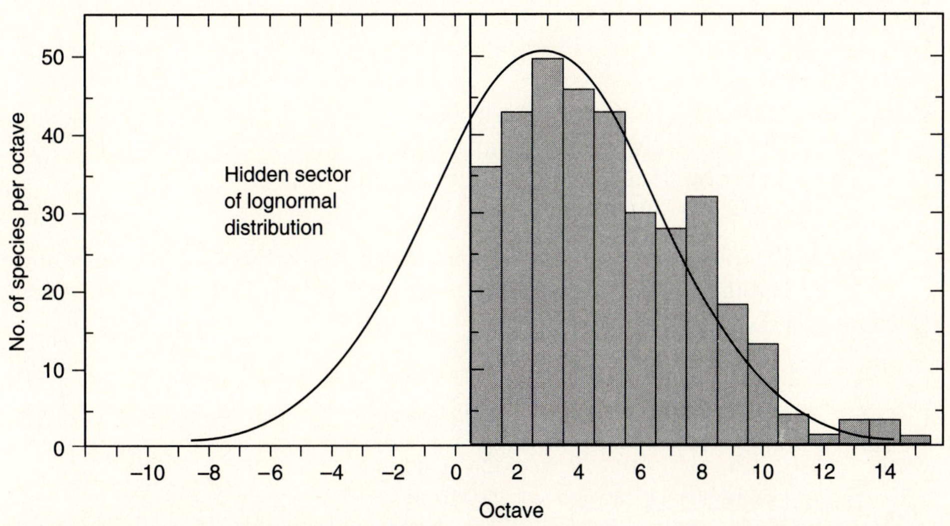

Figure 12.9 Species abundances in a collection of moths caught in a light trap (data from Preston 1948). The lognormal distribution is truncated at the point where species are represented by a single individual. More intensive sampling should cause the distribution to move to the right and to unveil the hidden sector of rare species. The left edge of the observed distribution is called the *veil line*. The abundance classes for each octave are listed in Table 12.4. (From Preston 1948.)

The lognormal distribution is a continuous statistical distribution, but species-abundance data are discrete in terms of individuals. Strictly speaking, the species-abundance data should be treated as Poisson variates, and one should fit the *Poisson lognormal* (= discrete lognormal) to most community data (Pielou 1975, 49; Bulmer 1974). The Poisson lognormal is difficult to compute and Bulmer (1974) has discussed methods of evaluating it. For practical purposes the ordinary lognormal is usually fitted to species-abundance data, using the maximum likelihood methods devised by Cohen (1959, 1961) and described by Pielou (1975, 50–53). Gauch and Chase (1974) discussed a non-linear regression method for fitting the lognormal distribution to species-abundance data, but Hansen and Zeger (1978) showed that this regression method was not appropriate for species-abundance data, and recommended the method of Cohen.

To fit a lognormal distribution to species-abundance data by the methods of Cohen (1959, 1961), proceed as follows:

Step 1. Transform all the observed data (number of individuals, or biomass, or other measure of species importance) logarithmically:

$$x_i = \log n_i \tag{12.18}$$

where n_i = Observed number of individuals of species i in sample
i = Species counter ($i = 1, 2, 3, \cdots, S_0$)*
x_i = Transformed value for lognormal distribution

Any base of logarithms can be used, as long as you are consistent. I will use log-base 10.

Step 2. Calculate the observed mean and variance of x_i by the usual statistical formulas (Appendix 1). Sample size is S_0, the observed number of species.

Step 3. Calculate the parameter y:

$$y = \frac{s^2}{(\bar{x} - x_0)^2} \tag{12.19}$$

where y = Parameter of lognormal distribution
s^2 = Observed variance (calculated in step 2)
$\bar{x}$ = Observed mean (calculated in step 2)
x_0 = $\log(0.5) = -0.30103$ if using $\log_{10}$

Step 4. From Table 12.5 obtain the estimate of θ corresponding to this estimate of y.

Step 5. Obtain corrected estimates of the mean and variance of the lognormal distribution from the following equations:

$$\hat{\mu} = \bar{x} - \theta(\bar{x} - x_0) \tag{12.20}$$

$$\hat{\sigma}^2 = s^2 + \theta(\bar{x} - x_0)^2 \tag{12.21}$$

where $\hat{\mu}$ = Estimate of true mean of the lognormal
$\bar{x}, s^2$ = Observed mean and variance from step 2
θ = Correction factor from Table 12.5 (step 4)
x_0 = Truncation point of observed data = $\log(0.5)$
$\hat{\sigma}^2$ = Estimate of true variance of the lognormal

*To avoid confusion but maintain traditional symbols, I use S_0 for the number of species observed and s for the standard deviation.

TABLE 12.5 VALUES OF THE ESTIMATION FUNCTION θ CORRESPONDING TO VALUES OF y OBTAINED IN EQUATION (12.19)[a]

y	.000	.001	.002	.003	.004	.005	.006	.007	.008	.009	y
0.05	.00000	.00000	.00000	.00001	.00001	.00001	.00001	.00001	.00002	.00002	0.05
0.06	.00002	.00003	.00003	.00003	.00004	.00004	.00005	.00006	.00007	.00007	0.06
0.07	.00008	.00009	.00010	.00011	.00013	.00014	.00016	.00017	.00019	.00020	0.07
0.08	.00022	.00024	.00026	.00028	.00031	.00033	.00036	.00039	.00042	.00045	0.08
0.09	.00048	.00051	.00055	.00059	.00063	.00067	.00071	.00075	.00080	.00085	0.09
0.10	.00090	.00095	.00101	.00106	.00112	.00118	.00125	.00131	.00138	.00145	0.10
0.11	.00153	.00160	.00168	.00176	.00184	.00193	.00202	.00211	.00220	.00230	0.11
0.12	.00240	.00250	.00261	.00272	.00283	.00294	.00305	.00317	.00330	.00342	0.12
0.13	.00355	.00369	.00382	.00396	.00410	.00425	.00440	.00455	.00470	.00486	0.13
0.14	.00503	.00519	.00536	.00553	.00571	.00589	.00608	.00627	.00646	.00665	0.14
0.15	.00685	.00705	.00726	.00747	.00769	.00791	.00813	.00835	.00858	.00882	0.15
0.16	.00906	.00930	.00955	.00980	.01006	.01032	.01058	.01085	.01112	.01140	0.16
0.17	.00168	.01197	.01226	.01256	.01286	.01316	.01347	.01378	.01410	.01443	0.17
0.18	.01476	.01509	.01543	.01577	.01611	.01646	.01682	.01718	.01755	.01792	0.18
0.19	.01830	.01868	.01907	.01946	.01986	.02026	.02067	.02108	.02150	.02193	0.19
0.20	.02236	.02279	.02323	.02368	.02413	.02458	.02504	.02551	.02599	.02647	0.20
0.21	.02695	.02744	.02794	.02844	.02895	.02946	.02998	.03050	.03103	.03157	0.21
0.22	.03211	.03266	.03322	.03378	.03435	.03492	.03550	.03609	.03668	.03728	0.22
0.23	.03788	.03849	.03911	.03973	.04036	.04100	.04165	.04230	.04296	.04362	0.23
0.24	.04429	.04497	.04565	.04634	.04704	.04774	.04845	.04917	.04989	.05062	0.24
0.25	.05136	.05211	.05286	.05362	.05439	.05516	.05594	.05673	.05753	.05834	0.25
0.26	.05915	.05997	.06080	.06163	.06247	.06332	.06418	.06504	.06591	.06679	0.26
0.27	.06768	.06858	.06948	.07039	.07131	.07224	.07317	.07412	.07507	.07603	0.27
0.28	.07700	.07797	.07896	.07995	.08095	.08196	.08298	.08401	.08504	.08609	0.28
0.29	.08714	.08820	.08927	.09035	.09144	.09254	.09364	.09476	.09588	.09701	0.29
0.30	.09815	.09930	.10046	.10163	.10281	.10400	.10520	.10641	.10762	.10885	0.30
0.31	.1101	.1113	.1126	.1138	.1151	.1164	.1177	.1190	.1203	.1216	0.31
0.32	.1230	.1243	.1257	.1270	.1284	.1298	.1312	.1326	.1340	.1355	0.32
0.33	.1369	.1383	.1398	.1413	.1428	.1443	.1458	.1473	.1488	.1503	0.33
0.34	.1519	.1534	.1550	.1566	.1582	.1598	.1614	.1630	.1647	.1663	0.34
0.35	.1680	.1697	.1714	.1731	.1748	.1765	.1782	.1800	.1817	.1835	0.35
0.36	.1853	.1871	.1889	.1907	.1926	.1944	.1963	.1982	.2001	.2020	0.36
0.37	.2039	.2058	.2077	.2097	.2117	.2136	.2156	.2176	.2197	.2217	0.37
0.38	.2238	.2258	.2279	.2300	.2321	.2342	.2364	.2385	.2407	.2429	0.38
0.39	.2451	.2473	.2495	.2517	.2540	.2562	.2585	.2608	.2631	.2655	0.39
0.40	.2678	.2702	.2726	.2750	.2774	.2798	.2822	.2827	.2871	.2896	0.40
0.41	.2921	.2947	.2972	.2998	.3023	.3049	.3075	.3102	.3128	.3155	0.41
0.42	.3181	.3208	.3235	.3263	.3290	.3318	.3346	.3374	.3402	.3430	0.42
0.43	.3459	.3487	.3516	.3545	.3575	.3604	.3634	.3664	.3694	.3724	0.43
0.44	.3755	.3785	.3816	.3847	.3878	.3910	.3941	.3973	.4005	.4038	0.44
0.45	.4070	.4103	.4136	.4169	.4202	.4236	.4269	.4303	.4338	.4372	0.45
0.46	.4407	.4442	.4477	.4512	.4547	.4583	.4619	.4655	.4692	.4728	0.46
0.47	.4765	.4802	.4840	.4877	.4915	.4953	.4992	.5030	.5069	.5108	0.47
0.48	.5148	.5187	.5227	.5267	.5307	.5348	.5389	.5430	.5471	.5513	0.48
0.49	.5555	.5597	.5639	.5682	.5725	.5768	.5812	.5856	.5900	.5944	0.49
0.50	.5989	.6034	.6079	.6124	.6170	.6216	.6263	.6309	.6356	.6404	0.50

continues

TABLE 12.5 (Cont.)

y	.000	.001	.002	.003	.004	.005	.006	.007	.008	.009	y
0.51	.6451	.6499	.6547	.6596	.6645	.6694	.6743	.6793	.6843	.6893	0.51
0.52	.6944	.6995	.7046	.7098	.7150	.7202	.7255	.7308	.7361	.7415	0.52
0.53	.7469	.7524	.7578	.7633	.7689	.7745	.7801	.7857	.7914	.7972	0.53
0.54	.8029	.8087	.8146	.8204	.8263	.8323	.8383	.8443	.8504	.8565	0.54
0.55	.8627	.8689	.8751	.8813	.8876	.8940	.9004	.9068	.9133	.9198	0.55
0.56	.9264	.9330	.9396	.9463	.9530	.9598	.9666	.9735	.9804	.9874	0.56
0.57	.9944	1.001	1.009	1.016	1.023	1.030	1.037	1.045	1.052	1.060	0.57
0.58	1.067	1.075	1.082	1.090	1.097	1.105	1.113	1.121	1.129	1.137	0.58
0.59	1.145	1.153	1.161	1.169	1.177	1.185	1.194	1.202	1.211	1.219	0.59
0.60	1.228	1.236	1.245	1.254	1.262	1.271	1.280	1.289	1.298	1.307	0.60
0.61	1.316	1.326	1.335	1.344	1.353	1.363	1.373	1.382	1.392	1.402	0.61
0.62	1.411	1.421	1.431	1.441	1.451	1.461	1.472	1.482	1.492	1.503	0.62
0.63	1.513	1.524	1.534	1.545	1.556	1.567	1.578	1.589	1.600	1.611	0.63
0.64	1.622	1.634	1.645	1.657	1.668	1.680	1.692	1.704	1.716	1.728	0.64
0.65	1.740	1.752	1.764	1.777	1.789	1.802	1.814	1.827	1.840	1.853	0.65
0.66	1.866	1.879	1.892	1.905	1.919	1.932	1.946	1.960	1.974	1.988	0.66
0.67	2.002	2.016	2.030	2.044	2.059	2.073	2.088	2.103	2.118	2.133	0.67
0.68	2.148	2.163	2.179	2.194	2.210	2.225	2.241	2.257	2.273	2.290	0.68
0.69	2.306	2.322	2.339	2.356	2.373	2.390	2.407	2.424	2.441	2.459	0.69
0.70	2.477	2.495	2.512	2.531	2.549	2.567	2.586	2.605	2.623	2.643	0.70
0.71	2.662	2.681	2.701	2.720	2.740	2.760	2.780	2.800	2.821	2.842	0.71
0.72	2.863	2.884	1.905	2.926	2.948	2.969	2.991	3.013	3.036	3.058	0.72
0.73	3.081	3.104	3.127	3.150	3.173	3.197	3.221	3.245	3.270	3.294	0.73
0.74	3.319	3.344	3.369	3.394	3.420	3.446	3.472	3.498	3.525	3.552	0.74
0.75	3.579	3.606	3.634	3.662	3.690	3.718	3.747	3.776	3.805	3.834	0.75
0.76	3.864	3.894	3.924	3.955	3.986	4.017	4.048	4.080	4.112	4.144	0.76
0.77	4.177	4.210	4.243	4.277	4.311	4.345	4.380	4.415	4.450	4.486	0.77
0.78	4.52	4.56	4.60	4.63	4.67	4.71	4.75	4.79	4.82	4.86	0.78
0.79	4.90	4.94	4.99	5.03	5.07	5.11	5.15	5.20	5.24	5.28	0.79
0.80	5.33	5.37	5.42	5.46	5.51	5.56	5.61	5.65	5.70	5.75	0.80
0.81	5.80	5.85	5.90	5.95	6.01	6.06	6.11	6.17	6.22	6.28	0.81
0.82	6.33	6.39	6.45	6.50	6.56	6.62	6.68	6.74	6.81	6.87	0.82
0.83	6.93	7.00	7.06	7.13	7.19	7.26	7.33	7.40	7.47	7.54	0.83
0.84	7.61	7.68	7.76	7.83	7.91	7.98	8.06	8.14	8.22	8.30	0.84
0.85	8.39	8.47	8.55	8.64	8.73	8.82	8.91	9.00	9.09	9.18	0.85

Source: Cohen 1961.

[a] These values are used to fit the lognormal distribution to species-abundance data.

Step 6. Calculate the standardized normal deviation corresponding to the truncation point:

$$z_0 = \frac{x_0 - \mu}{\sigma} \tag{12.22}$$

Step 7. From tables of the standard normal distribution (e.g., Rohlf and Sokal 1995, 78 or Zar 1996, Table B.2), find the area (p_0) under the tail of the normal curve to the left of z_0. Then:

$$\hat{S}_T = \frac{S_0}{1 - p_0} \qquad (12.23)$$

where $\hat{S}_T$ = Estimated number of species in the community (including those to the left of the veil line, e.g., Figure 12.9)
S_0 = Observed number of species in sample
p_0 = Area of standard normal curve to left of z_0

In the notation of equation (12.16), note that

$$\hat{a} = \frac{1}{\sqrt{2\hat{\sigma}^2}} \qquad (12.24)$$

where $\hat{a}$ = Parameter measuring the spread of the lognormal distribution
$\hat{\sigma}^2$ = True variance of the lognormal (equation [12.21])

The variance of these estimates of the parameters of the lognormal distribution can be estimated, following Cohen (1961), as

$$\text{var}(\hat{\mu}) = \frac{\mu_{11}\hat{\sigma}^2}{s_0} \qquad (12.25)$$

where $\text{var}(\hat{\mu})$ = Estimated variance of mean of the lognormal
μ_{11} = Constant from Table 12.6
$\hat{\sigma}^2$ = Estimate of true variance of lognormal (equation 12.21)
s_0 = Observed number of species in sample

The variance of the standard deviation of the lognormal is given by

$$\text{var}(\hat{\sigma}) = \frac{\mu_{22}\hat{\sigma}^2}{s_0} \qquad (12.26)$$

where $\text{var}(\hat{\sigma})$ = Variance of estimated standard deviation of the lognormal
μ_{22} = Constant from Table 12.6
$\hat{\sigma}^2$ = True variance of lognormal
s_0 = Observed number of species in the sample

These two variances may be used to set confidence limits for the mean and standard deviation in the usual way. The goodness-of-fit of the calculated lognormal distribution can be determined by a chi-squared test (example in Pielou 1975, 51) or by a nonparametric Kolmolgorov-Smirnov test.

Unfortunately there is no estimate available of the precision of $\hat{S}_T$ (equation [12.23]), and this is the parameter of the lognormal we are most interested in (Pielou 1975; Slocomb and Dickson 1978). Simulation work on artificial diatom communities by Slocomb and Dickson (1978) showed that unreliable estimates of S_T were a serious problem unless sample sizes were very large (> 1000 individuals) and the species in the sample were 80% or more of the total species in the community. Such large-scale sampling is rare in the most species-rich communities that we might wish to fit to the lognormal distribution.

TABLE 12.6 FACTORS FOR ESTIMATING THE VARIANCE OF THE MEAN AND STANDARD DEVIATION OF A LOGNORMAL DISTRIBUTION[a]

	For truncated samples			For truncated samples	
z_0	μ_{11}	μ_{22}	z_0	μ_{11}	μ_{22}
−4.0	1.00054	.502287	0.0	22.1875	4.03126
−3.5	1.00313	.510366	0.1	27.1403	4.46517
−3.0	1.01460	.536283	0.2	33.1573	4.94678
−2.5	1.05738	.602029	0.3	40.4428	5.48068
−2.4	1.07437	.622786	0.4	49.2342	6.07169
−2.3	1.09604	.646862	0.5	59.8081	6.72512
−2.2	1.12365	.674663	0.6	72.4834	7.44658
−2.1	1.15880	.706637	0.7	87.6276	8.24204
−2.0	1.20350	.743283	0.8	105.66	9.11780
−1.9	1.26030	.785158	0.9	127.07	10.081
−1.8	1.33246	.832880	1.0	152.40	11.138
−1.7	1.42405	.887141	1.1	182.29	12.298
−1.6	1.54024	.948713	1.2	217.42	13.567
−1.5	1.68750	1.01846	1.3	258.61	14.954
−1.4	1.87398	1.09734	1.4	306.78	16.471
−1.3	2.10982	1.18642	1.5	362.91	18.124
−1.2	2.40764	1.28690	1.6	428.11	19.922
−1.1	2.78311	1.40009	1.7	503.57	21.874
−1.0	3.25557	1.52746	1.8	591.03	24.003
−0.9	3.84879	1.67064	1.9	691.78	26.311
−0.8	4.59189	1.83140	2.0	807.71	28.813
−0.7	5.52036	2.01172	2.1	940.38	31.511
−0.6	6.67730	2.21376	2.2	1091.4	34.405
−0.5	8.11482	2.43990	2.3	1265.4	37.575
−0.4	9.89562	2.69271	2.4	1458.6	40.858
−0.3	12.0949	2.97504	2.5	1677.8	44.392
−0.2	14.8023	3.28997			
−0.1	18.1244	3.64083			

Source: Cohen 1961.

[a] This table is entered with a value of z_0 as calculated in equation (12.22).

Program-group SPECIES DIVERSITY (Appendix 2) fits a truncated lognormal distribution to species-abundance data and calculates an expected distribution, using the approach outlined in Pielou (1975). Box 12.4 illustrates these calculations for a lognormal distribution.

12.4.3 Simpson's Index

Partly because of the complexity of the logarithmic series and the lognormal distribution, and the lack of a theoretical justification for these statistical approaches, ecologists have turned to a variety of nonparametric measures of heterogeneity that make no assumptions about the shape of species-abundance curves. The first nonparametric measure of diversity was proposed by Simpson (1949). Simpson suggested that diversity was inversely related

Box 12.4 Fitting a Truncated Lognormal to Species-Abundance Data

Kempton and Taylor (1974) provided moth data for site 49, Fort Augustus, Scotland, in 1969; 4534 individuals were collected in 165 species:

Individuals per species	Midpoint of interval, x	Observed no. of species, f_x
1	1	24
2–3	2.5	22
4–7	5.5	30
8–15	11.5	22
16–31	23.5	30
32–63	47.5	21
64–127	99.5	9
128–255	191.5	7

1. Calculate the mean and variance of the transformed data (log base 10) using the formulas for grouped data:

$$\bar{x} = \frac{\Sigma\, xf_x}{\Sigma\, f_x}$$

$$= \frac{(\log 1)(24) + (\log 2.5)(22) + (\log 5.5)(30) + \cdots}{24 + 22 + 30 + 22 + 30 + 21 + 9 + 7}$$

$$= \frac{164.5991}{165}$$

$$= 0.99757$$

$$s = \frac{\Sigma\, x^2 f_x - (\Sigma\, xf_x)^2/n}{n - 1}$$

$$= 0.41642$$

2. Estimate the parameter y from equation (12.19):

$$y = \frac{s^2}{(\bar{x} - x_0)^2}$$

$$= \frac{0.41642}{[0.99757 - (-0.30103)]^2}$$

$$= 0.24693$$

3. From Table 12.5, interpolating between y of 0.246 and 0.247,

$$\theta = 0.04912$$

4. Correct the observed mean and variance for the effects of truncation from equations (12.20) and (12.21):

$$\hat{\mu} = \overline{x} - \theta(\overline{x} - x_0)$$
$$= 0.99757 - (0.04912)[0.99757 - (-0.30103)]$$
$$= 0.93378$$

$$\hat{\sigma}^2 = s^2 + \theta(\overline{x} - x_0)^2$$
$$= 0.41642 + (0.04912)[0.99757 - (-0.30103)]^2$$
$$= 0.49925$$

5. Calculate the standard normal deviation corresponding to the truncation point from equation (12.22):

$$z_0 = \frac{x_0 - \hat{\mu}}{\hat{\sigma}}$$
$$= \frac{-0.30103 - 0.93378}{\sqrt{0.49925}}$$
$$= -1.7476$$

6. From Table 11 of Rohlf and Sokal (1995), obtain the area under the normal curve to the left of z_0:

$$\hat{p}_0 = 0.02005$$

7. From equation (12.23), calculate the estimated number of species in the whole community:

$$S_T = \frac{S_0}{1 - \hat{p}_0}$$
$$= \frac{165}{1 - 0.02005}$$
$$= 168.4 \text{ species}$$

Kempton and Taylor (1974) cautioned that this fitting procedure may give inexact parameter estimates compared with Bulmer's (1974) procedure.

to the probability that two individuals picked at random belong to the same species. For an infinite population, this is given by

$$D = \sum p_i^2 \tag{12.27}$$

where D = Simpson's index
p_i = Proportion of species i in the community

To convert this probability to a measure of diversity, most workers have suggested using the complement of Simpson's original measure:

$$\text{Simpson's index of diversity} = \left\{ \begin{array}{c} \text{Probability of picking two} \\ \text{organisms at random that} \\ \text{are different species} \end{array} \right\}$$

$$= 1 - \left\{ \begin{array}{c} \text{Probability of picking two} \\ \text{organisms that are the} \\ \text{same species} \end{array} \right\}$$

Thus,

$$1 - D = 1 - \sum (p_i)^2 \tag{12.28}$$

where $(1 - D)$ = Simpson's index of diversity
p_i = Proportion of individuals of species i in the community

Strictly speaking, this formula can be used to estimate Simpson's index only for an infinite population. Pielou (1969) showed that for a finite population, the appropriate estimator is

$$1 - \hat{D} = 1 - \sum_{i=1}^{s} \left[\frac{n_i(n_i - 1)}{N(N - 1)} \right] \tag{12.29}$$

where n_i = Number of individuals of species i in the sample
N = Total number of individuals in the sample $= \sum n_i$
s = Number of species in the sample

Note that this formula (12.29) can be used only when there are counts of individuals in the samples. When cover, biomass, or productivity are used as measures of species importance, equation (12.28) must be used. In practice, with a large sample there is almost no difference between these two equations.

There is some confusion in the literature over what should be called Simpson's index. Washington (1984) argues strongly for maintaining Simpson's original formulation, in which case equations (12.28) and (12.29) are the *complement* of Simpson's diversity. To confuse matters further, Williams (1964) and MacArthur (1972) used the reciprocal of Simpson's original formulation:

$$\frac{1}{D} = \frac{1}{\Sigma \, p_i^2} \tag{12.30}$$

where $\frac{1}{D}$ = Simpson's reciprocal index (= Hill's N_2)
p_i = Proportion of species i in the community

Hill (1973) called this reciprocal N_2.

Simpson's index $(1 - D)$ ranges from 0 (low diversity) to almost $1 (1 - 1/s)$. The reciprocal of Simpson's original formulation $(1/D)$ varies from 1 to s, the number of species in the sample. In this form Simpson's diversity can be most easily interpreted as the number of equally common species required to generate the observed heterogeneity of the sample.

Diversity is almost always measured by a sample from a community, and it is virtually impossible for an ecologist to obtain a simple random sample (Pielou 1969; Routledge 1980a). One way around this problem is to treat the community sample as a *collection*, or a complete statistical "universe," and to make inferences about this finite collection (Pielou 1966). Another approach is to use sampling units such as quadrats for plants or nets for insects, and to estimate diversity using a jackknife procedure. Zahl (1977) was the first to propose using this procedure to provide confidence estimates for Simpson's diversity measure. Routledge (1980a) showed that small samples (<30 quadrats) could give biased estimates for Simpson's diversity ($1 - D$ is underestimated), especially when less than 10 quadrats were counted. Heltshe and Forrester (1985) suggested that the jackknife estimate of confidence limits for Simpson's diversity ($1 - D$) was too large when applied to clumped populations, causing excessively wide confidence limits when more than 40 quadrats were sampled in their artificial populations. This overestimation depended on the exact shape of the species-abundance curves.

Jackknife procedures for estimating Simpson's index of diversity and its confidence limits from quadrat samples are outlined clearly in Routledge (1980a). Lyons and Hutcheson (1986) proposed an alternative method for estimating confidence limits for Simpson's diversity using Pearson curves, but there was little improvement over the jackknife procedure.

12.4.4 Shannon-Wiener Function

The most popular measures of species diversity are based on information theory. The main objective of information theory is to try to measure the amount of *order* (or disorder) contained in a system (Margalef 1958). Four types of information might be collected regarding *order* in the community: (1) the number of species, (2) the number of individuals in each species, (3) the places occupied by individuals of each species, and (4) the places occupied by individuals as separate individuals. In most community work only data of types 1 and 2 are obtained.

Information theory, Margalef suggested, provides one way to escape some of the difficulties of the lognormal curve and the logarithmic series. We ask the question, *How difficult would it be to predict correctly the species of the next individual collected?* This is the same problem faced by communication engineers interested in predicting correctly the next letter in a message. This uncertainty can be measured by the Shannon-Wiener function:*

$$H' = \sum_{i=1}^{s} (p_i)(\log_2 p_i) \tag{12.31}$$

where H' = Information content of sample (bits/individual)
= Index of species diversity
s = Number of species
p_i = Proportion of total sample belonging to ith species

Information content is a measure of the amount of uncertainty, so that the larger the value of H', the greater the uncertainty. A message such as bbbbbbb (or a community with only

*This function was derived independently by Shannon and Wiener and is sometimes mislabeled the Shannon-Weaver function.

one species in it) has no uncertainty in it, and $H' = 0$. Any base of logarithms can be used for this index, since they are all convertible to one another by a constant multiplier:

$$H' \text{(base 2 logs)} = 3.321928\, H' \text{(base 10 logs)}$$

$$H' \text{(base } e \text{ logs)} = 2.302585\, H' \text{(base 10 logs)}$$

If base 2 logs are used, the units of H' are in *bits per individual*; if base e logs, *nits*; and if base 10 logs, *decits*.

Strictly speaking, the Shannon-Wiener measure of information content should be used only on random samples drawn from a large community in which the total number of species is known (Pielou 1966). For most community samples this is not the case, and Pielou (1966) thus recommends using the more appropriate Brillouin index (see Section 12.4.5).

The Shannon-Wiener measure H' increases with the number of species in the community and in theory can reach very large values. In practice, for biological communities H' does not seem to exceed 5.0 (Washington 1984). The theoretical maximum value is $\log(S)$, and the minimum value (when $N \gg S$) is $\log[N/(N - S)]$ (Fager 1972).

Many workers have used H' as a measure of species diversity, but the information theoretic approach has been heavily criticized by Hurlbert (1971) and by Washington (1984). The decision to use H' as a measure of species diversity should be made more on empirical grounds than on theoretical grounds. For example, Taylor et al. (1976) showed that α of the logarithmic series was a better diversity statistic than H' because α varied less in replicate samples of moths taken at the same site over several years.

Sampling distributions for the Shannon-Wiener index H' have been determined by Good (1953) and Basharin (1959), but these standard errors of H' are valid only if you have a simple random sample from the community. This is never the case in field data where nets, traps, quadrats or transects are used for sampling (Kempton 1979). Adams and McCune (1979) showed that estimates of H' from field data are usually biased, so that observed H' is less than true H', and that the jackknife technique could be used to reduce this bias and to estimate standard errors for H' so that confidence limits might be calculated. Zahl (1977) and Routledge (1980a) presented jackknife estimators for the Shannon-Wiener function when data are collected by quadrat sampling. Adams and McCune (1979) have prepared a computer program for jackknifing the Shannon-Wiener function.

The Shannon-Wiener index may be expressed in another form (MacArthur 1965), in units of numbers of species as

$$N_1 = e^{H'} \tag{12.32}$$

where e = 2.71828 (base of natural logs)
H' = Shannon-Wiener function (calculated with base e logs)
N_1 = Number of equally common species that would produce the same diversity as H'

If a different base of logs is used, replace e with the base of the logs used. Hill (1973) recommends using N_1 rather than H' because the units (number of species) are more clearly understandable to ecologists. Peet (1974) recommends N_1 as the best heterogeneity measure that is sensitive to the abundances of the rare species in the community.

12.4.5 Brillouin Index

Many community samples should be treated as collections rather than as random samples from a large biological community, according to Pielou (1966). In any case in which we can assume the data are a finite collection, and sampling is done without replacement, the appropriate information-theoretic measure of diversity is Brillouin's formula:

$$\hat{H} = \frac{1}{N} \log \left(\frac{N!}{n_1! n_2! n_3! \cdots} \right) \tag{12.33}$$

where $\hat{H}$ = Brillouin's index
N = Total number of individuals in entire collection
n_1 = Number of individuals belonging to species 1
n_2 = Number of individuals belonging to species 2

Any base of logarithms may be used, as with the Shannon function. If base 2 logs are used, the units of H are *bits* per individual. Margalef (1958) was the first to propose using Brillouin's index as a measure of diversity.

There is much argument in the literature about whether the Brillouin index or the Shannon-Wiener function is a better measure of species diversity (Peet 1974; Washington 1984). In practice, this argument is irrelevant to field ecologists because H and H' are nearly identical for most ecological samples (when N is large). Legendre and Legendre (1983) also point out that Brillouin's index cannot be used when biomass, cover, or productivity is used as a measure of species importance in a community. Only the *number* of individuals can be used in equation (12.33). If the Brillouin index is applied to quadrats, the mean and standard error of the Brillouin index can be estimated by the jackknife procedure (Heltshe and Forrester 1985).

The Brillouin index is like the Shannon function in being most sensitive to the abundances of the rare species in the community. Peet (1974) recognized two categories of diversity indices. *Type I* indices are most sensitive to changes in the rare species in the community sample. The Shannon-Wiener index is an example of a Type I index, like the Brillouin index. *Type II* indices are most sensitive to changes in the more abundant species. Simpson's index is an example of a Type II index. The choice of what heterogeneity measure to use on your data should be made on this basis: Are you more interested in emphasizing the dominant or the rare species in your community? Box 12.5 illustrates the calculation of Simpson's index, the Shannon-Wiener function, and Brillouin's index for a forest community.

12.5 EVENNESS MEASURES

Many different measures of evenness (or *equitability*) have been proposed, and the literature is most confusing about which measure is best. Smith and Wilson (1996) have recently reviewed 14 indices of evenness with respect to the criterion that evenness measures must be independent of species richness. The most common approach has been to scale one of the heterogeneity measures relative to its maximal value when each species

Box 12.5 Calculation of Simpson's Index, the Shannon-Wiener Function, and Brillouin's Index of Species Diversity

Hough (1936) tallied the abundance of large trees in a virgin forest in Pennsylvania:

Tree species	No. of individuals n_i	Proportional abundance p_i
Hemlock	1940	0.521
Beech	1207	0.324
Yellow birch	171	0.046
Sugar maple	134	0.036
Black birch	97	0.026
Red maple	93	0.025
Black cherry	34	0.009
White ash	22	0.006
Basswood	15	0.004
Yellow poplar	7	0.002
Magnolia	4	0.001
Total	3724	1.000

Simpson's Index

From equation (12.27),

$$\begin{aligned}\hat{D} &= \sum p_i^2 \\ &= 0.521^2 + 0.324^2 + 0.046^2 + 0.036^2 + \cdots \\ &= 0.381\end{aligned}$$

The two indices of diversity follow from equations (12.28) and (12.30):

$$1 - \hat{D} = 1 - 0.381 = 0.619$$

This measure is the probability that two individuals chosen at random will be different species.

$$\frac{1}{\hat{D}} = \frac{1}{0.381} = 2.623 \text{ species}$$

This is the number of equally common species required to produce the observed value of D.

Note that with this large sample, the finite-population formula of equation (12.29) gives results identical to equation (12.28).

Shannon-Wiener Function

From equation (12.31),

$$H' = \sum_{i=1}^{s} (p_i)(\log_2 p_i)$$

$$= (0.521)(\log_2 0.521) + (0.324)(\log_2 0.324) + (0.046)(\log_2 0.046) + \cdots$$

$$= 1.829 \text{ bits per individual}$$

Note that $\log_2(x) = 3.321981 \log_{10}(x)$.

From equation (12.32),

$$\hat{N}_1 = e^{\hat{H}'} \qquad \text{(base } e \text{ logs)}$$

$$= 2^{\hat{H}'} \qquad \text{(base 2 logs)}$$

$$= 2^{1.829} = 3.55 \text{ species}$$

Brillouin's Index

From equation (12.33),

$$\hat{H} = \frac{1}{N} \log\left(\frac{N!}{n_1!n_2!n_3! \cdots}\right)$$

$$= \frac{1}{3724} \log_2\left(\frac{3724!}{1940!\ 1207!\ 171!\ 134!\ 97! \cdots}\right)$$

$$= 1.818 \text{ bits per individual}$$

Note that this is virtually identical to $\hat{H}'$.

Program-group SPECIES DIVERSITY (Appendix 2) can do these calculations.

in the sample is represented by the same number of individuals. Two formulations are possible:

$$\text{Evenness} = \frac{D}{D_{\text{MAX}}}$$

$$\text{Evenness} = \frac{D - D_{\text{MIN}}}{D_{\text{MAX}} - D_{\text{MIN}}}$$

where D = Observed index of species diversity

D_{MAX} = Maximum possible index of diversity, given S species and N individuals

D_{MIN} = Minimum possible index of diversity, given S and N

These two measures (labeled V' and V by Hurlbert 1971) are convergent for large samples, and the first type of evenness measures (V') are most commonly used in the literature. All these evenness measures should range from 0 to 1. Unfortunately many of the indices of evenness based on this approach are not independent of species richness, and other indices of evenness are needed. Smith and Wilson (1996) prefer the four indices described in the following sections.

Simpson's Measure of Evenness For Simpson's measure of heterogeneity, maximum diversity is obtained when all abundances are equal ($p = 1/S$), so in a very large population:

$$\hat{D}_{\text{MAX}} = \frac{1}{s} \tag{12.34}$$

where $\hat{D}_{\text{MAX}}$ = Maximum possible value for Simpson's index (equation [12.27])
s = Number of species in the sample

It follows from this that the maximum possible value of the reciprocal of Simpson's index ($1/D$) is always equal to the number of species observed in the sample. This leads to a simple definition of Simpson's index of evenness:

$$E_{1/D} = \frac{1/\hat{D}}{s} \tag{12.35}$$

where $E_{1/\hat{D}}$ = Simpson's measure of evenness
$\hat{D}$ = Simpson's index (equation [12.27])
s = Number of species in the sample

This index ranges from 0 to 1 and is relatively unaffected by the rare species in the sample.

Camargo's Index of Evenness Camargo (1993) proposed a new index of evenness that is unaffected by species richness and is simple to compute:

$$E' = 1.0 - \left(\sum_{i=1}^{s} \sum_{j=i+1}^{s} \left[\frac{|p_i - p_j|}{s} \right] \right) \tag{12.36}$$

where E' = Camargo's index of evenness
p_i = Proportion of species i in total sample
p_j = Proportion of species j in total sample
s = Number of species in total sample

This index, like Simpson's, is relatively unaffected by the rare species in the sample.

Smith and Wilson's Index of Evenness Smith and Wilson (1996) invented a new index of evenness based on the variance in abundance of the species. The variance is measured over the log of the abundances in order to use proportional differences rather than absolute differences in abundance. The new index is defined as

$$E_{\text{var}} = 1 - \left[\frac{2}{\pi \arctan \left\{ \sum_{i=1}^{s} \left(\log_e(n_i) - \sum_{j=1}^{s} \log_e(n_j)/s \right)^2 \Big/ s \right\}} \right] \tag{12.37}$$

where the arctangent is measured as an angle in radians, and

E_{var} = Smith and Wilson's index of evenness
n_i = Number of individuals in species i in sample ($i = 1, 2, 3, 4, \ldots s$)
n_j = Number of individuals in species j in sample ($j = 1, 2, 3, 4, \ldots s$)
s = Number of species in entire sample

This is the best available index of evenness, according to Smith and Wilson (1996) because it is independent of species richness and is sensitive to both rare and common species in the community.

Modified Nee Index of Evenness Nee et al. (1992) suggested using the slope of the Whittaker dominance-diversity relationship (Figure 12.5) to measure evenness, but the index they proposed ranged from $-\infty$ to zero. Smith and Wilson (1996) improved and modified the Nee index to provide a new index defined as follows:

$$E_Q = \frac{2\arctan(b)}{\pi} \tag{12.38}$$

where the arctangent is measured as an angle in radians, and

E_Q = Modified Nee index of evenness
b = Slope of the Whittaker dominance relationship

Note that the slope is obtained from the regression of $\log_e$ abundances (x-axis) on the scaled rank of abundance (y-axis) such that the most rare species has rank $1/S$ and the most common species has rank 1.0. This index ranges from 0–1 and is independent of species richness. It is sensitive to both common and rare species in the sample.

Which measure of evenness is best? The key ecological decision you must make is whether or not you wish to weight rare and common species equally. Some ecologists like Routledge (1983) argue that we should treat rare and common species similarly. Other ecologists like Alatalo (1981) argue that rare species are often poorly sampled and often missed, so that it is best not to put much weight on the abundance of rare species. Once you make this critical ecological decision, you can use the key provided by Smith and Wilson (1996) to select which evenness index is best for your data:

- If rare and common species are to be given equal weight in the sample,
  (a) and a minimum of zero with any number of species is needed, use $E_{1/D}$
  (b) and a wide range of evenness is being measured, use Camargo's E'
- If common species are to be emphasized over rare species in the sample,
  (c) and very skewed distributions are to be expected, use Nee's index E_Q
  (d) and for most data sets, the best overall index is Smith and Wilson's E_{var}

There is a general problem with all measures of evenness: they all assume you know the total number of species in the whole community (Pielou 1969). But this number is almost always impossible to determine for species-rich communities. Since observed species numbers must always be less than true species numbers in the community, the evenness ratios are always *overestimated* (Sheldon 1969). Peet (1974, 1975) and Routledge (1983) recommend that evenness measures should not be used in ecological work unless the number of species in the whole community is known. This is a very stringent restriction and is probably too purist. It may be possible to know the total number of species in some temperate-zone communities and in well-studied tropical communities, but this will be rare in most insect groups and in other invertebrate taxa. Program-group SPECIES DIVERSITY (Appendix 2) calculates all these evenness measures for species-abundance data.

12.6 RECOMMENDATIONS

Community ecology has paid more attention to the measurement of species diversity than to almost any other parameter. There is thus an enormous literature on diversity, full of contradictory recommendations. In some respects the problems of measuring diversity are similar to the problems of measuring similarity. Following Southwood (1978) and Routledge (1979), it appears best to take an empirical approach, as follows:

1. *Construct Whittaker plots of log abundance on species rank* (Figure 12.5). The shape of the dominance-diversity curve will indicate which models of species-abundance relations might be applied to the data. These graphs may themselves be useful for publication.
2. *Estimate species richness using the rarefaction method* (Figure 12.2). This will permit the comparison of species richness among several communities sampled with different intensity.
3. *Fit the logarithmic series or the lognormal curve to the data*, if the Whittaker plot indicates this is reasonable. The logarithmic series α may be a useful index of diversity even for communities that deviate from the logarithmic distribution.
4. *Use the reciprocal of Simpson's index (equation [12.30]) or the exponential form of the Shannon-Wiener function (equation [12.32]) to describe heterogeneity.* Decide beforehand whether you wish to weight the common species more (Simpson's) or the rare species more (Shannon's) in your community analysis.
5. *Use Smith and Wilson's index of evenness (equation [12.37]) to estimate evenness for the community sample*, unless you have good data on both the rare species and the common ones (in which case use Camargo's E').

Practical wisdom is still accumulating in community ecology about which indices of diversity are most useful for specific applications (Taylor et al. 1976; James and Rathbun 1981; Washington 1984; Smith and Wilson 1996). It is clear no single universal approach can be recommended for all communities, and much more empirical work is required.

12.7 SUMMARY

Species diversity is a dual concept that includes the number of species in the community and the evenness with which the individuals are divided among the species. Many ways of measuring species diversity exist, and there is much controversy about which indices of diversity are best.

Species richness, or the number of species in the community, is easy to determine only in easily censused communities with few species. In all other cases, the larger the sample size, the longer the species list. The rarefaction technique allows one to adjust a series of samples to a common sample size (number of individuals) so that species richness can be compared among samples. For quadrat sampling, a jackknife estimate of species richness can be made, based on the number of species that occur in only one quadrat.

Heterogeneity measures confound species richness and evenness in a single index of diversity. Two statistical distributions have commonly been fitted to species-abundance data: the logarithmic series and the lognormal distribution. Nonparametric measures of heterogeneity are commonly used because they assume no statistical distribution. Type I heterogeneity measures place most weight on the rare species in the sample, and the Shannon-Wiener function is an example of these measures. Type II heterogeneity measures place most weight on the common species, and Simpson's index is an example of these. Confidence limits can be obtained for heterogeneity measures by jackknife procedures.

Evenness can be estimated in many different ways, and the key concept is to relate observed species abundances to maximum possible heterogeneity, when all species have an equal number of individuals. Good measures of evenness are now available, but to use them you must first have decided whether to emphasize or de-emphasize the rare species in your community samples.

SELECTED READING

Heltshe, J. F., and Forrester, N. E. 1985. Statistical evaluation of the jackknife estimate of diversity when using quadrat samples. *Ecology* 66: 107–111.

Hughes, R. G. 1986. Theories and models of species abundance. *American Naturalist* 128: 879–899.

James, F. C., and Rathbun, S. 1981. Rarefaction, relative abundance, and diversity of avian communities. *Auk* 98: 785–800.

Koch, C. F. 1987. Prediction of sample size effects on the measured temporal and geographic distribution patterns of species. *Paleobiology* 13: 100–107.

Loehle, C. 1990. Proper statistical treatment of species-area data. *Oikos* 57: 143–145.

Longino. J. T., and Colwell, R. K. 1997. Biodiversity assessment using structured inventory: capturing the ant fauna of a tropical rain forest. *Ecological Applications* 7: 1263–1277.

Peet, R. K. 1974. The measurement of species diversity. *Annual Review of Ecology and Systematics* 5: 285–307.

Routledge, R. D. 1980. Bias in estimating the diversity of large, uncensused communities. *Ecology* 61: 276–281.

Smith, B., and Wilson, J. B. 1996. A consumer's guide to evenness indices. *Oikos* 76: 70–82.

Taylor, L. R., Kempton, R. A., and Woiwod, I. P. 1976. Diversity statistics and the log-series model. *Journal of Animal Ecology* 45: 255–272.

Washington, H. G. 1984. Diversity, biotic and similarity indices: a review with special relevance to aquatic ecosystems. *Water Research* 18: 653–694.

QUESTIONS AND PROBLEMS

12.1. Calculate a rarefaction curve for a sample of a subalpine forest community containing 278 individual trees in 7 species as follows: ES, 126; SF, 103; LP, 27; PC, 12; AL, 6; DF, 2; and AF, 2. Estimate a 95% confidence interval for species richness at a sample size of 100 individuals.

12.2. Use the jackknife estimator of species richness to estimate the total richness of herbs in an Indiana oak-hickory forest from the following set of 17 quadrats:

	Quadrat 1																
Species	1	2	3	4	5	6	7	8	9	10	11	12	13	14	15	16	17
A	13		5		1		7		12		2	2	4	6	7	1	4
B	4	1	1	6			2		4		1	7	8	9	12	2	7
C		2	7	1	3	8	1	2	1			7	1	2	2	1	1
D	1																
E	6		5	1	7	12	6	7		2	3	1	4	1	1	6	5
F		1															
G									1								
H	3		1	7	6	1	2		3	4	2	1	1	7	1	4	2
I	6																
J	1	4	12	1	3	8	6	4	2	1	1	3	2	1	7	4	2

12.3. Fit the logarithmic series to Patrick's (1968) diatom data for Box 7 given in Table 12.1. Graph the theoretical and the observed species-abundance data in a Whittaker plot, and do a chi-squared goodness-of-fit test on these data.

12.4. Fit the lognormal distribution to the diatom data of Problem 12.3, and discuss whether the canonical lognormal is a good description of this data set.

12.5. Fit the lognormal distribution to the data on moths given in Kempton and Taylor (1974, Table 4). Compare the estimated parameters with those given in that paper, and discuss why they differ.

12.6. Estimate Simpson's index of diversity for the small-mammal data used in Box 12.3. Calculate the maximum and the minimum possible values of Simpson's index for these data (1287 individuals, 10 species).

12.7. Calculate the Shannon-Wiener function H' and the Brillouin index H for the seabird communities on St. Paul and St. George Island (Table 11.1). Estimate evenness for these two communities. Are these measures of evenness biased?

12.8. Calculate Simpson's index of diversity and the Shannon-Wiener index of diversity for the following sets of hypothetical data:

	Proportion of species in community			
Species	W	X	Y	Z
1	0.143	0.40	0.40	0.40
2	0.143	0.20	0.20	0.20
3	0.143	0.15	0.15	0.15
4	0.143	0.10	0.10	0.10
5	0.143	0.05	0.025	0.01
6	0.143	0.05	0.025	0.01
7	0.143	0.05	0.025	0.01
8			0.025	0.01
9			0.025	0.01
10			0.025	0.01
11				0.01

Species	Proportion of species in community W	X	Y	Z
12				0.01
13				0.01
14				0.01
15				0.01
16				0.01
17				0.01
18				0.01
19				0.01
	1.00	1.00	1.00	1.00

What do you conclude about the sensitivity of these measures?

12.9. Plot and calculate species-area curves for the following set of data for bird species on the Channel Islands of California (Wright 1981, 743). Use three regressions: species (Y) on area (X), species on log area, and log species on log area. Which regression describes these data best? Why are the slope values for these three regressions not identical? See Loehle (1990) for an evaluation.

Island	Area (km²)	Bird species
Santa Barbara	2.6	10
Anacapa	2.8	14
San Miguel	36	15
San Nicholas	57	11
San Clemente	145	24
Santa Catalina	194	34
Santa Rosa	218	25
Santa Cruz	249	37

12.10. Calculate evenness indices for the four recommended evenness measures for the hypothetical two-species communities suggested by Alatalo (1981) and used by Molinari (1989):

Community	Species *X* abundance	Species *Y* abundance
A	999	1
B	900	100
C	800	200
D	700	300
E	600	400
F	500	500

The expectation is that the index of evenness should respond in a reasonable way to this gradual change in evenness. Plot the shapes and discuss which indices fulfill this expectation. Smith and Wilson (1996, 79) discuss this criterion.

CHAPTER 13

Niche Measures and Resource Preferences

The analysis of community dynamics depends in part on the measurement of how organisms utilize their environment. One way to do this is to measure the *niche* parameters of a population and to compare the niche of one population with that of another. Since food is one of the most important dimensions of the niche, the analysis of animal diets is closely related to the problem of niche specifications. In this chapter I review niche metrics and the related measurement of dietary overlap and dietary preferences.

Before you decide on the appropriate measures of niche size and dietary preference, you must think carefully about the exact questions you wish to answer with these measures. The hypothesis must drive the measurements and the ways in which the raw data will be summarized. As in all ecological work, it is important to think before you leap into analysis.

13.1 WHAT IS A RESOURCE?

The measurement of niche parameters is fairly straightforward, once the decision about what resources to include has been made. The question of defining a resource state can be subdivided into three questions (Colwell and Futuyma 1971). First, what *range* of resource states should be included? Second, how should samples be taken across this range? And third, how can nonlinear niche dimensions be analyzed? Figure 13.1 illustrates some of these questions graphically.

Resource states may be defined in a variety of ways:

1. *Food resources*: the taxonomic identity of the food taken may be used as a resource state, or the size category of the food items (without regard to taxonomy) could be defined as the resource state.
2. *Habitat resources*: habitats for animals may be defined botanically or from physical-chemical data into a series of resource states.
3. *Natural sampling units*: sampling units like lakes or leaves or individual fruits may be defined as resource states.
4. *Artificial sampling units*: a set of random quadrats may be considered to comprise different resource states.

The idea of a resource state is very broad and depends on the type of organism being studied and the purpose of the study. Resource states based on clearly significant resources like food or habitat seem preferable to more arbitrarily defined states (like 3 and 4 above).

In analyzing the comparative use of resource states by a group of species, it is important to include the extreme values found for all the species combined as upper and lower bounds for your measurements (Colwell and Futuyma 1971). Only if the complete range of possible resource states is used will the niche measurements be valid on an absolute scale. Conversely, you should not measure beyond the extreme values for the set of species, or you will waste time and money in measuring resource states that are not occupied.

If samples are taken across the full range of resource states, there is still a problem of spacing. Compare the sampling at hypothetical sites I, III, and IV in Figure 13.1. All these sampling schemes range over the same extreme limits of soil moisture, but niche breadths calculated for each species would differ depending on the spacing of the samples. If all resource states are ecologically distinct to the same degree, the problem of spacing is not serious, but this is rarely the case in a community in nature. The important point is to sample evenly across all the resource states as much as possible.

Resource states may be easily quantified on an absolute scale if they are physical or chemical parameters like soil moisture. But the effects of soil moisture or any other physical-chemical parameter on the abundance of a species is never a simple straight line (Hanski 1978; Green 1979). Colwell and Futuyma (1971) made the first attempt to weight resource states by their level of distinctness. Hanski (1978) provided a second method for

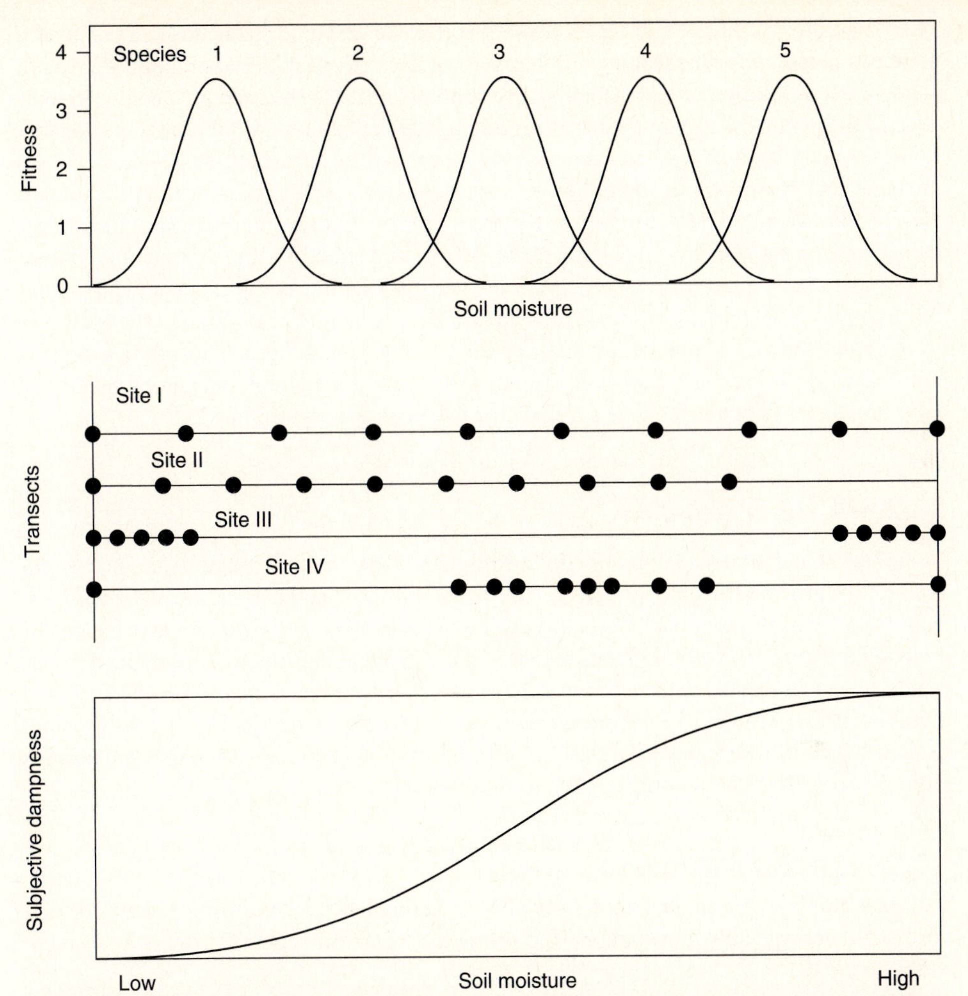

Figure 13.1 Hypothetical example to illustrate some problems in the measurement of niche breadth and niche overlap. The horizontal axis is a gradient of soil moisture from dry hillside (left) to moist stream bank (right). Consider 5 species of plants, each adapted to different soil moisture levels (top). The same moisture gradient exists at each of four different study sites (I, II, III, IV), but the sampling quadrats (large dots) are placed in different patterns relative to soil moisture. Estimates of niche breadth and overlap will vary at the different sites because of the patterns of spacing of the quadrats and the total range of soil moisture covered. (From Colwell and Futuyma 1971.)

weighting resource states, but neither of these two methods seems to have solved the problem of nonlinear niche dimensions. In practice we can do little to correct for this problem except to recognize that it is present in our data.

A related problem is how resource states are recognized by organisms and by field ecologists. If an ecologist recognizes more resource states than the organism, there is no problem in calculating niche breadth and overlap, assuming a suitable niche metric (Abrams 1980). Many measures of niche overlap show increased bias as the number of resource states increases (see page 472), so that one must be careful in picking a suitable niche measure. But if the ecologist does not recognize resource states that organisms do, there is a potential for misleading comparisons of species in different communities. There is no simple resolution of this difficulty, and it points again to the need for detailed knowledge of the natural history of the organisms being studied in order to minimize such distortions. In most studies of food niches, food items can be classified to species, and we presume that herbivores or predators do not subdivide resources more finely, although clearly animals may select different age-classes or growth-stages within a species. Microhabitat resource states are more difficult to define. Schoener (1970), for example, on the basis of detailed natural history observations, recognized three perch-diameter classes and three perch-height classes in sun or in shade for *Anolis* lizards in Bermuda. These classes were sufficient to show microhabitat segregation in *Anolis* and thus a finer subdivision was not necessary. There is an important message here: You must know the natural history of your organisms to quantify niche parameters in an ecologically useful way.

13.2 NICHE BREADTH

Some plants and animals are more specialized than others, and measures of *niche breadth* attempt to measure this quantitatively. Niche breadth has also been called *niche width* or *niche size* by ecologists. Niche breadth can be measured by observing the distribution of individual organisms within a set of resource states. The table formed by assigning species to the rows and resource states to the columns is called the *resource matrix* (Colwell and Futuyma 1971). Table 13.1 illustrates a resource matrix for lizards in the southwestern United States in which microhabitats are divided into 14 resource states. Three measures of niche breadth are commonly applied to the resource matrix.

13.2.1 Levins's Measure

Levins (1968) proposed that niche breadth be estimated by measuring the uniformity of distribution of individuals among the resource states. He suggested one way to measure this:

$$B = \frac{1}{\Sigma\, p_j^2} \tag{13.1}$$

which can also be written as

$$B = \frac{Y^2}{\Sigma\, N_j^2}$$

where B = Levins's measure of niche breadth

p_j = Proportion of individuals found in or using resource state j, or fraction of items in the diet that are of food category j $\left(\text{estimated by } \frac{N_j}{Y}\right)\left(\Sigma\, p_j = 1.0\right)$

N_j = Number of individuals found in or using resource state j

$Y = \Sigma\, N_j$ = Total number of individuals sampled

TABLE 13.1 EXAMPLE OF A RESOURCE MATRIX FOR LIZARD MICROHABITATS IN THE SOUTHWESTERN UNITED STATES

	Terrestrial										Arboreal					
Species	Open sun	Grass sun	Bush sun	Tree sun	Other sun	Open shade	Grass shade	Bush shade	Tree shade	Other shade	Low sun	Low shade	High sun	High shade	Sample size, *N*	Levins's *B* (microhabitat niche breadth)
Cnemidophorus tigris	47.5	2.5	34.6	2.5	2.5	0.0	0.0	7.8	0.0	2.5	0.1	0.0	0.0	0.0	1801	2.82
Uta stansburiana	27.1	4.6	41.3	4.6	4.6	0.8	0.0	6.0	0.0	3.9	5.3	0.1	1.8	0.0	768	3.87
Phrymosoma platyrhinos	92.6	0.0	2.5	0.0	0.0	2.5	0.0	2.5	0.0	0.0	0.0	0.0	0.0	0.0	121	1.17
Crotaphytus wislizeni	59.1	0.7	22.6	0.7	0.7	2.2	0.0	9.5	0.0	0.0	2.2	0.0	2.2	0.0	137	2.43
Callisaurus daraconoides	80.7	2.0	4.3	3.9	2.0	0.7	0.4	0.9	0.4	0.4	3.2	0.0	1.1	0.0	538	1.52
Sceloporus magister	0.0	0.0	1.2	1.2	3.5	0.0	0.0	3.5	1.2	1.2	10.0	5.9	38.2	34.1	85	3.58
Urosaurus graciosus	0.0	0.0	2.2	0.0	0.0	0.0	0.0	0.0	0.0	0.0	7.7	1.1	39.6	49.5	46	2.45
Dipsosaurus dorsalis	46.3	1.5	40.3	1.5	1.5	0.0	0.0	9.0	0.0	0.0	0.0	0.0	0.0	0.0	67	2.60
Uma scoparia	80.5	0.8	7.3	0.8	0.8	2.4	0.0	7.3	0.0	0.0	0.0	0.0	0.0	0.0	41	1.52
Coleonyx variegatus	0.0	0.0	0.0	0.0	0.0	95.3	0.0	4.7	0.0	0.0	0.0	0.0	0.0	0.0	43	1.10
Xantusia vigilis	0.0	0.0	0.0	0.0	0.0	0.0	0.0	0.0	100.0	0.0	0.0	0.0	0.0	0.0	27	1.00

Source: Pianka 1986, 160–161.

[a]The percentage utilization of 14 microhabitats by 11 lizard species from the deserts of the southwestern United States was determined by recording where each individual was sitting when first sighted. This table is based on 3674 sightings.

Note that B is the reciprocal of Simpson's index of diversity (Chapter 12, Section 12.4.3). Like the reciprocal of Simpson's index, B is maximum when an equal number of individuals occurs in each resource state, so that the species does not discriminate among the resource states and has the broadest possible niche. Levins's B is minimal when all the individuals occur in only one resource state (minimum niche breadth, maximum specialization). The range of B is from 1 to n, where n is the total number of resource states.

It is often useful to standardize niche breadth to express it on a scale from 0 to 1.0. This can be done easily for Levins's measure by dividing B by the total number of resource states after correcting for a finite number of resources. Hurlbert (1978) suggests the following measure for standardized niche breadth:

$$B_A = \frac{B - 1}{n - 1} \tag{13.2}$$

where B_A = Levins's standardized niche breadth
B = Levins's measure of niche breadth
n = Number of possible resource states

Box 13.1 illustrates the calculation of Levins's niche breadth.

Levins's measure of niche breadth does not allow for the possibility that resources vary in abundance. Hurlbert (1978) argues that in many cases ecologists should allow for the fact that some resources are very abundant and common, and other resources are uncommon or rare. The usage of resources ought to be scaled to their availability. If we add to the resource matrix a measure of the proportional abundance of each resource state, we can use the following measure of niche breadth:

$$B' = \frac{1}{\Sigma\, (p_j^2 / a_j)} \tag{13.3}$$

where B' = Hurlbert's niche breadth
p_j = Proportion of individuals found in or using resource j ($\Sigma\, p_j = 1.0$)
a_j = Proportion of the total available resources consisting of resource j ($\Sigma\, a_j = 1.0$)

B' can take on values from $1/n$ to 1.0 and should be standardized for easier comprehension. To standardize Hurlbert's niche breadth to a scale of 0–1, use the following equation:

$$B'_A = \frac{B' - a_{\min}}{1 - a_{\min}} \tag{13.4}$$

where B'_A = Hurlbert's standardized niche breadth
B' = Hurlbert's niche breadth (equation [13.3])
$a_{\min}$ = Smallest observed proportion of all the resources (minimum a_j)

Note that when all resource states are equally abundant, the a_j are all equal to $1/n$, and Levins's standardized niche breadth (equation [13.2]) and Hurlbert's standardized niche breadth (equation [13.4]) are identical.

The variance of Hurlbert's niche breadth can be estimated by the delta method (Smith 1982), as follows:

$$\text{var}(B') = \frac{4B'^4[\Sigma\, (p_j^3 / a_j^2) - (1/B')^2]}{Y} \tag{13.5}$$

Box 13.1 Calculation of Niche Breadth for Desert Lizards

Pianka (1986) gives the percentage utilization of 19 food sources for two common lizards of southwestern United States as follows:

	Cnemidophorus tigris (whiptail lizard)	*Uta stansburiana* (side-blotched lizard)
Spiders	1.9	3.9
Scorpions	1.3	0
Solpugids	2.1	0.5
Ants	0.4	10.3
Wasps	0.4	1.3
Grasshoppers	11.1	18.1
Roaches	4.8	1.5
Mantids	1.0	0.9
Ant lions	0.3	0.4
Beetles	17.2	23.5
Termites	30.0	14.7
Hemiptera and Homopters	0.6	5.8
Diptera	0.4	2.3
Lepidoptera	3.8	1.0
Insect eggs and pupae	0.4	0.1
Insect larvae	18.1	7.4
Miscellaneous arthropods	2.6	6.5
Vertebrates	3.6	0.2
Plants	0.1	1.6
Total	100.1	100.0
No. of individual lizards	1975	944

Levins's Measure of Niche Breadth

For the whiptail lizard, from equation (13.1),

$$B = \frac{1}{\Sigma p_j^2}$$

$$= \frac{1}{0.019^2 + 0.013^2 + 0.021^2 + 0.004^2 + 0.004^2 + 0.111^2 + \cdots}$$

$$= \frac{1}{0.171567} = 5.829$$

To standardize this measure of niche breadth on a scale of 0 to 1, calculate Levins's measure of standardized niche breadth (equation [13.2]):

$$B_A = \frac{B - 1}{n - 1}$$

$$= \frac{5.829 - 1}{19 - 1} = 0.2683$$

Shannon-Wiener Measure

For the whiptail lizard, using equation (13.7) and logs to the base e,

$$H' = -\sum p_j \log p_j$$

$$= -[(0.019)\log 0.019 + (0.013)\log 0.013 + (0.021)\log 0.021 + (0.004)\log 0.004 + (0.004)\log 0.004 + (0.111)\log 0.111 + \cdots$$

$$= 2.103 \text{ nits per individual}$$

To express this in the slightly more familiar units of *bits,*

$$H' \text{ (bits)} = 1.442695\ H' \text{ (nits)}$$

$$= (2.103)(1.442695) = 3.034 \text{ bits/individual}$$

To standardize this measure, calculate evenness from equation (13.8):

$$J' = \frac{H'}{\log n}$$

$$= \frac{2.103}{\log(19)} = 0.714$$

Smith's Measure

For the whiptail lizard data, by the use of equation (13.9),

$$FT = \sum (\sqrt{p_j a_j})$$

and assuming all 19 resources have equal abundance (each as a proportion 0.0526)

$$FT = \sqrt{(0.019)(0.0526)} + \sqrt{(0.013)(0.0526)} + \sqrt{(0.021)(0.0526)} + \cdots$$

$$= 0.78$$

The 95% confidence interval is given by equations (13.10) and (13.11):

$$\text{Lower 95\% confidence limit} = \sin\left[\arcsin(0.78) - \frac{1.96}{2\sqrt{1975}}\right]$$

$$= \sin(0.8726) = 0.766$$

$$\text{Upper 95\% confidence limit} = \sin\left[\arcsin(0.78) + \frac{1.96}{2\sqrt{1975}}\right]$$

$$= \sin(0.9167) = 0.794$$

Number of Frequently Used Resources

If we adopt 5% as the minimum cutoff, the whiptail lizard uses four resources frequently (grasshoppers, beetles, termites, and insect larvae).

These measures of niche breadth can all be calculated by Program-group NICHE MEASURES (Appendix 2).

where $\text{var}(B')$ = Variance of Hurlbert's measure of niche breadth (B')
p_j = Proportion of individuals found in or using resource state j ($\Sigma\, p_j = 1.0$)
a_j = Proportion resource j is of the total resources ($\Sigma\, a_j = 1.0$)
Y = Total number of individuals studied = $\Sigma\, N_j$

This variance, which assumes a multinomial sampling distribution, can be used to set confidence limits for these measures of niche breadth, if sample sizes are large, in the usual way:

$$B' \pm 1.96\sqrt{\text{var}(B')} \tag{13.6}$$

would give an approximate 95% confidence interval for Hurlbert's niche breadth.

In measuring niche breadth or niche overlap for food resources, an ecologist typically has two counts available: the number of *individual animals* and the number of *resource items*. For example, a single lizard specimen might have several hundred insects in its stomach. The sampling unit is usually the individual animal, and one must assume these individuals constitute a random sample. It is this sample size, of individual animals, that is used to calculate confidence limits (equation [13.6]). The resource items in the stomach of each individual are *not* independent samples, and they should be counted only to provide an estimate of the dietary proportions for that individual. If resource items are pooled over all individuals, the problem of sacrificial pseudoreplication occurs (see Chapter 10, page 349). If one fox has eaten one hare and a second fox has eaten 99 mice, the diet of foxes is 50% hares, not 1% hares.

13.2.2 Shannon-Wiener Measure

Colwell and Futuyma (1971) suggested using the Shannon-Wiener formula from information theory to measure niche breadth. Given the resource matrix, the formula is

$$H' = -\sum p_j \log p_j \tag{13.7}$$

where H' = Shannon-Wiener measure of niche breadth
p_j = Proportion of individuals found in or using resource j ($j = 1, 2, 3, \ldots n$)
n = Total number of resource states

and any base of logarithms may be used (see page 445). Since the Shannon-Wiener measure can range from 0 to ∞, one may wish to standardize it on a 0–1 scale. This can be done simply by using the evenness measure J':

$$J' = \frac{\text{Observed Shannon measure of niche breadth}}{\text{Maximum possible Shannon measure}} = \frac{H'}{\log n} \tag{13.8}$$

where J' = Evenness measure of the Shannon-Wiener function
n = Total number of possible resource states

and the same base of logarithms is used in equations (13.7) and (13.8). The Shannon-Wiener function is used less frequently than Levins's measure for niche breadth. Hurlbert (1978) argues against the use of the Shannon-Wiener measure because it has no simple ecological interpretation; he prefers Levins's measure of niche breadth. The former gives relatively more weight to the rare resources used by a species; conversely, the Levins measure gives more weight to the abundant resources used. Box 13.1 illustrates the calculation of the Shannon-Wiener measure of niche breadth.

13.2.3 Smith's Measure

Smith (1982) proposed another measure of niche breadth. It is similar to Hurlbert's measure (equation [13.3]) in that it allows you to take resource availability into account. The measure is

$$FT = \sum (\sqrt{p_j a_j}) \tag{13.9}$$

where FT = Smith's measure of niche breadth
p_j = Proportion of individuals found in or using resource state j
a_j = Proportion resource j is of the total resources
n = Total number of possible resource states

For large sample sizes, an approximate 95% confidence interval for FT can be obtained using the arcsine transformation as follows:

$$\text{Lower 95\% confidence limit} = \sin\left[x - \frac{1.96}{2\sqrt{y}}\right] \tag{13.10}$$

$$\text{Upper 95\% confidence limit} = \sin\left[x + \frac{1.96}{2\sqrt{y}}\right] \tag{13.11}$$

where x = Arcsin (FT)
y = Total number of individuals studied = ΣN_j

and the arguments of the trigonometric functions are in radians (not degrees).

Smith's measure of niche breadth varies from 0 (minimal) to 1.0 (maximal) and is thus a standardized measure. It is a convenient measure to use because its sampling distribution is known (Smith 1982).

Smith argues that his measure FT is the best measure of niche breadth that takes resource availability into account. Hurlbert's measure B' (equation [13.3]) is very sensitive to the selectivity of rare resources, which are more heavily weighted in the calculation of B'. Smith's FT measure is much less sensitive to selectivity of rare resources.

All niche breadth measures that consider resource availability estimate the overlap between the two frequency distributions of use and availability. The choice of the niche breadth measure to use in these situations depends upon how you wish to weight the differences. One simple measure is the percentage similarity measure (*PS*; see Chap. 11, Section 11.1.4), suggested as a measure of niche breadth by Feinsinger et al. (1981) and Schluter (1982). The percentage similarity measure of niche breadth is the opposite of Hurlbert's B' because it gives greater weight to the abundant resources and little weight to rare resources. For this reason Smith (1982) recommended against the use of percentage similarity as a measure of niche breadth. The decision about which measure is best depends completely on whether you wish for ecological reasons to emphasize dominant or rare resources in the niche breadth measure. Box 13.1 illustrates the calculation of Smith's measure of niche breadth.

13.2.4 Number of Frequently Used Resources

The simplest way to measure niche breadth is to count the number of resources used more than some minimal amount (Schluter, pers. comm.). The choice of the cutoff for frequent resource use is completely arbitrary, but if it is too high (>10%) the number of frequently used resources is constrained to be small. A reasonable value for the cutoff for many species might be 5%, so that the number of frequently used resources would always be 20 or less.

This simple measure of niche breadth may be adequate for many descriptive purposes. Figure 13.2 illustrates how closely it is correlated with Levins's measure of niche breadth for some of Pianka's lizard food-niche data.

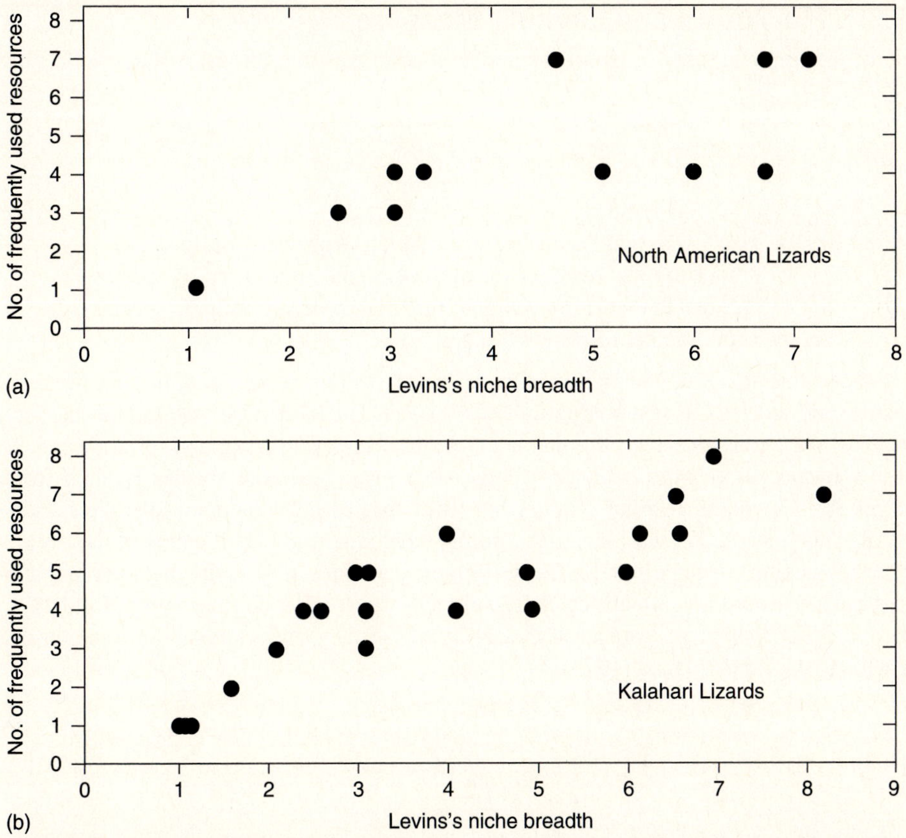

Figure 13.2 Relationship between two measures of niche breadth for desert lizard communities. (a) Diet niche breadths for North American lizards, 11 species, $r = 0.74$. (b) Diet niche breadths for Kalahari lizards, 21 species, $r = 0.84$ (data from Pianka 1986). The simple measure of number of frequently used resources is highly correlated with the more complex Levins's niche breadth measure.

If resources are subdivided in great detail, the minimal cutoff for the calculation of the number of frequently used resources must be reduced. As a rule of thumb, the cutoff should be approximately equal to the reciprocal of the number of resources, but never above 10%.

13.3 NICHE OVERLAP

One step in understanding community organization is to measure the overlap in resource use among the different species in a community guild. The most common resources measured in order to calculate overlap are *food* and *space* (or microhabitat). Several measures of niche overlap have been proposed, and there is considerable controversy about which is best (Hurlbert 1978, Abrams 1980, Linton et al. 1981, Loreau 1990, Manly 1990, Liebold 1995). The general problem of measuring niche overlap is very similar to the problem of measuring similarity (Chapter 11, Section 11.1), and some of the measures of niche overlap are identical to those we have already discussed for measuring community similarity.

13.3.1 MacArthur and Levins's Measure

One of the first measures proposed for niche overlap was that of MacArthur and Levins (1967):

$$M_{jk} = \frac{\sum_{i}^{n} p_{ij} p_{ik}}{\sum p_{ij}^2} \tag{13.12}$$

where M_{jk} = MacArthur and Levins's niche overlap of species k on species j
p_{ij} = Proportion that resource i is of the total resources used by species j
p_{ik} = Proportion that resource i is of the total resources used by species k
n = Total number of resource states

Note that the MacArthur-Levins measure of overlap is *not* symmetrical between species j and species k as you might intuitively expect. The MacArthur and Levins measure estimates the extent to which the niche space of species k overlaps that of species j. If species A specializes on a subset of foods eaten by a generalist species B, then from species A's viewpoint, overlap is total, but from species B's viewpoint, overlap is only partial. This formulation was devised to mimic the competition coefficients of the Lotka-Volterra equations (MacArthur 1972). Since most ecologists now agree that overlap measures cannot be used as competition coefficients (Hurlbert 1978; Abrams 1980; Holt 1987), the MacArthur-Levins measure has been largely replaced by a very similar but symmetrical measure first used by Pianka (1973):

$$O_{jk} = \frac{\sum_{i}^{n} p_{ij} p_{ik}}{\sqrt{\sum_{i}^{n} p_{ij}^2 \sum_{i}^{n} p_{ik}^2}} \tag{13.13}$$

where O_{jk} = Pianka's measure of niche overlap between species j and species k
p_{ij} = Proportion resource i is of the total resources used by species j
p_{ik} = Proportion resource i is of the total resources used by species k
n = Total number of resource states

This is a symmetric measure of overlap, so that overlap between species A and species B is identical to the overlap between species B and species A. The measure ranges from 0 (no resources used in common) to 1.0 (complete overlap). It has been used by Pianka (1986) for his comparison of desert lizard communities. Box 13.2 illustrates the calculation of niche overlap with the MacArthur-Levins measure and the Pianka measure.

Box 13.2 Calculation of Niche Overlap for African Finches

Dolph Schluter measured the diet of two seed-eating finches in Kenya in 1985 from stomach samples and obtained the following results, expressed as number of seeds in stomachs and proportions (in parentheses):

Seed species	Green-winged Pytilia (*Pytilia melba*)	Vitelline masked weaver (*Ploceus velatus*)
Sedge # 1	7 (0.019)	0 (0)
Sedge # 2	1 (0.003)	0 (0)
Setaria spp. (grass)	286 (0.784)	38 (0.160)
Grass # 2	71 (0.194)	24 (0.101)
Amaranth spp.	0 (0)	30 (0.127)
Commelina # 1	0 (0)	140 (0.591)
Commelina # 2	0 (0)	5 (0.021)
Total	365 food items	237 food items

MacArthur and Levins's Measure

From equation (13.12),

$$M_{jk} = \frac{\sum_{i}^{n} p_{ij}p_{ik}}{\sum p_{ij}^2}$$

$$\sum p_{ij}p_{ik} = (0.019)(0) + (0.003)(0) + (0.784)(0.160) + (0.194)(0.101) + (0)(0.127) + \cdots = 0.1453325$$

$$\sum p_{ij}^2 = 0.019^2 + 0.003^2 + 0.784^2 + 0.194^2 = 0.652662$$

$$\sum p_{ik}^2 = 0.160^2 + 0.101^2 + 0.127^2 + 0.591^2 + 0.021^2 = 0.401652$$

$$M_{jk} = \frac{0.14533}{0.6527} = 0.223 \quad \left(\text{extent to which } \textit{Pytilia} \text{ is overlapped by } \textit{Ploceus}\right)$$

$$M_{kj} = \frac{0.14533}{0.40165} = 0.362 \quad \left(\text{extent to which } \textit{Ploceus} \text{ is overlapped by } \textit{Pytilia}\right)$$

Note that these overlaps are not symmetrical, and for this reason this measure is rarely used by ecologists.

Pianka's modification of the MacArthur-Levins measure gives a symmetric measure of overlap that is preferred (equation [13.13]):

$$O_{jk} = \frac{\sum_{i}^{n} p_{ij}p_{ik}}{\sqrt{\sum_{i}^{n} p_{ij}^2 \sum_{i}^{n} p_{ik}^2}}$$

$$O_{jk} = \frac{0.1453325}{\sqrt{(0.6527)(0.40165)}} = 0.284$$

Note that this measure of overlap is just the geometric mean of the two MacArthur and Levins overlaps:

$$\text{Pianka's } O_{jk} = \sqrt{\text{MacArthur and Levins's } M_{jk}M_{kj}}$$

Percentage Overlap

From equation (13.14),

$$P_{jk} = \left[\sum_{i=1}^{n} (\text{minimum } p_{ij}, p_{ik})\right] 100$$
$$= (0 + 0 + 0.1603 + 0.1013 + 0 + 0 + 0)\,100 = 26.2\%$$

Morisita's Measure

From equation (13.15),

$$C = \frac{2\,\Sigma_i^n p_{ij}p_{ik}}{\Sigma_i^n p_{ij}[(n_{ij} - 1)/(N_j - 1)] + \Sigma_i^n p_{ik}[(n_{ik} - 1)/(N_k - 1)]}$$

From the calculations given above,

$$\sum p_{ij}p_{ik} = 0.14533$$

$$\sum p_{ij}[(n_{ij} - 1)/(N_j - 1)] = 0.019\left(\frac{7 - 1}{365 - 1}\right) + 0.003\left(\frac{1 - 1}{365 - 1}\right) + 0.784\left(\frac{286 - 1}{365 - 1}\right) + \cdots = 0.6514668$$

$$\sum p_{ik}[(n_{ik} - 1)/(N_k - 1)] = 0.160\left(\frac{38 - 1}{237 - 1}\right) + 0.101\left(\frac{24 - 1}{237 - 1}\right) + 0.127\left(\frac{30 - 1}{237 - 1}\right) + \cdots = 0.3989786$$

$$C = \frac{2(0.14533)}{0.6514668 + 0.3989786} = 0.277$$

Simplified Morisita Index

From equation (13.16),

$$C_H = \frac{2\,\Sigma_i^n p_{ij}p_{ik}}{\Sigma_i^n p_{ij}^2 + \Sigma_i^n p_{ik}^2}$$

These summation terms were calculated above for the MacArthur and Levins measures; thus,

$$C_H = \frac{2(0.1453325)}{0.652662 + 0.401652} = 0.276$$

Note that the simplified Morisita index is very nearly equal to the original Morisita measure and to the Pianka modification of the MacArthur-Levins measure.

Horn's Index

From equation (13.17),

$$R_o = \frac{\Sigma\,(p_{ij} + p_{ik})\log(p_{ij} + p_{ik}) - \Sigma\, p_{ij}\log p_{ij} - \Sigma\, p_{ik}\log p_{ik}}{2\log 2}$$

Using logs to base e:

$$\begin{aligned}
\sum (p_{ij} + p_{ik})\log(p_{ij} + p_{ik}) &= (0.019 + 0)\log(0.019 + 0) + (0.003 + 0)\log(0.003 + 0) \\
&\quad + (0.784 + 0.160)\log(0.784 + 0.160) + \cdots \\
&= -1.16129
\end{aligned}$$

$$\begin{aligned}
\sum p_{ij}\log p_{ij} &= (0.019)\log 0.019 + (0.003)\log 0.003 + (0.784)\log 0.784 + \cdots \\
&= -0.60165
\end{aligned}$$

$$\begin{aligned}
\sum p_{ik}\log p_{ik} &= (0.160)\log 0.160 + (0.101)\log 0.101 + (0.127)\log 0.12784 + \cdots \\
&= -1.17880
\end{aligned}$$

$$R_o = \frac{-1.16129 + 0.60165 + 1.17880}{2\log 2} = 0.4466$$

Hurlbert's Index

From equation (13.18),

$$L = \sum_{i}^{n} (p_{ij}p_{ik}/a_i)$$

If we assume that all seven seed species are equally abundant (a_i = 1/7 for all), we obtain

$$L = \frac{(0.019)(0)}{0.1429} + \frac{(0.003)(0)}{0.1429} + \frac{(0.784)(0.160)}{0.1429} + \cdots = 1.015$$

These calculations can be carried out in Program-group NICHE MEASURES (Appendix 2).

13.3.2 Percentage Overlap

Percentage overlap is identical with the *percentage similarity* measure proposed by Renkonen (1938) and is one of the simplest and most attractive measures of niche overlap. This measure is calculated as a *percentage* (see Chapter 11, Section 11.1.4) and is given by

$$P_{jk} = \left[\sum_{i=1}^{n} (\text{minimum } p_{ij}, p_{ik}) \right] 100 \qquad (13.14)$$

where P_{jk} = Percentage overlap between species j and species k
p_{ij} = Proportion resource i is of the total resources used by species j
p_{ik} = Proportion resource i is of the total resources used by species k
n = Total number of resource states

Percentage overlap is the simplest measure of niche overlap to interpret because it is a measure of the actual area of overlap of the resource utilization curves of the two species. This overlap measure was used by Schoener (1970) and has been labeled the Schoener overlap index (Hurlbert 1978). It would seem preferable to call it the Renkonen index or, more simply, percentage overlap. Abrams (1980) recommends this as the best of the measures of niche overlap. One strength of the Renkonen measure is that it is not sensitive to how one divides up the resource states. Human observers may recognize resource categories that animals or plants do not; conversely, organisms may distinguish resources lumped together by human observers. The first difficulty will affect the calculated value of MacArthur and Levins's measure of overlap, but should not affect the percentage overlap measure if sample size is large. The second difficulty is implicit in all niche measurements and emphasizes the need for sound natural history data on the organisms under study. Box 13.2 illustrates the calculation of the percentage overlap measure of niche overlap.

13.3.3 Morisita's Measure

Morisita's index of similarity (Chapter 11, Section 11.1.4), first suggested by Morisita (1959), can also be used as a measure of niche overlap. It is calculated from the following formula:

$$C = \frac{2 \sum_i^n p_{ij} p_{ik}}{\sum_i^n p_{ij}[(n_{ij} - 1)/(N_j - 1)] + \sum_i^n p_{ik}[(n_{ik} - 1)(N_k - 1)]} \qquad (13.15)$$

where C = Morisita's index of niche overlap between species j and k
p_{ij} = Proportion resource i is of the total resources used by species j
p_{ik} = Proportion resource i is of the total resources used by species k
n_{ij} = Number of individuals of species j that use resource category i
n_{ik} = Number of individuals of species k that use resource category i
N_j, N_k = Total number of individuals of each species in sample
$\sum_{i=1}^n n_{ij} = N_j, \sum_{i=1}^n n_{ik} = N_k$

Box 13.2 illustrates the calculation of the Morisita index of niche overlap.

Morisita's measure was formulated for counts of individuals and not for other measures of usage like proportions or biomass. If your data are not formulated as numbers of individuals, you can use the next measure of niche overlap, the simplified Morisita index, which is very similar to Morisita's original measure.

13.3.4 Simplified Morisita Index

The simplified Morisita index proposed by Horn (1966) is another similarity index that can be used to measure niche overlap. It is sometimes called the Morisita-Horn index. It is calculated as outlined in Box 13.2, from the following formula:

$$C_H = \frac{2 \sum_i^n p_{ij} p_{ik}}{\sum_i^n p_{ij}^2 + \sum_i^n p_{ik}^2} \tag{13.16}$$

where C_H = Simplified Morisita index of overlap (Horn 1966) between species j and species k
p_{ij} = Proportion resource i is of the total resources used by species j
p_{ik} = Proportion resource i is of the total resources used by species k
n = Total number of resource states ($i = 1, 2, 3, \ldots, n$)

The simplified Morisita index is very similar to the Pianka modification of the MacArthur and Levins measure of niche overlap, as can be seen by comparing equations (13.13) and (13.16). Linton et al. (1981) showed that for a wide range of simulated populations, the values obtained for overlap were nearly identical for the simplified Morisita and the Pianka measures. In general for simulated populations, Linton et al. found that the Pianka measure was slightly less precise (larger standard errors) than the simplified Morisita index in replicated random samples from two hypothetical distributions, and they recommended the simplified Morisita index as better.

13.3.5 Horn's Index

Horn (1966) suggested an index of similarity or overlap based on information theory. It is calculated as outlined in Chapter 11 (Section 11.1.4).

$$R_o = \frac{\sum (p_{ij} + p_{ik}) \log(p_{ij} + p_{ik}) - \sum p_{ij} \log p_{ij} - \sum p_{ik} \log p_{ik}}{2 \log 2} \tag{13.17}$$

where R_o = Horn's index of overlap for species j and k
p_{ij} = Proportion resource i is of the total resources utilized by species j
p_{ik} = Proportion resource i is of the total resources utilized by species k

and any base of logarithms may be used. Box 13.2 illustrates the calculation for Horn's index of overlap.

13.3.6 Hurlbert's Index

None of the previous four measures of niche overlap recognize that the resource states may vary in abundance. Hurlbert (1978) defined niche overlap as "the degree to which the frequency of encounter between two species is higher or lower than it would be if each species utilized each resource state in proportion to the abundance of that resource state." The appropriate measure of niche overlap that allows resource states to vary in size is as follows:

$$L = \sum_i^n \left(\frac{p_{ij} p_{ik}}{a_i} \right) \tag{13.18}$$

where L = Hurlbert's measure of niche overlap between species j and species k
p_{ij} = Proportion resource i is of the total resources utilized by species j
p_{ik} = Proportion resource i is of the total resources utilized by species k
a_i = Proportional amount or size of resource state i ($\Sigma\ a_i = 1.0$)

Hurlbert's overlap measure is not like other overlap indices in ranging from 0 to 1. It is 1.0 when both species utilize each resource state in proportion to its abundance, 0 when the two species share no resources, and >1.0 when the two species both use certain resource states more intensively than others and the preferences of the two species for resources tend to coincide.

Hurlbert's index L has been criticized by Abrams (1980) because its value changes when resource states used by neither one of the two species are added to the resource matrix. Hurlbert (1978) considers this an advantage of his index because it raises the critical question of what resource states one should include in the resource matrix. Box 13.2 illustrates the calculation of Hurlbert's index of overlap.

13.3.7 Which Overlap Index Is Best?

The wide variety of indices available to estimate niche overlap has led many ecologists to argue that the particular index used is relatively unimportant, since they all give the same general result (Pianka 1974). One way to evaluate overlap indices is to apply them to artificial populations with known overlaps. Three studies have used simulation techniques to investigate the bias of niche overlap measures and their sensitivity to sample size. Ricklefs and Lau (1980) analyzed four indices of niche overlap, as did Linton et al. (1981). Smith and Zaret (1982) analyzed seven measures of niche overlap. I will not discuss the conclusions of Linton et al. (1981), which suffer from the methodological flaw of defining true overlap as the percentage overlap measure (equation [13.14]) and then comparing all the other measures to it—a circular argument.

Ricklefs and Lau (1980) showed by computer simulation that the sampling distribution of all measures of niche overlap are strongly affected by sample size (see Figure 13.3). When niche overlap is complete, there is a negative bias in the percentage overlap measure that is reduced but not eliminated as sample size increases (Figure 13.3). This negative bias at high levels of overlap seems to be true of all measures of niche overlap (Ricklefs and Lau 1980; Linton et al. 1981).

Smith and Zaret (1982) have presented the most penetrating analysis of bias in estimating niche overlap. Figure 13.4 shows the results of their simulations. Bias (= true value − estimated value) increases in all measures of overlap as the number of resources increases, and overlap was always underestimated. This effect is particularly strong for the percentage overlap measure. The amount of bias can be quite large when the number of resource categories is large, even if sample size is reasonably large ($N = 200$; see Figure 13.4a). All niche overlap measures show decreased bias as sample size goes up (Figure 13.4b). Increasing the smaller sample size has a much greater effect than increasing the larger sample size. The bias is minimized when both species are sampled equally ($N_1 = N_2$). As evenness of resource use increases, bias increases only for the percentage overlap measure and the simplified Morisita index (Figure 13.4c).

The percentage overlap measure (equation [13.14]) and the simplified Morisita index (equation [13.16]) are the two most commonly used measures of niche overlap, yet they

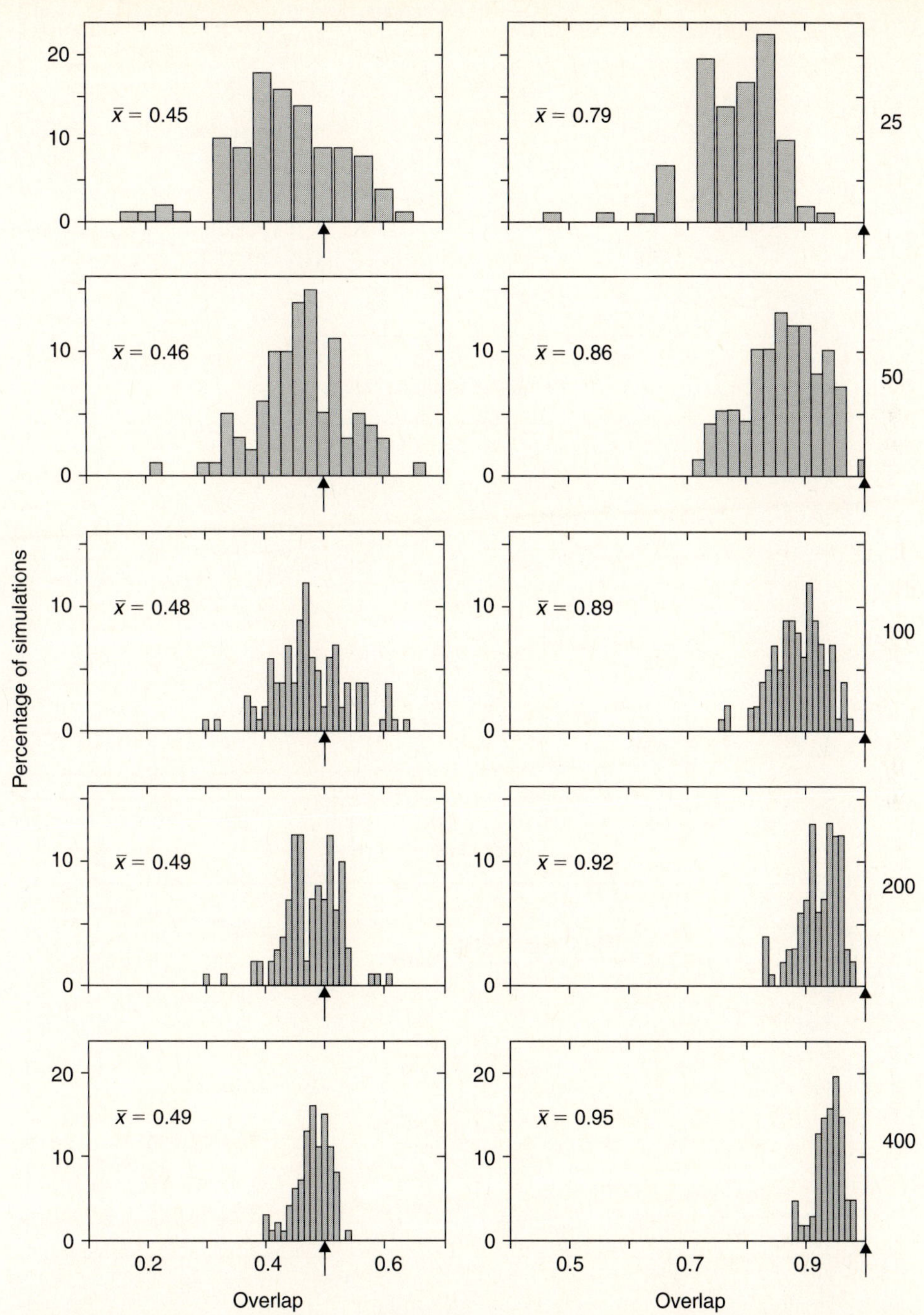

Figure 13.3 Sampling distributions of the percentage overlap measure of niche breadth (equation [13.14]) for probability distributions having five resource categories. Sample sizes were varied from $n = 25$ at top to $n = 400$ at the bottom. The expected values of niche overlap are marked by the arrows (0.5 for the left side and 1.0 for the right side). Simulations were done 100 times for each example. (After Ricklefs and Lau 1980.)

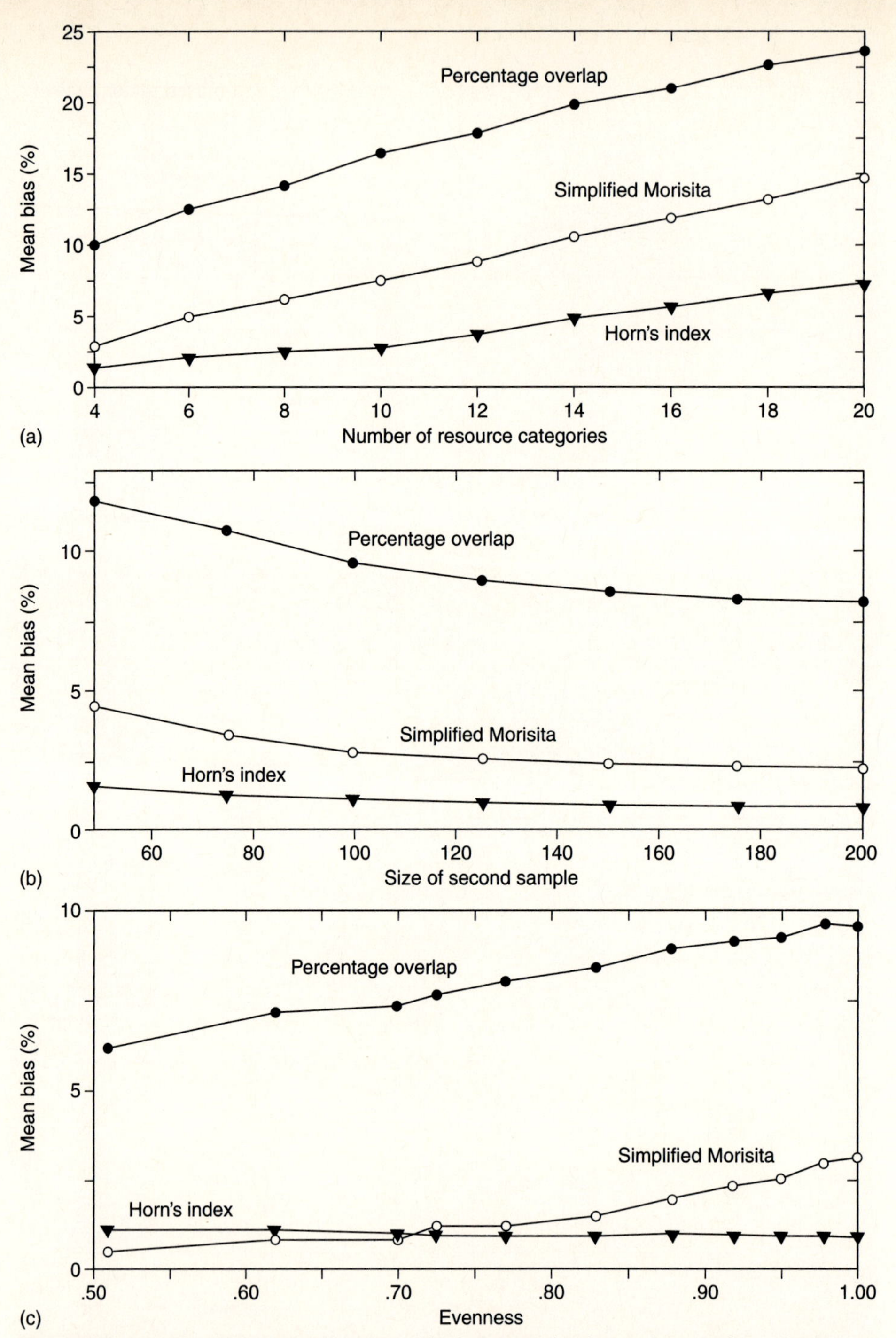

Figure 13.4 Bias in niche overlap measures. Bias is measured as (true value − estimated value) expressed as a percentage. All overlap bias results in an underestimation of the overlap. (a) Effect of changing the number of resource categories on the bias of the percentage overlap measure (eq. [13.14]), the simplified Morisita measure (eq. [13.16]), and the Horn index (eq. [13.17]). Simulations were run with equal sample sizes for the two species. (b) Effect of the size of the second sample on bias. The first sample was $n_1 = 100$, and four resource categories were used. (c) Effect of evenness of resource use on the bias of measures of niche overlap. Evenness is 1.0 when all four resource categories are used equally. Sample sizes for both species were 100 in these simulations. (From Smith and Zaret 1982.)

are the two measures that Smith and Zaret (1982) found to be most biased under changing numbers of resources, sample size, and resource evenness (Figure 13.4). The best overlap measure found by Smith and Zaret is Morisita's measure (equation [13.15]), which is not graphed in Figure 13.4 because it has nearly zero bias at all sample sizes and also when there are a large number of resources. The recommendation to minimize bias is thus to use Morisita's measure to assess niche overlap. If resource use cannot be expressed as numbers of individuals (which Morisita's measure requires), the next best measure of overlap appears to be Horn's index (Smith and Zaret 1982; Ricklefs and Lau 1980).

If confidence intervals or tests of significance are needed for niche overlap measures, two approaches may be used: first, obtain replicate sets of samples, calculate the niche overlap for each set, and calculate the confidence limits or statistical tests from these replicate values (Horn 1966; Hurlbert 1978). Alternatively, use statistical procedures to estimate standard errors for these indices.

Three statistical procedures can be used to generate confidence intervals for measures of niche overlap (Mueller and Altenberg 1985; Maurer 1982; Ricklefs and Lau 1980): the delta method, the jackknife method, and the bootstrap. The delta method is the standard analytical method used in mathematical statistics for deriving standard errors of any estimated parameter. Standard errors estimated by the delta method are not always useful to ecologists because they are difficult to derive for complex ecological measures, and they cannot be used to estimate confidence limits that are accurate when sample sizes are small and variables do not have simple statistical distributions. For this reason the jackknife and bootstrap methods—resampling methods most practical with a computer—are of great interest to ecologists (see Chapter 15, Section 15.5). Mueller and Altenberg (1985) argue that in many cases the populations being sampled may be composed of several unrecognized subpopulations (e.g., based on sex or age differences). If this is the case, the bootstrap method is the best to use. See Mueller and Altenberg (1985) for a discussion of the application of the bootstrap to generating confidence limits for niche overlap measures.

The original goal of measuring niche overlap was to infer interspecific competition (Schoener 1974), but the relationship between niche overlap and competition is poorly defined in the literature. The particular resources being studied may not always be limiting populations, and species may overlap with no competition. Conversely, MacArthur (1968) pointed out that zero niche overlap did not mean that interspecific competition was absent. Abrams (1980) pointed out that niche overlap does not always imply competition, and that in many cases niche overlap should be used as a descriptive measure of community organization. The relationship between competition and niche overlap is complex (Holt 1987).

13.4 MEASUREMENT OF HABITAT AND DIETARY PREFERENCES

If an animal is faced with a variety of possible food types, it prefers some and avoids others. We ought to be able to measure preference very simply by comparing usage and availability. Manly et al. (1993) have recently reviewed the problem of resource selection by animals, and they provide a detailed statistical discussion of the problems of measuring preferences. Note that resource selection may involve habitat preferences, food preferences, or nest site preferences. We discuss here more cases of diet preference, but the principles are the same for any resource selection problem.

Three general study designs for measuring preferences are reviewed by Manly et al. (1993):

1. **Design I**: With this design all measurements are made at the population level and individual animals are not recognized. Used and unused resources are sampled for the entire study area with no regard for individuals. For example, fecal pellets are recorded as present or absent on a series of quadrats.
2. **Design II**: Individual animals are identified and the use of resources measured for each individual, but the availability of resources is measured at the population level for the entire study zone. For example, stomach contents of individuals can be measured and compared with the food available on the study area.
3. **Design III**: Individual animals are measured as in Design II, but in addition the resources available for each individual are measured. For example, habitat locations can be measured for a set of radio-collared individuals, and these can be compared to the habitats available within the home range of each individual.

Clearly Designs II and III are most desirable, since they allow us to measure resource preferences for each individual. If the animals studied are a random sample of the population, we can infer the average preference of the entire population, as in Design I, but we can also ask questions about preferences of different age or sex groups in the population.

The key problem in all these designs is to estimate the resource selection probability function, defined as the probability of use of resources of different types. Often we cannot estimate these probabilities in an absolute sense, but we can estimate a set of preferences that are proportional to these probabilities. We can thus conclude, for example, that moose prefer to eat the willow *Salix glauca* without knowing exactly what fraction of the willows are in fact eaten by the moose in an area.

Several methods have been suggested for measuring preferences (Chesson 1978; Cock 1978; Johnson 1980; Manly et al. 1993). The terminology for all the indices of preference is the same, and can be described as follows: Assume an array of n types of food (or other resources) in the environment, and that each food type has m_i items or individuals ($i = 1, 2 \ldots, n$), and the total abundance of food items is

$$M = \sum_{i=1}^{n} m_i \tag{13.19}$$

where M = Total number of food items available

Let u_i be the number of food items of species i in the diet, so that we have a second array of items selected by the species of interest. The total diet is given by

$$U = \sum_{i=1}^{n} u_i \tag{13.20}$$

where U = Total number of food items in the diet

In most cases the array of items in the environment and the array of food items in the diet are expressed directly as proportions or percentages. Table 13.2 gives an example of dietary data in meadow voles.

TABLE 13.2 INDEX OF ABUNDANCE OF 15 GRASSES AND HERBS IN AN INDIANA GRASSLAND AND PERCENTAGE OF THE DIET OF THE MEADOW VOLE *Microtus pennsylvanicus* (BY VOLUME) ESTIMATED FROM STOMACH SAMPLES

	Grassland		Diet	
Plant	Index of abundance	Proportion of total	Percent of total volume	Percent of plant volume
Poa	6.70	26.8	32.1	36.9
Muhlengergia	6.30	25.2	14.6	16.8
Panicum	2.90	11.6	24.7	28.4
Achillea	2.90	11.6	4.7	5.4
Plantago	2.25	9.0	5.8	6.7
Daucus	0.70	2.8	0	0
Aster	0.55	2.2	0	0
Solidago	0.55	2.2	0	0
Bromus	0.50	2.0	0	0
Ambrosia	0.40	1.6	0	0
Rumex	0.35	1.4	0	0
Taraxacum	0.30	1.2	2.1	2.4
Phleum	0.20	0.8	1.4	1.6
Asclepias	0.20	0.8	0	0
Oxalis	0.20	0.8	1.6	1.8
Total	25.00	100.0	87.0	100.0
Other items (roots, fungi, insects)			13.0	

Source: Zimmerman 1965.

Most measures of preference assume that the density of the food items available is constant. Unless food items are replaced as they are eaten, food densities will decrease. If any preference is shown, the relative proportion of the food types will thus change under exploitation. If the number of food items is very large, or if food can be replaced as in a laboratory test, this problem of exploitation is unimportant. Otherwise you must be careful in choosing a measure of preference (Cock 1978).

How can we judge the utility of measures of preference? Cock (1978) has suggested that three criteria should be considered in deciding on the suitability of an index of preference:

1. *Scale* of the index: it is best to have both negative and positive preference scales of equal size, symmetric about 0.
2. *Adaptability* of the index: it is better to have the ability to include more than two food types in the index.
3. *Range* of the index: it is best if maximum index values are attainable at all combinations of food densities.

We now consider several possible measures of preference and evaluate their suitability on the basis of these criteria.

13.4.1 Forage Ratio

The simplest measure of preference is the *forage ratio* first suggested by Savage (1931) and by Williams and Marshall (1938):

$$w_i = \frac{o_i}{p_i} \tag{13.21}$$

where w_i = Forage ratio for species i (Index 2 of Cock 1978)
o_i = Proportion or percentage of species i in the diet
p_i = Proportion or percentage of species i available in the environment

The forage ratio is more generally called the *selection index* by Manly et al. (1993), since not all resource selection problems involve food. One example will illustrate the forage ratio. Lindroth and Batzli (1984) found that *Bromus inermis* comprised 3.6% of the diet of meadow voles in bluegrass fields when this grass comprised 0.3% of the vegetation in the field. Thus the forage ratio or selection index for *Bromus inermis* is

$$\hat{w}_i = \frac{3.6}{0.3} = 12.0$$

Selection indices above 1.0 indicate preference; values less than 1.0 indicate avoidance. Selection indices may range from 0 to ∞, which is a nuisance; consequently, Manly et al. (1993) suggest presenting forage ratios or selection indices as standardized ratios that sum to 1.0 for all resource types:

$$B_i = \frac{\hat{w}_i}{\sum_{i=1}^{n} \hat{w}_i} \tag{13.22}$$

where B_i = Standardized selection index for species i
$\hat{w}_i$ = Forage ratio for species i (equation [13.19])

Standardized ratios of (1/number of resources) indicate no preference. Values below this indicate relative avoidance, and values above indicate relative preference.

Table 13.3 illustrates data on selection indices from a Design I–type study of habitat selection by moose in Minnesota.

Statistical tests of selection indices depend on whether the available resources are censused completely or estimated with a sample. In many cases of habitat selection, air photo maps are used to measure the area of different habitat types, so that there is a complete census with (in theory) no errors of estimation. Or in laboratory studies of food preference, the exact ratios of the foods made available are set by the experimenter. In other cases sampling is carried out to estimate the available resources, and these availability ratios are subject to sampling errors.

Let us consider the first case of a complete census of available resources, so that there is no error in the proportions available (p_i). To test the null hypothesis that animals are selecting resources at random, Manly et al. (1993) recommend the G-test:

$$\chi^2 = 2 \sum_{i=1}^{n} \left[u_i \ln\left(\frac{u_i}{Up_i}\right) \right] \tag{13.23}$$

TABLE 13.3 SELECTION INDICES FOR MOOSE TRACKS IN FOUR HABITAT TYPES ON 134 SQUARE KILOMETERS OF THE LITTLE SIOUX BURN IN MINNESOTA DURING THE WINTER OF 1971–72.[a]

Habitat	Proportion available (p_i)	No. of moose tracks (u_i)	Proportion of tracks in habitat (o_i)	Selection index (w_i)	Standardized selection index[b] (B_i)
Interior burn	0.340	25	0.214	0.629	0.110
Edge burn	0.101	22	0.188	1.866	0.326
Forest edge	0.104	30	0.256	2.473	0.433
Forest	0.455	40	0.342	0.750	0.131
Total	1.000	117	1.000	5.718	1.000

[a] Data from Neu et al. 1974.

[b] Standardized selection indices above (1/number of resources), or 0.25 in this case, indicate preference.

where u_i = Number of observations using resource i
U = Total number of observations of use = Σu_i
χ^2 = Chi-squared value with $(n - 1)$ degrees of freedom (H_0: random selection)
n = Number of resource categories

The standard error of a selection ratio can be approximated by

$$s_{\bar{w}_i} = \sqrt{\frac{o_i(1 - o_i)}{Up_i^2}} \tag{13.24}$$

where $s_{\bar{w}_i}$ = Standard error of the selection ratio for resource i
o_i = Observed proportion of use of resource i
U = Total number of observations of use
p_i = Proportion of type i resources that are available in study area

The confidence limits for a single selection ratio are found as usual:

$$\hat{w}_i \pm z_\alpha s_{\bar{w}_i} \tag{13.25}$$

where z_α is the standard normal deviate (1.96 for 95% confidence, 2.576 for 99%, and 1.645 for 90% confidence).

Two selection ratios can be compared to see if they are significantly different with the following test from Manly et al. (1993, 48):

$$\chi^2 = \frac{(\hat{w}_i - \hat{w}_j)^2}{\text{variance}(\hat{w}_i - \hat{w}_j)}$$
$$\text{variance}(\hat{w}_i - \hat{w}_j) = \frac{o_i(1 - o_i)}{Up_i^2} - \frac{2o_io_j}{Up_ip_j} + \frac{o_j(1 - o_j)}{Up_j^2} \tag{13.26}$$

where χ^2 = Chi-squared value with 1 degree of freedom (H_0: $\hat{w}_i = \hat{w}_j$)
o_i, o_j = Observed proportion of use of resource i or j
U = Total number of observations of use
p_i, p_j = Proportion of type i or j resources that are available in study area

Box 13.3 illustrates the calculation of selection indices and their confidence limits for the moose data in Table 13.3.

In the second case in which the resources available must be estimated from samples (and hence have some possible error), the statistical procedures are slightly altered. To test

Box 13.3 Calculation of Selection Indices for Moose in Four Habitat Types

Neu et al. (1974) provided these data for moose in Minnesota:

Habitat	Proportion available (p_i)	No. of moose tracks (u_i)	Proportion of tracks in habitat (o_i)
Interior burn	0.340	25	0.214
Edge burn	0.101	22	0.188
Forest edge	0.104	30	0.256
Forest	0.455	40	0.342
Total	1.000	117	1.000

This is an example of Design I data in which only population-level information is available. Since the proportion available was measured from aerial photos, we assume it is measured exactly without error.

The selection index is calculated from equation (13.21):

$$w_i = \frac{o_i}{p_i}$$

For the interior burn habitat,

$$w_1 = \frac{o_1}{p_1} = \frac{25/117}{0.340} = 0.629$$

Similarly for the other habitats,

$$w_2 = \frac{22/117}{0.101} = 1.866$$

$$w_3 = 2.473$$

$$w_4 = 0.750$$

We can test the hypothesis of equal use of all four habitats with equation (13.23):

$$\begin{aligned}\chi^2 &= 2\sum_{i=1}^{n}[u_i \ln(u_i/Up_i)]\\ &= 2[25(\log_e(25/117(0.340)) + 22\{\log_e(22/117(0.101))\} + \cdots]\\ &= 35.40 \text{ with 3.d.f. } (p < 0.01)\end{aligned}$$

This value of chi-squared is considerably larger than the critical value of 3.84 at $\alpha = 5\%$, so we reject the null hypothesis that moose use all these habitats equally.

We can compute the confidence limits for these selection indices from equations (13.24) and (13.25). For the interior burn habitat,

$$s_{\bar{w}_1} = \sqrt{\frac{o_1(1 - o_1)}{Up_1^2}} = \sqrt{\frac{0.214(1 - 0.214)}{117(0.340^2)}} = 0.112$$

The 95% confidence limits for this selection index must be corrected using the Bonferroni correction for (α/n) are given by

$$\hat{w}_i \pm z_\alpha s_{\bar{w}_1}$$

$$0.629 \pm z_{0.0125}(0.112), \text{ or } 0.629 \pm 2.498(0.112), \text{ or } 0.350 \text{ to } 0.907$$

The Bonferroni correction corrects for multiple comparisons to maintain a consistent overall error rate by reducing the α value to α/n and thus in this example using $z_{0.0125}$ instead of $z_{0.05}$. Sokal and Rohlf (1995, 240) discuss the Bonferroni method.

Similar calculations can be done for the confidence limits for the edge burn (0.968 to 2.756). Since these two confidence belts do not overlap we suspect that these two habitats are selected differently.

We can now test if the selection index for the interior burn habitat differs significantly from that for the edge burn habitat. From equation (13.26),

$$\chi^2 = \frac{(\hat{w}_i - \hat{w}_j)^2}{\text{variance}(\hat{w}_i - \hat{w}_j)}$$

The variance of the difference is calculated from

$$\text{variance}(\hat{w}_i - \hat{w}_j) = \frac{o_i(1 - o_i)}{Up_i^2} - \frac{2o_io_j}{Up_ip_j} + \frac{o_j(1 - o_j)}{Up_j^2}$$

$$= \frac{0.214(1 - 0.214)}{117(0.340^2)} - \frac{2(0.214)(0.188)}{117(0.340)(0.101)} + \frac{0.188(1 - 0.188)}{117(0.101^2)}$$

$$= 0.1203$$

Thus,

$$\chi^2 = \frac{(\hat{w}_i - \hat{w}_j)^2}{\text{variance }(\hat{w}_i - \hat{w}_j)} = \frac{(0.629 - 1.866)^2}{0.1203} = 12.72 \qquad (p < 0.01)$$

This chi-squared has 1 d.f., and we conclude that these two selection indices are significantly different, edge burn being the preferred habitat.

These calculations can be carried out by Program-group NICHE MEASURES (Appendix 2).

the null hypothesis of no selection, we compute the G-test in a manner similar to that of equation (13.23):

$$\chi^2 = 2 \sum_{i=1}^{n} \left[u_i \ln\left(\frac{u_i}{Up_i}\right) + m_i \ln\left(\frac{m_i}{(m_i + u_i M/(U + M))}\right)\right] \quad (13.27)$$

where u_i = Number of observations using resource i
m_i = Number of observations of available resource i
U = Total number of observations of use = $\Sigma\, u_i$
M = Total number of observations of availability = $\Sigma\, m_i$
χ^2 = Chi-squared value with $(n - 1)$ degrees of freedom (H_0: random selection)
n = Number of resource categories

Similarly, to estimate a confidence interval for the selection ratio when availability is sampled, we estimate the standard error of the selection ratio as

$$s_{\bar{w}_i} = \sqrt{\frac{(1 - o_i)}{Uo_i} + \frac{(1 - p_i)}{p_i M}} \quad (13.28)$$

where all the terms are as defined above. Given this standard error, the confidence limits for the selection index are determined in the usual way (equation [13.25]).

We have discussed so far only Design I studies. Design II studies, in which individuals are recognized, allow a finer level of analysis of resource selection. The general principles are similar to those just provided for Design I studies, and details of the calculations are given in Manly et al. (1993, Chapter 4).

When a whole set of confidence intervals are to be computed for a set of proportions of habitats utilized, one should not use the simple binomial formula, because the anticipated confidence level (e.g., 95%) is often not achieved in this multinomial situation (Cherry 1996). Cherry showed through simulation that acceptable confidence limits for the proportions of resources utilized could be obtained with Goodman's (1965) formulas as follows:

$$L_{o_i} = \frac{C + 2(u_i - 0.5) - \sqrt{C(C + 4(u_i - 0.5)(U - u_i + 0.5))/U}}{2(C + U)} \quad (13.29)$$

$$U_{o_i} = \frac{C + 2(u_i + 0.5) - \sqrt{C(C + 4(u_i + 0.5)(U - u_i - 0.5))/U}}{2(C + U)} \quad (13.30)$$

where L_{o_i} = Lower confidence limit for the proportion of habitat i used
U_{o_i} = Upper confidence limit for the proportion of habitat i used
C = Upper $\frac{\alpha}{n}$ percentile of the χ^2 distribution with 1 d.f.
u_i = Number of observations of resource i being used
U = Total number of observation of resource use
n = Number of habitats available or number of resource states

These confidence limits can be calculated in Program-group EXTRAS (Appendix 2).

13.4.2 Murdoch's Index

A number of indices of preference are available for the two-prey case in which an animal is choosing whether to eat prey species *a* or prey species *b*. Murdoch (1969) suggested the index *C* such that

$$\frac{r_a}{r_b} = C\left(\frac{n_a}{n_b}\right) \quad \text{or} \quad C = \left(\frac{r_a}{r_b}\right)\left(\frac{n_b}{n_a}\right) \tag{13.31}$$

where C = Murdoch's index of preference (Index 4 of Cock 1978)
r_a, r_b = Proportion of prey species *a*, *b* in diet
n_a, n_b = Proportion of prey species *a*, *b* in the environment

Murdoch's index is similar to the instantaneous selective coefficient of Cook (1971) and the survival ratio of Paulik and Robson (1969), and was used earlier by Cain and Sheppard (1950) and Tinbergen (1960). Murdoch's index is limited to comparisons of the relative preference of two prey species, although it can be adapted to the multiprey case by pooling prey into two categories: species A and all other species.

Murdoch's index has the same scale problem as the forage ratio: ranging from 0 to 1.0 for negative preference and from 1.0 to infinity for positive preference. Jacobs (1974) pointed out that by using the logarithm of Murdoch's index, we can achieve symmetrical scales for positive and negative preference. Murdoch's index has the desirable attribute that actual food densities do not affect the maximum attainable value of the index C.

13.4.3 Manly's α

A simple measure of preference can be derived from probability theory using the probability of encounter of a predator with a prey and the probability of capture upon encounter (Manly et al. 1972; Chesson 1978). Chesson argues strongly against Rapport and Turner (1970), who attempted to separate *availability* and *preference*. Preference, according to Chesson, reflects any deviation from random sampling of the prey, and therefore includes all the biological factors that affect encounter rates and capture rates, including availability.

Two situations must be distinguished to calculate Manly's α as a preference index (Chesson 1978).

Constant Prey Populations When the number of prey eaten is very small in relation to the total (or when replacement prey are added in the laboratory), the formula for estimating α is

$$\alpha_i = \frac{r_i}{n_i}\left(\frac{1}{\Sigma\ (r_j/n_j)}\right) \tag{13.32}$$

where α_i = Manly's α (preference index) for prey type *i*
r_i, r_j = Proportion of prey type *i* or *j* in the diet (i and $j = 1, 2, 3, \ldots, m$)
n_i, n_j = Proportion of prey type *i* or *j* in the environment
m = Number of prey types possible

Note that the α values are normalized so that

$$\sum_{i=1}^{m} \alpha_i = 1.0 \tag{13.33}$$

When selective feeding does not occur, $\alpha_i = 1/m$ (m = total number of prey types). If α_i is greater than $(1/m)$, then prey species i is preferred in the diet. Conversely, if α_i is less than $(1/m)$, prey species i is avoided in the diet.

Given estimates of Manly's α for a series of prey types, it is easy to eliminate one or more prey species and obtain a relative preference for those remaining (Chesson 1978). For example, if four prey species are present, and you desire a new estimate for the alphas without species 2 present, the new alpha values are simply

$$\alpha_1 = \frac{\alpha_1}{\alpha_1 + \alpha_3 + \alpha_4} \tag{13.34}$$

and similarly for the new α_3 and α_4. One important consequence of this property is that you should not compare α values from two experiments with different numbers of prey types, since their expected values differ.

Variable Prey Populations When a herbivore or a predator consumes a substantial fraction of the prey available, or in laboratory studies in which it is not possible to replace prey as they are consumed, one must take into account the changing numbers of the prey species in estimating α. This is a much more complex problem than that outlined above for the constant prey case. Chesson (1978) describes one method of estimation. Manly (1974) showed that an approximate estimate of the preference index for experiments in which prey numbers are declining is given by

$$\alpha_i = \frac{\log p_i}{\sum_{i=1}^{m} p_j} \tag{13.35}$$

where
α_i = Manly's α (preference index) for prey type i
p_i, p_j = Proportion of prey i or j remaining at the end of the experiment $(i = 1, 2, 3, \ldots, m)$ $(j = 1, 2, 3, \ldots, m) = e_i/n_i$
e_i = Number of prey type i remaining uneaten at end of experiment
n_i = Initial number of prey type i in experiment
m = Number of prey types

Any base of logarithms may be used and will yield the same final result. Manly (1974) gives formulas for calculating the standard error of these α values. Manly suggested that equation (13.35) provided a good approximation to the true values of α when the number of individuals eaten *and* the number remaining uneaten at the end were all larger than 10.

There is a general problem in experimental design in estimating Manly's α for a variable prey population. If a variable prey experiment is stopped before too many individuals have been eaten, it may be analyzed as a constant prey experiment with little loss in accuracy. Otherwise the stopping rule should be that for all prey species, both the number eaten and the number left uneaten should be greater than 10 at the end of the experiment.

Manly's α is also called Chesson's index in the literature. Manly's α is strongly affected by the values observed for rare resource items and is affected by the number of resource types used in a study (Confer and Moore 1987). Box 13.4 illustrates the

Box 13.4 Calculation of Manly's α as an Index of Preference

Constant Prey Population

Three color-phases of prey were presented to a fish predator in a large aquarium. As each prey was eaten, it was immediately replaced with another individual. One experiment produced these results:

	Type 1 prey	Type 2 prey	Type 3 prey
No. of prey present in the environment at all times	4	4	4
Proportion present, n_i	0.333	0.333	0.333
Total number eaten during experiment	105	67	28
Proportion eaten, r_i	0.525	0.335	0.140

From equation (13.32),

$$\alpha_i = \frac{r_i}{n_i}\left(\frac{1}{\Sigma\,(r_j/n_j)}\right)$$

$$\alpha_1 = \frac{0.525}{0.333}\left[\frac{1}{(0.525/0.333) + (0.335/0.333) + (0.140/0.333)}\right] = 0.52$$

(preference for type 1 prey)

$$\alpha_2 = \frac{0.335}{0.333}\left(\frac{1}{3.00}\right) = 0.34 \text{ (preference for type 2 prey)}$$

$$\alpha_3 = \frac{0.140}{0.333}\left(\frac{1}{3.00}\right) = 0.14 \text{ (preference for type 3 prey)}$$

The α values measure the probability that an individual prey item is selected from a particular prey class when all prey species are equally available.

Since there are 3 prey types, α values of 0.33 indicate no preference. In this example prey type 1 is highly preferred, and prey type 3 is avoided.

Variable Prey Population

The same color-phases were presented to a fish predator in a larger aquarium, but no prey items were replaced after one was eaten, so the prey numbers declined during the experiment. The results obtained follow.

	Type 1 prey	Type 2 prey	Type 3 prey
No. of prey present at start of experiment, n_i	98	104	54
No. of prey alive at end of experiment, e_i	45	66	43
Proportion of prey alive at end of experiment, p_i	0.459	0.635	0.796

From equation (13.35),

$$\alpha_i = \frac{\log p_i}{\sum_{j=1}^{m} p_j}$$

Using logs to base e, we obtain

$$\alpha_1 = \frac{\log_e(0.459)}{\log_e(0.459) + \log_e(0.635) + \log_e(0.796)} = \frac{-0.7783}{-1.4608} = 0.53$$

$$\alpha_2 = \frac{\log_e(0.635)}{-1.4608} = 0.31$$

$$\alpha_3 = \frac{\log_e(0.796)}{-1.4608} = 0.16$$

As in the previous experiment, since there are 3 prey types, α values of 0.33 indicate no preference. Prey type 1 is highly preferred and prey type 3 is avoided.

Manly (1974) shows how the standard errors of these α values can be estimated.

Program-group NICHE MEASURES (Appendix 2) can do these calculations.

calculation of Manly's alpha. Program-group NICHE MEASURES (Appendix 2) does these calculations.

13.4.4 Rank Preference Index

The calculation of preference indices is critically dependent upon the array of resources that the investigator includes as part of the available food supply (Johnson 1980). Table 13.4 illustrates this with a hypothetical example. When a common but seldom-eaten food species is included or excluded in the calculations, a complete reversal of which food species are preferred may arise.

TABLE 13.4 HYPOTHETICAL EXAMPLE OF HOW THE INCLUSION OF A COMMON BUT SELDOM-USED FOOD ITEM WILL AFFECT THE PREFERENCE CLASSIFICATION OF A FOOD SPECIES

Food species	Percent in diet	Percent in environment	Classification
	Case A: species x included		
x	2	60	Avoided
y	43	30	Preferred
z	55	10	Preferred
	Case B: species x not included		
y	44	75	Avoided
z	56	25	Preferred

Source: Johnson 1980.

One way to avoid this problem is to *rank* both the utilization of a food resource and the availability of that resource, and then to use the difference in these ranks as a measure of relative preference. Johnson (1980) emphasized that this method will produce only a ranking of *relative* preferences, and that all statements of absolute preference should be avoided. The major argument in favor of the ranking method of Johnson is that the analysis is usually not affected by the inclusion or exclusion of food items that are rare in the diet. Johnson's method is applied to individuals and can thus be applied to Design II and Design III studies of how individuals select resources.

To calculate the rank preference index, proceed as follows:

Step 1. Determine for each individual the rank of usage (r_i) of the food items from 1 (most used) to m (least used), where m is the number of species of food resources.

Step 2. Determine for each individual the rank of availability (s_i) of the m species in the environment; these ranks might be the same for all individuals or specific for each individual.

Step 3. Calculate the rank difference for each of the m species:

$$t_i = r_i - s_i \tag{13.36}$$

where t_i = Rank difference (measure of relative preference)
r_i = Rank of usage of resource type i ($i = 1, 2, 3, \ldots, m$)
s_i = Rank of availability of resource type i

Step 4. Average all the rank differences across all the individuals sampled, and rank these averages to give an order of relative preference for all the species in the diet.

Box 13.5 illustrates these calculations of the rank preference index. Program RANK provided by Johnson (1980) can do these calculations.

13.4.5 Rodgers's Index for Cafeteria Experiments

In cafeteria experiments an array of food types is presented to an animal in equal abundance so that availability does not enter directly into the measurement of preference. But in many cases food types cannot be easily replenished as they are eaten; consequently, the most preferred foods are eaten first. Figure 13.5 illustrates one such cafeteria experiment on the collared lemming (*Dicrostonyx groenlandicus*). Rodgers and Lewis (1985) argued that if the total amount eaten at the end of the trial is used (as in Manly's α, page 484), a misleading preference score can be obtained because species not eaten until later in the trial will have equal preference scores to those eaten first. Rodgers suggests that the most appropriate measure of preference is the *area* under each of the cumulative consumption curves in Figure 13.5, standardized to a maximum of 1.0. Rodgers's index is calculated as follows:

Step 1. Measure the area under the curve for each of the species in the cafeteria trial. This can be done most simply by breaking the curve up into a series of triangles and trapezoids, and summing the area of these.

Step 2. Standardize the preference scores to the range 0–1.0 by the following formula:

$$R_i = \frac{A}{\max(A_i)} \tag{13.37}$$

Box 13.5 Calculation of Rank Preference Indices

Johnson (1980) gave data on the habitat preferences of two mallard ducks, as follows:

	Bird 5198		Bird 5205	
Wetland class	Usage[a]	Availability[b]	Usage	Availability
I	0.0	0.1	0.0	0.4
II	10.7	1.2	0.0	1.4
III	4.7	2.9	21.0	3.5
IV	20.1	0.8	0.0	0.4
V	22.1	20.1	5.3	1.2
VI	0.0	1.4	10.5	4.9
VII	2.7	12.6	0.0	1.0
VIII	29.5	4.7	15.8	5.1
IX	0.0	0.0	10.5	0.7
X	2.7	0.2	36.8	1.8
XI	7.4	1.1	0.0	1.2
Open water	0.0	54.9	0.0	78.3
Total	99.9	100.0	99.9	99.9

[a]Usage = percentage of recorded locations in each wetland class (r_i).

[b]Availability = percentage of wetland area in a bird's home range in each wetland class (s_i).

1. Rank the usage values (r_i) from 1 (most used) to 12 (least used), assigning average ranks to ties. Thus wetland class VIII has the highest usage score (29.5) for duck 5198, so it is assigned rank 1. Wetland classes I, VI, IX, and open water all have 0.0 for duck 5198, so these tie for ranks 9, 10, 11, and 12 and hence are assigned average rank of 10.5. The results are given below.

2. Rank the available resources (s_i) in the same manner from open water (rank 1) to the lowest availability (class IX for bird 5198, class I for bird 5205). Note that because this is a Design III study, in which individuals are scored for both availability and usage, we must rank each individual separately. The resulting ranks for these two mallards are as follows:

	Rank			
	Bird 5198		Bird 5205	
Wetland class	Usage	Availability	Usage	Availability
I	10.5	11	9.5	12
II	4	7	9.5	6
III	6	5	2	4
IV	3	9	9.5	11
V	2	2	6	7.5
VI	10.5	6	4.5	3
VII	7.5	3	9.5	9
VIII	1	4	3	2
IX	10.5	12	4.5	10
X	7.5	10	1	5
XI	5	8	9.5	7.5
Open water	10.5	1	9.5	1

3. Calculate the differences in ranks for each individual to get a relative measure of preference, $t_i = r_i - s_i$

Wetland class	Bird 5198	Bird 5205	Average rank difference
I	−0.5	−2.5	−1.5
II	−3	+3.5	+0.25
III	+1	−2	−0.5
IV	−6	−1.5	−3.7
V	0	−1.5	−0.7
VI	+4.5	+1.5	+3.0
VII	+4.5	+0.5	+2.5
VIII	−3	+1	−1.0
IX	−1.5	−5.5	−3.5
X	−2.5	−4	−3.2
XI	−3	+2	−0.5
Open water	+9.5	+8.5	+9.0

The smallest average rank indicates the most preferred resource. For these ducks we can rank the wetland habitats as follows:

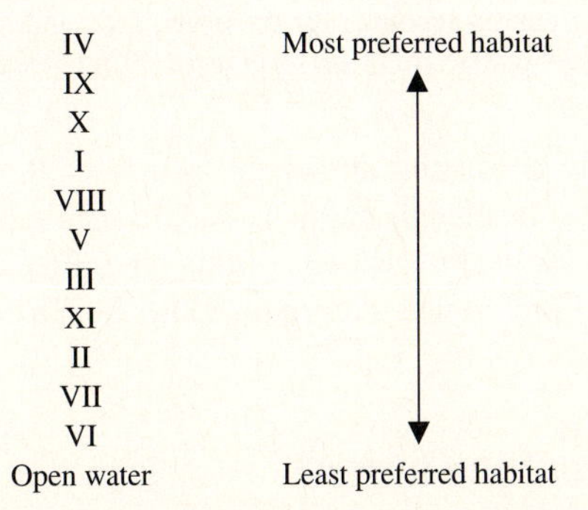

Johnson (1980) discusses a method by which the significance of these preferences can be evaluated. Note that statistical tests cannot be done unless the number of individuals is equal to or greater than the number of preference categories (habitats). This simple case with two ducks was used here for illustration only.

These rank preferences can be obtained with Program RANK provided by Johnson (1980).

where R_i = Rodgers's index of preference for cafeteria experiments for species i
A_i = Area under the cumulative proportion eaten curve for species i
$\max(A_i)$ = Largest value of the A_i

Cafeteria-type experiments could also be analyzed using Manly's α for variable prey populations, but this is not recommended. Note that Manly's α was designed for experiments in which some fraction of *each* of the food items remains uneaten at the end of the experiment, while Rodgers's index can be applied to experiments in which one or more

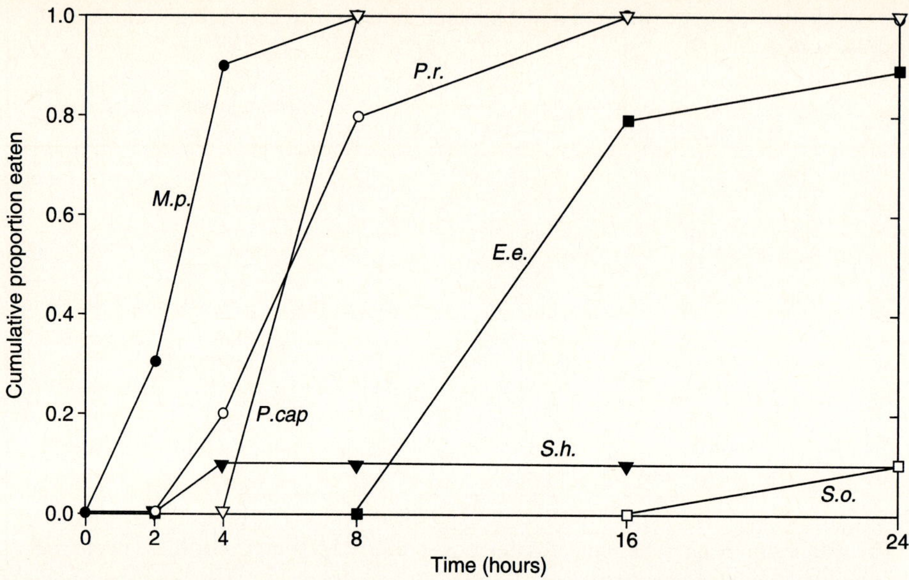

Figure 13.5 Cafeteria test with seven species of herbs given to a collared lemming for 24 hours. The area under each of these curves accounts for the order, rate, and total amount of each species eaten and is the most appropriate measure of performance. (From Rodgers 1984.)

food types are completely consumed during the course of the experiment. Manly et al. (1993, Chapter 6) provide a detailed discussion of additional, more comprehensive methods that can be used for time-series observations of this type. Box 13.6 illustrates the calculation of Rodgers's index of preference. Program-group NICHE MEASURES (Appendix 2) can do these calculations.

13.4.6 Which Preference Index?

No general agreement exists in the literature about which of these indices is the best measure of preference. No one has done a simulation study to look at the properties of these indices under a variety of situations. Until this is done, I can only make some tentative recommendations.

The selection index (forage ratio), Manly's α, and the rank preference index would appear to be the best indices of preference for most situations. The only exception would seem to be the variable prey case (cafeteria experiment) in which some prey types are eaten entirely; in these cases, use Rodgers's index of preference or the methods of Manly et al. (1993, Chapter 6).*

Measures of resource preference are usually presented and discussed in the literature without any method of estimating standard errors of these measures or their confidence limits. General methods of estimating confidence limits can be applied to these measures, under the usual assumptions of approximate normality. Because of the complex nature of these measures, the best approach to estimating probable errors would seem to be re-

*Manly et al. (1993) provide a computer program, RSF (*Resource Selection Functions*), for the IBM PC for the more complex designs to estimate preferences.

Box 13.6 Calculation of Rodgers's Indices of Preference from Cafeteria Experiments

In one trial of a cafeteria experiment with snowshoe hares, we obtained these results:

	Proportion eaten		
Time (hr)	*Betula glandulosa*	*Salix glauca*	*Picea glauca*
6	0.20	0.10	0.00
12	0.45	0.25	0.05
18	0.55	0.40	0.05
24	0.90	0.55	0.10
36	1.00	0.75	0.15
48	1.00	0.85	0.20

I will illustrate the calculations only for bog birch (*Betula glandulosa*).

1. Calculate the area under the cumulative consumption curve as a series of triangles and trapezoids:

(a) 0–6 hr:

$$\text{Triangle area} = \frac{1}{2}(\text{base})(\text{height})$$

$$= \frac{1}{2}(6)(0.2) = 0.6$$

(b) 6–12 hr: Using the equation for the area of a trapezoid,

$$\text{Area} = \text{base} \times \text{average height of trapezoid}$$

$$= (12 - 6)\left(\frac{0.2 + 0.45}{2}\right) = 1.95$$

(c) 12–18 hr: by the same rule,

$$\text{Area} = (18 - 12)\left(\frac{0.45 + 0.55}{2}\right) = 3.00$$

(d) 18–24 hr: by the same rule,

$$\text{Area} = (24 - 18)\left(\frac{0.55 + 0.90}{2}\right) = 4.35$$

(e) 24–36 hr: by the same rule,

$$\text{Area} = (36 - 24)\left(\frac{0.90 + 1.00}{2}\right) = 11.40$$

(f) 36–48 hr: for one piece,

$$\text{Rectangular area} = (48 - 36)(1.00) = 12.0$$

$$\text{Total area under the curve for } \textit{Betula} = 0.6 + 1.95 + 3.0 + 4.35 + 11.40 + 12.00$$
$$= 33.3 \text{ units of area (hours} \times \text{proportion)}$$

By a similar set of calculations, we obtain

$$\text{Total area for } \textit{Salix} = 23.55 \text{ units of area}$$

$$\text{Total area for } \textit{Picea} = 4.50 \text{ units of area}$$

2. Standardize the preference scores by the use of equation (13.37):

$$R_i = \frac{A_i}{\max(A_i)}$$

$$R_1 = \frac{33.3}{33.3} = 1.00$$

$$R_2 = \frac{23.55}{33.3} = 0.71$$

$$R_3 = \frac{4.50}{33.3} = 0.14$$

The most preferred food has preference score 1.0, and the smaller the standardized preference score, the less preferred the food. In this experiment hares prefer to eat *Betula glandulosa,* and they tend to avoid *Picea glauca* when they have a choice.

These calculations can be done by Program-group NICHE MEASURES (Appendix 2).

sampling schemes such as the jackknife and bootstrap methods discussed in Chapter 15 Section 15.5.

Dietary preference is one component of diet selection models (Ellis et al. 1976). There is now an elegant body of theory on optimal foraging in animals (Stephens and Krebs 1986) within which measures of preference also reside. Combining field measures of diet preference with theoretical insights from optimal foraging theory is a major challenge for ecologists.

Food preference is only one form of preference, and it is important to note that any resource axis can be analyzed by means of these same preference indices. For example, habitat preferences, temperature preferences, or nest site preferences could be analyzed using the approaches just outlined.

13.5 SUMMARY

How organisms utilize their environment can be quantified by measuring the *niche*. Organisms may be *generalists* and utilize a wide spectrum of resources, or they may be *special-*

ists and use only a few resources. *Niche breadth* is usually measured by the same general formulas used to measure species diversity (heterogeneity measures). Levins's measure (= Simpson's diversity) and the Shannon-Wiener measure have both been used to estimate niche breadth, although neither of these measures takes account of possible differences in resource abundance. The choice of a measure of niche breadth depends upon whether you choose to emphasize resources used often or those used less frequently, and whether you wish to take resource availability into account.

Niche overlap is important in analyzing community organization, and several measures of overlap are in common use. Bias can be severe in two of the commonly used indices of niche overlap (percentage overlap and simplified Morisita index). Only Morisita's original measure is free from bias over most of the range of possible overlap, and this overlap measure is recommended.

Resource selection by animals can occur for food, habitat, or other resources like nest sites. Several measures of preference are commonly used. The selection index (forage ratio), Manly's α, and the rank preference index appear to be the most easily understood measures of preference, and are recommended for most studies of preference. Habitat preference can be analyzed using the same methods as dietary preferences. Defining a *resource*, determining the resources that are available to animals, and sampling the array of resources actually used by organisms are all major problems to which ecologists must address more attention.

SELECTED READING

Abrams, P. 1980. Some comments on measuring niche overlap. *Ecology* 61: 44–49.

Arthur, S. M., Manly, B. F. J., McDonald, L. L., and Garner, G. W. 1996. Assessing habitat selection when availability changes. *Ecology* 77: 215–227.

Cherry, S. 1996. A comparison of confidence interval methods for habitat use-availability studies. *Journal of Wildlife Management* 60: 653–658.

Chesson, J. 1978. Measuring preference in selective predation. *Ecology* 59: 211–215.

Cock, M. J. W. 1978. The assessment of preference. *Journal of Animal Ecology* 47: 805–816.

Colwell, R. K., and Futuyma, D. J. 1971. On the measurement of niche breadth and overlap. *Ecology* 52: 567–576.

Hurlbert, S. H. 1978. The measurement of niche overlap and some relatives. *Ecology* 59: 67–77.

Johnson, D. H. 1980. The comparison of usage and availability measurements for evaluating resource preference. *Ecology* 61: 65–71.

Manly, B. F. J., McDonald, L. L., and Thomas, D. L. 1993. *Resource selection by animals: Statistical design and analysis for field studies*. Chapman and Hall, London.

Mueller, L. D., and Altenberg, L. 1985. Statistical inference on measures of niche overlap. *Ecology* 66: 1204–1210.

Smith, E. P., and Zaret, T. M. 1982. Bias in estimating niche overlap. *Ecology* 63: 1248–1253.

QUESTIONS AND PROBLEMS

13.1. Calculate dietary niche breadth for meadow voles from the data given in Table 13.2. How does niche breadth change when you use a measure of niche breadth that allows for differences in abundance of the food resources, as given in Table 13.2?

13.2. Calculate the niche breadth for two hypothetical species utilizing four resources:

	Resource type			
	A	B	C	D
Relative availability of resource	0.40	0.40	0.10	0.10
Relative utilization				
Species 1	0.40	0.40	0.10	0.10
Species 2	0.25	0.25	0.25	0.25

Which of these two species is a generalist, and which is a specialist? Compare your evaluation with that of Petraitis (1979).

13.3. Calculate niche overlap for two separate species pairs from the following hypothetical data (percent utilization):

	Resource type				
	A	B	C	D	E
Community X					
Species 1	60	30	10	0	0
Species 2	0	0	10	30	60
Community Y					
Species 3	0	90	10	0	0
Species 4	0	0	10	90	0

Would you expect a measure of niche overlap to give equal overlap values for these two communities? Compare your answer with the comments in Hurlbert (1978, 68).

13.4. What measure of niche overlap would you recommend using for the data in Table 13.1? Review the recommendations in Linton et al. (1981) regarding sample size and discuss whether the data in Table 13.1 meet Linton's criteria for adequate sampling for all the species pairs.

13.5. Ivlev (1961, Table 18) gives data on the diet of carp during a depletion experiment over 5 days as follows:

	Biomass of food remaining (grams)				
Food type	Day 0	Day 1	Day 2	Day 3	Day 4
Chironomid larvae	10.0	6.92	4.21	1.64	0.32
Amphipods	10.0	7.63	5.59	3.72	2.26
Freshwater isopods	10.0	8.50	7.13	5.95	4.94
Mollusks	10.0	8.95	7.98	6.95	6.25

Calculate Manly's α for these four food types, and compare these values with those calculated using Rodgers's technique.

13.6. Arnett et al. (unpublished, 1989; data in Manly et al. 1993) reported on habitat selection in radio-collared bighorn sheep in Wyoming. Ten habitat types were mapped and the locations of individual sheep were obtained during autumn 1988 as follows:

Habitat	Available proportion	Habitat locations for bighorn sheep number						
		A	B	C	D	E	F	Total
Riparian	0.06	0	0	0	0	0	0	0
Conifer	0.13	0	2	1	1	0	2	6
Shrub type A	0.16	0	1	2	3	2	1	9
Aspen	0.15	2	2	1	7	2	4	18
Rock outcrop	0.06	0	2	0	5	5	2	14
Sagebrush	0.17	16	5	14	3	18	7	63
Ridges	0.12	5	10	9	6	10	6	46
Shrub type B	0.04	14	10	8	9	6	15	62
Burnt areas	0.09	28	35	40	31	25	19	178
Clearcuts	0.02	8	9	4	9	0	19	49
Total	1.00	73	76	79	74	68	75	445

(a) What type of design is this study?

(b) Calculate the selection indices for each of the 6 sheep. Which habitats are preferred, and which are avoided?

(c) What two methods could be used to obtain average selection ratios for these sheep? Which method is better? Manly et al. (1993, 55) discusses this problem.

(d) How would you test the hypothesis that these 6 sheep use habitats in the same manner? Do they? What assumptions must you make to do this test?

13.7. Calculate the selection index (forage ratio) for the preference data in Box 13.5, and compare these values to those estimated in Box 13.5 using Johnson's (1980) rank preference index.

PART FIVE

Ecological Miscellanea

The first four parts of this book present statistical methods that every ecologist should be able to use for measuring ecological systems. But many other statistical problems arise in ecology, and we now enter the gray zone of things you might wish to know ecologically and things you should have been taught by your statistics professor.

Population ecologists must estimate both survival and reproduction if they are to understand the demographic equation. Reproductive parameters are species- or group-specific, so the statistical methods for measuring reproductive output in Cladocera are quite different from those for measuring the reproductive rate of deer. For this reason I do not try to discuss the specific methods for estimating reproductive rates in this book.

But survival rates have some general methodological problems in their estimation, and in Chapter 14 I discuss some of the methods for estimating survival that can be applied to many types of animals and plants. There are some idiosyncratic choices even here—for example, wildlife ecologists need a good discussion of the problem of estimating survival from radiotelemetry data, so I have presented a synopsis here.

Some statistical procedures like transformations are so obvious to statisticians that they are covered all too quickly in most statistics texts. Other statistical methods like the bootstrap are most useful to ecologists but are often deemed too sophisticated for introductory statistics courses. In Chapter 15 I present a few of the statistical methods that you may need to know before you take on the real world with all its problems.

CHAPTER 14

Estimation of Survival Rates

Survival is a critical population process, and the estimation of survival rates is one important part of measuring the parameters of the demographic equation. There are many different approaches to estimating survival rates, and we have already discussed the mark-recapture method in Chapter 2. In this chapter I discuss some of the more general techniques for estimating survival rates. There is a large literature on the estimation of survival rates in wild populations, and in this chapter we only scratch the surface of the methods available. There is a bias in the methods discussed here toward the larger vertebrates.

14.1 FINITE AND INSTANTANEOUS RATES

The simplest measures of survival are obtained by following a group of individuals over time. This gives us the *finite* survival rate:

$$\text{Finite survival rate} = \frac{\text{Number of individuals alive at end of time period}}{\text{Number of individuals alive at start of time period}} \quad (14.1)$$

$$\hat{S}_0 = \frac{N_t}{N_0}$$

If the time interval is one year, this will give a finite *annual* survival rate. Finite survival rates can range from 0 to 1, and they always apply to some specific time period.

One common problem in calculating survival is to convert observed rates to a standardized time base. For example, you might observe a 0.95 finite survival rate over 38 days and wish to express this as a 30-day rate or as an annual rate. This is easily done with finite rates as follows:

$$\begin{Bmatrix}\text{Adjusted finite}\\ \text{survival rate}\end{Bmatrix} = \begin{pmatrix}\text{Observed finite}\\ \text{survival rate}\end{pmatrix}^{(t_s/t_0)} \quad (14.2)$$

where t_s = Standardized time interval (e.g., 30 days)
t_0 = Observed time interval (e.g., 38 days)

Box 14.1 illustrates this conversion procedure. In general you should not extrapolate time periods too much when doing these conversions. If you measure survival rates over 3 days it is somewhat misleading to convert this to an annual survival rate.

Survival rates can be expressed as finite rates or as instantaneous rates. If the number of deaths in a short time period is proportional to the total number of individuals at that time, the rate of drop in numbers can be described by the geometric equation

$$\frac{dN}{dt} = iN \quad (14.3)$$

where N = Number of individuals
i = Instantaneous mortality rate (always a negative number)
t = Time

Figure 14.1 illustrates geometric population decline. In integral form equation (14.3) becomes

$$N_t = N_0 e^{it} \quad (14.4)$$

where N_t = Number of individuals at time t
N_0 = Number of individuals at start of time period

If $t = 1$ time unit and we take logs, we obtain

$$\log_e\left(\frac{N_t}{N_0}\right) = i \quad (14.5)$$

Since (N_t/N_0) is the finite survival rate (from equation [14.1]), we have obtained the following simple relationship:

$$\log_e(\text{finite survival rate}) = \text{Instantaneous mortality rate} \quad (14.6)$$

Box 14.1 Conversion of Survival Rates to a Standard Time Base

1. Keith et al. (1984) observed in snowshoe hares a finite survival rate of 0.384 for a time span of 346 days. To convert this survival rate to a standard year (365 days) proceed as follows:

Using equation (14.2),

$$\begin{Bmatrix}\text{Adjusted annual}\\ \text{survival rate}\end{Bmatrix} = \begin{pmatrix}\text{Observed finite}\\ \text{survival rate}\end{pmatrix}^{(t_s/t_0)}$$

$$= (0.384)^{365/346} = (0.384)^{1.0549} = 0.364$$

The equivalent annual survival rate for these hares is thus 36.4%.

To express this survival rate as a weekly survival rate,

$$\begin{Bmatrix}\text{Adjusted weekly}\\ \text{survival rate}\end{Bmatrix} = \begin{pmatrix}\text{Observed finite}\\ \text{survival rate}\end{pmatrix}^{(t_s/t_0)}$$

$$= (0.384)^{7/346} = (0.384)^{0.10693} = 0.903$$

The equivalent weekly survival rate is thus 90.3% in this example.

2. Perennial lupine plants in the southern Yukon suffer a mortality rate of 22% per year (Turkington, pers. comm.). What fraction of these plants will survive for 10 years if the mortality rate does not change? For this example,

$$\begin{Bmatrix}\text{Adjusted 10 year}\\ \text{survival rate}\end{Bmatrix} = \begin{pmatrix}\text{Observed one year}\\ \text{survival rate}\end{pmatrix}^{(t_s/t_0)}$$

$$= (1.0 - 0.22)^{10/1} = (0.78)^{10} = 0.083$$

or about 8% of plants would live for 10 years in this population.

These calculations can be carried out by Program-group SURVIVAL (Appendix 2).

Note that there is always some *time unit* that must be specified for both finite and instantaneous rates. Instantaneous mortality rates are always negative numbers because they describe the slope of the lines illustrated in Figure 14.1. Because the finite mortality rate is defined as

$$\text{Finite mortality rate} = (1.0 - \text{finite survival rate}) \tag{14.7}$$

we can convert between finite and instantaneous rates of mortality by the following equation:

$$\text{Finite mortality rate} = 1.0 - e^{\text{instantaneous mortality rate}} \tag{14.8}$$

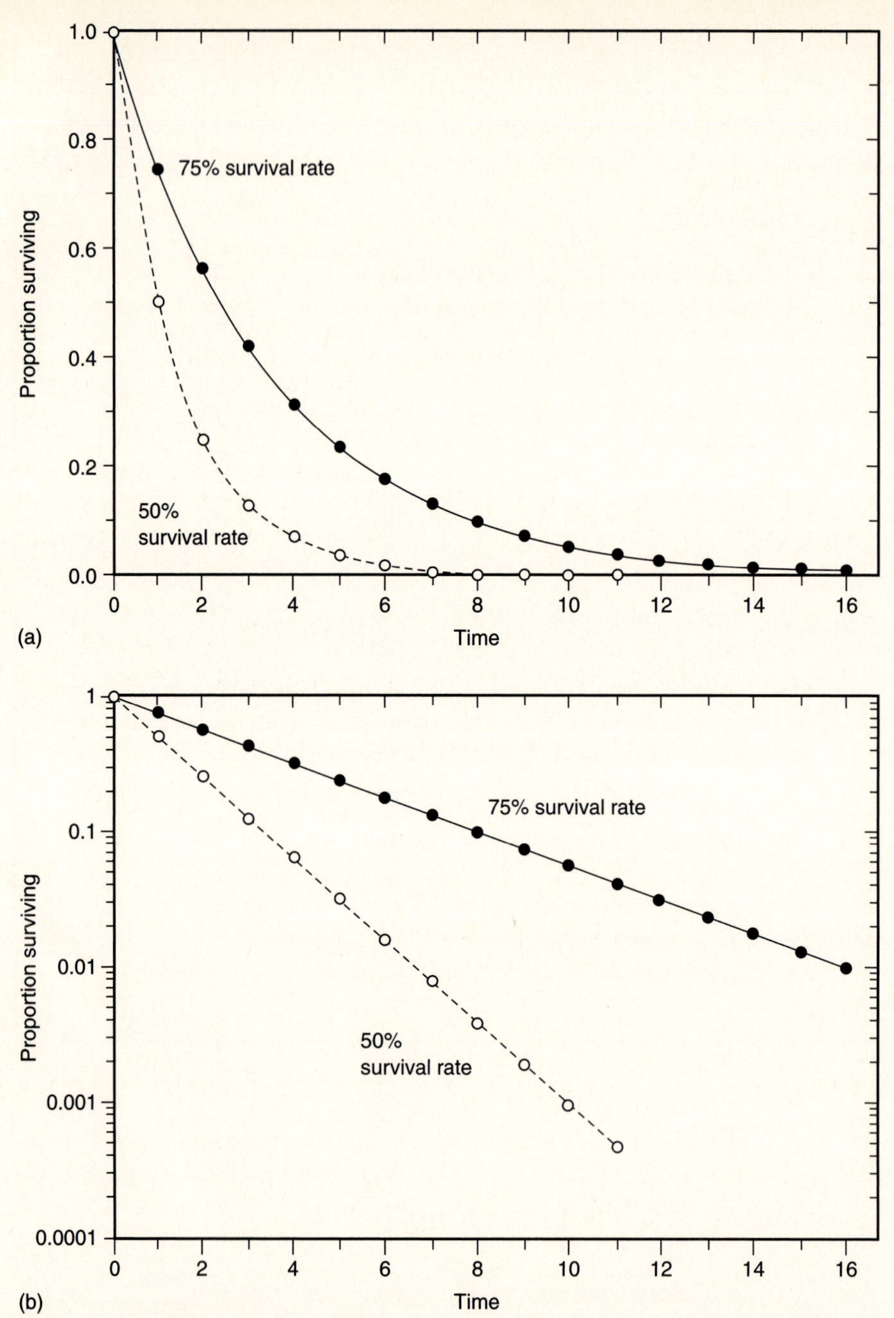

Figure 14.1 Illustration of geometric population decline. Many survival analyses assume that a cohort of animals or plants decreases at a constant survival rate. (a) Geometric declines at 50% per time period and 75%. (b) On a semilogarithmic plot, in which the proportion surviving is expressed on a log scale, these same declines are linear. The slope of these lines is the instantaneous mortality rate defined in equation (14.3). This graph illustrates why *geometric* and *logarithmic* are used interchangeably to describe proportional changes in abundance.

The following table illustrates the conversion between these three measures of survival:

Finite survival rate	Finite mortality rate	Instantaneous mortality rate
1.0	0.0	0.0
0.95	0.05	−0.051
0.90	0.10	−0.105
0.75	0.25	−0.287
0.50	0.50	−0.693
0.25	0.75	−1.386
0.10	0.90	−2.303
0.01	0.99	−4.605

Note that finite rates of survival and mortality can range only from 0 to 1, but instantaneous rates of mortality can range from 0 to $-\infty$. It is important to remember that both finite and instantaneous rates of survival and mortality always refer to a specific time period such as one month or one year.

It is clearly easiest to think about finite survival rates or finite mortality rates, and survival data should usually be reported in this form. Instantaneous rates are useful mathematically, but they are difficult to conceptualize. Survival is a multiplicative process, and we obtain average survival rates by chain multiplication. For example, to calculate total generation survival for an insect population,

$$S_G = S_1 \times S_2 \times S_3 \times S_4 \times S_5 \tag{14.9}$$

where S_G = Generation survival rate
S_1 = Survival of egg state
S_2 = Survival of small larval stage
S_3 = Survival of large larval stage
S_4 = Survival of pupal stage
S_5 = Survival of adult stage

The same general principle applies to any type of survival average. The important point is that finite survival rates are not averaged arithmetically but as a geometric mean (as in equation [14.9]).

14.2 ESTIMATION FROM LIFE TABLES

The mortality schedule operating on a population can be described succinctly by a *life table*. A life table is an age-specific summary of the mortality rates operating on a population. A population may often be usefully subdivided into groups like males and females, and different life tables may be calculated for each group. Table 14.1 gives an example of a life table for female African buffalo.

Life tables have been presented in several formats, and the following symbols are commonly used to summarize the data in the columns:

TABLE 14.1 LIFE TABLE FOR FEMALE AFRICAN BUFFALO (*Syncerus caffer*) IN THE SERENGETI REGION OF EAST AFRICA

Age, x (years)	No. alive at start of age interval, n_x	Proportion surviving at start of age interval, l_x	No. dying within age interval x to $x + 1$, d_x	Finite rate of mortality, q_x	Finite rate of survival, p_x
0	1000	1.000	330	0.330	0.670
1	670	0.670	94	0.140	0.860
2	576	0.576	11	0.019	0.981
3	565	0.565	12	0.021	0.979
4	553	0.553	21	0.038	0.962
5	532	0.532	16	0.030	0.970
6	516	0.516	29	0.056	0.944
7	487	0.487	20	0.041	0.959
8	467	0.467	35	0.075	0.925
9	432	0.432	52	0.120	0.880
10	380	0.380	44	0.116	0.884
11	336	0.336	73	0.217	0.783
12	263	0.263	67	0.255	0.745
13	196	0.196	56	0.286	0.714
14	140	0.140	49	0.350	0.650
15	91	0.091	43	0.473	0.527
16	48	0.048	26	0.542	0.458
17	22	0.022	15	0.682	0.318
18	7	0.007	7	1.000	0.000
19	0	0.000	—	—	—

Source: Sinclair 1977.

x = Age interval
n_x = Number of individuals of a cohort alive at the *start* of age interval x
l_x = Proportion of individuals surviving at the start of age interval x
d_x = Number of individuals of a cohort dying *during* the age interval x to $x + 1$
q_x = Finite rate of mortality during the age interval x to $x + 1$
p_x = Finite rate of survival during the age interval x to $x + 1$
e_x = Mean expectation of life for individuals alive at *start* of age x

Be aware, however, that not every ecologist uses these symbols in the same way.

To set up a life table, you must first decide on the time interval for the data. With longer-lived animals and plants this is often 1 year, but for voles it might be 1 week and for some insects 1 day. By making the time interval shorter, you will increase the detail of the mortality picture shown by the life table. But the shorter the time interval, the more data you will need for the life table.

The first point to be noted about life tables is that, given any one of the columns, you can calculate all the rest. There is nothing new in each of the six columns of the life table. They are just different ways of viewing the same data. The essential conversion formulas are as follows:

$$n_{x+1} = n_x - d_x \tag{14.10}$$

TABLE 14.2 FORMULAS FOR CONVERTING BETWEEN DIFFERENT COLUMNS OF THE LIFE TABLE. TO USE THIS TABLE, DETERMINE WHAT TYPE OF DATA YOU HAVE (LEFT SIDE OF TABLE) AND MOVE ACROSS THE ROW TO THE COLUMN OF THE TYPE OF DATA YOU WOULD LIKE TO HAVE.

	n_x	l_x	d_x	q_x	p_x
n_x	n_x	$\frac{n_x}{n_0}$	$n_{x+1} - n_x$	$\frac{n_{x+1}}{n_x} - 1$	$\frac{n_{x+1}}{n_x}$
l_x	$l_x n_0$	l_x	$(l_{x+1} - l_x)n_0$	$1 - \frac{l_{x+1}}{l_x}$	$\frac{l_{x+1}}{l_x}$
d_x	$\sum_{y=x}^{\infty} d_y$	$\frac{\Sigma_{y=x}^{\infty} d_y}{n_0}$	d_x	$\left(\frac{d_x}{\Sigma_{y=x}^{\infty} d_y}\right)$	$1 - \left(\frac{d_x}{\Sigma_{y=x}^{\infty} d_y}\right)$
q_x	$n_0 \prod_{y=0}^{x-1} (1 - q_y)$	$\frac{\Pi_{y=0}^{x-1} (1 - q_y)}{n_0}$	$q_x \prod_{y=0}^{x-1} (1 - q_y)$	q_x	$1 - q_x$
p_x	$n_0 \prod_{y=0}^{x-1} p_y$	$\frac{\Pi_{y=0}^{x-1} p_y}{n_0}$	$1 - q_x \prod_{y=0}^{x-1} (1 - q_y)$	$1 - p_x$	p_x

Source: Caughley 1977a.

$$q_x = \frac{d_x}{n_x} \tag{14.11}$$

$$l_x = \frac{n_x}{n_0} \tag{14.12}$$

$$e_x = \frac{T_x}{n_x} \tag{14.13}$$

where $T_x = \sum_{i=0}^{x} L_i$

$L_x = (n_x + n_{x+1})/2$

and all other symbols are defined as above. Table 14.2 combines these equations to illustrate the relationships among the life table variables.

14.2.1 Methods of Collecting Life Table Data

At least six types of data can be collected to calculate a life table. Caughley (1977a) has reviewed these methods and cautions the ecologist to pick the method of maximal efficiency for the organism being studied.

Method 1: Age at Death Recorded Directly The number of individuals dying in successive intervals of time is recorded for a group of individuals born at the same time. This is the most precise type of data available because it is based on a single cohort followed through time. Every individual must be known to be alive or dead at each time period. The observed data are the d_x column of the life table. The other columns can be calculated from the formulas in Table 14.2.

Method 2: Cohort Size Recorded Directly The number of individuals alive in successive intervals of time is recorded for a cohort of individuals. These data are similar to those obtained with Method 1, except that those *surviving* are tallied, not those dying. These data are also precise and specific to the cohort studied. The observed data are the n_x column of the life table. The other parameters of the life table follow from Table 14.2.

Method 3: Age at Death Recorded for Several Cohorts Individuals are marked at birth, and their age at death is recorded, as in Method 1. But several different cohorts are pooled from different years or seasons. These data are usually treated as if the individuals were members of one cohort, and the analysis of Method 1 is applied.

Method 4: Age Structure Recorded Directly The number of individuals aged x in a population is compared with the number of these that die before reaching age $x + 1$. The number of deaths in that age interval, divided by the number alive at the start of the age interval, gives an estimate of q_x directly. Note that the numbers must be counted directly, and the same result cannot be achieved by estimating the proportions in different age classes.

Each of these four methods is completely general and requires no assumptions to be made about population stability or stability of the age structure. The next two methods are much more specific because they require you to know the rate of increase of the population and that the age distribution is stable.

Method 5: Ages at Death Recorded for a Population with a Stable Age Distribution and Known Rate of Increase Often it is possible to find skulls or other remains that give the age at death of an individual. These data can be tallied into a frequency distribution of deaths and thus give d_x directly. To correct for the fact that the population is growing (or declining), each d_x value is corrected as follows:

$$d'_x = d_x e^{rx} \tag{14.14}$$

where d'_x = Corrected estimate of number dying within the age interval x to $x + 1$
d_x = Observed number of deaths within the age interval x to $x + 1$
r = Instantaneous rate of population growth
x = Age class (0, 1, 2, . . .)
e = 2.71828 . . .

The life table can then be calculated from Table 14.2 using the corrected d'_x values. If the population is constant, note that e^{rx} will be 1.0, so no correction is needed. Box 14.2 illustrates these calculations.

Method 6: Age Distribution Recorded for a Population with a Stable Age Distribution and Known Rate of Increase In this case the age distribution of the population is measured directly by sampling. The number of individuals born is calculated from fecundity rates. The rate of increase of the population must be known. In this case the l_x schedule is estimated directly:

$$l_x = \frac{n_x e^{rx}}{n_0} \tag{14.15}$$

where l_x = Proportion surviving at start of age x
n_x = Number observed in age class x to $x + 1$
n_0 = Number of births estimated from fecundity data
r = Instantaneous rate of population growth

Box 14.3 illustrates the use of Method 6.

Methods 5 and 6 both assume that the rate of population growth is known and that the age distribution is stable. Caughley (1977a) cataloged the numerous ways in which life tables have been calculated incorrectly in the literature by means of Methods 5 and 6. Most of these pitfalls involve biased sampling. For example, if only winter-kills are located, there are no data on summer mortality and a life table cannot be computed. If the conspicuousness of carcasses varies with age, the observed ages at death are a biased frequency

Box 14.2 Life Table Construction for Method 5: Age at Death Recorded in a Growing Population

Sinclair (1977) picked up skulls of male African buffalo in the southern Serengeti and obtained these data:

Age (years)	No. of skulls found of this age class
2	2
3	5
4	5
5	6
6	18
7	17
8	20
9	17
10	15
11	16
12	18
13	15
14	14
15	8
16	5
17	1
18	0
19	1
	183

Sinclair estimated that 48.5% of the male buffalo died in their first year, and 12.9% of the original cohort died in their second year, but these young skulls disintegrated and so were not represented in the sample. The simplest way to add these two ages into the observed data is to construct a hypothetical cohort of 1000 skulls:

Age x	Hypothetical cohort of 1000 skulls	Corrected estimate of number dying within the age interval x to $x + 1$, d'_x
0	485	485
1	129	139.3
2	4.2	4.9
3	10.5	13.2
4	10.5	14.3
5	12.7	18.7
6	38.0	60.3
7	35.9	61.5
8	42.2	78.1
9	35.9	71.8
10	31.6	68.2
11	33.7	78.6
12	38.0	95.7
13	31.6	86.0
14	29.5	86.7
15	16.9	53.6
16	10.5	36.0
17	2.1	7.8
18	0.0	0.0
19	2.1	9.1
Totals	999.9	1468.8

The original data for ages 2–19 are multiplied by the constant 2.109 to make the new cohort sum to 1000. This hypothetical cohort must now be corrected for the fact that this buffalo population has been increasing at $r = 0.077$ per year. From equation (14.14),

$$d'_x = d_x e^{rx}$$

$$d'_0 = 485e^{(0.077(0))} = 485$$

$$d'_1 = 129e^{(0.077(1))} = 139.3$$

$$d'_2 = 4.2e^{(0.077(2))} = 4.9$$

and these estimates are listed in the table above. We can now treat these corrected estimates (d'_x) as d_x estimates and calculate the life table in the usual way from the formulas in Table 14.2. For example, to calculate the survivorship curve (l_x) function,

$$l_x = \frac{\sum_{y=x}^{\infty} d_y}{n_0}$$

$$l_0 = \frac{1468.8}{1468.8} = 1.000$$

$$l_1 = \frac{983.8}{1468.8} = 0.670$$

$$l_2 = \frac{844.5}{1468.8} = 0.575$$

and the final results are tabulated below:

Age, x	Proportion surviving to start of age x, l_x
0	1.000
1	0.670
2	0.575
3	0.572
4	0.563
5	0.553
6	0.540
7	0.499
8	0.457
9	0.404
10	0.355
11	0.308
12	0.254
13	0.189
14	0.130
15	0.070
16	0.034
17	0.011
18	0.006
19	0.006

Program-group SURVIVAL (Appendix 2) can do these calculations.

distribution and should not be converted to a life table. Animals shot by hunters often represent a very biased age distribution.

A life table calculated from age distribution data as in Method 6 will give appropriate estimates only when the frequency of each age class is the same as or less than that of the preceding age class. If this is not the case, the d_x values will be negative, which is impossible. In these cases Caughley (1977a) recommends smoothing the age distribution. Many computer packages for data analysis and statistics have smoothing functions that can also be used to smooth age distributions. Cleveland (1994) illustrates the use of cubic splines to smooth a data series.

Box 14.3 Life Table Construction for Method 6: Age Distribution Recorded in a Growing Population

Caughley (1970) obtained a sample of Himalayan thar in New Zealand by shooting and got these results:

Age, x (yr)	Sampled age distribution, n_x	Correction factor, $e^{rx}(r = 0.12)$	Corrected frequency $n_x e^{rx}$	Smoothed frequency, n'_x	Estimates of life table parameters l_x	d_x	q_x
0	43	1.00	43.0	43	1.00	0.37	0.37
1	25	1.13	28.3	27	0.63	0.02	0.03
2	18	1.27	22.9	26	0.61	0.03	0.04
3	18	1.43	25.8	25	0.58	0.03	0.05
4	19	1.62	30.8	23	0.55	0.05	0.09
5	11	1.82	20.0	22	0.50	0.05	0.10
6	12	2.05	24.7	19	0.45	0.06	0.13
7	8	2.32	18.6	17	0.39	0.06	0.15
8	2	2.61	5.2	14	0.33	0.06	0.18
9	3	2.94	8.8	11	0.27	0.06	0.21
10	4	3.32	13.3	9	0.21	0.05	0.23
	163						

The age distribution at birth (age 0) was estimated from fecundity rates. The analysis proceeds as follows:

1. Calculate the corrected age distribution from the equation:

$$n'_x = n_x e^{rx}$$

where n'_x = Corrected number of individuals in age class x
n_x = Observed number of individuals in age class x
r = Instantaneous rate of population growth

For example,

$$n'_3 = 18e^{(0.12)(3)} = 25.8$$

$$n'_6 = 12e^{(0.12)(6)} = 24.7$$

These values are given in column 4 of the table above.

2. Smooth the corrected age distribution either by some smoothing function (moving averages, splines) or by eye. This step is necessary because the age distribution must *decline* with age (i.e., a monotonic, decreasing function). For example, it is logically impossible for there to be more 3-year-old thar than 2-year-olds if the population is growing smoothly. In this particular case, smoothing was done by eye, and the results are given in column 5 of the table above.

3. The smoothed age distribution can now be used as n_x values to calculate the life table functions, as in Table 14.2. For example, to calculate the survivorship function (l_x) from equation (14.12),

$$l_x = \frac{n_x}{n_0}$$

Thus we obtain:

$$l_1 = \frac{27}{43} = 0.628$$

$$l_2 = \frac{26}{43} = 0.605$$

$$l_3 = \frac{25}{43} = 0.581$$

and these values are given in the above table.

Program-group SURVIVAL (Appendix 2) can do these calculations from a smoothed age distribution.

There is little discussion in the literature about the sample sizes needed for the construction of a life table. Caughley (1977a) recommends at least 150 individuals as a minimum when age distributions are used in the analysis. Age determinations may also be crude in many species, and if age is not measured carefully, the resulting life table may be very inaccurate. I have seen no discussion of power analysis applied to life table data.

14.2.2 Key Factor Analysis

Morris (1959) developed key factor analysis as a technique for determining the cause of population outbreaks in the spruce budworm, which periodically defoliates large areas of balsam fir forests in eastern Canada. The methods described by Morris (1959, 1963) have been criticized and replaced by a series of methods developed by Varley and Gradwell (1960, 1963, 1965). Dempster (1975) and Southwood (1978, 376–84) describe these methods in detail, and Stiling (1988) and Royama (1996) discuss the application of key factor analysis.

Key factor analysis was designed for insect populations with discrete generations in which a series of mortality factors operate in a linear sequence with no interaction. For example, if two parasites and one disease kill larval insects, key factor analysis assumes that (for example) parasite A acts first to kill a sample, then parasite B works, and finally disease C acts to kill some that remain. All factors operate independently of one another, so that for this example parasites A and B are independent agents. These assumptions are very simplistic and are useful only as a starting point in analyzing population dynamics. In particular the application of these methods to long-lived vertebrates must be done with caution.

The Varley and Gradwell method of key factor analysis proceeds as follows:

Step 1. *Maximum natality* is obtained for the population by multiplying the number of breeding females by the mean fecundity per female. There may be some disagreement about what level of fecundity to assume—the maximum observed or the mean value. In most cases the value used is not critical for the analysis that follows.

Step 2. *Population size* estimates for each age or stage in the generation are entered in the table (Table 14.3). These estimates are basically the n_x column of the life table, with the addition of known mortality from predation or parasitism.

Step 3. To convert these absolute numerical losses into relative (proportional) losses, the population size estimates are converted into logarithms. Base 10 logs are normally used, but any base will give the same results.

Step 4. *Age-specific mortality* is calculated by subtracting each logarithm of population size from the previous one:

$$k_x = \log(N_x) - \log(N_{x+1}) \tag{14.16}$$

where k_x = Age-specific mortality for age or stage x
N_x = Population size at age x
N_{x+1} = Population size at the next age $(x + 1)$

Step 5. *Total generation mortality* is determined by adding the k values:

$$K = k_0 + k_1 + k_2 + k_3 + k_4 + \cdots = \sum_{x=0}^{n} k_x \tag{14.17}$$

TABLE 14.3 LIFE TABLE FOR THE WINTER MOTH IN WYTHAM WOODS, NEAR OXFORD, ENGLAND, FOR 1955–1956

	Percent of previous stage killed	No. killed (per m²)	No. alive (per m²)	Log of no. alive (per m²)	k-value
Adult stage					
Females climbing trees in 1955			**4.39**		
Egg stage					
Females × 150			658.0	2.82	**0.84** = k_1
Larval stage					
Full-grown larvae	86.9	551.6	**96.4**	1.98	0.03 = k_2
Attacked by *Cyzenis*	6.7	**6.2**	90.2	1.95	0.01 = k_3
Attacked by other parasites	2.3	**2.6**	87.6	1.94	
Infected by microsporidian	4.5	**4.6**	83.0	1.92	0.02 = k_4
Pupal stage					**0.47** = k_5
Killed by predators	66.1	54.6	28.4	1.45	
Killed by *Cratichneumon*	46.3	**13.4**	15.0	1.18	0.27 = k_6
Adult stage					
Females climbing trees in 1956			**7.5**		

Source: Varley et al. 1973.

Note: The figures in bold are those actually measured. The rest of the life table is derived from these.

where K = Total generation mortality
$k_0, k_1, \ldots$ = Age-specific mortalities from equation (14.16)

As a check on the calculations, you can also calculate total generation mortality as

$$K = \log(\text{maximum natality}) - \log(\text{reproductive adults}) \qquad (14.18)$$

Box 14.4 illustrates the use of key factor analysis on data from the cinnabar moth. These steps are repeated for several generations (a minimum of 5–10), and one then proceeds to the more useful and interesting part of key factor analysis.

Recognition of the Key Factor Morris (1963) defined the *key factor* as that *k*-factor which was of the greatest predictive value in forecasting future population changes. The key factor is seen most easily by plotting the total generation mortality and all the individual *k*-values against time (or generation number). Figure 14.2 illustrates these plots for red deer on Rhum from 1971 to 1983. For both males and females, the key factor is clearly k_4, winter mortality, since most of the changes in total mortality K are associated with changes in k_4. Changes in reproductive output (k_1) are also closely correlated with total mortality K. Summer mortality (k_2) and emigration (k_3), which occurred only in males, are both unimportant in causing changes in total K (Clutton-Brock et al. 1985).

In some cases key factors are not immediately obvious by graphical inspection (see Figure 14.2). Podoler and Rogers (1975) suggested a simple graphical method to determine key factors. Plot each *k*-value (*y*-axis) against total generation mortality K (*x*-axis). Calculate the slope of these lines, and the *k*-value with the largest slope is the key factor. Figure 14.3 illustrates a simple example of this method for the great tit (*Parus major*).

The *k*-values calculated in key factor analysis are closely related to the instantaneous mortality rate defined in equation (14.3). The conversion equation is

$$k = \frac{1}{\log_{10}(1 - i)} \qquad (14.19)$$

Box 14.4 Key Factor Analysis of the Cinnabar Moth (*Tyria jacobaeae*)

Dempster (1975) gave the following census data for the cinnabar moth for 1967 on an area of heathland in Norfolk:

	No. entering stage
Adults of 1966	109
% females	55%
Fecundity	285.2 eggs per female
Eggs	17,110
Larvae I and II	16,244
Larvae III and IV	5,623
Larvae V	3,439
Pupae	1,746
Adults of 1967	362

The maximum fecundity of the cinnabar moth is 600 eggs per female (measured by Dempster), and thus potential production for 1967 was:

$$109 \text{ adults} \times 0.55 \times 600 = 35{,}970 \text{ eggs}$$

Convert these population estimates into logarithms (base 10):

	$\log_{10}$ (number)	k-value	
Maximum fecundity	4.5560		
		0.3228	k_0
Actual eggs laid	4.2332		
		0.0225	k_1
Larval hatching	4.2107		
		0.4608	k_2
Third instar larvae	3.7499		
		0.2135	k_3
Fifth instar larvae	3.5364		
		0.2937	k_4
Pupae	3.2427		
		0.6840	k_5
Adults	2.5587		

From these logarithms, the k values can be calculated by equation (14.16):

$$k_x = \log(N_x) = \log(N_{x+1})$$

For example, for the first k-value measuring the loss of fecundity,

$$k_0 = \log(35{,}970) - \log(17{,}110)$$
$$= 4.5560 - 4.2332 = 0.3228$$

Total generation mortality can be calculated from equation (14.17):

$$K = k_0 + k_1 + k_2 + k_3 + k_4 + k_5$$
$$= 0.3228 + 0.0225 + 0.4608 + 0.2135 + 0.2937 + 0.6840$$
$$= 1.9973$$

As a check on our arithmetic, use equation (14.18):

$$K = \log(\text{maximum natality}) - \log(\text{reproductive adults})$$
$$= 4.5560 - 2.5587 = 1.9973$$

These calculations are repeated for several generations (years), and Dempster (1975) shows the resulting analysis.

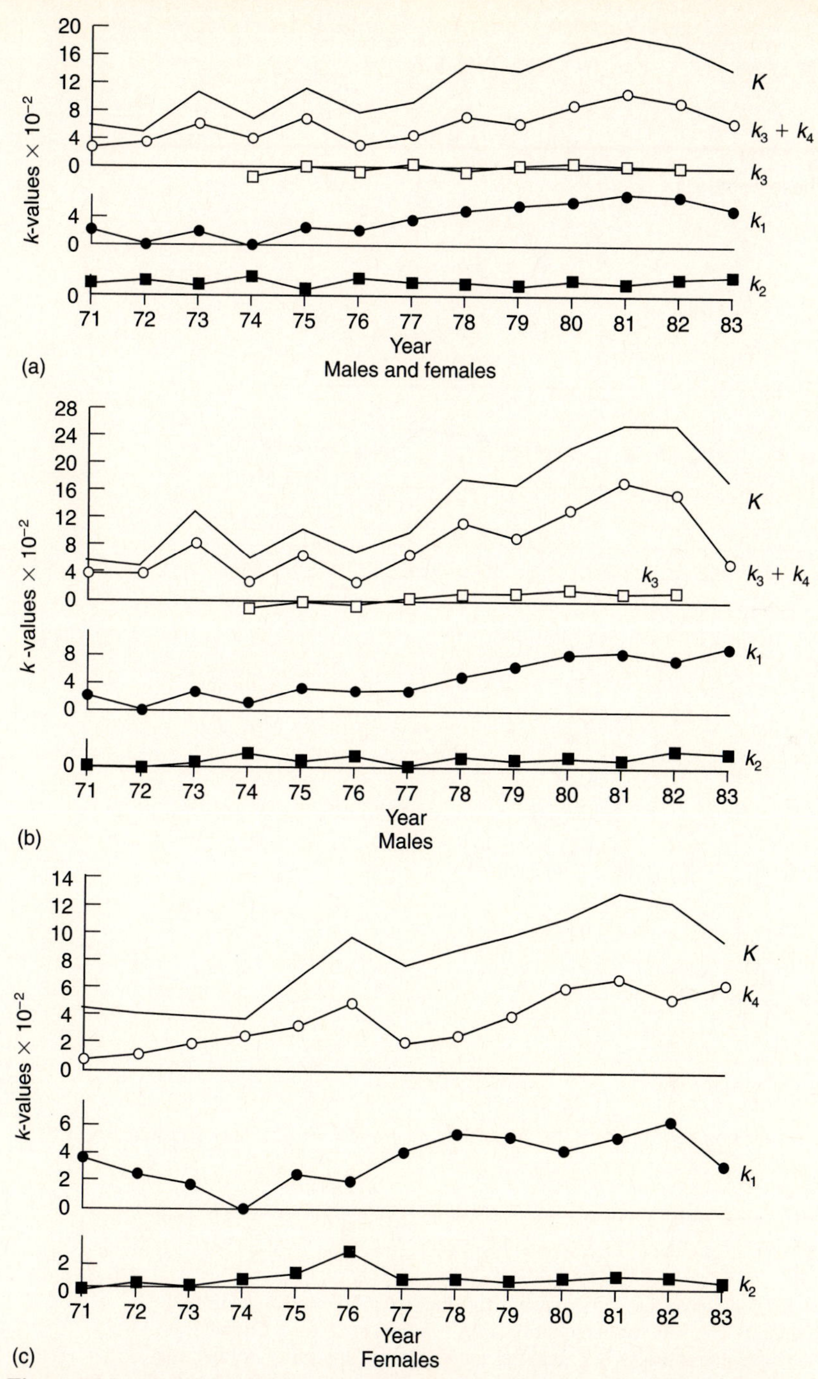

Figure 14.2 Key factor analysis for red deer (*Cervus elaphus*) on the Isle of Rhum, Scotland. The three plots show the analysis for (a) males and females combined, (b) males only, and (c) females only. k_1 fecundity; k_2 summer mortality; k_3 emigration; k_4 winter mortality. (From Clutton-Brock et al. 1985.)

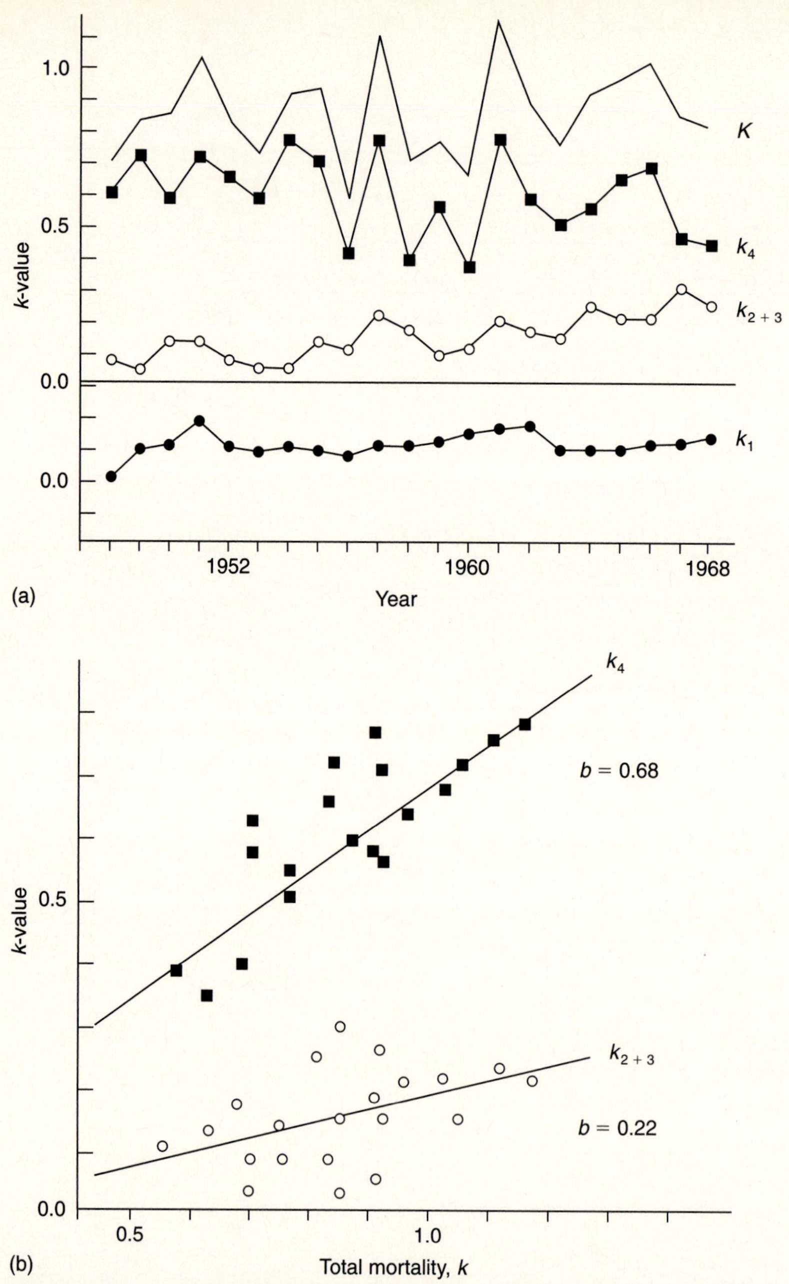

Figure 14.3 Key factor identification using the method of Podoler and Rogers (1975). (a) Key factor analysis for the great tit (*Parus major*) at Oxford, England. k_1, Clutch size; k_2, hatching success; k_3, nestling mortality; k_4, mortality outside the breeding season. (b) Podoler and Rogers plot of k versus K for the great tit; k_4 has the largest slope value ($b = 0.68$) and is the key factor in this population. (From Podoler and Rogers 1975.)

where k = k-value from key factor analysis (equation [14.16])
i = Instantaneous mortality rate (equation [14.3])

Testing for Density-Dependence One of the major uses of key factor analysis has been to look for evidence of density-dependent mortality in population studies (Dempster 1975). To do this, one plots each k_x value against the density or number of individuals entering the stage or age interval (N_x). The simple interpretation is that if this regression is significant and positive, you have evidence of direct density dependence (Varley and Gradwell 1960). This is not correct, as pointed out by Maelzer (1970) and St. Amant (1970). Ito (1972), Kuno (1971, 1973), and Bulmer (1975) provide a detailed critique of the problem of testing for density dependence with key factor analysis. Errors arise for two reasons: (1) There is serial correlation between population sizes in a series of years, so that k_x and N_x are not independent variables. (2) Population size is often estimated, and the errors of estimation are not equal for all stages of the life cycle (Kuno 1971). Bulmer (1975) provides statistical tests for the density-dependence hypothesis that make allowance for these two sources of error. He points out, however, that a large number of observations is needed to have a good chance of detecting density dependence. In practice this means more than 25 generations or years of data, and often more than 50 generations, so that almost no ecological data can be used. The recommendation is thus that you should not use key factor analysis to look for evidence of density dependence. Dennis and Taper (1994) provide a new likelihood ratio test for density dependence in a time series of population abundances that has good power if the number of time periods (years) is 16 or more. Holyoak (1994) details the difficulties of testing for delayed density dependence in time-series data. The problem is not easy, and experimental methods should be preferred (Sinclair and Pech 1996).

14.2.3 Expectation of Further Life

Life table data may be summarized in yet another form by calculating the expectation of further life. The mean expectation of further life can be used as one way of compressing an entire life table into one number. But this compression is done at a price of leaving out the details of the life table that may be of greatest interest to an ecologist.

To calculate mean expectation of life, proceed as follows:

Step 1. Calculate the *life-table age structure* L_x:

$$L_x = \frac{n_x + n_{x+1}}{2} \tag{14.20}$$

where L_x = Number of individuals alive on average during the age interval x to $x + 1$
n_x = Number of individuals alive at start of age interval x

This value is only approximate if numbers do not fall linearly within the age interval, but for most animal and plant populations, this is a reasonable approximation. If the time interval is long or mortality rates are high, a better estimate of L_x might be the geometric mean:

$$L_x = \sqrt{n_x(n_{x+1})} \tag{14.21}$$

For example, if $n_3 = 50$ and $n_4 = 10$, equation (14.20) would give $L_3 = 30$, while equation (14.21) would give $L_3 = 22.4$.

Step 2. Calculate the cumulative function for each age class T_x:

$$T_x = \sum_{i=x}^{m} L_i \qquad (14.22)$$

where T_x = A cumulative function of (individuals)(time units) for age x individuals
L_x = Life table age structure at age x
m = Maximum age class observed

Step 3. Estimate the *mean expectation of further life* from the following equation:

$$e_x = \frac{T_x}{n_x} \qquad (14.23)$$

where e_x = Mean expectation of further life for individuals alive at start of age interval x
T_x = Cumulative function of (individuals)(time units) for age x (equation [14.22])
n_x = Number of individuals alive at start of age interval x

Leslie et al. (1955) point out that the special case of the mean expectation of life for an individual alive at the start of the life table is approximately

$$\hat{e}_0 = 0.5 + l_1 + l_2 + l_3 + \cdots + l_{m-1} + 0.5l_m \qquad (14.24)$$

where $\hat{e}_0$ = Mean expectation of life for an individual alive at the start of the life table
l_x = Proportion of individuals surviving to the start of age interval x
l_m = Proportion of individuals surviving to the start of the last age interval m

The variance of the estimate e_0 is given by Leslie et al. (1955) as

$$\text{var}(\hat{e}_0) = \sum_{x=0}^{m-1} \left[\frac{S_{x+1}^2 q_x}{p_x(n_x - 0.5a_x)} \right] \qquad (14.25)$$

where $\text{var}(\hat{e}_0)$ = Variance of the mean expectation of life for an individual alive at the origin of the life table
q_x = Finite rate of mortality during the age interval x to $x + 1$ (equation [14.11])
p_x = Finite rate of survival during the age interval x to $x + 1 = 1.0 - q_x$
n_x = Number of individuals known to be alive at start of age interval x
a_x = Number of accidental deaths (or removals) during the interval x to $x + 1$
$S_x = l_x + l_{x+1} + \cdots + l_{m-1} + 0.5\, l_m$
l_x = Proportion of individuals surviving to the start of age interval x
m = Number of age groups in the data

From this variance you can obtain confidence intervals in the usual way:

$$\hat{e}_0 \pm t_\alpha(\text{S.E.}) \qquad (14.26)$$

where $\hat{e}_0$ = Mean expectation of life from age 0
t_α = Student's t-value for $(n - 1)$ degrees of freedom
S.E. = Standard error of the estimate $\hat{e}_0 = \sqrt{\text{var}(\hat{e}_0)}$ given above

Box 14.5 illustrates these calculations for expectation of life.

14.3 ESTIMATION OF SURVIVAL FROM AGE COMPOSITION

Fisheries scientists have developed a whole series of methods for estimating survival rates from data on age composition. Ricker (1975, Chap. 2) gives a detailed exposition of these methods, which I summarize here.

When it is possible to age individual organisms, like fish or trees, one can estimate survival rates directly from the ratio of numbers in each successive age group:

$$\hat{S}_t = \frac{N_{t+1}}{N_t} \qquad (14.27)$$

where S_t = Finite annual survival rate of individuals in age class t
N_{t+1} = Number of individuals in age class $t + 1$
N_t = Number of individuals in age class t

This simple approach to estimating survival can be used only when three assumptions are satisfied:

1. The survival rate is constant for each age group.
2. All year-classes are recruited at the same abundance.
3. All ages are sampled equally by the sampling gear.

These assumptions are very restrictive and cannot be correct for many populations.

If the survival rate is constant over a period of time, a combined estimate of average survival can be made from formulas given by Robson and Chapman (1961):

$$\hat{S} = \frac{T}{R + T - 1} \qquad (14.28)$$

where $\hat{S}$ = Finite annual survival rate estimate
$T = N_1 + 2N_2 + 3N_3 + 4N_4 + \cdots$
$R = \sum_{t=0}^{m} N_t$
N_t = Number of individuals in age group t

This estimate of survival has a variance of

$$\text{var}(\hat{S}) = \hat{S}\left[\hat{S} - \left(\frac{T - 1}{R + T - 2}\right)\right] \qquad (14.29)$$

Box 14.5 Calculation of Expectation of Further Life in Snowshoe Hares

A cohort of 63 snowshoe hares was followed at 6-month intervals with the following results:

Age class, x	Age (months)	No. alive, n_x	Proportion alive, l_x	Mortality rate, q_x	Accidental deaths, a_x
0	0	63	1.00	0.33	0
1	6	42	0.67	0.14	0
2	12	36	0.57	0.33	1
3	18	24	0.38	0.38	0
4	24	15	0.24	0.53	0
5	30	7	0.11	0.71	0
6	36	2	0.03	1.00	0
7	42	0	0.00	—	0

The first step is to calculate the L_x, T_x, and S_x functions needed to solve equations (14.20) to (14.23). Fill in this table as follows:

Age class	L_x	T_x	S_x
0	52.5	157.5	2.9850
1	39	105	1.9850
2	30	66	1.3183
3	19.5	36	0.7469
4	11	16.5	0.3659
5	4.5	5.5	0.1279
6	1	1	0.0167

From equation (14.20):

$$L_x = \frac{n_x + n_{x+1}}{2}$$

$$\hat{L}_0 = \frac{63 + 42}{2} = 52.5$$

$$\hat{L}_1 = \frac{42 + 36}{2} = 39$$

and so on. These results are listed above.

From equation (14.22),

$$T_x = \sum_{x=0}^{\infty} L_x$$

$$\hat{T}_0 = 52.5 + 39 + 30 + 19.5 + 11 + 4.5 + 1 = 157.5$$

$$\hat{T}_1 = 39 + 30 + 19.5 + 11 + 4.5 + 1 = 105$$

and so on. These results are listed in the table above.

$$S_x = l_x + l_{x+1} + l_{x+2} + \cdots + l_{m-1} + 0.5l_m$$

$$\hat{S}_0 = 1.00 + 0.67 + 0.57 + 0.38 + 0.24 + 0.11 + (0.5)(0.03) = 2.985$$

and so on, with the results listed in the table above.

From these summations the mean expectation of life at birth follows from equation (14.23):

$$\hat{e}_0 = \frac{T_0}{n_0} = \frac{157.5}{63} = 2.5 \text{ time units}$$

Since one time unit in this example is 6 months, this can be converted directly to

$$\hat{e}_0 = (2.5)(6) = 15 \text{ months}$$

The variance of this estimate from equation (14.25) is

$$\text{var}(\hat{e}_0) = \sum_{x=0}^{m-1}\left[\frac{S_{x+1}^2 q_x}{p_x(n_x - 0.5a_x)}\right]$$

$$= \frac{(1.985^2)(0.3333)}{0.6667(63 - 0)}$$

$$+ \frac{(1.3183^2)(0.1429)}{0.8571(42 - 0)}$$

$$+ \frac{(0.7469^2)(0.3333)}{0.6667(36 - 0.5)} + \cdots = 0.0524$$

The standard error of the estimate of e_0 is

$$\text{S.E.} = \sqrt{\text{var}(e_0)}$$

$$= \sqrt{0.0524} = 0.2290 \text{ time units}$$

or, expressed in months,

$$\text{S.E.} = (0.229)(6) = 1.374 \text{ months}$$

The 95% confidence limits are obtained from Student's t-distribution with 62 degrees of freedom:

$$\hat{e}_0 \pm t_\alpha(\text{S.E.})$$

$$15.0 \pm 2(1.374)$$

or 12.3 to 17.7 months for these hares.

Program-group SURVIVAL (Appendix 2) can do these calculations.

Ricker (1975) gives the following age data on Antarctic fin whales from 1947 to 1953 to illustrate these calculations:

Age	0	1	2	3	4	5	6+
Frequency (%)	0.3	2.3	12.7	17.2	24.1	14.1	29.5

For these whales, ages 0–3 are not sampled representatively by the whaling ships, so these data must be omitted from analysis. From equation (14.27) for ages 4 to 5,

$$\hat{S} = \frac{14.1}{24.1} = 0.585 \text{ per year}$$

A combined estimate from Robson and Chapman's method gives, from equation (14.28),

$$\hat{T} = 14.1 + 2(29.5) = 73.1$$

$$\hat{R} = 24.1 + 14.1 + 29.5 = 67.7$$

$$\hat{S} = \frac{73.1}{67.7 + 73.1 - 1} = 0.523 \text{ per year}$$

with the following variance:

$$\text{var}(\hat{S}) = 0.523\left[0.523 - \left(\frac{73.1 - 1}{67.7 + 73.1 - 2}\right)\right] = 0.00133$$

Fisheries scientists discovered in 1908 that a plot of size-frequency data often formed a dome-shaped curve with a long descending right limb. Baranov (1918) called these *catch curves*, and most fishery scientists now apply this name to plots of log frequency-at-age (y-axis) against age (x-axis). The analysis of catch curves can be useful in fisheries management. Figure 14.4 illustrates a catch curve for the petrale sole off the coast of western Canada.

The ascending left part of a catch curve and the dome of the curve represent age groups that are not adequately sampled by the gear. In a commercial fishery young fish may live in different habitats or not be caught in the nets used to fish for the adults. The more interesting part of the curve is the descending right limb. Baranov (1918) showed that the right limb is a survivorship curve that is both age-specific and time-specific if the following assumptions are correct:

1. The mortality rate is uniform with age.
2. The mortality rate is constant over time.
3. The sample is taken randomly from the age groups involved.
4. Recruitment is constant for all age groups.

If these assumptions are correct, the finite rate of survival can be estimated from the antilog of the slope of the right limb of the catch curve (for the linear regression $Y = \log_e$

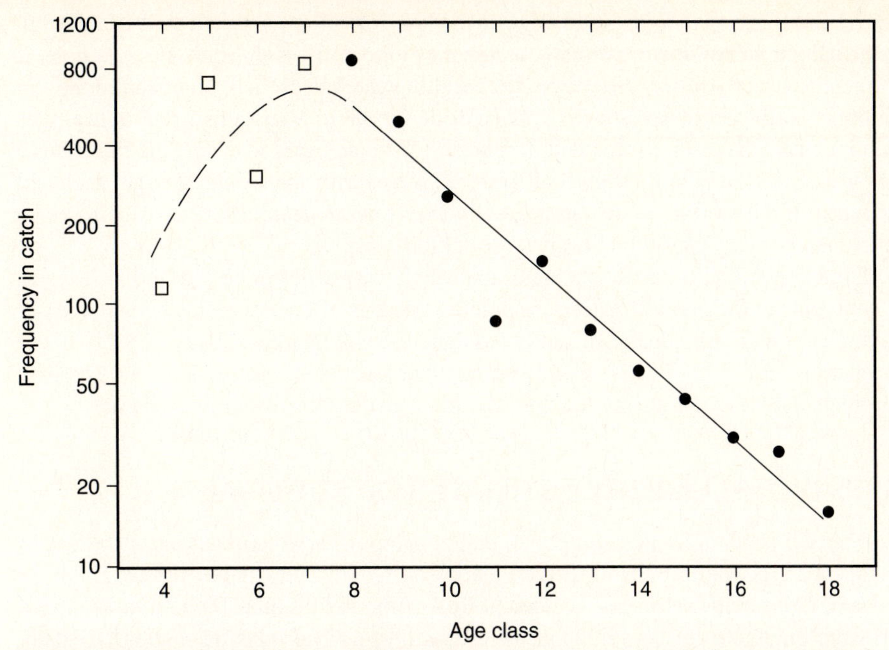

Figure 14.4 Catch curve for the 1945 year class of female petrale sole (*Eopsetta jordani*) off western Canada. The slope of the right limb of the catch curve (ages 8 to 18 years, •) is −0.36. (Data from Ketchen and Forrester 1966.)

[frequency-at-age] and x = age class). Alternatively, for two age classes t and $t + 1$, the survival rate is (using base e logs)

$$i = \log_e(f_t) - \log_e(f_{t+1}) \tag{14.30}$$

$$\hat{S} = e^{-i} \tag{14.31}$$

where f_t = Frequency of age t individuals
f_{t+1} = Frequency of age $t + 1$ individuals
$\hat{S}$ = Estimate of finite survival rate between age t and $t + 1$

For example, in Figure 14.4 the slope of the right limb of the curve for $\log_e$ (frequency) versus age is −0.36056. The finite survival rate is thus:

$$\hat{S} = e^{-0.36056} = 0.70 \text{ per year}$$

Some of these assumptions can be relaxed in certain situations. For example, if a fishery is sampled every year, a single year-class can be followed throughout its life span (e.g., the 1975 year class can be sampled in 1984 when it is age IX, in 1985 when it is age X, etc.). In this case the assumption of constant recruitment can be dropped, and the catch curve for that year-class is a cohort survivorship curve for which the average survival rate can be calculated.

If these four assumptions are violated, the right limb of the catch curve will often not be a straight line. Ricker (1975) discusses in detail the interpretation of nonlinear catch curves from commercial fisheries. Variations in the rate of recruitment will usually not make a catch curve nonlinear, but will change the slope of the right limb. For example, if recruitment is steadily falling over time, the resulting catch curve will be linear but have a lower slope, so the estimated survival rate will be too high. Variations in mortality rates will tend to make the catch curve bend. If mortality rate increases with age, the catch curve will show a convex right limb. If mortality rate decreases with age, a concave right limb will be produced. If the mortality rate varies greatly from year to year, catch curves will not be linear and are not very useful for survival estimation.

If age determination is not accurate, the catch curve will not be affected, as long as there is no bias in the age estimation. In practice there is often a negative bias in age determination—older individuals are usually assigned to younger ages (Ricker 1975). If age determination is strongly biased, estimated survival rates from catch curves are too low. The problem can be overcome by making sure that age determinations are accurate.

14.4 RADIOTELEMETRY ESTIMATES OF SURVIVAL

One use of radio collars for the study of animal populations is to estimate survival rates. In studies of this type, radio transmitters are placed on several individuals that are followed until death, radio malfunction, or accidental loss of the radio collar. Trent and Rongstad (1974) were the first to discuss methods of estimating survival rates from telemetry data. They proposed the following approach. The average daily survival rate is estimated by

$$\hat{S} = \frac{x - y}{x} \tag{14.32}$$

where $\hat{S}$ = Estimate of finite daily survival rate
x = Total number of radio-days observed over the period
y = Total number of deaths observed over the period

For example, Trent and Rongstad (1974) observed 31 cottontail rabbits for a total of 1660 radio-rabbit-days in September and October 1971, and in these two months observed 6 deaths. Thus,

$$\hat{S} = \frac{1660 - 6}{1660} = 0.99638 \text{ (survival rate per day)}$$

To convert these daily rates to weekly, monthly, or annual rates of survival, use the following equation:

$$\hat{p} = \hat{S}^n \tag{14.33}$$

where p = Estimate of finite survival rate per n days
S = Finite daily survival rate from equation (14.32)
n = Number of days you wish to convert to

For example, for these cottontails a 28-day survival rate would be

$$\hat{p} = 0.99638^{28} = 0.9036 \text{ (per 28 days)}$$

This method of estimating survival rates assumes that each rabbit-day is a binomial event in which the animal lives or dies, so that the distribution of deaths should follow a binomial distribution. In order to analyze the data this way, you must assume that each radio-day is an independent event and that there is a constant mortality rate over the whole time period. The method assumes that every death is recorded exactly on the day it occurs without guesswork.

To obtain confidence limits for these estimates, Trent and Rongstad (1974) use the Poisson approximation to the binomial, and thus used the methods outlined in Chapter 4 (page 122). Another approach would be to use the binomial distribution directly to obtain confidence limits (Program-group EXTRAS, Appendix 2).

14.4.1 Maximum Likelihood Method

Problems arise with the Trent and Rongstad (1974) method when the exact time of death is not known. Mayfield (1961, 1975) discussed the problem of estimating nest success in birds in which nests are visited periodically. This is exactly the same problem of estimation that Trent and Rongstad faced with radio-collared rabbits. Mayfield (1975) assumed that each nest loss occurred exactly halfway between visits. For example, if a nest was present on day 15 and then found destroyed at the next visit on day 19, one would assume it failed on day 17. Johnson (1979) recognized that the Mayfield method led to biased estimates of nest success; if nests were visited at irregular intervals, estimated nest success values would be too high. Hensler (1985) further discusses the possible biases of the Mayfield method.

These problems of estimation were resolved by Bart and Robson (1982), who suggested a maximum likelihood estimator of the survival rate calculated as follows:

Step 1. Compile the survival data in a frequency table, as in Table 14.4, in which interval length between observations is given for each observation of survival or mortality.

Step 2. Calculate the Mayfield (1975) estimator of the daily survival rate:

$$\hat{S} = 1 - \left(\frac{\text{Number of deaths}}{\Sigma_L \, L(n_{LS} + 0.5n_{LF})} \right) \tag{14.34}$$

TABLE 14.4 SURVIVAL DATA FOR NESTLING MOURNING DOVES FROM PERIPHERAL NESTING LOCATIONS, TALLIED IN A FREQUENCY DISTRIBUTION FOR CALCULATION OF THE MAXIMUM LIKELIHOOD ESTIMATOR OF FINITE DAILY SURVIVAL RATES[a]

Interval between subsequent visits to nests (days) (L)	Total no. of nests sampled with this interval (n_L)	No. of nests surviving intact (n_{LS})	No. of nests with mortality (n_{LF})
1	12	12	0
2	8	7	1
3	15	12	3
4	18	17	1
5	26	24	2
6	33	31	2

[a] Data from Bart and Robson 1982.

where $\hat{S}$ = Mayfield estimate of finite daily rate of survival
L = Interval length in days (1, 2, 3, . . .)
n_{LS} = Number of intervals of length L in which no mortality occurred
n_{LF} = Number of intervals of length L in which some mortality occurred

Step 3. Use the Mayfield estimate S as a preliminary estimate for the maximum likelihood estimate from the following equations:

(a) Calculate A and B from

$$A = \sum_L \left[\frac{L}{\hat{S}} \left(n_{LS} - \frac{n_{Lf} \hat{S}^L}{1 - \hat{S}^L} \right) \right] \qquad (14.35)$$

$$B = \sum_L \frac{L}{\hat{S}} \left[n_{LS} + \left(\frac{n_{LF} \hat{S}^L (L - 1 + \hat{S}^L)}{(1 - \hat{S}^L)^2} \right) \right] \qquad (14.36)$$

where all terms are as defined above.

(b) The maximum likelihood estimate of $\hat{S}$ is given by

$$\hat{S}_M = \hat{S} + \frac{A}{B} \qquad (14.37)$$

where $\hat{S}_M$ = Maximum likelihood estimate of survival rate
$\hat{S}$ = Mayfield estimate of survival rate

and A, B are as defined above.

Use this estimate as a trial value of $\hat{S}$, go back to step 3 and repeat steps (a) and (b) a second time. Bart and Robson (1982) found that two or three iterations were adequate for convergence to a final estimate of the daily survival rate. Box 14.6 gives an example of these calculations.

To obtain confidence limits for this estimate of survival, Bart and Robson (1982) showed that a square root transformation was required. The calculations are somewhat tedious and proceed as follows:

Step 1. Calculate the transformed survival rate:

$$\hat{S}_t = \sqrt{1 - \hat{S}_M^x} \qquad (14.38)$$

where $\hat{S}_t$ = Transformed daily survival rate
$\hat{S}_M$ = Maximum likelihood estimate of survival, equation (14.37)

x = Average interval length $= \dfrac{\Sigma \, Ln_L}{\Sigma \, n_L}$

$n_L = n_{LS} + n_{Lf}$ = Total number of observations in interval L
L = Interval length in days

Step 2. Calculate the standard error of $\hat{S}_t$:

$$\hat{S}_{pt} = \frac{x(\hat{S}_M^{x-1})\hat{S}_p}{2\sqrt{1 - \hat{S}_M^x}} \qquad (14.39)$$

where $\hat{S}_{pt}$ = Standard error of the transformed daily survival rate
x = Average interval length (defined above)
$\hat{S}_M$ = Maximum likelihood estimate of survival

$$\hat{S}_p = \left[\sum_L \left(\frac{n_L L^2(\hat{S}_M^{L-2})}{1 - \hat{S}_M^L}\right)\right] \quad (14.40)$$

Program-group SURVIVAL (Appendix 2) can do these calculations.

Box 14.6 Estimation of Survival Rate from Radiotelemetry Data

Red squirrels were fitted with radio collars and monitored over a 2-month period with these results:

Intervals between relocations (days), L	No. of red squirrels relocated, n_L	No. of survivors, n_{LS}	No. of deaths, n_{Lf}
1	47	45	2
2	23	22	1
3	36	33	3
4	12	12	0

Trent and Rongstad Estimator

From equation (14.32):

$$x = \text{Total number of radio-days observed}$$

$$= (47)(1) + (23)(2) + (36)(3) + (12)(4) = 249 \text{ radio-days}$$

$$\hat{S} = \frac{x - y}{x}$$

$$= \frac{249 - 6}{249} = 0.9759 \qquad \text{(finite survival rate per day)}$$

Mayfield Estimator

From equation (14.34):

$$\sum_L L(n_{LS} + 0.5n_{LF}) = 1[45 + (0.5)(2)] + 2[22 + (0.5)(1)]$$

$$+ 3[33 + (0.5)(3)] + 4[12 + 0] = 242.5 \text{ radio-days}$$

$$\hat{S} = 1 - \left(\frac{\text{number of deaths}}{\Sigma L(n_{LS} + 0.5n_{Lf})}\right)$$

$$= 1 - \left(\frac{6}{242.5}\right) = 0.9526 \qquad \text{(finite survival rate per day)}$$

Bart and Robson (1982) Maximum Likelihood Estimator

From equations (14.35), (14.36), and (14.37),

$$A = \sum_L \left[\frac{L}{\hat{S}} \left(n_{LS} - \frac{n_{Lf}\hat{S}^L}{1 - \hat{S}^L} \right) \right]$$

$$= \frac{1}{0.97526} \left[45 - \frac{2(0.97526)}{1 - 0.97526} \right]$$

$$+ \frac{2}{0.97526} \left[22 - \frac{1(0.97526^2)}{1 - 0.97526^2} \right] + \cdots = 3.0183154$$

$$B = \sum_L \frac{L}{\hat{S}^2} \left[n_{LS} + \frac{n_{Lf}\hat{S}^L(L - 1 + \hat{S}^L)}{(1 - \hat{S}^L)^2} \right]$$

$$= \frac{1}{(0.97526)^2} \left[45 + \frac{2(0.97526)(1 - 1 + 0.97526)}{(1 - 0.97526)^2} \right]$$

$$+ \frac{2}{(0.97526)^2} \left[22 + \frac{1(0.97526^2)(1 - 1 + 0.97526^2)}{(1 - 0.97526^2)^2} \right] + \cdots = 10{,}059.89$$

$$\hat{S}_M = \hat{S} + \frac{A}{B}$$

$$= 0.97526 + \frac{3.0183154}{10059.89} = 0.975558 \qquad \text{(finite daily survival rate)}$$

A second iteration using this estimate of 0.975558 for $\hat{S}$ gave

$$\hat{S}_M = 0.975555$$

and a third iteration gave the same result:

$$\hat{S}_M = 0.975555$$

so there was no need to continue the iterations. This is the best estimate of the daily survival rate for these red squirrels.

These calculations, along with estimates of the 95% confidence intervals, can be carried out by Program-group SURVIVAL (Appendix 2).

This transformation can also be used to estimate the sample size needed to get a specified precision in an estimate of the daily survival rate. Bart and Robson (1982) give this procedure:

1. Guess the likely value of the finite daily survival rate (S) (e.g., 0.98).
2. Specify the desired lower confidence limit (S_L) (e.g., 0.95) on the assumption that the finite daily survival rate given in (1) is correct.
3. Guess the average interval length for your data (A) (e.g., 2 days).
4. Decide on the probability you wish to tolerate that the lower estimate will be *less than* the specified limit (e.g., 0.025).

From these values, the sample size you need is given by

$$\hat{n} = \left(\frac{z\ S^L}{2\ \Delta}\right)^2 \tag{14.41}$$

where
n = Estimated total sample size needed
z = Standard normal deviate for the probability level chosen in (4) (e.g., if $p = .025$, z will be 1.96; if $p = .05$, z will be 1.645)
S = Guessed value of the daily survival rate
L = Guessed value of the average interval length / 2.0
$\Delta = \sqrt{1 - S_L^A} - \sqrt{1 - S^A}$
A = Average interval length (guessed)
S_L = Desired lower confidence limit for the probability level chosen

To be conservative, you should underestimate the daily survival rate slightly.

One restriction on these estimates is that survival rates are assumed to be constant within the time unit specified. Heisey and Fuller (1985) describe a computer program that allows one to combine data for several time periods and test the hypothesis that survival was constant throughout the whole time period. White (1983) describes a completely general survival program that computes maximum likelihood estimates for any type of radiotelemetry, band-recovery, or nest success data. It is important to recognize that survival rates in natural populations are rarely constant from year to year or equal in the two sexes. White (1983) warns against uncritical acceptance of the null hypothesis of equal survival rates when sample sizes are small and the ability to specify narrow confidence limits does not exist. There is no substitute for reasonable sample sizes in any estimation of survival rates.

14.4.2 Kaplan-Meier Method

An alternative approach to estimating survival rates from radio-collared animals has been proposed by Pollock et al. (1989a). This approach centers on the time of death of each individual in the sample. Individuals may live through the time period of interest or lose their radio during the period. The survival rate from the start of the period until the day of death of the last recorded death is given by

$$\hat{S}_K = \prod_{i=1}^{n} \left[1 - \left(\frac{d_i}{r_i}\right)\right] \tag{14.42}$$

where
$\hat{S}_K$ = Kaplan-Meier estimate of finite survival rate for the period
d_i = Number of deaths recorded at time i
r_i = Number of individuals alive and at risk at time i
n = Number of time checks for possible deaths

Radio-tagged individuals may be checked every day or week, every second day or week, or at irregular time periods for this method. Table 14.5 illustrates the type of data obtained on radio-tagged bobwhite quail.

The Kaplan-Meier method has two important advantages for field ecologists. First, newly tagged animals may be added to the sample at any time, so that the cohort being

TABLE 14.5 SURVIVAL DATA ON RADIO-TAGGED BOBWHITE QUAIL, SUMMARIZED FOR KAPLAN-MEIER SURVIVAL ESTIMATION

Sample week (i)	Time of checks	No. birds with radios (r_i)	No. found dead (d_i)	No. censored
1	17–23 Nov.	20	0	0
2	24–30 Nov.	21	0	0
3	1–7 Dec.	22	2	1
4	8–14 Dec.	19	5	0
5	15–21 Dec.	14	3	0
6	22–28 Dec.	11	0	0
7	29 Dec.–4 Jan.	11	0	0
8	5–11 Jan.	11	2	0
9	12–18 Jan.	9	1	0
10	19–25 Jan.	8	0	1
11	26 Jan.–1 Feb.	7	0	0
12	2–8 Feb.	10	0	0
13	9–15 Feb.	16	4	0
14	16–22 Feb.	22	4	0
15	23 Feb.–1 March	23	4	1
16	2–8 March	24	4	0
17	9–15 March	20	2	0

Source: Pollock et al. 1989a, Table 2.

studied can be maintained at a large size even if deaths are occurring throughout the study. This is called the *staggered entry design*, and Table 14.5 illustrates this with data on bobwhite quail. The previous methods all assume that a cohort of individuals is marked at one time and that the subsequent history of that cohort is studied without any possible additions. It is important in any survival study to keep the sample size as large as possible, so that confidence limits are relatively narrow. Second, animals may be lost to the study inadvertently without affecting the estimated survival rate. These individuals represent *censored data* in statistical jargon; they have dropped out of the study, and we do not know their fate (Figure 14.5). Animals may lose their radio tags, or radios may fail electronically. The Kaplan-Meier method accounts for censored data by adjusting the number of individuals at risk. Note that we do *not* assume that censored animals have died but that they go on living without our knowing their subsequent fate. We make use of their data up to the point at which they were lost from the sample.

The variance of the Kaplan-Meier estimate of the survival rate can be calculated in two ways. I present here Greenwood's formula (Pollock et al. 1989a):

$$\text{var}(\hat{S}_K) = \hat{S}_K^2 \left[\sum_{i=1}^{n} \left(\frac{d_i}{r_i(r_i - d_i)} \right) \right] \tag{14.43}$$

where $\text{var}(\hat{S}_K)$ = Greenwood's estimate of the variance of the Kaplan-Meier survival rate

d_i = Number of deaths recorded at time i

r_i = Number of radio-tagged animals at risk at time i

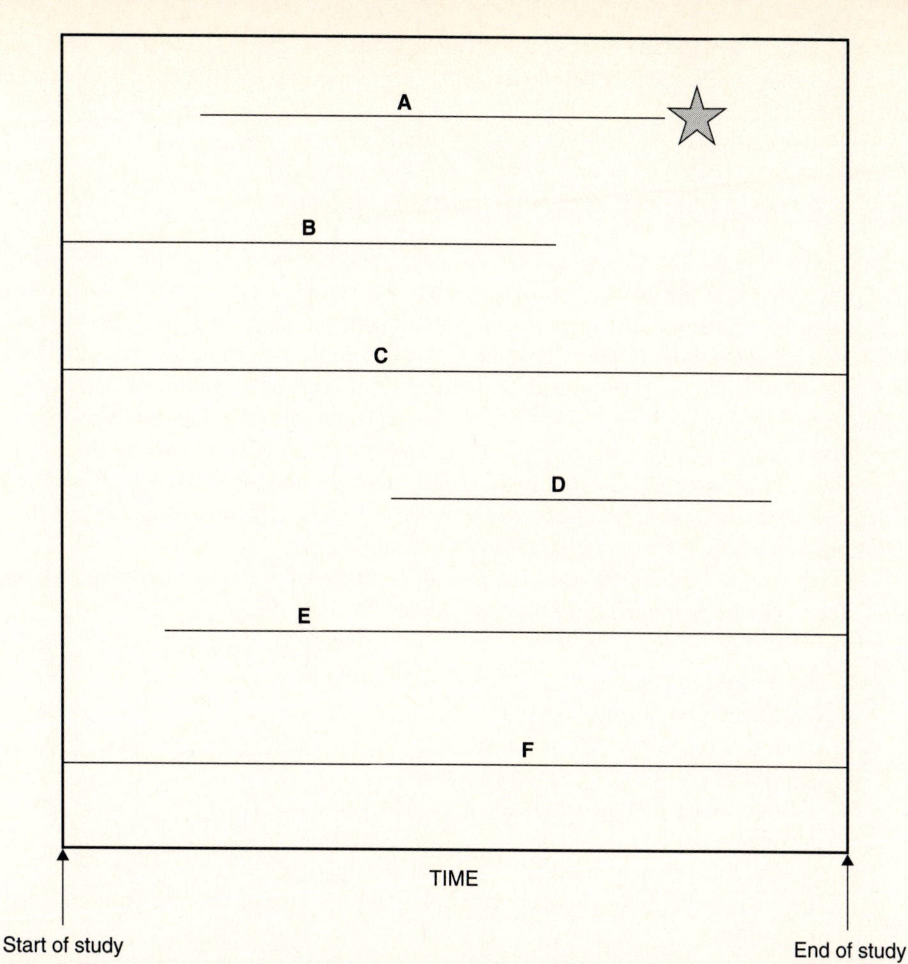

Figure 14.5 Illustration of censored survival data in a mark-recapture or radiotelemetry study. The *x*-axis represents time, which is a defined study period. In this example 6 individuals (A to F) enter the study at different times when they are first captured (the *staggered-entry* design). Individuals alive at the end of the study (C, E, and F) are all *right-censored* since we know they lived that long but not how much longer. During the study animals may lose their tags or radio collars and thus be lost to observation. In this example animals B and D are *censored* observations, and we use their data to the point of disappearance. One animal (A) died during the study period.

The standard error of this estimate is the square root of this variance, and confidence limits may be obtained in the usual way:

$$\hat{S}_K \pm t_\alpha \sqrt{\text{var}(\hat{S}_K)} \tag{14.44}$$

The *t*-value can be based conservatively on the number of radio-tagged individuals alive at the end of the time period or more liberally on the average number of tagged individuals during the study period.

Pollock et al. (1989b) discuss the advantages of this estimator over the Mayfield (1975) and Bart and Robson (1982) estimators. Program-group SURVIVAL (Appendix 2) can do these calculations.

14.5 ESTIMATION OF BIRD SURVIVAL RATES

A common technique for studying survival in bird populations has been to band young birds in the nest year after year and to record the recoveries of these birds in subsequent years. Table 14.6 illustrates the type of data often obtained in these studies. Anderson et al. (1985) and Lebreton et al. (1993a) have discussed the statistical problems associated with survival data of this type. If the recovery data are from dead birds, there is clearly an incomplete registration of deaths, since not all dead, tagged birds are located. Similarly, if the recovery data is of survivors, there is an incomplete registration of surviving animals. The probability of survival may vary both in time and with respect to the age of the birds. To separate time- and age-dependence of survival, bird-ringing schemes should mark both adult birds and young birds each year (Anderson et al. 1985).

If only young birds in the nest are ringed, eight assumptions are necessary before survival rates can be estimated from the data (Anderson et al. 1985):

1. The ringed sample is representative of the population.
2. There is no loss of rings.
3. Survival is not affected by ringing.
4. The year of recovery is correctly tabulated.
5. The fate of each ringed individual is independent of the fates of other ringed individuals.
6. Survival rates are age-specific and do not depend on the calendar year.
7. The reporting rate for ringed birds is the same for all age classes and in all years of the study.
8. There are no subgroups in the population that have different survival rates.

The difficulty is that many of these assumptions are not valid for birds, and the statistical

TABLE 14.6 RINGING AND RECOVERY DATA FOR YOUNG BIRDS RINGED OVER k YEARS AND RECOVERED OVER l YEARS ($l > k$).[a] THIS APPROACH IS NOT RECOMMENDED FOR SURVIVAL ESTIMATION.

		Recoveries (R_{ij}) in year j						
Ringing year	Number ringed	1	2	3	. . .	k	. . .	l
1	N_1	R_{11}	R_{12}	R_{13}	. . .	R_{1k}	. . .	R_{1l}
2	N_2		R_{22}	R_{23}	. . .	R_{2k}	. . .	R_{2l}
3	N_3			R_{33}	. . .	R_{3k}	. . .	R_{3l}
⋮	⋮					⋮		⋮
k	N_k					R_{kk}	. . .	R_{kl}

[a] The recovery rate in any given year can depend on both the age of the bird and the year and variation in the chances of being found in that year, so this design is confounded unless you make very restrictive assumptions.

methods available for estimating survival rates are very sensitive to these assumptions. The main point is that this simple, biologically appealing approach to estimating survival by ringing baby birds is unacceptable statistically, and thus should not be used.

There is an extensive literature now on estimating survival in bird populations (Brownie et al. 1985; Conroy et al. 1989; Lebreton et al. 1992) and a set of computer programs (see Lebreton et al. 1993b) are available for carrying out the tedious calculations that are necessary to allow for time- and age-varying survival rates. The general procedure is to link year-to-year variation in ecological factors like temperature, rainfall, or population density with individual factors like sex and age in a complex survival model so that one can determine which factors are significant in explaining survival differences. Factors that are not significant can be dropped out of the overall model, and one searches for the best simple model to fit the data.

Survival rates estimated from a long sampling program by mark-recapture models like the Jolly-Seber model are not independent estimates and should not be analyzed by analysis of variance or regression as though they are independent (Lebreton et al. 1993a). This is one of the main reasons that specialized methods have been developed during the last 30 years for the analysis of survival.

14.6 TESTING FOR DIFFERENCES IN SURVIVAL RATES

Survival differences among different cohorts or between the two sexes in a population need to be analyzed statistically to see if they are significant. A whole array of statistical tests are available to answer the general question of whether the life span of individuals in population A is longer or shorter on average that that of individuals in population B. The statistical tests needed to answer these questions have been provided from medical statistics (Lee 1992), where they are used to decide about the effectiveness of new medical treatments, and from engineering, where they are used to test hazard or failure rates of equipment. Pyke and Thompson (1986), Pollock et al. (1989b), and Hutchings et al. (1991) have summarized some of the most useful parametric and nonparametric survival tests for ecologists and given examples to illustrate their utility for analyzing survivorship curves from life tables. We will illustrate two of these tests here.

14.6.1 Log-Rank Test

Two survival curves can be compared to test the null hypothesis that they are random samples from the same underlying survivorship curve. The log-rank test is one of many that can be applied to such data. It has the advantage of being a simple nonparametric test and the disadvantage of having relatively low statistical power. The log-rank test is a chi-squared test that compares the observed and the expected number of deaths that occur in each time interval of study. The null hypothesis is that the two groups have the same survival pattern. We expect each of the two groups to have the same number of deaths, corrected for differences in sample size of the groups. The expected number of deaths in the first sample can be estimated by

$$E(d_{1j}) = \frac{d_j r_{1j}}{r_j} \tag{14.45}$$

where $E(d_{1j})$ = Expected number of deaths in group 1 in time j
r_{1j} = Number of animals at risk in group 1 at sample time j

d_j = Total number of deaths in both groups = $\sum_{j=1}^{k} (d_{1j} + d_{2j})$

r_j = Total number at risk in both groups = $\sum_{j=1}^{k} (r_{1j} + r_{2j})$

The variance of the number of deaths is obtained from the following equation:

$$\text{var}(d_{1j}) = \frac{r_{1j} r_{2j} d_j}{r_j^2} \quad (14.46)$$

where $\text{var}(d_{1j})$ = Variance of number of deaths in group 1 at time j

and other terms are as defined above.

Given these expectations and their variance, the log-rank test statistic is:

$$\chi^2 = \frac{\left[\sum_{j=1}^{k} d_{1j} - \sum_{j=1}^{k} E(d_{1j}) \right]^2}{\sum_{j=1}^{k} \text{var}(d_{1j})} \quad (14.47)$$

where d_{1j} = Number of deaths in group 1 in time period j
$E(d_{1j})$ = Expected number of deaths in group 1 in time j (equation [14.45])
$\text{var}(d_{1j})$ = Variance of number of deaths in group 1 at time j (equation [14.46])

The log-rank chi-squared test has one degree of freedom and at the usual $\alpha = 5\%$ level, the critical value is 3.84. Box 14.7 illustrates the use of the log-rank test for a staggered entry design.

The log-rank test assumes that survival times are independent for the different animals. If you have banding or tagging data, it assumes that the tags or bands do not reduce survival. If you are dealing with radiotelemetry data, it assumes that carrying a radio tag does not reduce survival. Censored data in radiotelemetry or mark-recapture studies represent a particular problem for survival estimation, since we assume that censored individuals go on living a normal life span. But a radio failure may be caused by a predator that destroys the radio as it kills the individual. In this case an error would be introduced, and the survival rate would be lower than estimated. Upper and lower limits of errors from censoring can be estimated by assuming that every censored individual was in reality a death (to give a lower bound) and by assuming that every censored animal lived until the end of the study period (to give an upper bound).

14.6.2 Likelihood Ratio Test

The log-rank test is a nonparametric test, and in some cases it is more useful to use a parametric test. Parametric tests for survival differences assume a specific survival distribution model, like the exponential distribution shown in Figure 14.1. If survival distributions for two groups are the same, and less than 20% of the observations are censored, and sample size is large ($n_1 + n_2 > 50$), then the likelihood ratio test has greater power than the log-rank test (Lee 1992). It is important to check that the survival distribution is the same for the two groups being compared before using this test.

Box 14.7 Comparing Survival Distributions with the Log-Rank Test

Bobwhite quail were radio-tagged in North Carolina in the autumn of 1985 and again in the autumn of 1986. More deaths occurred in 1986, and the wildlife biologist wishes to know if the survival patterns shown in these two years differ significantly.

First, combine the data for the two samples:

	Fall 1985		Fall 1986		Combined data	
Week	No. at risk r_{1j}	No. of deaths, d_{1j}	No. at risk r_{2j}	No. of deaths, d_{2j}	No. at risk r_j	No. of deaths, d_j
1	7	1	7	0	14	1
2	6	0	6	0	12	0
3	8	0	11	1	19	1
4	13	0	10	0	23	0
5	18	0	16	1	34	1
6	18	0	15	0	33	0
7	18	0	15	1	33	1
8	18	0	14	0	32	0
9	18	0	14	3	32	3
Total		1		6		7

Then proceed as follows to compute the log-rank test statistic:

1. Calculate for each of the time periods in the study the number of deaths expected under the null hypothesis that the two samples have the same survival pattern. From equation (14.45),

$$E(d_{1j}) = \frac{d_j r_{1j}}{r_j}$$

where d_{1j} = Number of deaths occurring in sample 1 at time j
r_{1j} = Number of animals at risk in sample 1 at time j
d_j = Number of deaths occurring in combined data at time j
r_j = Number at risk in combined data at time j

$$E(d_{11}) = \frac{1(7)}{14} = 0.5$$

$$E(d_{12}) = \frac{0(6)}{12} = 0$$

$$E(d_{13}) = \frac{1(8)}{19} = 0.421$$

and the resulting values are given in the table below.

2. Calculate the variance of the number of deaths from equation (14.46). Note that if no deaths are observed in a time period, there is no variance (as in week 2):

$$\text{var}(d_{1j}) = \frac{r_{1j} r_{2j} d_j}{r_j^2}$$

$$\text{var}(d_{11}) = \frac{7(7)(1)}{14^2} = 0.25$$

$$\text{var}(d_{13}) = \frac{8(11)(1)}{19^2} = 0.244$$

Expected no. of deaths in first sample	Variance of expected no. of deaths
0.500	0.250
0	0
0.421	0.244
0	0
0.529	0.249
0	0
0.545	0.248
0	0
1.687	0.738
3.683	1.729

3. Calculate the chi-squared test statistic for the log-rank test from equation (14.47):

$$\chi^2 = \frac{[\Sigma_{j=1}^{k} d_{1j} - \Sigma_{j=1}^{k} \text{E}(d_{1j})]^2}{\Sigma_{j=1}^{k} \text{var}(d_{1j})} = \frac{[1 - 3.683]^2}{1.729} = 4.16$$

This chi-squared statistic tests the null hypothesis that the two survival distributions are the same and has one degree of freedom. At $\alpha = 5\%$, the critical value of χ^2 is 3.84, and thus we reject the null hypothesis for these quail survival data. Survival was significantly higher in fall 1985.

These calculations can be carried out in Program-group SURVIVAL (Appendix 2).

The likelihood ratio test proceeds in six steps:

Step 1. Sum the censored and uncensored life spans of individuals in the two treatments:

$$x_i = \sum_{j=2}^{k} (c_{ij} j) + s_i k \qquad (14.48)$$

$$y_i = \sum_{j=2}^{k} d_{ij} j$$

where x_i = Sum of censored life spans for group $i (i = 1, 2)$
y_i = Sum of uncensored life spans for group i
c_{ij} = Number of censored individuals in group i at time period j
d_{ij} = Number of deaths in group i at time period j
s_i = Number of survivors to the end of the study for group $i (i = 1, 2)$
k = Last time period of the study $(j = 1, 2, 3, \ldots k)$

Step 2. Compute the maximum likelihood estimates of the death rate per time interval:

$$\hat{\lambda}_i = \frac{d_i}{x_i + y_i} \tag{14.49}$$

where λ_i = Finite death rate for group i per unit time for the entire study

and the other terms are defined above.

Step 3. Calculate the combined individual likelihood functions for the two groups under the assumption that they have significantly different survival rates:

$$L(\hat{\lambda}_1, \hat{\lambda}_2) = \lambda_1^{d_1} \lambda_2^{d_2} (e^{[-\hat{\lambda}_1(x_1+y_1)-\hat{\lambda}_2(x_2+y_2)]}) \tag{14.50}$$

Step 4. Repeat these calculations of the death rate for the entire sample combined under the null hypothesis of equal survival in the two groups:

$$\hat{\lambda} = \frac{d_1 + d_2}{x_1 + y_1 + x_2 + y_2} \tag{14.51}$$

Step 5. Repeat the calculation of the likelihood function for the combined sample under the null hypothesis of equal survival:

$$L(\hat{\lambda}, \hat{\lambda}) = \hat{\lambda}^{d_1+d_2} (e^{[-\hat{\lambda}(x_1+y_1+x_2+y_2)]}) \tag{14.52}$$

Step 6. Calculate the test statistic for the likelihood ratio as follows:

$$\chi^2 = -2 \log_e \left(\frac{L(\hat{\lambda}, \hat{\lambda})}{L(\hat{\lambda}_1, \hat{\lambda}_2)} \right) \tag{14.53}$$

This chi-squared value for the likelihood ratio test has one degree of freedom and tests the null hypothesis of equal survival distributions in the two groups. Program-group SURVIVAL (Appendix 2) can calculate the log rank test and the likelihood ratio test for the comparison of survival in two groups.

14.6.3 Temporal Differences in Mortality Rates

Once you have determined that your two groups differ in mortality rates, it may be useful to try to pinpoint the exact time or age at which the differences occur. The simplest approach to this problem is to compute confidence limits for the time-specific mortality rates (Hutchings et al. 1991).

$$\hat{m}_j = \frac{d_j}{r_j - 0.5(d_j)} \tag{14.54}$$

where $\hat{m}_j$ = Finite rate of mortality during the time interval j to $j + 1$
d_j = Number of individuals dying during the time interval j to $j + 1$
r_j = Number of individuals alive at start of time interval j

The variance of this mortality rate is given by

$$\text{var}(\hat{m}_j) = \frac{(\hat{m}_j)^2}{d_j}\left\{1 - \left(\frac{\hat{m}_j}{2}\right)^2\right\} \tag{14.55}$$

This variance can be used to estimate confidence limits around the mortality rates for each time period in the usual manner (e.g., equation [14.42]). The precision of survival rates will be poor unless the sample size is 20 or more in each time period.

14.7 SUMMARY

Survival rates can be calculated in many different ways, and this chapter summarizes some of the more common methods of analysis. Survival rates can be expressed as finite or as instantaneous rates. Finite rates are easier to comprehend, but instantaneous rates are more convenient for mathematical manipulations.

Life tables are a convenient method of summarizing the age-specific mortality schedule of a population. Six methods have been used for constructing life tables, and each makes specific assumptions about the population. Key factor analysis can utilize life tables gathered over several years to determine the best way to predict population changes. It is useful in indicating the critical age or stage in the life cycle where numbers change.

Survival rates can be estimated from age-composition data, if survival rates are constant. Fisheries managers have used age-composition data most effectively to estimate the survival components of commercially harvested species.

Radiotelemetry data can be used to estimate survival rates with high precision if enough individuals are followed closely. The estimation problem is identical to that for estimating nesting success in birds from periodic visits to nests. Survival rate estimation from mark-recapture kinds of data are more difficult to estimate because not all individuals are recaptured or found dead. Maximum likelihood estimators have been developed, and computer programs are available to do the tedious calculations necessary to answer demographic questions about variation in survival rates in space and time among different sex and age groups in the population.

Statistical comparisons of survival between two groups can be done with a variety of parametric and nonparametric tests. The commonly used log-rank test has low statistical power unless samples are large. There is no substitute for large ($n > 40$) samples if you wish to have good precision in survival studies.

SELECTED READING

Bart, J., and Robson, D. S. 1982. Estimating survivorship when the subjects are visited periodically. *Ecology* 63: 1078–1090.

Caughley, G. 1977. *Analysis of Vertebrate Populations*. Chapter 8, Mortality, 85–106. John Wiley and Sons, London.

Heisey, D. M., and Fuller, T. K. 1985. Evaluation of survival and cause-specific mortality rates using telemetry data. *Journal of Wildlife Management* 49: 668–674.

Hutchings, M. J., Booth, K. D., and Waite, S. 1991. Comparison of survivorship by the logrank test: criticisms and alternatives. *Ecology* 72: 2290–2293.

Lachin, J. M., and Foulkes, M. A. 1986. Evaluation of sample size and power for analyses of survival with allowance for nonuniform patient entry, losses to follow-up, noncompliance, and stratification. *Biometrics* 42: 507–519.

Lebreton, J. D., and North, P. M., eds. 1993. *Marked Individuals in the Study of Bird Population.* Birkhäuser Verlag, Basel.

Lebreton, J. D., Pradel, R., and Clobert, J. 1993. The statistical analysis of survival in animal populations. *Trends in Ecology and Evolution* 8: 91–94.

Podoler, H., and Rogers, D. 1975. A new method for the identification of key factors from life-table data. *Journal of Animal Ecology* 44: 85–114.

Pollock, K. H., Winterstein, S. R., Bunck, C. M., and Curtis, P. D. 1989. Survival analysis in telemetry studies: The staggered entry design. *Journal of Wildlife Management* 53: 7–15.

Pyke, D. A., and Thompson, J. N. 1986. Statistical analysis of survival and removal rate experiments. *Ecology* 67: 240–245.

Sibley, R. M., and Smith, R. H. 1998. Identifying key factors using λ-contribution analysis. *Journal of Animal Ecology* 67: 17–24.

QUESTIONS AND PROBLEMS

14.1. Keith and Windberg (1978) estimated the survival rate of young snowshoe hares for 265 days as follows:

1966–67	0.557
1967–68	0.430
1968–69	0.241
1969–70	0.339

Express these as annual finite rates of survival.

14.2. Geist (1971) gave the following data for age at death from skulls of bighorn rams (*Ovis canadensis*) from the Palliser Range, Banff National Park. Note that age 1 rams are between 1 and 2 years of age, and so on.

	Age at death (years)																	
	1	2	3	4	5	6	7	8	9	10	11	12	13	14	15	16	17	Σ
No. found	0	2	1	3	4	0	6	6	8	9	8	13	9	3	3	1	1	77

(a) What method of life table do these data represent? Calculate a life table for these rams on the assumption that (1) $r = 0$ and (2) $r = -0.02$. How do they differ?

(b) Calculate the mean expectation of life for an individual born into this population.

14.3. Smith and Polacheck (1981) reported the following age-composition data for female northern fur seals (*Callorhinus ursinus*) from 1958–1961:

Age (years)	1958	1959	1960	1961
3	39	43	18	84
4	42	93	36	96
5	70	114	55	68
6	99	118	45	62
7	103	143	66	95
8	102	164	105	107
9	81	108	144	114
10	97	96	129	112
11	113	98	136	82
12	134	76	106	71
13	110	56	120	76
14	92	70	107	67
15	71	87	67	68
16	56	69	53	55
17	36	36	46	24
18	22	27	23	25
19	14	16	19	10
20–22	5	17	12	9

(a) Can you calculate a life table from these data? Why or why not?
(b) Can you estimate survival rates from these data? Are survival rates constant with age?

14.4. Smith (1987) gives the age structure of the ringed seal (*Phoca hispida*) from the 1980 and 1981 catch at Holman Island in the western Canadian arctic as follows:

Age	1980	1981
0+	195	118
1	61	87
2	26	89
3	59	82
4	13	73
5	32	57
6	21	52
7	8	15
8	8	10
9	28	16
10	24	17
11	19	11
12	10	13
13	16	11
14	15	15
15	10	16
16	6	6
17	4	7
18	2	4
19	1	7
20	3	4
Totals	561	710

Calculate an average annual survival rate from these catch curves. What assumptions must be made to make these estimates?

14.5. Mayfield (1961) reported the following data on nest loss for Kirkland's warbler: 154 nests located; total exposure = 882.5 nest days, with a total of 35 nests lost during incubation. Estimate the probability of a nest surviving the 14-day incubation period for this species. How large a sample would be needed to keep the confidence limits to ±0.02 (rate per day)?

14.6. Twenty moose were radio-collared and followed through a winter (October–March). Two individuals died: one on November 17, and one on January 6. Two radio collars failed; no signal was received after December 27 on one and February 22 on the other. Estimate the survival rate for the six-month period.

14.7. How much precision would be lost by doing radiotelemetry checks every second day instead of every day? every fifth day? On what factors will this loss of precision depend? Formalize your intuition on this question and then read Bart and Robson (1982, 1085) for a discussion of this problem.

14.8. Waterfowl biologists have often estimated the success of duck nests by dividing the number of nests from which eggs hatch by the number of nests found (Johnson and Klett 1985). Discuss whether this procedure will give an unbiased estimate of nest success and under what conditions.

14.9. Mark O'Donoghue radio-collared Canada lynx in the Yukon during the winter of 1990–91 and obtained these results:

Week	No. at risk	No. deaths	No. censored	Week	No. at risk	No. deaths	No. censored
1	8	0	0	16	16	0	0
2	8	0	0	17	16	0	0
3	8	0	0	18	16	0	0
4	8	0	0	19	16	0	0
5	12	0	0	20	16	1	1
6	12	0	0	21	14	0	0
7	14	0	0	22	14	0	1
8	14	1	0	23	13	0	0
9	13	2	0	24	13	0	0
10	12	0	0	25	15	0	0
11	14	0	0	26	15	0	0
12	16	0	0	27	15	0	5
13	17	0	0	28	10	0	0
14	17	0	0	29	10	0	0
15	17	1	0	30	10	0	0

(a) Estimate the overwinter survival rate for lynx during this time period.
(b) Convert this survival estimate to an annual rate of survival.
(c) What difference would it make to your conclusions if the 5 lynx censored in week 27 had died at the same time their radios stopped functioning?
(d) Suggest two ways of improving the precision of the survival estimate.
(e) What would be gained in this study by checking the lynx daily instead of weekly?
(f) What is wrong with calculating the survival rate by noting that there were 22 lynx studied in total over the winter, and 5 deaths were observed, so the mortality estimate is 5/22 = 23% (77% survival rate)?

14.10. Recompute the expectation of further life data in Box 14.5 using equation (14.21) as a better estimate of L_x. Discuss the magnitude of the difference in the two estimates of e_0 and which estimate is likely to be more accurate for these data.

CHAPTER 15

The Garbage Can

Every book on methods must end with a series of miscellaneous items that do not fit easily into one of the more structured chapters. In this final chapter I discuss the statistical ideas of transformations and repeatability, the problems of estimating trend lines, how to measure temporal variability, and two new computer-intensive methods that hold much promise for ecological statistics.

15.1 TRANSFORMATIONS

A transformation is a change of numerical scale. When you convert inches to centimeters you are applying a transformation. Most familiar transformations are of this type, and are called *linear* transformations. With all linear transformations, the standard statistics can be readily converted. Given a multiplicative transformation of the type $X_T = cX_0$, it follows that:

$$\bar{x}_T = c\bar{x}_0 \tag{15.1}$$

where $\bar{x}_T$ = Transformed mean (e.g., cm)
$\bar{x}_0$ = Original mean (e.g., inches)
c = Constant of conversion (e.g., 2.54 to convert inches to centimeters)

Similarly 95% confidence limits can be converted with equation (15.1). To convert standard deviations or standard errors,

$$S_T = cS_0 \tag{15.2}$$

where S_T = Transformed standard error or standard deviation
S_0 = Original standard error or standard deviation
c = Constant of conversion

If a fixed amount is added to or subtracted from each observation, the mean will be increased or decreased by that amount, but the standard deviation and standard error are not changed at all. These simple transformations are not usually the type that statisticians write about.

Parametric methods of statistical analysis assume that the variable being measured has a normal distribution. Ecological variables often have skewed distributions that violate this assumption. In addition to this normality assumption, more complex parametric methods such as the analysis of variance (ANOVA) assume that all groups have the same variance and that any treatment effects are additive. All statistics books discuss these assumptions (e.g., Sokal and Rohlf 1995, Chap. 13; Zar 1996, Chap. 13).

There are four solutions to the problems of non-normality, heterogeneous variances, and non-additivity. The first solution is to use nonparametric methods of data analysis. There is a large literature on nonparametric methods that can be applied to much ecological data (Siegel and Castellan 1988; Tukey 1977; Chambers et al. 1983). Nonparametric methods rely on ranks rather than absolute values and thus lose the arithmetic precision many ecologists desire. Nonparametric methods have been waning in popularity in recent years (Underwood 1997; Day and Quinn 1989, 448).

The second solution is to transform the scale of measurement so that the statistical demands of parametric analyses are approximately satisfied. The advantage of this solution is that the whole array of powerful methods developed for parametric statistics can be employed on the transformed data. If you choose to use transformations on your data, you have to decide on exactly what transformation to use. There are two ways to do this: (1) use one of the four standard transformations discussed in every statistics book; or (2) choose a general transformation that can be tailor-made to your specific data. The most widely used general transformation is the *Box-Cox transformation*.

The third solution is to ignore the problem and to argue that parametric methods like ANOVA are robust to violations of their assumptions. This is not recommended, although some statisticians argue that if you have large sample sizes ($n > 30$ in all groups), you do not need to worry much about these assumptions. In most cases I would not recommend this solution because one of the other three is preferable.

The fourth solution is to use computer-intensive methods of data analysis. These are the most recent tools available to ecologists who must deal with data that are not normally distributed, and they are now available widely in statistics packages for desktop computers. Randomization tests are one simple type of computer-intensive method (Sokal and Rohlf 1995, 803). Students are referred to Noreen (1990) and Manly (1991), specialized books that discuss these procedures in detail.

15.1.1 Standard Transformations

The four most commonly used statistical transformations are the *logarithmic* transformation, the *square root* transformation, the *arcsine* transformation, and the *reciprocal* transformation. These are discussed in more detail by Hoyle (1973) and Thöni (1967).

Logarithmic Transformation The logarithmic transformation is commonly used in ecological data in which percentage changes or multiplicative effects occur. The original data are replaced by

$$X' = \log(X) \quad \text{or} \qquad (15.3)$$

$$X' = \log(X + 1) \quad \text{(if data contains 0 values)}$$

where X' = Transformed value of data
X = Original value of data

and logarithms may be to base 10 or base e (which differ only by a constant multiplier).

Asymmetrical or skewed distributions are common in ecological data. Figure 15.1 illustrates two examples. Positively skewed distributions have a tail of data pointing to the right, and negatively skewed distributions have a tail pointing to the left. A number of measures of skewness are available (Sokal and Rohlf 1995, 114) in standard statistical books and packages.

The logarithmic transformation will convert a positively skewed frequency distribution into a more nearly symmetrical one. Data of this type may be described by the *log-normal* distribution. Note that the logarithmic transformation is very strong in correcting positive skew, much stronger than the square root transformation. Figure 15.2 illustrates the impact of a logarithmic transformation on a positively skewed frequency distribution.

In all work with parametric statistics, you use the transformed (log) values, so that means, standard errors, and confidence limits are all expressed in logarithm units. If you wish to have the final values expressed in the original measurement units, remember the following three rules (which apply to all transformations):

1. Never convert *variances, standard deviations*, or *standard errors* back to the original measurement scale. They have no statistical meaning on the original scale of measurement.
2. Convert *means* and *confidence limits* back to the original scale by the inverse transformation. For example, if the transformation is log $(X + 1)$, the mean expressed in original units will be

$$\bar{X} = [\text{antilog}(\bar{X}')] - 1 = 10^{\bar{X}'} - 1 \qquad (15.4)$$

In the particular case of the logarithmic transformation, the mean expressed in original units is equivalent to the geometric mean of the original data. In some cases it may be desirable to estimate the arithmetic mean of the original data (Thöni 1967; Hoyle 1973). If you use the original data to estimate the mean, you usually obtain a very poor estimate (both inaccurate and with low precision) if sample size is relatively small ($n < 100$). An unbiased estimate of the mean in the original measurement units can be obtained from the

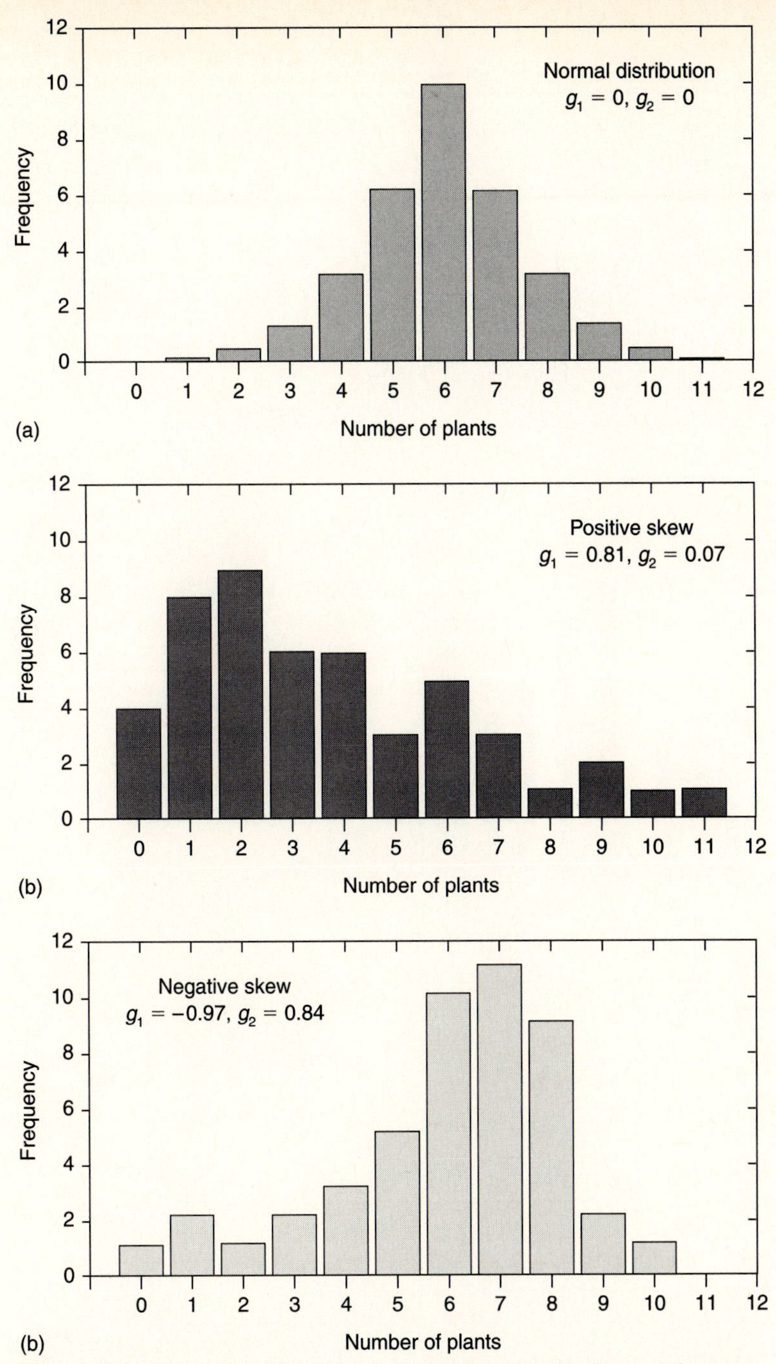

Figure 15.1 Illustration of positive and negative skew in observed frequency distributions. (a) The normal distribution, assumed in all of parametric statistics. (b) Positive skew illustrated by plant counts that fit a clumped distribution. (c) Negative skew illustrated by counts of highly clumped species. Skewed distributions can be normalized by standard transformations like the square root or the logarithmic transformation.

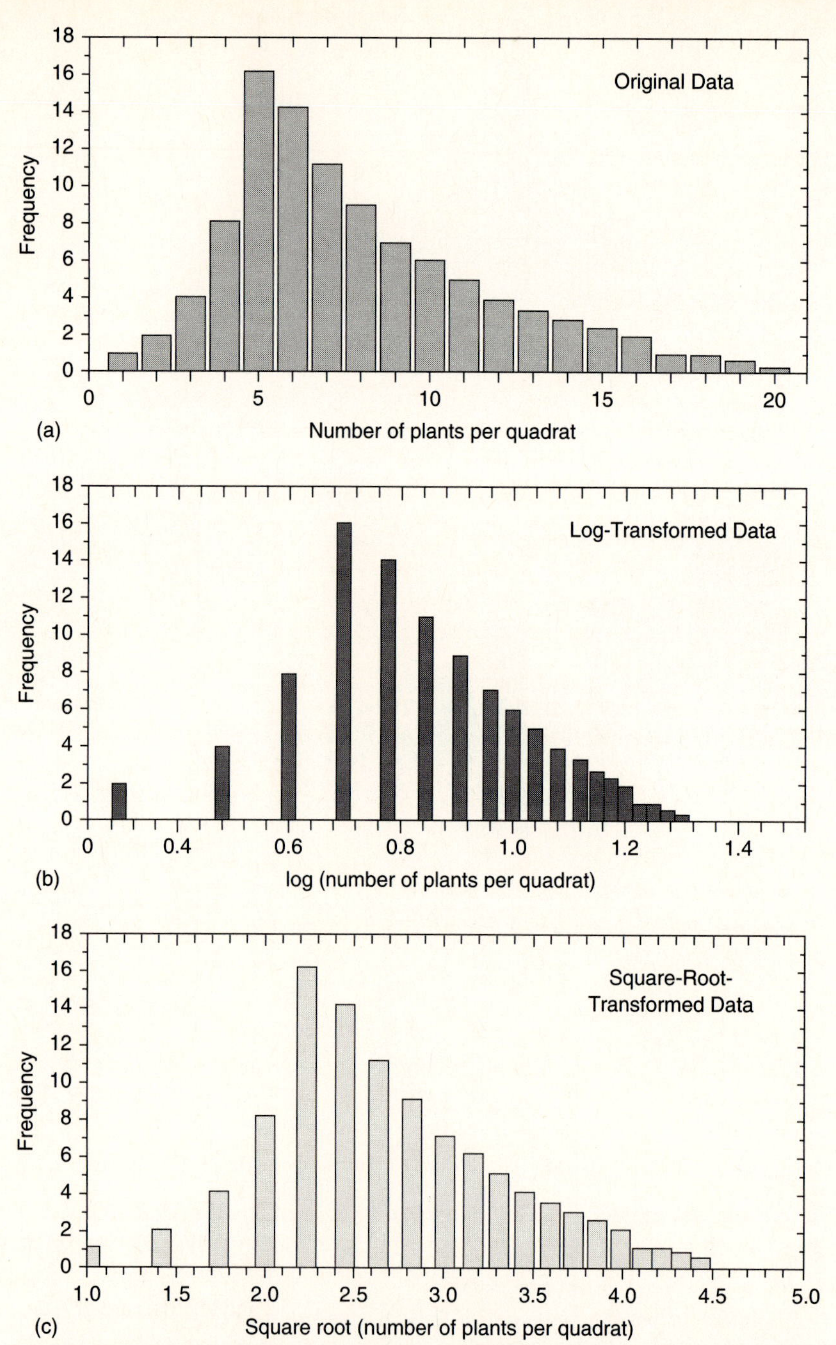

Figure 15.2 Illustration of the impact of a logarithmic transformation and a square root transformation on a positively skewed frequency distribution. Both these transformations make the original data more nearly normal in shape, but the square root transformation is slightly better in this example.

approximate equation given by Finney (1941):

$$\hat{\hat{X}} = (e^{\bar{X}' + 0.5s^2}) \left[1 - \frac{s^2(s^2 + 2)}{4n} + \frac{s^4(3s^4 + 44s^2 + 84)}{96n^2} \right] \quad (15.5)$$

where $\hat{\hat{X}}$ = Estimated arithmetic mean of original data
X' = Observed $\log_e$-transformed mean of data
s^2 = Observed variance of $\log_e$-transformed data
n = Sample size

Finney suggested that a minimum sample size of 50 is needed to get a reasonable estimate of the arithmetic mean with this procedure.

3. *Never* compare means calculated from untransformed data with means calculated from any transformation, reconverted back to the original scale of measurement. They are not comparable means. All statistical comparisons between different groups must be done using one common transformation for all groups.

When logarithmic transformations are used in linear regression on both the X and Y variables (log-log plots), the estimation of Y from X is biased if antilogs are used without a correction. Baskerville (1971) and Sprugel (1983) give the correct unbiased formulas for estimating Y in the original data units.

There is considerable discussion in the statistical literature about the constant to be used in logarithmic transformations. There is nothing magic about $\log(X + 1)$, and one could use $\log(X + 0.5)$ or $\log(X + 2)$. Berry (1987) discusses the statistical problem associated with choosing the value of this constant. Most ecologists ignore this problem and use $\log(X + 1)$, but Berry points out that the value of the constant chosen may greatly affect the results of parametric statistical tests like ANOVA. Berry argues that we should choose the constant c that minimizes the sum of skewness and kurtosis:

$$G_c = |g_1(c)| + |g_2(c)| \quad (15.6)$$

where G_c = Function to be minimized for a particular value of c
c = The constant to be added in a logarithmic transform
$g_1(c)$ = Estimate of skewness from a normal distribution for the chosen value of c
$g_2(c)$ = Estimate of kurtosis from a normal distribution for the chosen value of c

Figure 15.1 illustrates the concept of skewness in data. Kurtosis refers to the proportion of observations in the center of the frequency distribution in relation to the observations in the tails. Measures of both skewness (g_1) and kurtosis (g_2) are relative to the normal distribution in which g_1 and g_2 are zero. Methods to estimate skewness and kurtosis are given in most statistics books (Sokal and Rohlf 1995, 115; Zar 1996, 67):

$$g_1 = \frac{n \sum X^3 - 3(\sum X)(\sum X^2) + 2(\sum X)^3/n}{(n - 1)(n - 2)s^3} \quad (15.7)$$

where g_1 = Measure of skewness (= 0 for a normal distribution)
n = Sample size
$\sum X$ == Sum of all the observed X-values
$\sum X^2$ = Sum of all the observed X-values squared
s = Observed standard deviation

$$g_2 = \frac{(n+1)\{n\sum X^4 - 4(\sum X)(\sum X^3) + [6(\sum X)^2(\sum X^2)/n] - [3(\sum X)^4/n^2]\}}{(n-1)(n-2)(n-3)s^4} - \frac{3(n-1)^2}{(n-2)(n-3)} \tag{15.8}$$

where g_2 = Measure of kurtosis (= 0 for a normal distribution)
$\sum X^3$ = Sum of all the observed X-values cubed
$\sum X^4$ = Sum of all the observed X-values to fourth power

The procedure is to compute G_c (equation [15.6]) using the logarithmic transformation with a series of c values from (−) the minimum value observed in the raw data to 10 times the maximum value observed in the data (Berry 1987), as well as for the raw data with no transformation. This procedure is tedious and best done by computer. Program-group EXTRAS (Appendix 2) does these calculations. Plot the resulting G_c values on the y-axis against the c-values (x-axis) as in Figure 15.3, and choose the value of c that minimizes G_c.

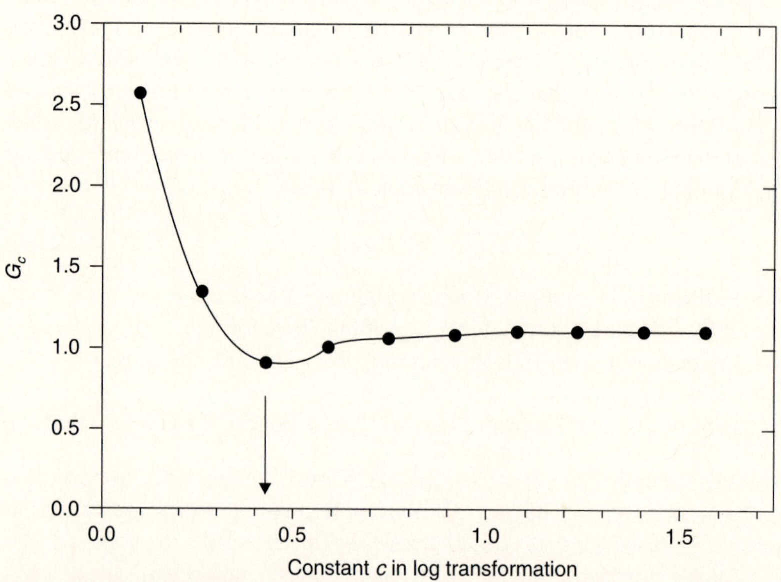

Figure 15.3 Estimation of the constant c to be used in logarithmic transforms of the type $\log(X + c)$ using the method of Berry (1987). Locate the minimum of the function G_c as defined in equation (15.6). Data from Box 4.2 on the abundance of black bean aphids was used for these calculations. There is a minimum in the G_c function around 0.43 (arrow), and the recommended logarithmic transform for these data would be $\log(X + 0.43)$.

Berry recommends using this value as the basis of the transformation:

$$X' = \log(X + c) \tag{15.9}$$

where c = Constant that minimizes the function G_c defined in equation (15.6)

In some cases c will be nearly 1.0, and the standard logarithmic transformation of equation (15.3) will be adequate. In other cases the G_c curve is relatively flat (as in Figure 15.3) and a broad range of possible c values will be equally good. In these cases it is best to use the smallest value of c possible.

Square Root Transformation The square root transformation is used when the variance is proportional to the mean, a common ecological situation (see Taylor's power law, Chapter 9, page 331). Any ecological data of counts fitting a Poisson distribution should be transformed with square roots before parametric statistics are applied. The original data are replaced by

$$X' = \sqrt{X + 0.5} \tag{15.10}$$

This transform is preferable to the straight square root transformation when the observed data are small numbers and include some zero values (Zar 1996, 279).

If you wish to obtain the mean and confidence limits in the original measurement units, you can reverse the transformation:

$$\bar{X} = (\bar{X}')^2 - 0.5 \tag{15.11}$$

This mean is slightly biased (Thöni 1967), and a more precise estimate is given by

$$\bar{X} = (\bar{X}')^2 - 0.5 + s^2\left(1 - \frac{1}{n}\right) \tag{15.12}$$

where $\bar{X}$ = Mean expressed in original measurement units
$\bar{X}'$ = Mean obtained from square root transformation (equation [15.10])
s^2 = Variance of square-root-transformed data
n = Sample size

Several variants of the square root transformation have been suggested to reduce the relationship between the variance and the mean. Anscombe (1948) suggested that

$$X' = \sqrt{X + 3/8} \tag{15.13}$$

was better than equation (15.10) for stabilizing the variance. Freeman and Tukey (1950) showed that when $X \leq 2$, a better transformation is

$$X' = \sqrt{X} + \sqrt{X + 1} \tag{15.14}$$

These variants of the square root transformation are rarely used in practice.

Arcsine Transformation Percentages and proportions form a binomial distribution when there are two categories, or a multinomial distribution when there are several categories, rather than a normal distribution. Consequently parametric statistics should not be computed for percentages or proportions without a transformation. In cases where the percentages range from 30% to 70%, there is no need for a transformation, but if any values are nearer to 0% or 100%, you should use an arcsine transformation. The term *arcsine* stands for the angle whose sine is a given value. Note that in mathematical jargon:

$$\text{arcsine} = \text{inverse sine} = \sin^{-1}$$

The recommended arcsine transformation is given by

$$X' = \arcsin\sqrt{p} \tag{15.15}$$

where X' = transformed value (measured in degrees)
p = observed proportion (range 0–1.0)

Transformed values may also be given in radians rather than degrees. The conversion factor is simply

$$1 \text{ radian} = \frac{180}{\pi} \text{ degrees} = 57.2957795 \text{ degrees}$$

To convert arcsine-transformed means back to the original scale of percentages or proportions, reverse the procedure:

$$\bar{p} = (\sin \bar{X}')^2 \tag{15.16}$$

where $\bar{p}$ = Mean proportion
X' = Mean of arcsine-transformed values

Mean proportions obtained in this way are slightly biased, and a better estimate from Quenouille (1950) follows:

$$\bar{p}_c = (\sin \bar{X}')^2 + 0.5 \cos[(2\bar{p})(1 - e^{-2s^2})] \tag{15.17}$$

where $\bar{p}_c$ = Corrected mean proportion
$\bar{p}$ = Mean proportion estimated from equation (15.16)
s^2 = Variance of the transformed values of X' from equation (15.15)

If you have the raw data, you can use a better transformation suggested by Anscombe (1948):

$$X' = \sqrt{n + 0.5}\left(\arcsin\sqrt{\frac{X + 3/8}{n + 3/4}}\right) \tag{15.18}$$

where X' = Transformed value
n = Sample size
X = Number of individuals with the attribute being measured

This transformation leads to a variable X' with expected variance 0.25 over all values of X and n. An alternative variant is suggested by Zar (1996, 240):

$$X' = 0.5\left(\arcsin\sqrt{\frac{X}{n + 1}} + \arcsin\sqrt{\frac{X + 1}{n + 1}}\right) \tag{15.19}$$

where all terms are as defined above.

This alternative may be slightly better than Anscombe's when most of the data are very large or very small proportions.

There is always a problem with binomial data because they are constrained to be between 0 and 1, and consequently no transformation can make binomial data truly normal in distribution. Binomial data can also be transformed by the logit transformation:

$$X' = \log_e \left(\frac{p}{q}\right) \tag{15.20}$$

where X' = logit transform of observed proportion p
$q = 1 - p$

Note that the logit transform is not defined for $p = 1$ or $p = 0$, and if your data include these values, you may add a small constant to the numbers of individuals with the attribute, as we did above in equation (15.18). The logit transformation will act to spread out the tails of the distribution and may help to normalize the distribution.

Reciprocal Transformation Some ecological measurements of rates show a relationship between the standard deviation and the square of the mean. In these cases, the reciprocal transformation is applied to achieve a nearly normal distribution:

$$X' = \frac{1}{X} \tag{15.21}$$

or if there are observed zero values, use

$$X' = \frac{1}{X + 1} \tag{15.22}$$

Thöni (1967) discusses the reciprocal transformation in more detail.

15.1.2 Box-Cox Transformation

In much ecological work using parametric statistics, transformations are applied using rules of thumb or tradition without any particular justification. When there is no strong reason for preferring one transformation over another, it is useful to use a more general approach. Box and Cox (1964) developed a general procedure for finding out the best transformation to use on a set of data in order to achieve a normal distribution. They used a family of power transformations of the following form:

$$X' = \frac{X^\lambda - 1}{\lambda} \qquad (\text{when } \lambda \neq 0) \tag{15.23}$$

or

$$X' = \log(X) \qquad (\text{when } \lambda = 0) \tag{15.24}$$

This family of equations yields most of the standard transformations as special cases, depending on the value of λ. For example, if $\lambda = 1$, there is effectively no transformation (since $X' = X - 1$); when $\lambda = 0.5$, you get a square root transformation; and when $\lambda = -1$, you get a reciprocal transformation.

To use the Box-Cox transformation, choose the value of λ that maximizes the log-likelihood function:

$$L = -\frac{\nu}{2}\log_e s_T^2 + (\lambda - 1)\frac{\nu}{n}\sum(\log_e X) \tag{15.25}$$

where L= Value of log-likelihood
ν= Number of degrees of freedom $(n - 1)$
s_T^2= Variance of transformed X-values, using equation (15.23)
λ= Provisional estimate of power transformation parameter
X= Original data values

This equation must be solved iteratively to find the value of λ that maximizes L. Since this is tedious, it is usually done by computer. Box 15.1 shows these calculations, and Figure 15.4 illustrates the resulting plot of the log-likelihood function for a set of data.

When the original data include zeros, equation (15.25) becomes insoluble because the log of 0 is negative infinity. For these cases, Box and Cox suggest adding a constant like 0.5 or 1.0 to each X value before doing this transformation. It is possible to use the log-likelihood function to search for the best value of this constant as well as λ, but in most cases $(X + 0.5)$ or $(X + 1.0)$ will be sufficient to correct for data that include zeros.

Box and Cox showed that confidence limits could be estimated for the λ parameter of the power transformation from the chi-squared distribution. If these confidence limits include $\lambda = 1.0$, the data may not need any transformation.

In practice, when one is collecting the same type of data from many areas or from many years, the Box-Cox procedure may be applied to several sets of data to see if a common value of λ may be estimated. This value of λ could then be used to specify the data transformation to be used in future data sets. As with other transformations, one should not change values of λ within one analysis, or the transformed means will not be comparable.

Box 15.1 Calculation of Box-Cox Transformation for Clip-Quadrat Data

A series of grassland plots were clipped at the height of the growing season to estimate production for one season, with these results:

Plot no.	Dry weight of grass (g)
1	55
2	23
3	276
4	73
5	41
6	97

1. Choose a trial value of λ, say $\lambda = -2.0$.
2. Transform each data point using equation (15.23) and $\log_e(X)$:

X	$\log_e(X)$	$\frac{X^\lambda - 1}{\lambda}$
55	4.00733	0.4998347
23	3.13549	0.4990548
276	5.62040	0.4999934
73	4.29046	0.4999062
41	3.71357	0.4997026
97	4.57471	0.4999469
Sum	25.34196	

3. Calculate the variance of the transformed weights (third column):

$$s_T^2 = \frac{\Sigma X^2 - (\Sigma X)^2/n}{n - 1}$$

$$= 12.29013 \times 10^{-8}$$

4. Calculate the log-likelihood function (equation [15.25]):

$$L = -\frac{\nu}{2}\log_e s_T^2 + \left[(\lambda - 1)\frac{\nu}{n}\sum (\log_e X)\right]$$

$$= -\frac{5}{2}[\log_e(0.0000001229)] + \left[(-2 - 1)\left(\frac{5}{6}\right)(25.34196)\right] = -23.58$$

5. Repeat (1) to (4) using different values of λ to get

λ	L
−3	−27.23
−1	−20.86
−0.5	−20.21
0	−20.28
+0.5	−21.13
1	−22.65
2	−26.90

These values are plotted in Figure 15.4. Clearly, there is a maximum between $\lambda = 0$ and $\lambda = -1.0$. By further application of steps (1) to (4) you can show that the maximum likelihood occurs at $\lambda = -0.29$.

Program-group EXTRAS (Appendix 2) can do these calculations and estimates confidence limits for the best value of λ.

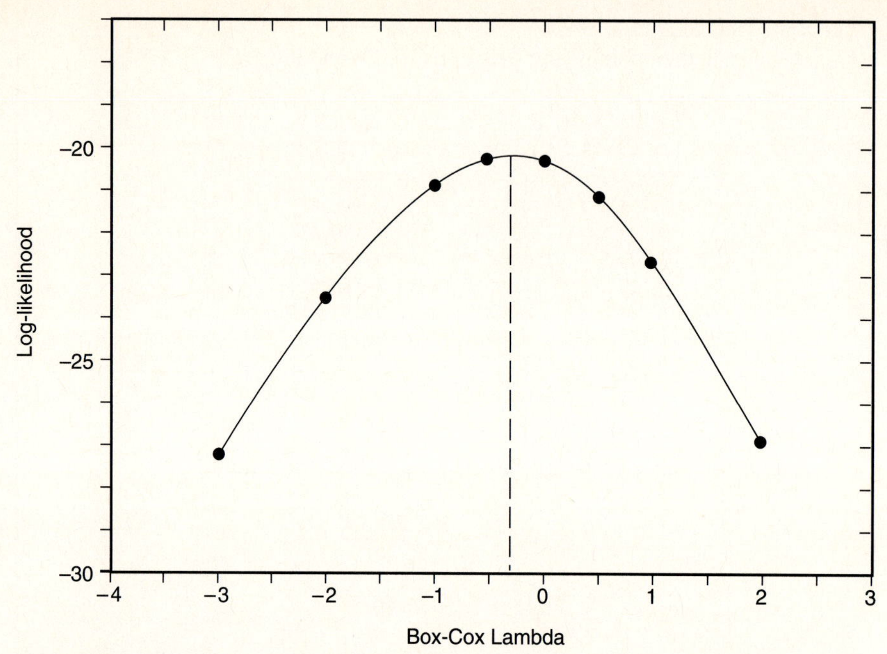

Figure 15.4 Log-likelihood function for the grassland plot data given in Box 15.1. Maximum likelihood occurs at $\lambda = -0.29$, and this exponent could be used in the Box-Cox transformation.

The Box-Cox procedure is a very powerful tool for estimating the optimal transformation to use for ecological data of a wide variety. Its application is limited only by the amount of computation needed to use it. Program-group EXTRAS (Appendix 2) does these calculations to estimate λ and gives 95% confidence limits for λ. Rohlf (1995) provides additional programs that employ this transformation.

15.2 REPEATABILITY

Ecologists usually assume all the measurements they take are highly precise and thus repeatable. If individual *a* counts 5 eggs in a nest, it is reasonable to assume that individual *b* will also count 5 eggs in the same nest. For continuous variables like weight, this assumption is less easily made, and in cases where some observer skill is required to take the measurement, one should not automatically assume that measurements are repeatable. Consequently, one of the first questions an ecologist should ask is, *Are these measurements repeatable*?

Repeatability is a measure ranging from 0 to 1.0 that shows how similar repeated measurements are on the same individual items. It is also known as the *intraclass correlation coefficient* (Sokal and Rohlf 1995, 213) and is calculated as follows. A series of individuals (or items) is measured repeatedly by the same individual or by several individuals. Table 15.1 gives an example from aerial census of moose. Data are cast in an ANOVA table for a one-way design:

TABLE 15.1 REPEATABILITY OF AERIAL COUNTS OF MOOSE IN SIX AREAS OF HABITAT IN BRITISH COLUMBIA

	Habitat Block					
	A	B	C	D	E	F
	16	8	27	43	4	14
	17	6	27	41	3	15
	15	8	24	44	4	13
	16	8	26	40	3	15
	18		25	41	4	14
	16		27	39		13
			26			15
No. of counts	6	4	7	6	5	7
$\sum X$	98	30	182	248	18	99
Mean count	16.3	7.5	26.0	41.3	3.6	14.1

Note: Several observers were used on sequential trips within the same 10-day period.

ANALYSIS OF VARIANCE TABLE

Source	Degrees of freedom	Sum of squares	Mean square
Among groups	$a - 1$	SS_A	MS_A
Within groups between measurements ('error')	$\Sigma (n_i - 1)$	SS_E	MS_E

where a = Total number of items being measured repeatedly
n_i = Number of repeated measurements made on item i

Formulas for calculating the sums of squares and mean squares in this table are given in every statistics book (e.g., Sokal and Rohlf 1995, 211; Zar 1996, 182). Box 15.2 works out one example. Repeatability is given by

$$R = \frac{s_A^2}{s_E^2 + s_A^2} \tag{15.26}$$

where R = Repeatability
s_A^2 = Variance among groups
s_E^2 = Variance within groups

R is thus the proportion of the variation in the data that occurs among groups. If measurements are perfectly repeatable, there will be zero variance within a group, and R will be 1.0. The two variance components are obtained directly from the ANOVA table above:

$$s_E^2 = \text{MS}_E \tag{15.27}$$

$$s_A^2 = \frac{\text{MS}_A - \text{MS}_E}{n_0} \tag{15.28}$$

Box 15.2 Calculation of Repeatability for Snowshoe Hare Hind Foot Length

A group of four field workers measured the right hind foot on the same snowshoe hares during one mark-recapture session with the following results:

	Observer					
Hare tag	A	B	C	D	n_i	$\sum X$
171	140	140			2	280
184	125	125			2	250
186	130	129	135		3	394
191	130	132	132		3	394
192	131	134			2	265
193	139	140	142		3	421
196	127	127			2	254
202	130	130	130	133	4	523
203	129	132	132		3	393
207	138	137	138	138	4	551
211	140	141	143	141	4	565
217	147	149	147		3	443
				Total	35	4733

Grand mean foot length = 135.2286 mm

1. Calculate the sums for each individual measured and the grand total, as shown above in column 7, and the sum of the individual items squared (i = hare, j = observer):

$$\sum X_{ij}^2 = 140^2 + 140^2 + 125^2 + \cdots = 641{,}433$$

2. Calculate the *sums of squares*. The sums of squares *among* hares is

$$SS_A = \sum \left[\frac{(\Sigma X_i)^2}{n_i} \right] - \frac{(\text{Grand total})^2}{\text{Total sample size}}$$

$$= \frac{280^2}{2} + \frac{250^2}{2} + \frac{394^2}{3} + \cdots + \frac{443^2}{3} - \frac{(4733)^2}{35}$$

$$= 641{,}379.5834 - 640{,}036.8286 = 1342.7548$$

The sum of squares *within* individual hares is

$$SS_E = \sum (X_{ij}^2) - \sum \left[\frac{(\Sigma X_i)^2}{n_i} \right]$$

$$= 641{,}433 - 641{,}379.5834 = 53.4166$$

3. Fill in the ANOVA table and divide the sums of squares by the degrees of freedom to get the mean squares:

Source	d.f.	Sum of squares	Mean square
Among hares	11	1342.7548	122.0686
Within individuals ("error")	23	53.4166	2.32246

4. Calculate the *variance components* from equations (15.27) to (15.29):

$$s_E^2 = \text{MS}_E = 2.32246$$

$$n_0 = \frac{1}{a-1}\left(\sum n_i - \frac{\sum n_i^2}{\sum n_i}\right)$$

$$= \frac{1}{12-1}\left(35 - \frac{2^2 + 2^2 + 3^2 + 3^2 + 2^2 + \cdots}{35}\right)$$

$= 2.8987$ (effective average number of replicates per individual hare)

$$s_A^2 = \frac{\text{MS}_A - \text{MS}_E}{n_0}$$

$$= \frac{122.0686 - 2.32246}{2.8987} = 41.31$$

5. Calculate *repeatability* from equation (15.26):

$$R = \frac{s_A^2}{s_A^2 + s_E^2}$$

$$= \frac{41.31}{41.31 + 2.32} = 0.947$$

6. Calculate the *lower confidence limit* for R from equation (15.32):

$$R_L = 1.0 - \frac{n_0 \text{MS}_E F}{\text{MS}_A + \text{MS}_E(n_0 - 1)(F)}$$

With $n_1 = 11$ d.f. and $n_2 = 23$ d.f from the F-table for $\alpha = 0.025$ we get $F = 2.62$ and thus:

$$R_L = 1.0 - \frac{(2.8987)(2.32246)(2.62)}{122.0686 + 2.32246(2.8987 - 1)(2.62)} = 0.868$$

7. Calculate the *upper confidence limit* for R from equation (15.33). To calculate F for $\alpha = 0.975$ for n_1 and n_2 degrees of freedom, note that

$$F_{.975}(n_1, n_2) = \frac{1.0}{F_{.025}(n_1, n_2)}$$

Since $F_{.025}$ for 23, and 11 d.f. is 3.17, $F_{.975}$ for 11, and 23 d.f. is 0.3155.

$$R_U = 1.0 - \frac{n_0 \text{MS}_E F}{\text{MS}_A + \text{MS}_E(n_0 - 1)(F)}$$

$$= 1.0 - \frac{(2.8987)(2.32246)(0.3155)}{122.0686 + 2.32246(2.8987 - 1)(0.3155)} = 0.983$$

These calculations can be carried out in Program-group EXTRAS (Appendix 2).

$$\text{where} \quad n_0 = \frac{1}{a - 1}\left(\sum n_i - \frac{\Sigma n_i^2}{\Sigma n_i}\right) \tag{15.29}$$

where a = Number of groups (items) being measured repeatedly
n_i = Number of repeated measurements made on item i

Lessells and Boag (1987) pointed out a recurring mistake in the literature on repeatabilities where incorrect values of R are obtained because of the mistake in confusing mean squares and variance components:

$$s_A^2 \neq \text{MS}_A \tag{15.30}$$

Since repeatability can be used in quantitative genetics to give an upper estimate of the heritability of a trait, these mistakes are serious and potentially confusing. Lessells and Boag (1987) give an approximate method for estimating the correct repeatability value from published ANOVA tables. They point out that if

$$\text{Published } R \text{ value} = \frac{F}{F + 1} \tag{15.31}$$

where F = F-ratio from a one-way ANOVA table

then the published repeatability must be *wrong*! This serves as a useful check on the literature.

Confidence limits on repeatability can be calculated as follows (Becker 1984):
Lower confidence limit:

$$R_L = 1.0 - \frac{n_0 \text{MS}_E F_{\alpha/2}}{\text{MS}_A + \text{MS}_E(n_0 - 1)(F)} \tag{15.32}$$

where $F_{\alpha/2}$ = value from F-table for $\alpha/2$ level of confidence (e.g., $F_{.025}$ for 95% confidence limits).

Upper confidence limit:

$$R_U = 1.0 - \frac{n_0 \text{MS}_E F_{1-\alpha/2}}{MS_A + MS_E(n_0 - 1)(F)} \tag{15.33}$$

where $F_{1-\alpha/2}$ = value from F-table for $(1 - \alpha/2)$ level of confidence (e.g., $F_{.975}$ for 95% confidence limits)

The F-table is entered with the degrees of freedom shown in the ANOVA table. These confidence limits are not symmetric about R.

Box 15.2 gives an example of repeatability calculations. Program-group EXTRAS (Appendix 2) will do these calculations.

15.3 CENTRAL TREND LINES IN REGRESSION

Linear regression theory, as developed in standard statistics texts, applies to the situation in which one independent variable (X) is used to predict the value of a dependent variable (Y). In many ecological situations, there may be no clear dependent or independent variable. For example, fecundity is usually correlated with body size, or fish length is related to fish weight. In these cases there is no clear causal relationship between the X and Y variables, and the usual regression techniques recommended in statistics texts are not appropriate. In these situations Ricker (1973, 1984) recommended the use of a central trend line to describe the data more accurately.

Central trend lines may also be useful in ecological situations in which both the X- and Y-variables have measurement errors. Standard regression theory assumes that the X-variable is measured without error, but many types of ecological data violate this simple assumption (Ricker 1973). If the X-variable is measured with error, estimates of the slope of the usual regression line will be biased toward a lower absolute value.

Figure 15.5 illustrates regressions between hind foot length and body weight in snowshoe hares. Two standard regressions can be calculated for these data:

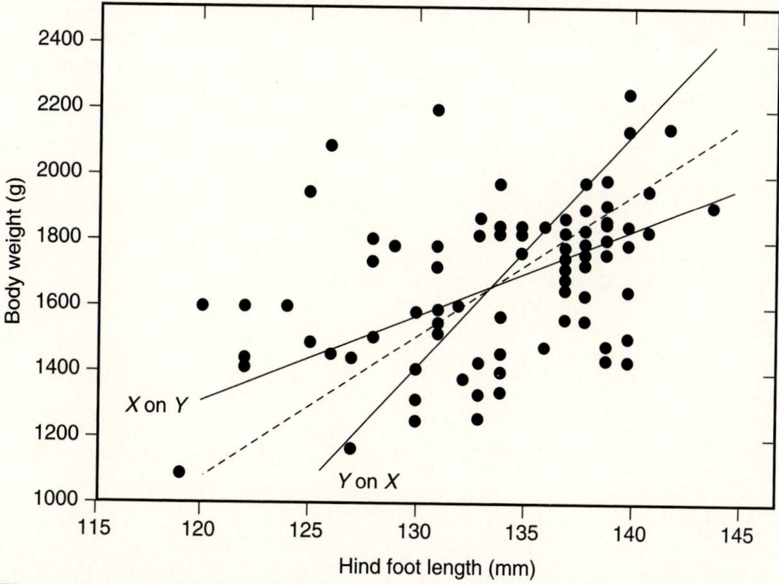

Figure 15.5 Regression of body weight on hind foot length for snowshoe hares. Three regressions can be calculated for these data. The usual regression of Y on X is shown as a solid line along with the less usual regression of X on Y. The functional or geometric mean regression (GMR) of Ricker (1973) is shown as the dashed line.

1. *Regression of hind foot length (Y) on body weight (X)*; this regression minimizes the vertical deviations (squared) from the regression line.
2. *Regression of body weight (Y) on hind foot length (X)*; this regression could alternatively be described as a regression of X on Y in terms of (1) above.

Ricker (1973) argues that a better description of this relationship is given by the central trend line called the *functional regression* or the *geometric mean regression* (GMR). The functional regression is simple to obtain, once the computations for the standard regression (1) have been made:

$$\hat{v} = \frac{\hat{b}}{r} = \sqrt{\frac{\hat{b}}{\hat{d}}} \tag{15.34}$$

where $\hat{v}$ = Estimated slope of the geometric mean regression
$\hat{b}$ = Estimated slope of the regression of Y on X
r = Correlation coefficient between X and Y
$\hat{d}$ = Estimated slope of the regression of X on Y

and $\hat{v}$ has the same sign as r and $\hat{b}$. The term *geometric mean regression* follows from the second way of estimating v given in equation (15.34). Box 15.3 illustrates the calculation of a central trend line.

Box 15.3 Calculation of a Geometric Mean Regression

Garrod (1967) gives the following data for fishing effort and total instantaneous mortality rate for ages 6 through 10 for the Arcto-Norwegian cod fishery:

Year	Fishing effort	Total mortality rate
1950–51	2.959	0.734
1951–52	3.551	0.773
1952–53	3.226	0.735
1953–54	3.327	0.759
1954–55	4.127	0.583
1955–56	5.306	1.125
1956–57	5.347	0.745
1957–58	4.577	0.859
1958–59	4.461	0.942
1959–60	4.939	1.028
1960–61	6.348	0.635
1961–62	5.843	1.114
1962–63	6.489	1.492
Totals	60.500	11.524
Means	4.6538	0.8865

The total mortality rate should depend on the fishing effort, and there is a large measurement error in estimating both of these parameters. To calculate the GMR, proceed as follows:

1. Calculate the usual statistical sums:

$$\sum X = 2.959 + 3.551 + 3.226 + \cdots = 60.500$$

$$\sum Y = 0.734 + 0.773 + 0.735 + \cdots = 11.524$$

$$\sum X^2 = 2.959^2 + 3.551^2 + 3.226^2 + \cdots = 298.40551$$

$$\sum Y^2 = 0.734^2 + 0.773^2 + 0.735^2 + \cdots = 10.965444$$

$$\sum XY = (2.959)(0.734) + (3.551)(0.773) + \cdots = 55.604805$$

2. Calculate the *sums of squares* and *sums of cross products*:

$$SS_X = \sum X^2 - \frac{(\Sigma X)^2}{n}$$

$$= 298.40551 - \frac{60.500^2}{13} = 16.847818$$

$$SS_Y = \sum Y^2 - \frac{(\Sigma Y)^2}{n}$$

$$= 10.965444 - \frac{11.524^2}{13} = 0.7498612$$

$$\text{Sum of cross products} = \sum XY - \frac{(\Sigma X)(\Sigma Y)}{n}$$

$$= 55.604805 - \frac{(60.5)(11.524)}{13} = 1.9738819$$

3. Calculate the parameters of the *standard regression of Y on X*:

$$Y = a + bX$$

$$\text{Slope} = \hat{b} = \frac{\text{Sum of cross products}}{\text{Sum of squares in } X}$$

$$= \frac{1.9738819}{16.847818} = 0.117159$$

$$Y \text{ intercept} = \hat{a} = \bar{y} - \hat{b}\bar{x}$$

$$= 0.8865 - (0.117159)(4.65385) = 0.34122$$

$$\text{Correlation coefficient} = r = \frac{\text{Sum of cross products}}{\sqrt{(\text{SS}_x)(\text{SS}_y)}}$$

$$= \frac{1.9738819}{\sqrt{(16.847818)(0.7498612)}} = 0.55534$$

$$\text{Variance about regression} = s^2_{yx}$$

$$= \frac{\text{SS}_y - ((\text{sum of cross products})^2/\text{SS}_x)}{n-2}$$

$$= \frac{0.7498612 - ((1.9738819)^2/16.847818)}{11}$$

$$= 0.0471457$$

$$\text{Standard error of slope} = s_b = \sqrt{\frac{s^2_{xy}}{\text{SS}_x}}$$

$$= \sqrt{\frac{0.0471457}{16.847818}} = 0.052899$$

4. Calculate the *geometric mean regression* from equations (15.34) and (15.35):

$$\text{Slope of GMR} = \hat{v} = \frac{\hat{b}}{r}$$

$$= \frac{0.117159}{0.55534} = 0.21097$$

$$Y \text{ intercept of GMR} = \hat{a}' = \bar{y} - \hat{v}\bar{x}$$

$$= 0.8865 - (0.21097)(4.6538) = -0.0953$$

The standard error of the slope of the GMR is the same as the standard error of $\hat{b}$.

Both regression lines can be plotted from the two points $(\bar{x}, \bar{y})$ and the Y-intercept $(\hat{a}')$, that is, the value of Y when $X = 0$.

Program-group EXTRAS (Appendix 2) can do these calculations for the geometric mean regression.

For the functional regression, the y-intercept is calculated in the usual way:

$$y \text{ intercept} = \bar{y} - \hat{v}\bar{x} \tag{15.35}$$

where $\bar{y}$ = Observed mean value of Y
$\bar{x}$ = Observed mean value of X
$\hat{v}$ = Estimated slope of the GMR from equation (15.34)

Ricker (1973) showed that the standard error of the slope of the GMR is the same as that of the slope $\hat{b}$ in a standard regression and that confidence limits for $\hat{v}$ can be obtained in the usual way from this standard error.

The central trend line described by the functional regression has three characteristics that are essential for any descriptive line (Ricker 1984):

1. The line must be *symmetrical*; thus, if X and Y are interchanged, there will be no change in the position of the line relative to the data.
2. The line must be *invariant to linear changes of scale*, so that its position among the data points does not depend on whether inches or centimeters are used in measuring.
3. The line should be *robust*, so that deviations from the statistical assumption of a bivariate normal distribution are not fatal.

Ordinary regressions discussed in all the statistics books are scale-invariant, but they are not symmetrical, nor are they robust (Schnute 1984). In contrast, the geometric mean regression has all of these traits.

Ricker (1984) and Jensen (1986) discuss several situations in which the functional regression is superior to ordinary regression procedures. Figure 15.6 illustrates a common ecological situation in which incomplete samples are taken from a population. The functional regression is obviously superior to the regression of Y on X in describing this relationship.

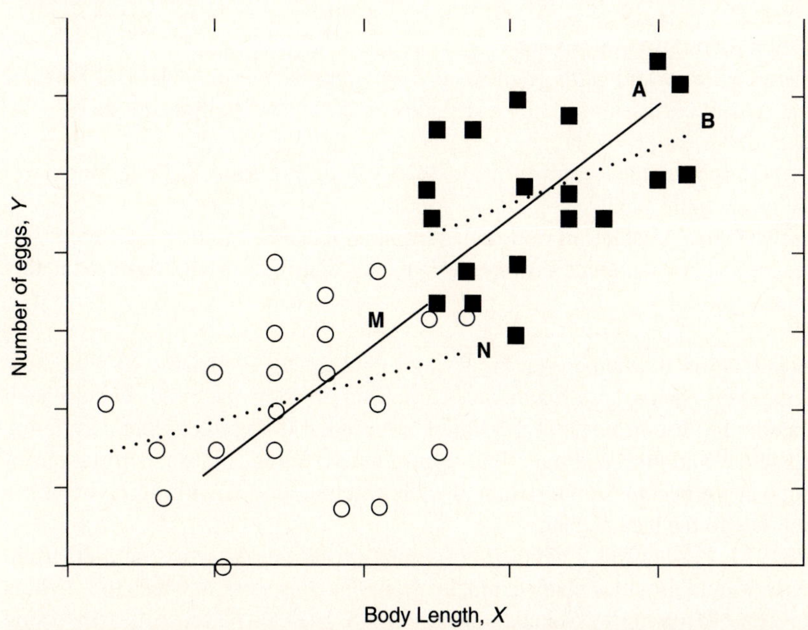

Figure 15.6 Hypothetical example of two incomplete samples from a population in which there is a linear regression between X and Y. Lines B and N (dashed lines) show the usual regression of Y on X for the two groups of data indicated by the two symbols. If you used these two regressions, you would probably conclude that these data come from two distinct statistical populations. Lines A and M show the geometric mean regressions for the two groups and suggest strongly that there is only one statistical population with a common relationship between body length and egg production. Clearly, the GMR is a better way of describing this relationship for this population. (Modified from Ricker 1984.)

TABLE 15.2 RICKER'S (1984) RECOMMENDATIONS REGARDING THE TYPE OF LINEAR REGRESSION TO USE FOR DIFFERENT KINDS OF DATA

A. *No serious measurement errors in the* X *or* Y *variables.*
   B. *Random sample from a bivariate normal population:*
      • Use GMR to describe general trend.
      • Use ordinary regression for prediction.
   B′. *Random sample from a population that is not bivariate normal:*
      • Use GMR or Schnute (1984) method for description and for prediction.
   B″. *Nonrandom sample:*
      • Use GMR method for description and for prediction.

A′. *Measurement errors in* Y *but not in* X.
   • Use ordinary regression for description and for prediction.

A″. *Measurement errors in both* Y *and* X *variables.*
      C. Error variances available for both *X* and *Y*:
         • Use Jolicoeur and Heusner (1971) method.
      C′. Error variances of *X* or *Y* unknown:
         • Use GMR method for description and prediction.

Note: GMR = geometric mean regression of Ricker (1973); ordinary regression = least-squares regression of *Y* on *X*.

When should you use the geometric mean regression? Ricker (1973, 1984) has put together some recommendations for guidance that are summarized in Table 15.2. The decision on what type of regression to choose depends on the answer to three questions:

1. Are there serious measurement errors in the *X*- or *Y*-variables? Or are measurement errors quite small?
2. Is the statistical population well-defined, so that it can be sampled randomly?
3. Does the statistical population show a frequency distribution that is approximately bivariate normal?

The main consideration is whether you are trying to predict the *Y*-variable with minimal error, or whether you wish to fit a functional relationship to show the trend. The decision as to which method to use in linear regression is not critical if there is a tight correlation between the *X* and *Y* variables (Jensen 1986). But in much ecological data, correlations of 0.5 to 0.7 occur, and it becomes important to decide which of the available regression lines is most appropriate to the data at hand.

Ricker (1973, 1984) gives a detailed discussion of the linear regression problem in ecological research and provides many examples from fisheries work in which the GMR is the appropriate regression to use. Sokal and Rohlf (1995, 541) also discuss this regression problem. Program-group EXTRAS (Appendix 2) can do these calculations for a geometric mean regression.

15.4 MEASURING TEMPORAL VARIABILITY OF POPULATIONS

Some populations fluctuate greatly in abundance from year to year and others are more stable. How can you measure this variability? This question arises in a number of ecological contexts, from theories that predict that species of smaller body size have more variable populations than larger species (Gaston and Lawton 1988), to suggestions that endangered species that fluctuate more are prone to extinction (Karr 1982). Given a set of population estimates for a species, it would seem to be a simple matter to estimate the variability of these estimates in the standard statistical manner. Unfortunately, this is not the case (McArdle et al. 1990), and it is instructive to see why.

Since populations typically change in size by proportions, rather than by constant amounts, the first suggestion we might make is to log-transform the density estimates. We can then use the standard deviation to measure variability with the usual formula first suggested by Lewontin (1966):

$$s = \sqrt{\frac{\sum_{t=1}^{k} (\log N_t - \log \bar{N})^2}{k - 1}} \tag{15.36}$$

where s = Standard deviation of log abundances = index of variability
N_t = Population size at time t
$\log \bar{N}$ = Mean of the logarithms of population size = $\sum \frac{(\log N_t)}{n}$
k = Number of observations in the time series (sample size)

This is the most commonly used measure of variability of populations, and it has serious problems that suggest it should never be used (McArdle et al. 1990; Stewart-Oaten et al. 1995). The most obvious problem is what to do when the population estimate is zero (McArdle and Gaston 1993), since the log of zero is not defined. Most authors sweep over this difficulty by adding a constant to the population estimate (e.g., $N + 1$) for each sample. But the value of s in fact changes with the value of the constant added, and this is not satisfactory. A second problem is that the variance of populations is typically related to the mean (see Taylor's power law, page 331), and unless the slope of Taylor's power law is 2, the log transformation is not entirely effective in removing this relationship. We need to search elsewhere for a better index of variability.

One measure that is not affected by zeros is the coefficient of variation. It is widely used in biometrics as a scale-independent measure of variability, and is defined in the usual way:

$$\text{CV}(N) = \frac{\text{standard deviation of } N}{\bar{N}} \tag{15.37}$$

where $\bar{N}$ = Mean population size for a series of time periods

The coefficient of variation is similar to the s statistic in being independent of population density if the slope of Taylor's power law is 2. Note that the usual measure of the coefficient of variation defined in equation (15.37) is slightly biased (Haldane 1955) and should be corrected by

$$\text{CV}(N) = \left(1 + \frac{1}{4n}\right)\left(\frac{\text{standard deviation of } N}{\bar{N}}\right) \tag{15.38}$$

where n is sample size.

The recommended procedure to measure population variability is as follows:

Step 1. Plot the regression of the log of the coefficient of variation (Y) vs. the log of population density (X) for the species of interest. This plot uses the same information as

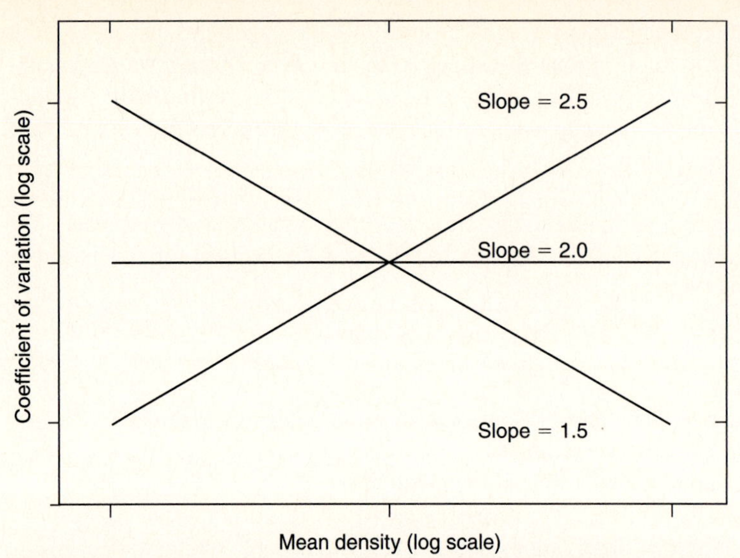

Figure 15.7 The problem of estimating variability in a population. The figure shows the relationship between the coefficient of variation (log scale) and the mean density of the population (log scale) that is expected under different slopes from Taylor's Power Law. Only if the Taylor slope = 2 is the measure of variability independent of population density, so that you could use the coefficient of variation as a measure of population variability.

Taylor's power law, and thus requires several samples of populations that have different average densities (a great deal of data!).

Step 2. If the variability does not depend on the mean (see Figure 15.7), you can use either the coefficient of variation or the standard deviation of log abundances (if no zeros) to measure variability in this population. If variability does not depend on mean density, Taylor's power law will have a slope of 2.0 (McArdle et al. 1990).

Step 3. If variability increases or decreases with mean density (Figure 15.7), you cannot compare population variability for this species with that of other species. A population at high density will of necessity be more or less variable than one at low density. Instead of comparing population variability, you should turn your attention to interpreting the pattern of relationships illustrated in Figure 15.7. Two situations are common.

Comparing Temporal Variability

When a number of independent sites are sampled repeatedly through time and the samples are taken far enough apart in time that they are independent statistically, we compare temporal variability. For each site, calculate the mean density and the coefficient of variation over time. Plot the data as in Figure 15.7 and calculate Taylor's power law (page 331).

1. If $b = 2$, the variability over time of the population is constant from site to site.
2. If $b > 2$, temporal variability is greater at good sites (= high-density sites).
3. If $b < 2$, temporal variability is greater at poor sites, and low-density populations are subject to more fluctuations than those in good sites.

Comparing Spatial Variability

In some cases interest centers on a series of populations in different sites (e.g., Perry 1988). As in the previous case, a number of independent sites are sampled repeatedly through time, and the samples are taken far enough apart in time and space that they are independent statistically. Calculate the mean and coefficient of variation over all the sites in a given year. Plot the data as in Figure 15.7 and calculate Taylor's power law for the spatial data.

1. If $b = 2$, the variability over space of the population is constant from year to year.
2. If $b > 2$, spatial variability is greater in good years (= high-density years).
3. If $b < 2$, spatial variability is greater in bad years, and in low-density years populations are more variable spatially. In good years densities tend to even out in space.

Clearly if we sample in the same sites at the same times in the same set of years, we can compare both temporal and spatial variability of our population.

Ecologists also wish to compare variability between different species, for example, to determine if bird populations fluctuate more than mammal populations. The same principles given above for single-species comparisons apply here. For simplicity, consider the case of two species. If for both of the species the slope of Taylor's power law is 2 for both temporal and spatial variability, then it is possible to use the coefficient of variation to compare population variability in the two species. Much background information is clearly needed before you can make this assumption. If you have only two sets of samples, one for each of the species, you cannot make any legitimate statement about relative variability. If the slope of Taylor's power law is not 2, comparisons are more difficult and must be limited to the observed range of densities (Figure 15.8). If many species are to be compared, one can only hope that the Taylor's slope is 2, and that you do not have to deal with a situation like that shown in Figure 15.8. Ecologists need to pay attention to these details to make proper statistical statements about variability of populations. Recent reviews have indeed suggested that most of the existing literature comparing population variability is invalid and the conclusions are artifacts of sampling (McArdle and Gaston 1992; Stewart-Oaten et al. 1995). It is important that we use better methods for future research on relative variability of populations.

15.5 JACKKNIFE AND BOOTSTRAP TECHNIQUES

The advent of modern computers has opened up a series of new statistical techniques that are of great importance to ecologists because they release us from two restrictive assumptions of parametric statistics: (1) that data conform to a normal frequency distribution, and (2) that statistical measures must have good theoretical properties, so that confidence limits can be derived mathematically. The price of giving up these traditional assumptions is a massive increase in computations. Computers are thus essential for all these new methods (Diaconis and Efron 1983).

Two computer-intensive methods have been particularly important in ecological statistics: the *jackknife* and the *bootstrap*. I will describe each of these briefly. We have already used these methods in Chapter 12 to estimate species richness (Sections 12.3.2 and 12.3.3)

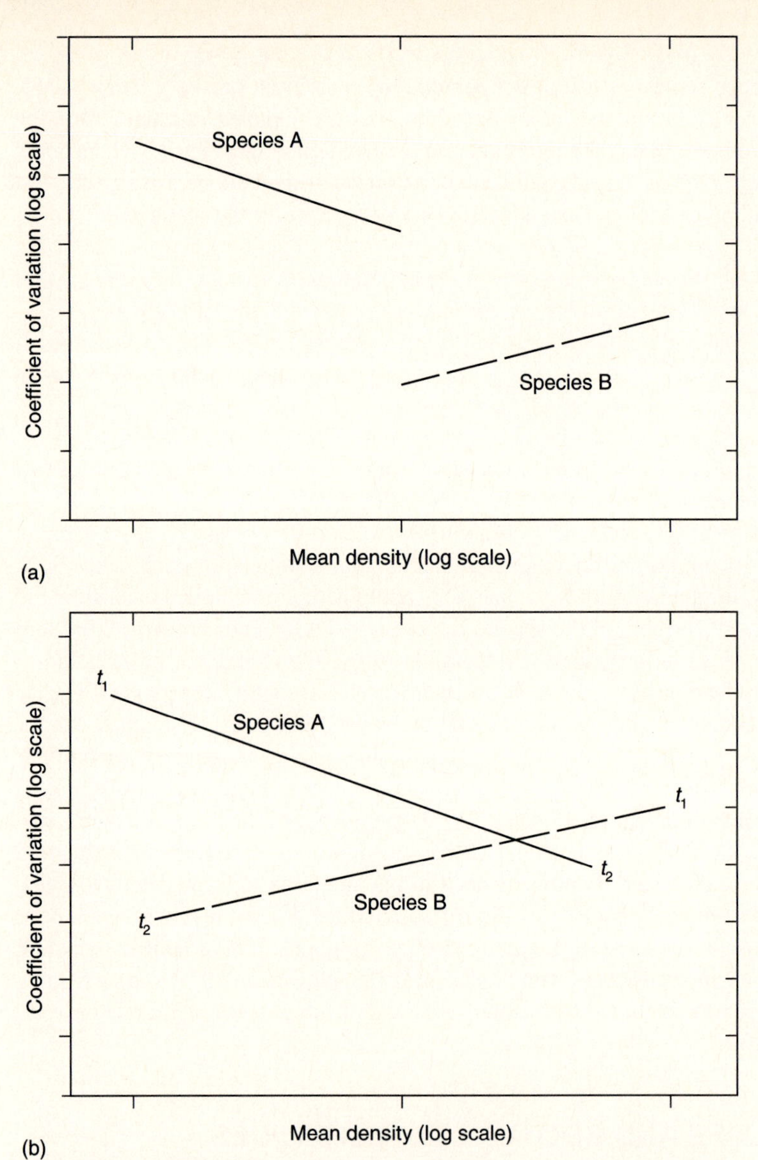

Figure 15.8 Problems of comparing temporal population variability between species. In all cases the slope of Taylor's power law differs from 2.0. (a) Different slopes are present, but species A is always more variable than species B, so they may be compared statistically. (b) If the species abundances are correlated, the species cannot be compared directly. In this instance the two species are negatively correlated, so that when the density of species A goes up from time t_1 to time t_2, the density of species B goes down. No general conclusions can be drawn from (b). (After McArdle et al. 1990.)

and pointed out their utility for estimating niche overlap in Chapter 13 (Section 13.3.7). Ecologists should be aware that these methods exist; they are sufficiently complicated, however, that you should consult a statistician before applying them uncritically to your data.

The *jackknife* technique was first suggested by Tukey (1958). We would like to know how much better our estimate would be if we had one more sample, but we do not have any more samples, so we ask the converse question: *How much worse would we be if we had one less sample*? Beginning with a set of n measurements, the jackknife is done as follows:

Step 1. *Recombine the original data*: We do this by omitting one of the n replicates from the jackknife sample.

Step 2. *Calculate pseudovalues of the parameter of interest* for each recombining of the data:

$$\Phi_i = nS - (n - 1)S_T \tag{15.39}$$

where Φ_i = Pseudovalue for jackknife estimate i
n = Original sample size
S = Original statistical estimate
S_T = Statistical estimate when original value i has been discarded from sample i = sample number $(1, 2, 3, \ldots, n)$

Step 3. *Estimate the mean and standard error of the parameter of interest* from the resulting pseudovalues.

The jackknife technique has been applied to several ecological parameters. One good example is the estimation of population growth rates (Meyer et al. 1986). Population growth rates can be estimated from the characteristic equation of Lotka:

$$1 = \sum_{x=0}^{\infty} e^{-rx} l_x m_x \tag{15.40}$$

where r = Per capita instantaneous rate of population growth
l_x = Probability of surviving to age x
m_x = Fecundity at age x

Meyer et al. (1986) had data on 10 individual *Daphnia pulex* females giving the reproductive output and the age at death. To calculate the jackknife estimate of r, they proceeded as indicated above, discarding in turn one female from each jackknife sample, to generate 10 pseudovalues of r. By averaging these 10 values and obtaining a standard error from them, they could estimate the r-value for each population of *Daphnia* studied.

The *bootstrap* technique was developed by B. Efron in 1977 (Efron 1982). The bootstrap method asks what another sample of the same size would look like if we had one. Since we do not, we pretend that the sample we have is a universe, and we sample from it with replacement to estimate the variability of the sampling process. The bootstrap method follows the same general procedure as for the jackknife:

1. *Recombine the original data*: The original data of n measurements are placed in a pool, and then n values are sampled *with replacement*. Thus any measurement in

the original data could be used once, twice, several times, or not at all in the bootstrap sample. Typically one repeats this bootstrap sampling at least 500 times and often several thousand times.

2. *Calculate the parameter of interest* from each bootstrap sample.
3. *Estimate the mean and standard error of the parameter of interest* from the replicate bootstrap estimates.

Bootstrap estimates of parameters are known to be biased, If the true population value for the mean (for example) is μ, and if the observed mean of the whole original sample is $\bar{x}_S$, the bootstrap estimate of the mean $\bar{x}_B$ will be biased because it estimates $\bar{x}_S$ rather than μ. The bias of $\bar{x}_B$ is defined as

$$\text{Bias}(\bar{x}_S) = \bar{x}_S - \mu \tag{15.41}$$

which can be estimated by:

$$\text{Bias}(\bar{x}_S) = \bar{x}_B - \bar{x}_S \tag{15.42}$$

Because of this bias, bootstrap estimates are usually bias-adjusted by combining equations (15.41) and (15.42):

$$\text{Bias-adjusted bootstrap mean} = 2\bar{x}_S - \bar{x}_B \tag{15.43}$$

where $\bar{x}_S$ = Observed mean of original sample
$\bar{x}_B$ = Bootstrap estimate of the mean

The precision of a bootstrap estimate will depend on how many times the original data are randomly recombined, and the bootstrap estimate will converge on a stable estimate as the number of recombinations becomes large. Note that repeated bootstrap calculations performed with the same number of recombinations and the same original data will vary somewhat because the items randomly chosen differ in every sample of recombinations.

Confidence limits for bootstrap estimates can be obtained in the usual way with standard errors calculated from the replicate bootstrap samples. Alternatively, Efron (1982) suggested measuring confidence limits directly from the frequency distribution of bootstrap estimates. If a large number (n > 500) of bootstrap estimates are done, the 2.5th and 97.5th percentile values of this frequency distribution delimit a confidence belt of 95%. This empirical approach, known as the *percentile method* of Efron, may produce somewhat biased confidence limits when sampling distributions of bootstrap estimates are skewed. Efron discusses ways of correcting for this bias.

There is as yet no general agreement on when jackknife estimates are better and when bootstrap methods are better, and more empirical work is needed for ecological measures. Meyer et al. (1986) found that the jackknife and the bootstrap were equally effective for estimating population growth rates (r) for *Daphnia pulex*. Since jackknife estimates require much less computing than bootstrap estimates (which typically need 500 or 1000 replicates), it may be more useful to use jackknife procedures when computing time is limited.

We have already seen cases in which jackknife estimators have been useful in ecological statistics (e.g., Chapter 12, Section 12.3.2, for species richness measures). There is no doubt that these computer-intensive techniques will be used more and more to improve the estimation of difficult ecological parameters. Box 15.4 illustrates the use of the

Box 15.4 Bootstrap Calculation of Food Niche Breadth for Coyotes

Mark O'Donoghue measured the frequency of 8 food sources for coyotes in the southwestern Yukon, Canada, in the winter of 1992–93 as follows:

	Number of scats with species	Proportions p_i
Snowshoe hares	91	0.421
Red squirrels	17	0.079
Arctic ground squirrels	21	0.097
Field voles	54	0.250
Red-backed voles	14	0.065
Least chipmunk	3	0.014
Moose	7	0.032
Willow ptarmigan	9	0.042
Totals	216	1.000

1. We calculate first the observed food niche breadth.

Levins's Measure of Niche Breadth
For the coyote data, from equation (13.1),

$$B = \frac{1}{\Sigma p_j^2}$$

$$= \frac{1}{0.421^2 + 0.079^2 + 0.097^2 + 0.250^2 + 0.065^2 + 0.014^2 + \cdots}$$

$$= \frac{1}{0.26267} = 3.807$$

2. We next resample the observed distribution of 216 observations at random with replacement to obtain another sample of observations with the same sample size. We obtained the following data in this first random resampling:

	Number of scats	Proportions
Snowshoe hares	85	0.394
Red squirrels	19	0.088
Arctic ground squirrels	26	0.120
Field voles	61	0.282
Red-backed voles	11	0.051
Least chipmunk	4	0.019
Moose	6	0.028
Willow ptarmigan	4	0.018
Totals	216	1.000

Use these resampled data to estimate niche breadth in the same manner as previously:

$$B = \frac{1}{\Sigma\, p_j^2}$$

$$= \frac{1}{0.394^2 + 0.088^2 + 0.120^2 + 0.282^2 + 0.051^2 + 0.019^2 + \cdots}$$

$$= \frac{1}{0.26076} = 3.835$$

3. We repeat this procedure in the computer 1000 times to obtain 1000 resampled estimates of niche breadth. From this sample of 1000 we calculate the mean niche breadth in the usual way to obtain

$$\bar{x} = \frac{\Sigma\, B}{1000} = 3.816$$

$$\text{Bias-adjusted bootstrap mean} = 2\bar{x}_S - \bar{x}_B$$

$$= 2(3.807) - 3.816 = 3.798$$

Since we have generated 1000 bootstrap estimates (in the computer), we can use the percentile method of Efron (1982) to provide estimates of confidence limits. For 95% confidence limits, we rank all the 1000 estimates and locate the 2.5% and the 97.5% points in the frequency distribution. These will be (in ranked form) the 25th and the 975th values, and for these particular data, these were found to be:

25th percentile = 3.772

97.5th percentile = 3.829

and these values can be used as empirical estimates of the 95% confidence limits for this estimate of niche breadth.

The bootstrap estimates can be obtained from a variety of statistical packages including those of Rohlf (1995) and Manly (1991).

bootstrap method to estimate niche breadth in mammals. Rohlf (1995) provides a computer program to calculate jackknife and bootstrap estimates.

15.6 SUMMARY

Transformations are essential for much ecological data that are not normally distributed and in which variances are not equal in all groups. Traditional transformations, like the logarithmic and square root, can be applied to known classes of data or used *ad hoc* to improve the fit of data to the assumptions of parametric statistics. Alternatively, one can use a general-purpose transformation like the Box-Cox transformation to produce a specially tailored exponential transformation for a particular data set.

Repeatability is a statistical measure of how similar duplicate measurements on the same individual are, in relation to differences among individuals. It can be a useful measure in quantitative genetics, where it is an upper bound on heritability, and in data analysis in which one is looking for the best way of measuring complex traits.

Linear regression data do not often satisfy the standard assumption that the X-variable is measured without error. Central trend lines may be a better description of the regression and give more accurate predictions. The *geometric mean regression* is one simple central trend line that should be used on more ecological data.

Estimating the relative variability of populations in time and space would appear to be a simple task, but it is fraught with statistical problems. The coefficient of variation should be used for estimates of variability rather than the standard deviation of log abundances. Only if the slope of Taylor's power law is 2 can one make simple comparisons among populations or between species.

Computer-intensive estimates of statistical parameters can be achieved in two general ways using *jackknife* or *bootstrap* methods. By discarding some of the original data or by sampling the original data with replacement, one can calculate nearly unbiased estimates of complex ecological parameters like the rate of population growth, along with confidence limits. This new area of statistics holds much promise for dealing with ecological parameters that have been intractable to normal statistical methods.

SELECTED READING

Berry, D. A. 1987. Logarithmic transformations in ANOVA. *Biometrics* 43: 439–456.

Diaconis, P., and Efron, B. 1983. Computer-intensive methods in statistics. *Scientific American* 248: 116–130.

Efron, B., and Tibshirani, R. J. 1993. *An Introduction to the Bootstrap.* Chapman and Hall, New York.

Hoyle, M. H. 1973. Transformations—an introduction and a bibliography. *International Statistical Review* 41: 203–223.

Jensen, A. L. 1986. Functional regression and correlation analysis. *Canadian Journal of Fisheries and Aquatic Sciences* 43: 1742–1745.

Lessells, C. M., and Boag, P. T. 1987. Unrepeatable repeatabilities: A common mistake. *Auk* 104: 116–121.

Manly, B. F. J. 1991. *Randomization and Monte Carlo Methods in Biology.* Chapman and Hall, New York.

McArdle, B. H., Gaston, K. J., and Lawton, J. H. 1990. Variation in the size of animal populations: Patterns, problems and artifacts. *Journal of Animal Ecology* 59: 439–454.

Meyer, J. S., Ingersoll, C. G., McDonald, L. L., and Boyce, M. S. 1986. Estimating uncertainty in population growth rates: Jackknife vs. bootstrap techniques. *Ecology* 67: 1156–1166.

Ricker, W. E. 1984. Computation and uses of central trend lines. *Canadian Journal of Zoology* 62: 1897–1905.

Stewart-Oaten, A., Murdoch, W. W., and Walde, S.J. 1995. Estimation of temporal variability in populations. *American Naturalist* 146: 519–535.

QUESTIONS AND PROBLEMS

15.1. Twelve samples of forest soil were cultured to estimate the abundance of a species of protozoa in the soil, with these results: 56, 160, 320, 640, 900, 900, 5200, 5200, 7200, 20800, 20800 and 59000. Do these data require transformation? What transformation do you recommend?

15.2. Aerial counts of caribou from photographs taken sequentially produced these counts for four adjacent flight paths:

Flight Path			
A	B	C	D
240	80	10	72
600	34	54	35
50	58	250	27
135	90	4	100
82	73	70	660
98	430	180	20
32	33	150	160
220	250	92	90

(a) Do these data require transformation?

(b) If a logarithmic transformation is to be used, what value of the constant c would you recommend?

15.3. J. N. M. Smith took three measurements repeatedly on bill size in song sparrows (*Melospiza melodia*) on Mandarte Island, B.C. with these results:

Bird no.	Bill length (mm)	Bill depth (mm)	Bill width (mm)
10993	8.5	6.2	6.6
	8.5	6.1	6.5
	8.5	6.1	6.7
	8.1	6.0	6.6
	8.6	6.2	6.5
	8.8	6.2	6.7
10994	8.0	5.9	6.6
	8.4	5.6	6.3
	8.5	5.7	6.3
10996	8.5	6.0	6.8
	8.5	5.8	6.5
	9.0	6.0	6.8
10999	8.2	5.6	6.5
	8.4	5.8	6.3
11000	9.1	5.9	6.9
	8.7	5.8	6.6
	8.7	5.8	6.5
10982	7.7	5.4	5.9
	7.9	5.3	6.0
	7.9	5.5	6.0

Calculate repeatability for each of these measures. Is one measurement better than the others, as far as repeatability is concerned?

15.4. Standard length (cm) and total weight (g) were measured for 34 herring caught in a purse seine off the coast of British Columbia. Age was determined from growth rings in the scales.

Age 1+		Age 2+		Age 3+		Age 4+		Age 5+	
SL	WT	SL	WT	SL	WT	SL	WT	SL	WT
17.2	74	19.7	119	19.3	115	20.8	159	22.0	177
16.8	62	19.0	106	19.7	125	20.1	120	22.0	164
16.2	64	18.8	101	19.2	118	19.5	115	20.7	146
16.3	52	18.2	92	19.7	118	21.0	145		
16.1	58	18.1	92	19.8	124	21.0	141		
15.4	48	18.8	91	19.3	115	21.5	156		
16.1	57	19.6	117	20.6	142	21.7	146		
16.4	55	18.7	100	19.7	119				
17.3	73	19.3	106	20.2	136				
		18.1	82	19.8	111				
		19.1	100	19.1	110				
		17.0	75	20.9	143				
		18.0	96	19.3	110				
		19.3	110	17.7	91				
		17.9	85	20.1	134				
				19.7	120				

Data from Ricker (1975, 213).

(a) Is length related to weight in these fish? Is a transformation necessary to do a linear regression?

(b) Calculate an ordinary regression and a functional regression for these data. Which is more appropriate if you wish to estimate weight from length?

15.5. Calculate repeatability for the data on aerial counts of moose in Table 15.1.

15.6. Compare the temporal variability of these 10 grey-sided vole (*Clethrionomys rufocanus*) populations on Hokkaido, Japan (Saitoh et al. 1997). Values are number caught per 100 trap-nights in each area in the autumn of each year.

	Area									
	1	2	3	4	5	6	7	8	9	10
1963	9.75	7.91	0.42	3.14	1.78	1.44	1.44	1.78	2.80	0.76
1964	9.32	6.21	8.25	7.23	3.48	8.93	1.10	7.23	6.89	2.29
1965	6.43	5.18	9.27	11.65	8.93	6.21	9.27	14.38	11.65	5.53
1966	6.97	7.57	3.82	12.68	10.29	0.08	1.78	0.76	2.80	3.48
1967	16.25	12.34	4.67	9.95	23.23	6.55	1.78	3.14	6.55	3.31
1968	5.87	0.42	1.10	4.16	0.42	2.46	6.89	1.10	10.29	1.44
1969	19.94	19.15	19.74	13.70	24.59	2.29	10.63	2.80	6.21	4.42
1970	17.78	18.46	16.88	3.14	14.38	12.34	36.17	14.38	9.61	16.08
1971	12.11	1.44	19.83	13.70	9.27	6.21	17.44	2.80	4.84	7.34
1972	18.69	15.06	18.12	5.18	19.49	4.84	6.21	0.76	3.48	6.55
1973	10.29	13.70	16.08	7.91	7.91	0.76	9.27	13.70	3.48	7.23
1974	18.92	15.74	6.89	4.84	5.53	4.16	9.95	8.25	11.31	6.72
1975	7.34	1.44	2.46	3.48	2.80	0.08	3.48	0.08	3.48	4.67
1976	20.28	8.93	9.61	11.65	15.74	4.16	8.25	2.80	5.18	1.27
1977	9.67	17.00	4.00	2.00	23.00	42.00	14.00	14.00	14.00	17.25
1978	23.80	30.00	11.33	39.00	10.00	45.00	11.00	30.00	12.50	26.67
1979	14.40	0.00	5.67	13.00	4.00	3.00	3.00	1.00	1.00	2.33

	Area									
	1	2	3	4	5	6	7	8	9	10
1980	12.00	3.00	4.33	6.00	7.00	1.00	3.00	6.00	1.50	3.00
1981	17.00	17.00	15.67	19.00	24.00	12.00	9.00	31.00	13.50	9.00
1982	5.80	5.00	16.00	10.00	14.00	8.00	16.00	15.00	3.00	9.33
1983	8.80	8.00	4.00	8.00	21.00	2.00	3.00	0.00	8.00	9.33
1984	5.20	0.00	2.33	0.00	3.00	1.00	0.00	0.00	1.50	0.33
1985	21.60	10.00	14.67	8.00	19.00	12.00	11.00	5.00	23.00	10.67
1986	1.80	3.00	2.00	2.00	6.00	3.00	15.00	6.00	19.50	16.33
1987	4.50	0.00	1.00	4.00	0.00	0.00	0.00	0.00	0.00	0.00
1988	17.75	10.00	10.50	14.00	8.00	10.00	11.00	7.00	12.00	9.67
1989	2.50	7.00	5.50	1.00	16.00	21.00	6.00	16.00	19.00	5.67
1990	6.25	6.00	0.00	6.00	1.00	0.00	3.00	1.00	0.00	2.00
1991	5.33	6.00	2.00	16.00	2.00	1.00	0.00	0.00	9.00	0.00
1992	2.00	4.00	1.00	0.00	2.00	1.00	1.00	1.00	4.00	0.00

(a) What is the slope of Taylor's power law for this population?
(b) Is population 1 more variable than population 4?

Appendices

APPENDIX 1 DEFINITIONS AND FUNCTIONS

For individual observations,

$$\sum X = X_1 + X_2 + X_3 + X_4 + \cdots$$

$$\sum X^2 = X_1^2 + X_2^2 + X_3^2 + X_4^2 + \cdots$$

$$\text{Mean} = \bar{x} = \frac{\Sigma X}{n}$$

$$\text{Variance} = s^2 = \frac{\Sigma X^2 - (\Sigma X)^2/n}{n - 1}$$

For data grouped into a frequency distribution,

$$\text{Mean} = \bar{x} = \frac{\Sigma f_x X}{n}$$

$$\text{Variance} = s^2 = \frac{\Sigma f_x X^2 - (\Sigma f_x X)^2/n}{n - 1}$$

Gamma Function: Can be approximated from Stirling's formula:

$$\Gamma(z) \approx e^{-z} z^{(z-0.5)}(\sqrt{2\pi}) \left[1 + \frac{1}{12z} + \frac{1}{288z^2} - \frac{139}{51{,}840z^3} - \frac{571}{2{,}488{,}320z^4} + \cdots \right]$$

Trigamma Function: Can be approximated by

$$\Psi'(z) = \frac{1}{z} + \frac{1}{2z^2} + \frac{1}{6z^3} - \frac{1}{30z^5} + \frac{1}{42z^7} - \frac{1}{30z^9}$$

APPENDIX 2 SOFTWARE PROGRAMS

The computations discussed in this book can be accomplished in several ways. Those who are familiar with Microsoft EXCEL or other spreadsheet programs can write functions that will calculate the necessary equations. Specialized software is becoming more and more available every year to achieve some of these calculations, as for mark-recapture work. I have given specific details in the chapters about software that is available for specialized computations. In addition I have written a set of programs in DELPHI 3 and FORTRAN for the PC-Windows environment to compute estimates for many of the methods discussed here. These programs are an updated version of the FORTRAN programs that appeared in the first edition of this book and that were for the DOS operating system (Krebs 1989). This appendix includes a brief listing of each of these new programs, which are organized into 13 program groups. They are available for IBM PC 32-bit Windows 95, Windows NT 4.0, or later operating systems. The set of programs can be obtained from Exeter Software, 47 Route 25A, Suite 2, Setauket, New York 11733-2870, who may be contacted through their web site at *http://www.exetersoftware.com* or through their e-mail at Sales@ExeterSoftware.com. The program groups follow.

1. **MARK-RECAPTURE.** This group contains 9 programs from Chapters 2 and 3 as follows:
   Petersen Method
   Schnabel-Schumacher Method
   Resight Estimation (Radiotelemetry)
   Jolly-Seber Full Model
   Zero-Truncated Poisson Test of Equal Catchability
   Leslie, Chitty, and Chitty Test for Marked Animals: Equal Catchability
   Caughley Closed Population Estimators
   Catch-Effort Models for Exploited Populations
   Change-in-Ratio Estimators for Exploited Populations
2. **QUADRAT SAMPLING.** This group contains 8 programs from Chapter 4 as follows:
   Optimal Quadrat Size and Shape
   Poisson Distribution
   Negative Binomial Distribution
   Goodness-of-Fit to a Negative Binomial Distribution
   Calculating a Theoretical Negative Binomial Distribution
   Testing for Equality Among Several Negative Binomial Samples
   Aerial Survey Methods
   Line Intercept Sampling
3. **LINE TRANSECTS.** This group contains one program to compute a line transect estimator as described in Chapter 5.
   Hayne Estimator
4. **DISTANCE METHODS.** This group contains 5 programs from Chapter 5.
   Byth and Ripley
   T-square
   Ordered Distance

Variable-Area Transect
Point Quarter

5. **SPATIAL PATTERN.** This group contains four programs from Chapter 6 to test for random spatial patterns as follows:
   Clark and Evans Test
   Campbell and Clarke Test
   Contiguous Quadrats (TTLQV)
   Indices of Dispersion (Quadrat Counts)
6. **SAMPLE SIZE.** This group contains 6 programs from Chapter 7 to estimate sample sizes needed as follows:
   Continuous Variables—Means
   Continuous Variables—Variances
   Discrete Variables—Proportions
   Discrete Variables—Poisson Counts
   Discrete Variables—Negative Binomial Counts
   Specialized Ecological Variables
7. **SAMPLING.** This group contains 5 programs from Chapter 8 on sampling designs as follows:
   Means of Ratios
   Stratified Sampling
   Construction of Strata
   Adaptive Sampling
   Multistage Sampling
8. **SEQUENTIAL SAMPLING.** This group contains 7 programs from Chapter 9 to estimate sequential designs as follows:
   Means—Normal Distributions
   Variances—Normal Distributions
   Proportions—Binomial Distributions
   Counts—Negative Binomial Distribution
   Stopping Rules
   Sequential Schnabel Population Estimates
   Iwao's Method for Quadrat Counts
9. **SIMILARITY COEFFICIENTS.** This group contains 6 programs from Chapter 11 to estimate similarity as follows:
   Binary Coefficients
   Distance Coefficients
   Correlation Coefficients
   Percentage Similarity
   Morisita's Index of Similarity
   Horn's Index
10. **SPECIES DIVERSITY.** This group contains 8 programs from Chapter 12 to estimate species diversity as follows:
    Species Richness—Rarefaction Method
    Species Richness—Jackknife Method for Quadrat Counts
    Logarithmic Series
    Lognormal Distribution

Simpson's Index of Diversity
Shannon-Wiener Measure
Brillouin's Index of Diversity
Evenness Measures

11. **NICHE MEASURES AND DIET PREFERENCE.** This group contains 4 programs from Chapter 13 as follows:
    Niche Breadth
    Niche Overlap
    Habitat and Dietary Preferences
    Rodgers's Index for Cafeteria Food Preferences
12. **SURVIVAL.** This group contains 7 programs from Chapter 14 as follows:
    Conversion of Survival Rates
    Life Tables
    Expectation of Life
    Radiotelemetry Survival Rates
    Kaplan-Meier Method
    Log-Rank Test
    Likelihood Ratio Test
13. **EXTRAS.** This group contains 10 miscellaneous programs from Chapter 15 and other chapters, as follows:
    Significant Digits for Recording Data (Chapter 1)
    Binomial Confidence Limits (Chapter 2)
    Multinomial Confidence Limits (Chapter 13)
    Poisson Confidence Limits (Chapters 2 and 4)
    Logarithmic Transformation Constant in log $(X + c)$ Transformations
    Correction for Means from Logarithmic Transformations
    Box-Cox Transformation
    Repeatability of Measurements
    Geometric Mean Regression
    Means from Frequency Distributions

Any of these programs can be selected in the Windows version of the ECOLOGICAL METHODOLOGY software (available from Exeter Software).

References

Abramowitz, M. and Stegun, I. A. 1964. *Handbook of Mathematical Functions with Formulas, Graphs, and Mathematical Tables.* National Bureau of Standards, Washington, D.C.

Abrams, P. 1980. Some comments on measuring niche overlap. *Ecology* 61: 44–49.

Adams, J. E. and McCune, E. D. 1979. Application of the generalized jackknife to Shannon's measure of information used as an index of diversity. In *Ecological Diversity in Theory and Practice*, eds. J. F. Grassle, G. P. Patil, W. Smith and C. Taillie, 117–131. International Cooperative Publishing House, Fairland, Maryland.

Alatalo, R. V. 1981. Problems in the measurement of evenness in ecology. *Oikos* 37: 199–204.

Allen, J., Gonzalez, D. and Gokhale, D. V. 1972. Sequential sampling plans for the bollworm, *Heliothis zea*. *Environmental Entomology* 1: 771–780.

Andersen, J. 1962. Roe-deer census and population analysis by means of modified marking and release technique. In *The Exploitation of Natural Animal Populations*, eds. E. D. LeCren and M. W. Holdgate, 72–82. Blackwell, Oxford.

Anderson, D. R., Burnham, K. P. and White, G. C. 1985. Problems in estimating age-specific survival rates from recovery data of birds ringed as young. *Journal of Animal Ecology* 54: 89–98.

Anderson, D. R., Burnham, K. P., White, G. C. and Otis, D. L. 1983. Density estimation of small-mammal populations using a trapping web and distance sampling methods. *Ecology* 64: 674–680.

Anderson, D. R., Laake, J. L., Crain, B. R. and Burnham, K. P. 1979. Guidelines for line transect sampling of biological populations. *Journal of Wildlife Management* 43: 70–78.

Anganuzzi, A. A. and Buckland, S. T. 1993. Post-stratification as a bias reduction technique. *Journal of Wildlife Management* 57: 827–834.

Anscombe, F. J. 1948. The transformation of Poisson, binomial and negative binomial data. *Biometrika* 35: 246–254.

Anscombe, F. J. 1950. Sampling theory of the negative binomial and logarithmic series distributions. *Biometrika* 37: 358–382.

Armitage, P. 1950. Sequential analysis with more than two alternative hypotheses, and its relation to discriminant function analysis. *Journal of the Royal Statistical Society, Series B* 12: 137–144.

Arnason, A. N. and Baniuk, L. 1980. A computer system for mark-recapture analysis of open populations. *Journal of Wildlife Management* 44: 325–332.

Arnason, A. N., Schwarz, C. J. and Gerrard, J. M. 1991. Estimating closed population size and number of marked animals from sighting data. *Journal of Wildlife Management* 55: 716–730.

Arthur, S. M., Manly, B. F. J., McDonald, L. L. and Garner, G. W. 1996. Assessing habitat selection when availability changes. *Ecology* 77: 215–227.

Atchley, W. R., Gaskins, C. T. and Anderson, D. 1976. Statistical properties of ratios. I. Empirical results. *Systematic Zoology* 25: 137–148.

Bailey, N. T. J. 1952. Improvements in the interpretation of recapture data. *Journal of Animal Ecology* 21: 120–127.

Baranov, F. I. 1918. On the question of the biological basis of fisheries. *Nauchn. Issled. Ikhtiologicheskii Inst. Izv.* 1: 81–128 (in Russian).

Barcikowski, R. S. and Robey, R. R. 1984. Decisions in single group repeated measures analysis: statistical tests and three computer packages. *American Statistician* 38: 148–150.

Barford, N. C. 1985. *Experimental Measurements: Precision, Error, and Truth*. Wiley, New York.

Baroni-Urbani, C. and Buser, M. W. 1976. Similarity of binary data. *Systematic Zoology* 25: 251–259.

Bart, J. and Robson, D. S. 1982. Estimating survivorship when the subjects are visited periodically. *Ecology* 63: 1078–1090.

Basharin, G. P. 1959. On a statistical estimate for the entropy of a sequence of independent random variables. *Theory of Probability and Its Application* 4: 333–336.

Baskerville, G. L. 1971. Use of logarithmic regression in the estimation of plant biomass. *Canadian Journal of Forestry* 2: 49–53.

Batcheler, C. L. 1971. Estimation of density from a sample of joint point and nearest-neighbor distances. *Ecology* 52: 703–709.

Batcheler, C. L. 1975. Development of a distance method for deer census from pellet groups. *Journal of Wildlife Management* 39: 641–652.

Bayliss, P. and Giles, J. 1985. Factors affecting the visibility of kangaroos counted during aerial surveys. *Journal of Wildlife Management* 49: 686–692.

Bayliss, P. and Yeomans, K. M. 1989. Correcting bias in aerial survey population estimates of feral livestock in northern Australia using the double-count technique. *Journal of Applied Ecology* 26: 925–935.

Beall, G. 1940. The fit and significance of contagious distributions when applied to observations on larval insects. *Ecology* 21: 460–474.

Bear, G. D., White, G. C., Carpenter, L. H., Gill, R. B. and Essex, D. J. 1989. Evaluation of aerial mark-resighting estimates of elk populations. *Journal of Wildlife Management* 53: 908–915.

Becker, W. A. 1984. *A Manual of Quantitative Genetics*. Academic Enterprises, Pullman, Washington.

Begon, M., Harper, J. L. and Townsend, C. R. 1996. *Ecology: Individuals, Populations, and Communities*. Blackwell Science Ltd., Oxford.

Berger, J. O. and Berry, D. A. 1988. Statistical analysis and the illusion of objectivity. *American Scientist* 76: 159–165.

Berry, D. A. 1987. Logarithmic transformations in ANOVA. *Biometrics* 43: 439–456.

Besag, J. and Gleaves, J. T. 1973. On the detection of spatial pattern in plant communities. *Bulletin of the International Statistical Institute* 45: 153–158.

Binns, M. R. 1994. Sequential sampling for classifying pest status. In *Handbook of Sampling Methods for Arthropods in Agriculture*, eds. L. P. Pedigo and G. D. Buntin, 137–174. CRC Press, Boca Raton, Florida.

Binns, M. R. and Nyrop, J. P. 1992. Sampling insect populations for the purpose of IPM decision making. *Annual Review of Entomology* 37: 427–453.

Blair, W. F. 1942. Size of home range and notes on life history of the woodland deer-mouse and eastern chipmunk in northern Michigan. *Journal of Mammalogy* 23: 27–36.

Bliss, C. I. and Fisher, R. A. 1953. Fitting the negative binomial distribution to biological data and note on the efficient fitting of the negative binomial. *Biometrics* 9: 176–200.

Bondrup-Nielsen, S. 1983. Density estimation as a function of live-trapping grid and home range size. *Canadian Journal of Zoology* 61: 2361–2365.

Bonnell, M. L. and Ford, R. G. 1987. California sea lion distribution: a statistical analysis of aerial transect data. *Journal of Wildlife Management* 51: 13–19.

Boonstra, R. and Krebs, C. J. 1978. Pitfall trapping of *Microtus townsendii*. *Journal of Mammalogy* 59: 136–148.

Bormann, F. H. 1953. The statistical efficiency of sample plot size and shape in forest ecology. *Ecology* 34: 474–487.

Boulanger, J. G. and Krebs, C. J. 1996. Robustness of capture-recapture estimators to sample biases in a cyclic snowshoe hare population. *Journal of Applied Ecology* 33: 530–542.

Bowden, D. C. and Kufeld, R. C. 1995. Generalized mark-sight population size estimation applied to Colorado Moose. *Journal of Wildlife Management* 59: 840–851.

Box, G. E. P. and Cox, D. R. 1964. An analysis of transformations. *Journal of the Royal Statistical Society Series B* 26: 211–252.

Box, G. E. P., Hunter, W. G. and Hunter, J. S. 1978. *Statistics for Experimenters*. Wiley, New York.

Braaten, D. O. 1969. Robustness of the DeLury population estimator. *Journal of the Fisheries Research Board of Canada* 26: 339–355.

Bray, J. R. and Curtis, J. T. 1957. An ordination of the upland forest communities of southern Wisconsin. *Ecological Monographs* 27: 325–349.

Broome, L. S. 1985. Sightability as a factor in aerial survey of bird species and communities. *Australian Wildlife Research* 12: 57–67.

Brownie, C., Anderson, D. R., Burnham, K. P. and Robson, D. S. 1985. Statistical inference from band recovery data—a handbook. *United States Fish and Wildlife Service, Resource Publication* 156.

Buckland, S. T., Anderson, D. R., Burnham, K. P. and Laake, J. L. 1993. *Distance Sampling. Estimating Abundance of Biological Populations*. Chapman & Hall, London.

Bulmer, M. G. 1974. On fitting the Poisson lognormal distribution to species-abundance data. *Biometrics* 30: 101–110.

Bulmer, M. G. 1975. The statistical analysis of density dependence. *Biometrics* 31: 901–911.

Burgess, J. W. 1983. Reply to a comment by R.L. Mumme et al. *Ecology* 64: 1307–1308.

Burgess, J. W., Roulston, D. and Shaw, E. 1982. Territorial aggregation: An ecological spacing strategy in acorn woodpeckers. *Ecology* 63: 575–578.

Burnham, K. P. and Anderson, D. R. 1976. Mathematical models for nonparametric inferences from line transect data. *Biometrics* 32: 325–336.

Burnham, K. P. and Anderson, D. R. 1984. The need for distance data in transect counts. *Journal of Wildlife Management* 48: 1248–1254.

Burnham, K. P., Anderson, D. R. and Laake, J. L. 1980. Estimation of density from line transect sampling of biological populations. *Wildlife Monographs* 72: 1–202.

Burnstein, H. 1971. *Attribute Sampling. Tables and Explanations*. McGraw-Hill, New York.

Bustard, H. R. 1969. The population ecology of the gekkonid lizard (*Gehyra variegata* (Dumeril and Bibron)) in exploited forests in northern New South Wales. *Journal of Animal Ecology* 38: 35–51.

Byth, K. 1982. On robust distance-based intensity estimators. *Biometrics* 38: 127–135.

Byth, K. and Ripley, B. D. 1980. On sampling spatial patterns by distance methods. *Biometrics* 36: 279–284.

Cain, A. J. and Sheppard, P. M. 1950. Selection in the polymorphic land snail *Cepaea nemoralis*. *Heredity* 4: 275–294.

Camargo, J. A. 1993. Must dominance increase with the number of subordinate species in competitive interactions? *Journal of Theoretical Biology* 161: 537–542.

Camargo, J. A. 1995. On measuring species evenness and other associated parameters of community structure. *Oikos* 74: 538–542.

Campbell, D. J. and Clarke, D. J. 1971. Nearest neighbour tests of significance for non-randomness in the spatial distribution of singing crickets (*Teliogryllus commodus* (Walker)). *Animal Behaviour* 19: 750–756.

Carothers, A. D. 1971. An examination and extension of Leslie's test of equal catchability. *Biometrics* 27: 615–630.

Carpenter, S. R., Frost, T. M., Heisey, D. and Kratz, T. K. 1989. Randomized intervention analysis and the interpretation of whole-ecosystem experiments. *Ecology* 70: 1142–1152.

Caughley, G. 1970. Eruption of ungulate populations, with emphasis on Himalayan thar in New Zealand. *Ecology* 51: 53–72.

Caughley, G. 1974. Bias in aerial survey. *Journal of Wildlife Management* 38: 921–933.

Caughley, G. 1977a. *Analysis of Vertebrate Populations*. Wiley, London.

Caughley, G. 1977b. Sampling in aerial survey. *Journal of Wildlife Management* 41: 605–615.

Caughley, G. and Grice, D. 1982. A correction factor for counting emus from the air, and its application to counts in Western Australia. *Australian Wildlife Research* 9: 253–259.

Caughley, G. and Sinclair, A. R. E. 1994. *Wildlife Ecology and Management*. Blackwell Scientific Publications, Boston.

Caughley, G., Sinclair, R. and Scott-Kemmis, D. 1976. Experiments in aerial survey. *Journal of Wildlife Management* 40: 290–300.

Chambers, J. M., Cleveland, W. S., Kleiner, B. and Tukey, P. A. 1983. *Graphical Methods for Data Analysis*. Wadsworth, Belmont, California.

Chao, A. 1988. Estimating animal abundance with capture frequency data. *Journal of Wildlife Management* 52: 295–300.

Chao, M. T. 1982. A general purpose unequal probability sampling plan. *Biometrika* 69: 653–656.

Chapman, D. G. 1952. Inverse, multiple, and sequential sample censuses. *Biometrics* 8: 286–306.

Chapman, D. G. 1954. The estimation of biological populations. *Annals of Mathematical Statistics* 25: 1–15.

Cheetham, A. H. and Hazel, J. E. 1969. Binary (presence-absence) similarity coefficients. *Journal of Paleontology* 43: 1130–1136.

Cherry, S. 1996. A comparison of confidence interval methods for habitat use-availability studies. *Journal of Wildlife Management* 60: 653–658.

Chesson, J. 1978. Measuring preference in selective predation. *Ecology* 59: 211–215.

Clapham, A. R. 1932. The form of the observational unit in quantitative ecology. *Journal of Ecology* 20: 192–197.

Clark, P. J. and Evans, F. C. 1954. Distance to nearest neighbor as a measure of spatial relationships in populations. *Ecology* 35: 445–453.

Cleveland, W. S. 1994. *The Elements of Graphing Data*. AT&T Bell Laboratories, Murray Hill, New Jersey.

Clifford, H. T. and Stephenson, W. 1975. *An Introduction to Numerical Classification*. Academic Press, New York.

Clutton-Brock, T. H., Major, M. and Guinness, F. E. 1985. Population regulation in male and female red deer. *Journal of Animal Ecology* 54: 831–846.

Cochran, W. G. 1953. *Sampling Techniques*, 1st edition. Wiley, New York.

Cochran, W. G. and Cox, G. M. 1957. *Experimental Designs*. Wiley, New York.

Cock, M. J. W. 1978. The assessment of preference. *Journal of Animal Ecology* 47: 805–816.

Cohen, A. C. J. 1959. Simplified estimators for the normal distribution when samples are singly censored or truncated. *Technometrics* 1: 217–237.

Cohen, A. C. J. 1961. Tables for maximum likelihood estimates: singly truncated and singly censored samples. *Technometrics* 3: 535–541.

Cohen, J. 1988. *Statistical Power Analysis for the Behavioral Sciences*, 2nd ed. Lawrence Erlbaum Associates, Hillsdale, New Jersey.

Cohen, J. 1992. A power primer. *Psychological Bulletin* 112: 155–159.

Colwell, R. K. 1979. Toward a unified approach to the study of species diversity. In *Ecological Diversity in Theory and Practice*, eds. J. F. Grassle, G. P. Patil, W. Smith and C. Taillie, 75–91. International Cooperative Publishing House, Fairland, Maryland.

Colwell, R. K. and Coddington, J. A. 1994. Estimating terrestrial biodiversity through extrapolation. *Philosophical Transactions of the Royal Society of London, Series B* 345: 101–118.

Colwell, R. K. and Futuyma, D. J. 1971. On the measurement of niche breadth and overlap. *Ecology* 52: 567–576.

Confer, J. L. and Moore, M. V. 1987. Interpreting selectivity indices calculated from field data or conditions of prey replacement. *Canadian Journal of Fisheries and Aquatic Sciences* 44: 1529–1533.

Conroy, M. J., Hines, J. E. and Williams, B. K. 1989. Procedures for the analysis of band-recovery data and user instructions for program MULT. *United States Fish and Wildlife Service, Resource Publication* 175.

Cook, L. M. 1971. *Coefficients of Natural Selection*. Hutchinson, London.

Cormack, R. M. 1966. A test for equal catchability. *Biometrics* 22: 330–342.

Cottam, G. and Curtis, J. T. 1956. The use of distance methods in phytosociological sampling. *Ecology* 37: 451–460.

Cottam, G., Curtis, J. T. and Hale, B. W. 1953. Some sampling characteristics of a population of randomly dispersed individuals. *Ecology* 34: 741–757.

Cox, D. R. 1958. *Planning of Experiments*. Wiley, New York.

Cox, P. A. 1982. Vertebrate pollination and the maintenance of dioecism in *Freycinetia*. *American Naturalist* 120: 65–80.

Crow, E. L. and Gardner, R. S. 1959. Confidence intervals for the expectation of a poisson variable. *Biometrika* 46: 441–453.

Cullen, J. J., Zhu, M. and Pierson, D. C. 1986. A technique to assess the harmful effects of sampling and containment for determination of primary production. *Limnology and Oceanography* 31: 1364–1373.

Czaplewski, R. L., Crowe, D. M. and McDonald, L. L. 1983. Sample sizes and confidence intervals for wildlife population ratios. *Wildlife Society Bulletin* 11: 121–128.

Dale, M. 1997. *Spatial Pattern Analysis in Plant Ecology*. Cambridge University Press, Cambridge.

Damon, R. A. J. and Harvey, W. R. 1987. *Experimental Design, ANOVA, and Regression*. Harper and Row, New York.

Davies, O. L. 1956. *Design and Analysis of Industrial Experiments*. Hafner, New York.

Davies, R. G. 1971. *Computer Programming in Quantitative Biology*. Academic Press, London.

Day, R. W. and Quinn, G. P. 1989. Comparisons of treatments after an analysis of variance in ecology. *Ecological Monographs* 59: 433–463.

DeLury, D. B. 1947. On the estimation of biological populations. *Biometrics* 3: 145–167.

Deming, W. E. 1975. On probability as a basis for action. *American Statistician* 29: 146–152.

Dempster, J. P. 1975. *Animal Population Ecology*. Academic Press, London.

Dennis, B. and Taper, M. L. 1994. Density dependence in time series observations of natural populations: estimation and testing. *Ecological Monographs* 64: 205–224.

Diaconis, P. and Efron, B. 1983. Computer-intensive methods in statistics. *Scientific American* 248: 116–130.

Diamond, J. and Case, T. J. 1986. *Community Ecology*. Harper and Row, New York.

Dickman, M. 1968. Some indices of diversity. *Ecology* 49: 1191–1193.

Digby, P. G. N. and Kempton, R. A. 1987. *Multivariate Analysis of Ecological Communities*. Chapman & Hall, London.

Diggle, P. J. 1975. Robust density estimation using distance methods. *Biometrika* 62: 39–48.

Diggle, P. J. 1983. *Statistical Analysis of Spatial Point Patterns*. Academic Press, London.

Dixon, W. J. and Massey, F. J. J. 1983. *Introduction to Statistical Analysis*, 4th edition. McGraw-Hill, New York.

Donnelly, K. 1978. Simulations to determine the variance and edge-effect of total nearest neighbor distance. In *Simulation Methods in Archaeology*, ed. I. Hodder, 91–95. Cambridge University Press, London.

Downing, J. A. 1979. Aggregation, transformation, and the design of benthos sampling programs. *Journal of the Fisheries Research Board of Canada* 36: 1454–1463.

Downing, J. A. and Anderson, M. R. 1985. Estimating the standing biomass of aquatic macrophytes. *Canadian Journal of Fisheries and Aquatic Sciences* 42: 1860–1869.

Downing, J. A., Perusse, M. and Frenette, Y. 1987. Effect of interreplicate variance on zooplankton sampling design and data analysis. *Limnology and Oceanography* 32: 673–680.

Dunnet, G. M. 1963. A population study of the quokka, *Setonix brachyurus* Quoy and Gaimard (Marsupalia). III. The estimation of population parameters by means of the recapture technique. *CSIRO Wildlife Research* 8: 78–117.

Dutilleul, P. 1993. Spatial heterogeneity and the design of ecological field experiments. *Ecology* 74: 1646–1658.

Eberhardt, L. L. 1967. Some developments in "distance sampling." *Biometrics* 23: 207–216.

Eberhardt, L. L. 1969. Population estimates from recapture frequencies. *Journal of Wildlife Management* 33: 28–39.

Eberhardt, L. L. 1978a. Appraising variability in population studies. *Journal of Wildlife Management* 42: 207–238.

Eberhardt, L. L. 1978b. Transect methods for population studies. *Journal of Wildlife Management* 42: 1–31.

Eberhardt, L. L. 1982. Calibrating an index by using removal data. *Journal of Wildlife Management* 46: 734–740.

Eberhardt, L. L. and Thomas, J. M. 1991. Designing environmental field studies. *Ecological Monographs* 61: 53–74.

Edwards, A. W. F. 1972. *Likelihood: An Account of the Statistical Concept of Likelihood and its Application to Scientific Inference*. Cambridge University Press, Cambridge.

Efron, B. 1982. *The Jackknife, the Bootstrap, and Other Resampling Plans*. Society of Industrial and Applied Mathematics, Philadelphia.

Efron, B. and Tibshirani, R. J. 1993. *An Introduction to the Bootstrap*. Chapman and Hall, New York.

Elliott, J. M. 1977. Some methods for the statistical analysis of samples of benthic invertebrates. *Freshwater Biological Station Association, Scientific Publication* No. 25: 1–142.

Ellis, J. E., Wiens, J. A., Rodell, C. F. and Anway, J. C. 1976. A conceptual model of diet selection as an ecosystem process. *Journal of Theoretical Biology* 60: 93–108.

Engeman, R. M., Sugihara, R. T. and Dusenberry, W. E. 1994. A comparison of plotless density estimators using Monte Carlo simulation. *Ecology* 75: 1769–1779.

Evans, D. A. 1953. Experimental evidence concerning contagious distributions in ecology. *Biometrika* 40: 186–211.

Fager, E. W. 1972. Diversity: a sampling study. *American Naturalist* 106: 293–310.

Fairweather, P. G. 1991. Statistical power and design requirements for environmental monitoring. *Australian Journal of Marine and Freshwater Research* 42: 555–567.

Faith, D. P. 1983. Asymmetric binary similarity measures. *Oecologia* 57: 287–290.

Feinsinger, P., Spears, E. E. and Poole, R. W. 1981. A simple measure of niche breadth. *Ecology* 62: 27–32.

Feoli, E. and Orloci, L. 1991. *Computer Assisted Vegetation Analysis*. Kluwer Academic, Boston.

Field, J. G. 1970. The use of numerical methods to determine benthic distribution patterns from dredgings in False Bay. *Transactions of the Royal Society of South Africa* 39: 183–200.

Finney, D. J. 1941. On the distribution of a variate whose logarithm is normally distributed. *Journal of the Royal Statistical Society Series B* 7: 155–161.

Finney, D. J. 1978. Reader Reaction: Testing the effect of an intervention in sequential ecological data. *Biometrics* 34: 706–707.

Fischler, K. J. 1965. The use of catch-effort, catch-sampling, and tagging data to estimate a population of blue crabs. *Transactions of the American Fisheries Society* 94: 287–310.

Fisher, R. A., Corbet, A. S. and Williams, C. B. 1943. The relation between the number of species and the number of individuals in a random sample of an animal population. *Journal of Animal Ecology* 12: 42–58.

Freeman, M. F. and Tukey, J. W. 1950. Transformations related to the angular and square root. *Annals of Mathematical Statistics* 21: 607–611.

Garrod, D. J. 1967. Population dynamics of the arcto-Norwegian cod. *Journal of the Fisheries Research Board of Canada* 24: 145–190.

Gasaway, W. C., Dubois, S. D., Reed, D. J. and Harbo, S. J. 1986. Estimating moose population parameters from aerial surveys. *Biological Papers of the University of Alaska No* 22: 1–108.

Gaston, K. J. and Lawton, J. H. 1988. Patterns in the distribution and abundance of insect populations. *Nature* 331: 709–712.

Gates, C. E. 1979. Line transect and related issues. In *Sampling Biological Populations*, ed. R. M. Cormack, 71–154. International Cooperative Publishing House, Fairfield, Maryland.

Gates, C. E. 1980. LINETRAN, A general computer program for analyzing line-transect data. *Journal of Wildlife Management* 44: 658–661.

Gauch, H. G. J. 1982. *Multivariate Analysis in Community Ecology*. Cambridge University Press, Cambridge.

Gauch, H. G. J. and Chase, G. B. 1974. Fitting the Gaussian curve to ecological data. *Ecology* 55: 1377–1381.

Geist, V. 1971. *Mountain Sheep: A Study in Behavior and Evolution*. University of Chicago Press, Chicago.

Giles, R. H. 1978. *Wildlife Management*. Freeman, San Francisco.

Gleason, H. A. 1922. On the relation between species and area. *Ecology* 3: 158–162.

Goldsmith, F. B. 1973. The vegetation of exposed sea cliffs at South Stack, Anglesey. I. The multivariate approach. *Journal of Ecology* 61: 787–818.

Goldsmith, F. B. and Harrison, C. M. 1976. Description and analysis of vegetation. In

Methods In Plant Ecology, ed. S. B. Chapman, 85–155. Blackwell Scientific, Oxford.

Goldstein, R. 1989. Power and sample size via MS/PCV-DOS computers. *American Statistician* 43: 253–260.

Good, I. J. 1953. The population frequencies of species and the estimation of population parameters. *Biometrika* 40: 237–264.

Goodman, L. A. 1953. Sequential sampling tagging for population size problems. *Annals of Mathematical Statistics* 24: 56–69.

Goodman, L. A. 1965. On simultaneous confidence intervals for multinomial proportions. *Technometrics* 7: 247–254.

Gower, J. C. and Legendre, P. 1986. Metric and Euclidean properties of dissimilarity coefficients. *Journal of Classification* 3: 5–48.

Green, R. G. and Evans, C. A. 1940. Studies on a population cycle of snowshoe hares on the Lake Alexander area. *Journal of Wildlife Management* 4: 347–358.

Green, R. H. 1966. Measurement of non-randomness in spatial distributions. *Researches in Population Ecology* 8: 1–7.

Green, R. H. 1970. On fixed precision level sequential sampling. *Researches on Population Ecology* 12: 249–251.

Green, R. H. 1979. *Sampling Design and Statistical Methods for Environmental Biologists*. Wiley, New York.

Greig-Smith, P. 1979. Pattern in vegetation. *Journal of Ecology* 67: 755–779.

Grier, J. W. 1982. Ban of DDT and subsequent recovery of reproduction in bald eagles. *Science* 218: 1232–1234.

Gurevitch, J. and Chester Jr., S. T. 1986. Analysis of repeated measures experiments. *Ecology* 67: 251–254.

Hagood, M. J. 1970. The notion of a hypothetical universe. In *The Significance Test Controversy*, eds. D. E. Morrison and R. E. Henkel, 65–78. Butterworth, London.

Haldane, J. B. S. 1955. The measurement of variation. *Evolution* 9: 484.

Hankin, D. G. 1984. Multistage sampling designs in fisheries research: applications in small streams. *Canadian Journal of Fisheries and Aquatic Sciences* 41: 1575–1591.

Hansen, J. and Zeger, S. 1978. A comparison of two methods for estimating the parameters of a truncated normal distribution. In *Biological Data in Water Pollution Assessment: Quantitative and Statistical Analyses*, eds. K. L. Dickson, J. J. Cairns and R. J. Livingston, 29–37. American Society for Testing and Materials, Philadelphia.

Hanski, I. 1978. Some comments on the measurement of niche metrics. *Ecology* 59: 168–174.

Hanson, W. R. 1963. Calculation of productivity, survival and abundance of selected vertebrates from sex and age ratios. *Wildlife Monographs* 9: 1–60.

Harris, M. P. and Lloyd, C. S. 1977. Variations in counts of seabirds from photographs. *British Birds* 70: 200–205.

Hayes, R. J. and Buckland, S. T. 1983. Radial-distance models for the line-transect method. *Biometrics* 39: 29–42.

Hayne, D. W. 1949a. An examination of the strip census method for estimating animal populations. *Journal of Wildlife Management* 13: 145–157.

Hayne, D. W. 1949b. Two methods for estimating populations from trapping records. *Journal of Mammalogy* 30: 399–411.

Hayne, D. W. 1978. Experimental designs and statistical analysis. In *Populations of Small*

Mammals Under Natural Conditions. Pymatuning Symposia in Ecology, ed. D. P. Snyder, 3–10. University of Pittsburgh, Pittsburgh.

Healy, W. M. and Welsh, C. J. E. 1992. Evaluating line transects to monitor gray squirrel populations. *Wildlife Society Bulletin* 20: 83–90.

Heck, K. L. J., Van Belle, G. and Simberloff, D. 1975. Explicit calculation of the rarefaction diversity measurement and the determination of sufficient sample size. *Ecology* 56: 1459–1461.

Heffner, R. A., Butler, M. J. I. and Reilly, C. K. 1996. Pseudoreplication revisited. *Ecology* 77: 2558–2562.

Heisey, D. M. and Fuller, T. K. 1985. Evaluation of survival and cause-specific mortality rates using telemetry data. *Journal of Wildlife Management* 49: 668–674.

Heltshe, J. F. and Forrester, N. E. 1983a. Estimating diversity using quadrat sampling. *Biometrics* 39: 1073–1076.

Heltshe, J. F. and Forrester, N. E. 1983b. Estimating species richness using the jackknife procedure. *Biometrics* 39: 1–11.

Heltshe, J. F. and Forrester, N. E. 1985. Statistical evaluation of the jackknife estimate of diversity when using quadrat samples. *Ecology* 66: 107–111.

Hendricks, W. A. 1956. *The Mathematical Theory of Sampling*. The Scarecrow Press, New Brunswick, New Jersey.

Hensler, G. L. 1985. Estimation and comparison of functions of daily nest survival probabilities using the Mayfield method. In *Statistics in Ornithology*, eds. B. J. T. Morgan and P. M. North, 289–301. Springer-Verlag, New York.

Hill, M. O. 1973. Diversity and evenness: a unifying notation and its consequences. *Ecology* 54: 427–432.

Hines, W. G. S. and Hines, R. J. 1979. The Eberhardt statistic and the detection of nonrandomness of spatial point distributions. *Biometrika* 66: 73–79.

Hoisaeter, T. and Matthiesen, A. 1979. *Report on Some Statistical Aspects of Marine Biological Sampling*. San Carlos Publications, University of San Carlos, Cebu City, Phillipines.

Holt, R. D. 1987. On the relation between niche overlap and competition: the effect of incommensurable niche dimensions. *Oikos* 48: 110–114.

Holyoak, M. 1994. Identifying delayed density dependence in time-series data. *Oikos* 70: 296–304.

Hopkins, B. 1954. A new method for determining the type of distribution of plant individuals. *Annals of Botany* 18: 213–227.

Horn, H. S. 1966. Measurement of "overlap" in comparative ecological studies. *American Naturalist* 100: 419–424.

Hough, A. F. 1936. A climax forest community on East Tionesta Creek in northwestern Pennsylvania. *Ecology* 17: 9–28.

Hoyle, M. H. 1973. Transformations: An introduction and a bibliography. *International Statistical Review* 41: 203–223.

Hughes, R. G. 1986. Theories and models of species abundance. *American Naturalist* 128: 879–899.

Huhta, V. 1979. Evaluation of different similarity indices as measures of succession in arthropod communities of the forest floor after clear-cutting. *Oecologia* 41: 11–23.

Hunt, G. L., Eppley, Z. A. and Schneider, D. C. 1986. Reproductive performance of seabirds: the importance of population size and colony size. *Auk* 103: 306–317.

Hurlbert, S. H. 1971. The non-concept of species diversity: a critique and alternative parameters. *Ecology* 52: 577–586.

Hurlbert, S. H. 1978. The measurement of niche overlap and some relatives. *Ecology* 59: 67–77.

Hurlbert, S. H. 1984. Pseudoreplication and the design of ecological field experiments. *Ecological Monographs* 54: 187–211.

Hurlbert, S. H. 1990. Spatial distribution of the montane unicorn. *Oikos* 58: 257–271.

Hutchings, M. J., Booth, K. D. and Waite, S. 1991. Comparison of survivorship by the logrank test: criticisms and alternatives. *Ecology* 72: 2290–2293.

Iachan, R. 1985. Optimum stratum boundaries for shellfish surveys. *Biometrics* 41: 1053–1062.

Ito, Y. 1972. On the methods for determining density-dependence by means of regression. *Oecologia* 10: 347–372.

Iwao, S. 1975. A new method of sequential sampling to classify populations relative to a critical density. *Researches in Population Ecology* 16: 281–288.

Jackson, C. H. N. 1933. On the true density of tsetse flies. *Journal of Animal Ecology* 2: 204–209.

Jackson, D. A., Somers, K. M. and Harvey, H. H. 1989. Similarity coefficients: measures of co-occurrence and and association or simply measures of occurrence? *American Naturalist* 133: 436–453.

Jacobs, J. 1974. Quantitative measurement of food selection. *Oecologia* 14: 413–417.

James, F. C. and McCulloch, C. E. 1985. Data analysis and the design of experiments in ornithology. *Current Ornithology* 2: 1–63.

James, F. C. and Rathbun, S. 1981. Rarefaction, relative abundance, and diversity of avian communities. *Auk* 98: 785–800.

Jensen, A. L. 1986. Functional regression and correlation analysis. *Canadian Journal of Fisheries and Aquatic Sciences* 43: 1742–1745.

Jessen, R. J. 1978. *Statistical Survey Techniques*. Wiley, New York.

Johnson, D. H. 1979. Estimating nest success: the Mayfield method and an alternative. *Ornithology* 96: 651–671.

Johnson, D. H. 1980. The comparison of usage and availability measurements for evaluating resource preference. *Ecology* 61: 65–71.

Johnson, D. H. and Grier, J. W. 1988. Determinants of breeding distributions of ducks. *Wildlife Monographs* 100: 1–37.

Johnson, D. H. and Klett, A. T. 1985. Quick estimates of success rates of duck nests. *Wildlife Society Bulletin* 13: 51–53.

Johnson, E. G. and Routledge, R. D. 1985. The line transect method: a nonparametric estimator based on shape restriction. *Biometrics* 41: 669–679.

Johnson, N. L. and Kotz, S. 1969. *Distributions in Statistics: Discrete Distributions*. Houghton Mifflin, Boston.

Jolicoeur, P. and Heusner, A. A. 1971. The allometry equation in the analysis of the standard oxygen consumption and body weight of the white rat. *Biometrics* 27: 841–855.

Jolly, G. M. 1965. Explicit estimates from capture-recapture data with both death and immigration—stochastic model. *Biometrika* 52: 225–247.

Jolly, G. M. 1969a. Sampling methods for aerial censuses of wildlife populations. *East African Agricultural and Forestry Journal* 34: 46–49.

Jolly, G. M. 1969b. The treatment of errors in aerial counts of wildlife populations. *East African Agricultural and Forestry Journal* 34: 50–55.

Jolly, G. M. 1982. Mark-recapture models with parameters constant in time. *Biometrics* 38: 301–321.

Jolly, G. M. and Dickson, J. M. 1983. The problem of unequal catchability in mark-recapture estimation of small mammal populations. *Canadian Journal of Zoology* 61: 922–927.

Joule, J. and Cameron, G. N. 1975. Species removal studies. I. Dispersal strategies of sympatric *Sigmodon hispidus* and *Reithrodontomys fulvescens* populations. *Journal of Mammalogy* 56: 378–396.

Kamil, A. C. 1988. Experimental design in ornithology. *Current Ornithology* 5: 313–346.

Karr, J. R. 1982. Population variability and extinction in the avifauna of a tropical land-bridge island. *Ecology* 63: 1975–1978.

Kastenbaum, M. A., Hoel, D. G. and Bowman, K. O. 1970. Sample size requirements: one-way analysis of variance. *Bimetrika* 57: 421–430.

Keith, L. B., Cary, J. R., Rongstad, O. J. and Brittingham, M. C. 1984. Demography and ecology of a declining snowshoe hare population. *Wildlife Monographs* 90: 1–43.

Keith, L. B. and Meslow, E. C. 1968. Trap response by snowshoe hares. *Journal of Wildlife Management* 32: 795–801.

Keith, L. B. and Windberg, L. A. 1978. A demographic analysis of the snowshoe hare cycle. *Wildlife Monographs* 58: 1–70.

Kelker, G. H. 1940. Estimating deer populations by a differential hunting loss in the sexes. *Utah Academy of Science and Arts Letters* 17: 6–69.

Kempton, R. A. 1979. The structure of species abundance and measurement of diversity. *Biometrics* 35: 307–321.

Kempton, R. A. and Taylor, L. R. 1974. Log-series and log-normal parameters as diversity discriminants for the Lepidoptera. *Journal of Animal Ecology* 43: 381–399.

Kenward, R. 1987. *Wildlife Radio Tagging*. Academic Press, London.

Koch, C. F. 1987. Prediction of sample size effects on the measured temporal and geographic distribution patterns of species. *Paleobiology* 13: 100–107.

Kraemer, H. C. and Thiemann, S. 1987. *How many subjects?* Sage Publications, Newbury Park, California.

Krebs, C. J. 1966. Demographic changes in fluctuating populations of *Microtus californicus*. *Ecological Monographs* 36: 239–273.

Krebs, C. J. 1989. *Ecological Methodology*. Harper Collins, New York.

Krebs, C. J. 1994. *Ecology: The Experimental Analysis of Distribution and Abundance*. Harper Collins, New York.

Krebs, C. J. and Wingate, I. 1976. Small mammal communities of the Kluane Region, Yukon Territory. *Canadian Field-Naturalist* 90: 379–389.

Kuno, E. 1969. A new method of sequential sampling to obtain the population estimates with a fixed level of precision. *Researches on Population Ecology* 11: 127–136.

Kuno, E. 1971. Sampling error as a misleading artifact in key-factor analysis. *Researches on Population Ecology* 13: 28–45.

Kuno, E. 1972. Some notes on population estimation by sequential sampling. *Researches in Population Ecology* 14: 58–73.

Kuno, E. 1973. Statistical characteristics of the density-independent population fluctuation and the evaluation of density-dependence and regulation in animal populations. *Researches on Population Ecology* 15: 99–120.

Kuno, E. 1991. Sampling and analysis of insect populations. *Annual Review of Entomology* 36: 285–304.

Lachin, J. M. and Foulkes, M. A. 1986. Evaluation of sample size and power for analyses of survival with allowance for nonuniform patient entry, losses to follow-up, noncompliance, and stratification. *Biometrics* 42: 507–519.

Lance, G. N. and Williams, W. T. 1967. Mixed-data classificatory programs. I. Agglomerative systems. *Australian Computer Journal* 1: 15–20.

Lebreton, J. D., Burnham, K. P., Clobert, J. and Anderson, D. R. 1992. Modeling survival and testing biological hypotheses using marked animals: a unified approach with case studies. *Ecological Monographs* 62: 67–118.

Lebreton, J. D. and North, P. M. 1993. *Marked Individuals in the Study of Bird Population*. Birkhauser Verlag, Basel.

Lebreton, J. D., Pradel, R. and Clobert, J. 1993a. The statistical analysis of survival in animal populations. *Trends in Ecology and Evolution* 8: 91–94.

Lebreton, J. D., Reboulet, A. M. and Banco, G. 1993b. An overview of software for terrestrial vertebrate population dynamics. In *Marked Individuals in the Study of Bird Population*, eds. J. D. Lebreton and P. M. North, 357–372. Birkhauser Verlag, Basel.

Lee, E. T. 1992. *Statistical Methods for Survival Data Analysis*. John Wiley & Sons, New York.

Legendre, L. and Legendre, P. 1983. *Numerical Ecology*. Elsevier, New York.

LeResche, R. E. and Rausch, R. A. 1974. Accuracy and precision in aerial moose censusing. *Journal of Wildlife Management* 38: 175–182.

Leslie, P. H. 1952. The estimation of population parameters from data obtained by means of the capture-recapture method. II. The estimation of total numbers. *Biometrika* 39: 363–388.

Leslie, P. H. and Chitty, D. 1951. The estimation of population parameters from data obtained by means of the capture-recapture method. I. The maximum likelihood equations for estimating the death-rate. *Biometrika* 38: 269–292.

Leslie, P. H., Chitty, D. and Chitty, H. 1953. The estimation of population parameters from data obtained by means of the capture-recapture method. III. An example of the practical applications of the method. *Biometrika* 40: 137–169.

Leslie, P. H. and Davis, D. H. S. 1939. An attempt to determine the absolute number of rats on a given area. *Journal of Animal Ecology* 8: 94–113.

Leslie, P. H., Tener, J. S., Vizoso, M. and Chitty, H. 1955. The longevity and fertility of the Orkney vole, *Microtus orcadensis*, as observed in the laboratory. *Proceedings of the Zoological Society of London* 125: 115–125.

Lessells, C. M. and Boag, P. T. 1987. Unrepeatable repeatabilities: a common mistake. *Auk* 104: 116–121.

Levings, S. C. and Franks, N. R. 1982. Patterns of nest dispersion in a tropical ground ant community. *Ecology* 63: 338–344.

Levins, R. 1968. *Evolution in Changing Environments: Some Theoretical Explorations*. Princeton University Press, Princeton, New Jersey.

Lewontin, R. 1966. On the measurement of relative variability. *Systematic Zoology* 15: 141–142.

Lieberman, G. J. and Owen, D. B. 1961. *Tables of the Hypergeometric Probability Distribution*. Stanford University Press, Stanford, California.

Liebold, M. A. 1995. The niche concept revisited: mechanistic models and community context. *Ecology* 76: 1371–1382.

Lincoln, F. C. 1930. Calculating waterfowl abundance on the basis of banding returns. *U.S. Department of Agriculture Circular* 118: 1–4.

Lindroth, R. L. and Batzli, G. O. 1984. Food habits of the meadow vole (*Microtus pennsylvanicus*) in bluegrass and prairie habitats. *Journal of Mammalogy* 65: 600–606.

Linton, L. R., Davies, R. W. and Wrona, F. J. 1981. Resource utilization indices: an assessment. *Journal of Animal Ecology* 50: 283–292.

Lloyd, M. and Ghelardi, R. J. 1964. A table for calculating the "equitability" component of species diversity. *Journal of Animal Ecology* 33: 217–225.

Loehle, C. 1990. Proper statistical treatment of species-area data. *Oikos* 57: 143–145.

Longino, J. T. and Colwell, R. K. 1997. Biodiversity assessment using structured inventory: capturing the ant fauna of a tropical rain forest. *Ecological Applications* 7: 1263–1277.

Loreau, M. 1990. The Colwell-Futuyma method for measuring niche breadth and overlap: a critique. *Oikos* 58: 251–253.

Ludwig, J. A. and Reynolds, J. F. 1988. *Statistical Ecology: A Primer on Methods and Computing*. John Wiley and Sons, New York.

Lyons, N. I. and Hutcheson, K. 1986. Estimation of Simpson's Diversity when counts follow a Poisson distribution. *Biometrics* 42: 171–176.

MacArthur, R. H. 1965. Patterns of species diversity. *Biological Reviews* 40: 510–533.

MacArthur, R. H. 1968. The theory of the niche. In *Population Biology and Evolution*, ed. R. C. Lewontin, 159–176. Syracuse University Press, Syracuse, New York.

MacArthur, R. H. 1972. *Geographical Ecology*. Harper and Row, New York.

MacArthur, R. H. and Levins, R. 1967. The limiting similarity, convergence, and divergence of coexisting species. *American Naturalist* 101: 377–385.

Mace, A. E. 1964. *Sample-Size Determination*. Reinhold, New York.

Maelzer, D. A. 1970. The regression of log N_{n+1} on log N_n as a test of density dependence: an exercise with computer-constructed density-independent populations. *Ecology* 51: 810–822.

Magurran, A. E. 1988. *Ecological diversity and its measurement*. Princeton University Press, Princeton, New Jersey.

Mahon, R., Smith, R. W., Bernstein, B. B. and Scott, J. S. 1984. Spatial and temporal patterns of groundfish distribution on the Scotian Shelf and in the Bay of Fundy, 1970–1981. *Canadian Technical Reports in Fisheries and Aquatic Sciences* No. 1300.

Manly, B. F. J. 1971. A simulation study of Jolly's method for analyzing capture- recapture data. *Biometrics* 27: 415–424.

Manly, B. F. J. 1974. A model for certain types of selection experiments. *Biometrics* 30: 281–294.

Manly, B. F. J. 1984. Obtaining confidence limits on parameters of the Jolly-Seber model for capture-recapture data. *Biometrics* 40: 749–758.

Manly, B. F. J. 1990. On the statistical analysis of niche overlap data. *Canadian Journal of Zoology* 68: 1420–1422.

Manly, B. F. J. 1991. *Randomization and Monte Carlo Methods in Biology*. Chapman and Hall, New York.

Manly, B. F. J., McDonald, L. L. and Thomas, D. L. 1993. *Resource Selection by Animals: Statistical Design and Analysis for Field Studies*. Chapman and Hall, London.

Manly, B. F. J., Miller, P. and Cook, L. M. 1972. Analysis of a selective predation experiment. *American Naturalist* 106: 719–736.

Manly, B. F. J. and Parr, M. J. 1968. A new method of estimating population size, survivorship, and birth rate from capture-recapture data. *Transactions for the Society of British Entomologists* 18: 81–89.

Margalef, D. R. 1958. Information theory in ecology. *General Systems* 3: 36–71.

Marsh, H. and Sinclair, D. F. 1989. Correcting for visibility bias in strip transect aerial surveys of aquatic fauna. *Journal of Wildlife Management* 53: 1017–1024.

Maurer, B. A. 1982. Statistical inference for MacArthur-Levins niche overlap. *Ecology* 63: 1712–1719.

May, R. M. 1975. Patterns of species abundance and diversity. In *Ecology and Evolution of Communities*, eds. M. L. Cody and J. M. Diamond, 81–120. Belknap Press, Harvard University, Cambridge, Mass.

Mayfield, H. 1961. Nesting success calculated from exposure. *Wilson Bulletin* 73: 255–261.

Mayfield, H. 1975. Suggestions for calculating nest success. *Wilson Bulletin* 87: 456–466.

McArdle, B. and Gaston, K. 1993. The temporal variability of populations. *Oikos* 67: 187–191.

McArdle, B. H. and Gaston, K. J. 1992. Comparing population variabilities. *Oikos* 64: 610–612.

McArdle, B. H., Gaston, K. J. and Lawton, J. H. 1990. Variation in the size of animal populations: patterns, problems and artifacts. *Journal of Animal Ecology* 59: 439–454.

McIntosh, R. P. 1967. An index of diversity and the relation of certain concepts to diversity. *Ecology* 48: 392–404.

McNeil, W. J. 1967. Randomness of distribution of pink salmon redds. *Journal of the Fisheries Research Board of Canada* 24: 1629–1634.

Meyer, J. S., Ingersoll, C. G., McDonald, L. L. and Boyce, M. S. 1986. Estimating uncertainty in population growth rates: Jackknife vs. Bootstrap techniques. *Ecology* 67: 1156–1166.

Milne, A. 1959. The centric systematic area-sample treated as a random sample. *Biometrics* 15: 270–297.

Molinari, J. 1989. A calibrated index for the measurement of evenness. *Oikos* 56: 319–326.

Morin, A. 1985. Variability of density estimates and the optimization of sampling programs for stream benthos. *Canadian Journal of Fisheries and Aquatic Sciences* 42: 1530–1534.

Morisita, M. 1957. A new method for the estimation of density by the spacing method applicable to non-randomly distributed populations. *Physiology and Ecology* 7: 134–144 (in Japanese, U.S.D.A. Forest Service Translation No. 11116).

Morisita, M. 1959. Measuring of interspecific association and similarity between communities. *Memoirs of the Faculty of Science Kyushu University Series E* 3: 65–80.

Morisita, M. 1962. Id-index, a measure of dispersion of individuals. *Researches in Population Ecology* 4: 1–7.

Morris, R. F. 1954. A sequential sampling technique for spruce budworm egg surveys. *Canadian Journal of Zoology* 32: 302–313.

Morris, R. F. 1959. Single-factor analysis in population dynamics. *Ecology* 40: 580–588.

Morris, R. F. 1963. The dynamics of epidemic spruce budworm populations. *Memoirs of the Entomological Society of Canada* 31: 1–332.

Morrison, D. E. and Henkel, R. E. 1970a. *The Significance Test Controversy*. Butterworth, London.

Morrison, D. E. and Henkel, R. E. 1970b. Significance tests reconsidered. In *The Significance Test Controversy*, eds. D. E. Morrison and R. E. Henkel, 182–198. Butterworth, London.

Mueller, L. D. and Altenberg, L. 1985. Statistical inference on measures of niche overlap. *Ecology* 66: 1204–1210.

Mumme, R. L., Koenig, W. D. and Pitelka, F. A. 1983. Are acorn woodpecker territories aggregated? *Ecology* 64: 1305–1307.

Myers, J. H. 1978. Selecting a measure of dispersion. *Environmental Entomology* 7: 619–621.

Naranjo, S. E. and Hutchison, W. D. 1997. Validation of arthropod sampling plans using a resampling approach: software and analysis. *American Entomologist* 43: 48–57.

Neal, A. K., White, G. C., Gill, R. B., Reed, D. F. and Otterman, J. H. 1993. Evaluation of mark-resight model assumptions for estimating mountain sheep numbers. *Journal of Wildlife Management* 57: 436–450.

Nee, S., Harvey, P. H. and Cotgreave, P. 1992. Population persistence and the natural relationship between body size and abundance. In *Conservation of Biodiversity for Sustainable Development*, eds. O. T. Sandlund, K. Hindar and A. D. H. Brown, 124–136. Scandinavian University Press, Oslo.

Neu, C. W., Byers, C. R. and Peek, J. M. 1974. A technique for analysis of utilization-availability data. *Journal of Wildlife Management* 38: 541–545.

Noreen, E. W. 1990. *Computer Intensive Methods for Testing Hypotheses: An Introduction*. John Wiley and Sons, New York.

Norton-Griffiths, M. 1978. *Counting Animals*, 2nd edition. African Wildlife Leadership Foundation, Nairobi.

Nyrop, J. P. and Binns, M. 1991. Quantitative methods for designing and analyzing sampling programs for use in pest management. In *Handbook of Pest Management in Agriculture*, ed. D. Pimentel, 67–132. CRC Press, Boca Raton, Florida.

Nyrop, J. P. and Simmons, G. A. 1984. Errors incurred when using Iwao's sequential decision rule in insect sampling. *Environmental Entomology* 13: 1459–1465.

Orloci, L. 1978. *Multivariate Analysis in Vegetation Research*, 2nd ed. Dr. W. Junk B.V., The Hague, Netherlands.

Osenberg, C. W., Schmitt, R. J., Holbrook, S. J., Abu-Saba, K. E. and Flegal, A. R. 1994. Detection of environmental impacts: natural variability, effect size, and power analysis. *Ecological Applications* 4: 16–30.

Osenberg, C. W. and Schmitt, R. J. 1994. Detecting human impacts in marine habitats. *Ecological Applications* 4: 1–2.

Ossiander, F. J. and Wedemeyer, G. 1973. Computer program for sample sizes required to determine disease incidence in fish populations. *Journal of the Fisheries Research Board of Canada* 30: 1383–1384.

Otis, D. L., Burnham, K. P., White, G. C. and Anderson, D. R. 1978. Statistical inference from capture data on closed animal populations. *Wildlife Monographs* 62: 1–135.

Packard, J. M., Summers, R. C. and Barnes, L. B. 1985. Variation of visibility bias during aerial surveys of manatees. *Journal of Wildlife Management* 49: 347–351.

Palmer, M. W. 1990. The estimation of species richness by extrapolation. *Ecology* 71: 1195–1199.

Palsbøll, P. J., Allen, J., Berube, M., Clapham, P. J., Feddersen, T. P., Hammond, P. S., Hudson, R. R., Jørgensen, H., Katona, S., Larsen, A. H., Larsen, F., Lien, J., Mattila, D. K., Sigurjonsson, J., Sears, R., Smith, T., Sponer, R., Stevick, P. and Øien, N. 1997. Genetic tagging of humpback whales. *Nature* 388: 767–769.

Parker, K. R. 1979. Density estimation by variable area transect. *Journal of Wildlife Management* 43: 484–492.

Patil, G. P., Pielou, E. C. and Walters, W. E. 1971. *Spatial Patterns and Statistical Distributions*. Pennsylvania State University Press, University Park, Pennsylvania.

Patrick, R. 1968. The structure of diatom communities in similar ecological conditions. *American Naturalist* 102: 173–183.

Paulik, G. J. 1961. Detection of incomplete reporting of tags. *Journal of the Fisheries Research Board of Canada* 18: 817–829.

Paulik, G. J. and Robson, D. S. 1969. Statistical calculations for change-in-ratio estimators of population parameters. *Journal of Wildlife Management* 33: 1–27.

Peet, R. K. 1974. The measurement of species diversity. *Annual Review of Ecology and Systematics* 5: 285–307.

Peet, R. K. 1975. Relative diversity indices. *Ecology* 56: 496–498.

Perry, J. N. 1988. Some models for spatial variability of animal species. *Oikos* 51: 124–130.

Perry, J. N. 1995a. Spatial analysis by distance indices. *Journal of Animal Ecology* 64: 303–314.

Perry, J. N. 1995b. Spatial aspects of animal and plant distribution in patchy farmland habitats. In *Ecology and Integrated Farming Systems*, eds. D. M. Glen, M. P. Greaves and H. M. Anderson, 221–242. Wiley, New York.

Perry, J. N. and Hewitt, M. 1991. A new index of aggregation for animal counts. *Biometrics* 47: 1505–1518.

Peterman, R. M. 1990. Statistical power analysis can improve fisheries research and management. *Canadian Journal of Fisheries and Aquatic Sciences* 47: 2–15.

Peters, R. H. 1991. *A Critique for Ecology*. Cambridge University Press, Cambridge, England.

Petraitis, P. S. 1979. Likelihood measures of niche breadth and overlap. *Ecology* 60: 703–710.

Pianka, E. R. 1973. The structure of lizard communities. *Annual Review of Ecology and Systematics* 4: 53–74.

Pianka, E. R. 1974. Niche overlap and diffuse competition. *Proceedings of the National Academy of the USA* 71: 2141–2145.

Pianka, E. R. 1986. *Ecology and Natural History of Desert Lizards*. Princeton University Press, Princeton, New Jersey.

Pielou, E. C. 1966. The measurement of diversity in different types of biological collections. *Journal of Theoretical Biology* 13: 131–144.

Pielou, E. C. 1969. *An Introduction to Mathematical Ecology*. Wiley, New York.

Pielou, E. C. 1975. *Ecological Diversity*. Wiley, New York.

Pielou, E. C. 1977. *Mathematical Ecology*. Wiley, New York.

Pielou, E. C. 1984. *The Interpretation of Ecological Data*. Wiley, New York.

Pieters, E. P. and Sterling, W. L. 1974. A sequential sampling plan for the cotton fleahopper, *Pseudatomoscelis seriatus*. *Environmental Entomology* 3: 102–106.

Podoler, H. and Rogers, D. 1975. A new method for the identification of key factors from life-table data. *Journal of Animal Ecology* 44: 85–114.

Pollard, J. H. 1971. On distance estimators of density in randomly distributed forests. *Biometrics* 27: 991-1002.

Pollock, K. H. 1978. A family of density estimates for line-transect sampling. *Biometrics* 34: 475–478.

Pollock, K. H. 1982. A capture-recapture design robust to unequal probability of capture. *Journal of Wildlife Management* 46: 752–757.

Pollock, K. H., Hines, J. E. and Nichols, J. D. 1985a. Goodness-of-fit tests for open capture-recapture models. *Biometrics* 41: 399–410.

Pollock, K. H. and Kendall, W. L. 1987. Visibility bias in aerial surveys; a review of estimation procedures. *Journal of Wildlife Management* 51: 502–510.

Pollock, K. H., Lancia, R. A., Conner, M. C. and Wood, B. L. 1985b. A new change-in-ratio procedure robust to unequal catchability of types of animal. *Biometrics* 41: 653–662.

Pollock, K. H., Nichols, J. D., Brownie, C. and Hines, J. E. 1990. Statistical inference for capture-recapture experiments. *Wildlife Monographs* 107: 1–97.

Pollock, K. H., Winterstein, S. R., Bunck, C. M. and Curtis, P. D. 1989a. Survival analysis in telemetry studies: the staggered entry design. *Journal of Wildlife Management* 53: 7–15.

Pollock, K. H., Winterstein, S. R. and Conroy, M. J. 1989b. Estimation and analysis of survival distributions for radio-tagged animals. *Biometrics* 45: 99–109.

Preston, F. W. 1948. The commonness and rarity of species. *Ecology* 29: 254–283.

Preston, F. W. 1962. The canonical distribution of commonness and rarity. *Ecology* 43: 185–215, 410–432.

Pringle, J. D. 1984. Efficiency estimates for various quadrat sizes used in benthic sampling. *Canadian Journal of Fisheries and Aquatic Sciences* 41: 1485–1489.

Pyburn, W. F. 1958. Size and movements of a local population of cricket frogs (*Acris crepitans*). *Texas Journal of Science* 10: 325–342.

Pyke, D. A. and Thompson, J. N. 1986. Statistical analysis of survival and removal rate experiments. *Ecology* 67: 240–245.

Quenouille, M. H. 1950. *Introductory Statistics*. Butterworth-Springer, London.

Rapport, D. J. and Turner, J. E. 1970. Determination of predator food preferences. *Journal of Theoretical Biology* 26: 365–372.

Rasmussen, D. I. and Doman, E. R. 1943. Census methods and their applications in the management of mule deer. *Transactions of the North American Wildlife Conference* 8: 369–379.

Renkonen, O. 1938. Statisch-okologische Untersuchungen uber die terrestiche kaferwelt der finnischen bruchmoore. *Ann. Zool. Soc. Bot. Fenn. Vanamo* 6: 1–231.

Rice, E. L. 1967. A statistical method for determining quadrat size and adequacy of sampling. *Ecology* 48: 1047–1049.

Rice, W. R. and Gaines, S. D. 1994. "Heads I win, tails you lose": testing directional alternative hypotheses in ecological and evolutionary research. *Trends in Ecology and Evolution* 9: 235–237.

Ricker, W. E. 1973. Linear regressions in fishery research. *Journal of the Fisheries Research Board of Canada* 30: 409–434.

Ricker, W. E. 1975. Computation and interpretation of biological statistics of fish populations. *Fisheries Research Board of Canada Bulletin* 191.

Ricker, W. E. 1984. Computation and uses of central trend lines. *Canadian Journal of Zoology* 62: 1897–1905.

Ricklefs, R. E. and Lau, M. 1980. Bias and dispersion of overlap indices: results of some Monte Carlo simulations. *Ecology* 61: 1019–1024.

Ripley, B. D. 1981. *Spatial Statistics*. Wiley, New York.

Ripley, B. D. 1985. Analyses of nest spacings. In *Statistics in Ornithology*, eds. B. J. T. Morgan and P. M. North, 151–158. Springer-Verlag, Berlin.

Robson, D. S. and Chapman, D. G. 1961. Catch curves and mortality rates. *Transactions of the American Fisheries Society* 90: 181–189.

Robson, D. S. and Regier, H. A. 1964. Sample size in Petersen mark-recapture experiments. *Transactions of the American Fisheries Society* 93: 215–226.

Rodgers, A. R. and Lewis, M. C. 1985. Diet selection in Arctic lemmings (*Lemmus sibericus* and *Dicrostonyx groenlandicus*): food preferences. *Canadian Journal of Zoology* 63: 1161–1173.

Roff, D. A. 1973. On the accuracy of some mark-recapture estimators. *Oecologia* 12: 15–34.

Rohlf, F. J. 1995. *BIOM. A Package of Statistical Programs to Accompany the Text Biometry*. Exeter Software, Setauket, New York.

Rohlf, F. J. and Sokal, R. R. 1995. *Statistical Tables*, 3rd edition. Freeman, New York.

Romesburg, H. C. 1984. *Cluster Analysis for Researchers*. Lifetime Learning Publications, Belmont, California.

Root, R. B. 1973. Organization of a plant-arthropod association in simple and diverse habitats: the fauna of collards (*Brassica oleracea*). *Ecological Monographs* 43: 95–124.

Rose, F. L. and Armentrout, D. 1974. Population estimates of *Ambystoma tigrinum* inhabiting two playa lakes. *Journal of Animal Ecology* 43: 671–679.

Rosenberg, D. K., Overton, W. S. and Anthony, R. G. 1995. Estimation of animal abundance when capture probabilities are low and heterogeneous. *Journal of Wildlife Management* 59: 252–261.

Routledge, R. D. 1979. Diversity indices: which ones are admissible? *Journal of Theoretical Biology* 76: 503–515.

Routledge, R. D. 1980a. Bias in estimating the diversity of large, uncensused communities. *Ecology* 61: 276–281.

Routledge, R. D. 1980b. The form of species-abundance distributions. *Journal of Theoretical Biology* 82: 547–558.

Routledge, R. D. 1982. The method of bounded counts: when does it work? *Journal of Wildlife Management* 46: 757–761.

Routledge, R. D. 1983. Evenness indices: are any admissible? *Oikos* 40: 149–151.

Routledge, R. D. and Fyfe, D. A. 1992a. Confidence limits for line transect estimates based on shape restrictions. *Journal of Wildlife Management* 56: 402–407.

Routledge, R. D. and Fyfe, D. A. 1992b. Wildlife Software: TRANSAN: line transect estimates based on shape restrictions. *Wildlife Society Bulletin* 20: 455–456.

Rowell, J. G. and Walters, D. E. 1976. Analysing data with repeated observations on each experimental unit. *Journal of Agricultural Science* 87: 423–432.

Royama, T. 1996. A fundamental problem in key factor analysis. *Ecology* 77: 87–93.

Russell, H. J. J. 1972. Use of a commercial dredge to estimate a hardshell clam population by stratified random sampling. *Journal of the Fisheries Research Board of Canada* 29: 1731–1735.

Saitoh, T., Stenseth, N. C. and Bjørnstad, O. N. 1997. Density dependence in fluctuating grey-sided vole populations. *Journal of Animal Ecology* 66: 14–24.

Samuel, E. 1969. Comparison of sequential rules for estimation of the size of a population. *Biometrics* 25: 517–535.

Sanders, H. L. 1968. Marine benthic diversity: a comparative study. *American Naturalist* 102: 243–282.

Savage, R. E. 1931. The relation between the feeding of the herring off the east coast of England and the plankton of the surrounding waters. *Fishery Investigations, Ministry of Agriculture, Food, and Fisheries, Series 2*, 12: 1–88.

Scheiner, S. M. 1993. Introduction: Theories, Hypotheses, and Statistics. In *Design and Analysis of Ecological Experiments*, eds. S. M. Scheiner and J. Gurevitch, 1–13. Chapman and Hall, New York.

Schluter, D. 1982. Seed and patch selection by Galapagos ground finches: relation to foraging efficiency and food supply. *Ecology* 63: 1106–1120.

Schnabel, Z. E. 1938. The estimation of the total fish population of a lake. *American Mathematician Monthly* 45: 348–352.

Schnute, J. 1983. A new approach to estimating populations by the removal method. *Canadian Journal of Fisheries and Aquatic Sciences* 40: 2153–2169.

Schnute, J. 1984. Linear mixtures: a new approach to bivariate trend lines. *Journal of the American Statistical Association* 79: 1–8.

Schoener, T. W. 1970. Nonsynchronous spatial overlap of lizards in patchy habitats. *Ecology* 51: 408–418.

Schoener, T. W. 1974. Some methods for calculating competition coefficients from resource-utilization spectra. *American Naturalist* 108: 332–340.

Schumacher, F. X. and Eschmeyer, R. W. 1943. The estimation of fish populations in lakes and ponds. *Journal of the Tennessee Academy of Sciences* 18: 228–249.

Schweigert, J. F., Haegele, C. W. and Stocker, M. 1985. Optimizing sampling design for herring spawn surveys in the Strait of Georgia, B.C. *Canadian Journal of Fisheries and Aquatic Sciences* 42: 1806–1814.

Schweigert, J. F. and Sibert, J. R. 1983. Optimizing survey design for determining age structure of fish stocks: an example from British Columbia Pacific herring (*Clupea harengus pallasi*). *Canadian Journal of Fisheries and Aquatic Sciences* 40: 588–597.

Seber, G. A. F. 1982. *The Estimation of Animal Abundance*, 2nd edition. Charles Griffin and Company, London.

Seber, G. A. F. 1992. A review of estimating animal abundance II. *International Statistical Review* 60: 129–166.

Selby, S. M. 1962. *Handbook of Mathematical Tables*. Chemical Rubber Publishing Company, Cleveland, Ohio.

Tanaka, R. 1951. Estimation of vole and mouse populations on Mount Ishizuchi and on the uplands of Southern Shikoku. *Journal of Mammalogy* 32: 450–458.

Taylor, L. R. 1961. Aggregation, variance and the mean. *Nature* 189: 732–735.

Taylor, L. R. 1984. Assessing and interpreting the spatial distributions of insect populations. *Annual Review of Entomology* 29: 321–357.

Taylor, L. R., Kempton, R. A. and Woiwod, I. P. 1976. Diversity statistics and the log-series model. *Journal of Animal Ecology* 45: 255–272.

Taylor, L. R., Woiwod, I. P. and Perry, J. N. 1979. The negative binomial as a dynamic ecological model for aggregation, and the density dependence of *k*. *Journal of Animal Ecology* 48: 289–304.

Taylor, W. D. 1980. Comment on "Aggregation, transformations, and the design of benthos sampling programs." *Canadian Journal of Fisheries and Aquatic Sciences* 37: 1328–1329.

Thomas, L. 1997. Retrospective power analysis. *Conservation Biology* 11: 276–280.

Thomas, L. and Krebs, C. J. 1997. A review of statistical power analysis software. *Bulletin of the Ecological Society of America* 78: 126–139.

Thompson, H. R. 1956. Distribution of distance to *n*th neighbour in a population of randomly distributed individuals. *Ecology* 37: 391–394.

Thompson, S. K. 1992. *Sampling*. John Wiley and Sons, New York.

Thoni, H. 1967. Transformations of variables used in the analysis of experimental and observational data. A review. Technical Report No. 7, Statistical Laboratory, Iowa State University, Ames: 1–61.

Tinbergen, L. 1960. The natural control of insects in pine woods. I. Factors influencing the intensity of predation by songbirds. *Archives Neederlanishes Zoologie* 13: 265–344.

Trent, T. T. and Rongstad, O. J. 1974. Home range and survival of cottontail rabbits in southwestern Wisconsin. *Journal of Wildlife Management* 38: 459–472.

Trumble, J. T., Edelson, J. V. and Story, R. N. 1987. Conformity and incongruity of selected dispersion indices in describing the spatial distribution of *Trichoplusia ni* (Huebner) in geographically separate cabbage plantings. *Researches in Population Ecology* 29: 155–166.

Tukey, J. 1958. Bias and confidence in not quite large samples. *Annals of Mathematical Statistics* 29: 614.

Tukey, J. W. 1960. Conclusions vs. decisions. *Technometrics* 2: 423–433.

Tukey, J. W. 1977. *Exploratory Data Analysis*. Addison-Wesley, Reading, Massachusetts.

Turk, T. R. 1978. Testing the effect of an intervention in sequential ecological data. *Biometrics* 34: 128–129.

Turner, F. B. 1960. Size and dispersion of a Louisiana population of the cricket frog, *Acris gryllus*. *Ecology* 41: 258–268.

Underwood, A. J. 1981. Techniques of analysis of variance in experimental marine biology and ecology. *Oceanography and Marine Biology Annual Review* 19: 513–605.

Underwood, A. J. 1994. On beyond BACI: sampling designs that might reliably detect environmental disturbances. *Ecological Applications* 4: 3–15.

Underwood, A. J. 1997. *Experiments in Ecology: Their Logical Design and Interpretation Using Analysis of Variance*. Cambridge University Press, Cambridge.

Van Belle, G. and Zeisig, T. 1978. Response: Testing the effect of an intervention in sequential ecological data. *Biometrics* 34: 134–137.

Van Valen, L. 1978. The statistics of variation. *Evolutionary Theory* 4: 33–43.

Van Valen, L. M. 1982. A pitfall in random sampling. *Nature* 295: 171.

Varley, G. C. and Gradwell, G. R. 1960. Key factors in population studies. *Journal of Animal Ecology* 29: 399–401.

Varley, G. C. and Gradwell, G. R. 1963. The interpretation of insect population changes. *Proceedings of the Ceylon Association for the Advancement of Science* 18: 142–156.

Varley, G. C. and Gradwell, G. R. 1965. Interpreting winter moth population changes. *Proceeding of the XII International Congress of Entomology* : 377–378.

Venzon, D. J. and Moolgavkar, S. H. 1988. A method for computing profile-likelihood based confidence intervals. *Applied Statistics* 37: 87–94.

Wald, A. 1947. *Sequential Analysis*. John Wiley, New York.

Walters, C. J. 1993. Dynamic models and large scale field experiments in environmental impact assessment and management. *Australian Journal of Ecology* 18: 53–62.

Walters, C. J., Collie, J. S. and Webb, T. 1988. Experimental designs for estimating transient responses to management disturbances. *Canadian Journal of Fisheries and Aquatic Sciences* 45: 530–538.

Washington, H. G. 1984. Diversity, biotic and similarity indices: a review with special relevance to aquatic ecosystems. *Water Research* 18: 653–694.

Waters, W. E. 1955. Sequential sampling in forest insect surveys. *Forest Science* 1: 68–79.

Watson, N. H. F. 1974. Zooplankton of the St. Lawrence Great Lakes—species composition, distribution, and abundance. *Journal of the Fisheries Research Board of Canada* 31: 783–794.

Weatherhead, P. J. 1986. How unusual are unusual events? *American Naturalist* 128: 150–154.

Wetherill, G. B. and Glazebrook, K. D. 1986. *Sequential Methods in Statistics*, 3rd ed. Chapman and Hall, London.

White, G. C. 1983. Numerical estimation of survival rates from band-recovery and biotelemetry data. *Journal of Wildlife Management* 47: 716–728.

White, G. C. 1996. NOREMARK: Population estimation from mark-resighting surveys. *Wildlife Society Bulletin* 24: 50–52.

White, G. C., Anderson, D. R., Burnham, K. P. and Otis, D. L. 1982. Capture-recapture and removal methods for sampling closed populations. *Los Alamos National Laboratory* LA-8787-NERP, 235 pp.

White, G. C., Bartmann, R. M., Carpenter, L. H. and Garrott, R. A. 1989. Evaluation of aerial line transects for estimating mule deer densities. *Journal of Wildlife Management* 53: 625–635.

White, G. C. and Garrott, R. A. 1990. *Analysis of Wildlife Radio-tracking Data*. Academic Press, New York.

Whittaker, R. H. 1956. Vegetation of the Great Smoky Mountains. *Ecological Monographs* 26: 1–80.

Whittaker, R. H. 1965. Dominance and diversity in land plant communities. *Science* 147: 250–260.

Whittaker, R. H. 1967. Gradient analysis of vegetation. *Biological Reviews* 42: 207–264.

Wiegert, R. G. 1962. The selection of an optimum quadrat size for sampling the standing crop of grasses and forbs. *Ecology* 43: 125–129.

Williams, C. B. 1964. *Patterns in the Balance of Nature*. Academic Press, London.

Williams, C. S. and Marshall, W. H. 1938. Duck nesting studies, Bear River Migratory Bird Refuge, Utah, 1937. *Journal of Wildlife Management* 2: 29–48.

Williams, D. A. 1976. Improved likelihood ratio tests for complete contingency tables. *Biometrika* 63: 33–37.

Wilson, K. R. and Anderson, D. R. 1985a. Evaluation of a nested grid approach for estimating density. *Journal of Wildlife Management* 49: 675–678.

Wilson, K. R. and Anderson, D. R. 1985b. Evaluation of two density estimators of small mammal population size. *Journal of Mammalogy* 66: 13–21.

Winer, B. J., Brown, D. R. and Michels, K. M. 1991. *Statistical Procedures in Experimental Design*, 3rd ed. McGraw-Hill, New York.

Wolda, H. 1981. Similarity indices, sample size and diversity. *Oecologia* 50: 296–302.

Wolda, H. 1983. Diversidad de la Entomofauna y como Medirla. *Congress Latino Americano de Zoologia* 9: 181–186.

Wood, G. W. 1963. The capture-recapture technique as a means of estimating populations of climbing cutworms. *Canadian Journal of Zoology* 41: 47–50.

Worlund, D. D. and Taylor, G. 1983. Estimation of disease incidence in fish populations. *Canadian Journal of Fisheries and Aquatic Sciences* 40: 2194–2197.

Wright, S. J. 1981. Intra-archipelago vertebrate distributions: the slope of the species-area relation. *American Naturalist* 118: 726–748.

Wywialowski, A. P. and Stoddart, L. C. 1988. Estimation of jack rabbit density: methodology makes a difference. *Journal of Wildlife Management* 52: 57–59.

Yates, F. 1964. Sir Ronald Fisher and the Design of Experiments. *Biometrics* 20: 307–321.

Yoccoz, N. G. 1990. Use, overuse, and misuse of significance tests in evolutionary biology and ecology. *Bulletin of the Ecological Society of America* 72: 106–111.

Zahl, S. 1977. Jackknifing an index of diversity. *Ecology* 58: 907–913.

Zar, J. H. 1996. *Biostatistical analysis*, Third edition. Prentice-Hall, London.

Zimmerman, E. G. 1965. A comparison of habitat and food of two species of *Microtus*. *Journal of Mammalogy* 46: 605–612.

Index